LIQUID CRYSTAL DEVICES

STATE OF THE ART REVIEW Series:

Vol. 1   *HOLOGRAPHY...1969 (1969)*

Vol. 2   *LIQUID CRYSTALS AND THEIR APPLICATIONS (1970)*

Vol. 3   *HOLOGRAPHY...1970 (1970)*

Vol. 4   *ACOUSTIC SURFACE WAVE AND ACOUSTO-OPTIC DEVICES (1971)*

Vol. 5   *HOLOGRAPHY...1971/72 (1972)*

Vol. 6   *ELECTRET DEVICES FOR AIR POLLUTION CONTROL (1972)*

Vol. 7   *LIQUID CRYSTAL DEVICES (1973)*

LIQUID CRYSTAL DEVICES

Edited by Thomas Kallard

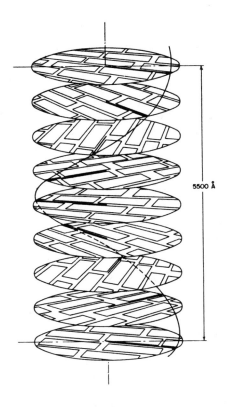

optosonic press

LIQUID CRYSTAL DEVICES

(STATE OF THE ART REVIEW, Vol. 7)

Copyright (c) 1973 by Optosonic Press

Published by Optosonic Press
Box 883, Ansonia Post Office
New York, N.Y. 10023

Additional copies may be procured
by addressing the publisher.

Library of Congress Catalog Card No. 73-78006

International Standard Book Number: 0-87739-007-X

PRINTED IN THE UNITED STATES OF AMERICA

PREFACE

This seventh volume in the STATE OF THE ART REVIEW Series is a continuation of Vol. 2: "Liquid Crystals and Their Applications," published in 1970. Several features have been added to make this volume even more useful than the previous one.

A large portion of the book is devoted to the detailed description of over 150 U.S. and foreign patents, including about 450 illustrations. The first paragraph of each patent description attempts to sum up briefly the concept of the invention. A list of patents is given covering the period March, 1970 through June, 1973, listed chronologically by the date of issue; each listing contains the patent number, country of issue and title. A separate index of inventors and patent holders is also included.

The bibliography lists approximately 1100 entries from world-wide sources, cross-referenced with an author index. The alphabetical list of periodicals includes only those journals regularly scanned, together with the latest issue indexed for the present volume.

A two-part catchword title index is included for readers mainly interested in certain specialized fields. The first part covers the patents, the second part the bibliography of papers.

In addition to the above-described reference material, the volume contains original papers by outstanding experts and manufacturers' announcements.

Our sincere gratitude to the authors of the four state-of-the-art papers for their valuable contributions to this volume.

The final "announcements" section describes some of the liquid crystal raw-materials, components, devices and services commercially available.

New York, N.Y.
September, 1973

Thomas Kallard
Editor

CONTENTS

PREFACE   v

CONTENTS   vii

PAPERS   ix

    LIQUID CRYSTAL DEVICES,
    Joseph A. Castellano, LCD, Inc.,   xi

    APPLICATIONS OF LIQUID CRYSTALS TO INFORMATION DISPLAY,
    FAULT DETECTION, AND MEDICAL THERMOGRAPHY,
    John M. Washick, International Liquid Crystal Co.,   xiii

    PERFORMANCE AND CHARACTERISTICS OF SMECTIC
    LIQUID CRYSTAL STORAGE DISPLAYS,
    F.J. Kahn, D. Maydan, G.N. Taylor, Bell Laboratories,   xvi

    MULTIPLEX OPERATION OF DYNAMIC SCATTERING
    LIQUID CRYSTAL DISPLAYS,
    C.R. Stein, General Electric Corporate R/D,   xxii

PATENT DESCRIPTIONS   1

PATENT INFORMATION   280

LIST OF PATENTS   281

INVENTOR INDEX   284

PATENT HOLDER INDEX   286

BIBLIOGRAPHY   287

AUTHOR INDEX   342

LIST OF PERIODICALS   350

CATCHWORD TITLE INDEX   351

ANNOUNCEMENTS   355

PAPERS

LIQUID CRYSTAL DEVICES

Dr. Joseph A. Castellano
LCD, Inc.
The Forrestal Industrial Center
Princeton, N. J. 08540

The number and type of applications for liquid crystals now appear to be limited only by the imagination. As of this writing, liquid crystals of one class or another have been used for: disposable thermometers, breast tumor detection, aerodynamic testing, digital readout on various instruments, wrist watch faces, electronic microcircuit testing, solvents in nuclear magnetic resonance spectroscopy, stationary phases in gas-liquid chromatography, toys and decorative materials.

The sharp changes in the color of cholesteric liquid crystals with small changes in temperature have resulted in a number of unique temperature sensing applications. One of these is the thermal mapping of electronic components (1). The technique merely involves coating of the component with the liquid crystal. Thermal mapping is thus used to determine heat flow and temperature distribution patterns in operating devices as well as hidden structural characteristics by localization of heat sources or conduction paths.

This technique has also been used to visualize two-dimensional temperature patterns developed during thermal nondestructive tests of aerospace components and structures (1). These temperature-sensitive materials have been used to evaluate the efficiency of heat exchangers, to detect nonuniformities in electrically resistive coatings, and to observe the conversion of mechanical-to-thermal energy in tensile and fatigue test specimens.

The medical applications of cholesteric liquid crystals have also received a great deal of attention recently. Continuous monitoring of skin temperature over an extended area with cholesteric liquid crystals provides the physician with a detailed and easily interpreted indication of circulatory improvements. It has also been possible to detect breast tumors by this technique (2) since the temperature of the skin in the vicinity of a tumor is higher than that of the surrounding area. Hoffmann-LaRoche has recently announced a line of disposable thermometers which use cholesteric liquid crystals on a thin film strip to detect changes in body temperature.

The discovery of dynamic scattering (3) in nematic liquid crystals has resulted in a whole new electronic display technology. The effect is produced in a parallel plate capacitor type cell that consists of a nematic liquid crystal sandwiched between two pieces of glass, each having a transparent conductive coating such as tin or indium oxide. Application of a DC or low frequency (1 kHz) AC signal across the plates changes the cell from an optically transparent to a translucent condition as a result of the creation of scattering centers. Other electro-optic effects in nematic liquid crystals have also been described. These field effects may use polarizers (4) or dichroic dyes (5) to produce various optical changes.

Numeric indicators have been fabricated with materials that exhibit the dynamic scattering mode or field effect. The complete range of numbers (0 to 9) is possible by photoetching a seven-segment pattern onto one of the conductively coated surfaces. In this way, it has been possible to fabricate numerical displays for electronic clocks and other digital instruments. Since the liquid crystal cell has low-power and flat construction characteristics, it also has been possible to extend the concept to electronic wrist watches. A number of companies are now manufacturing and marketing these products.

The use of liquid crystal numerics in compact, portable calculating machines is an obvious extension of the work on wrist watches. The low power required for the display and compatibility with ultra-low power integrated circuits now make it possible to produce machines that use throwaway batteries.

Cholesteric (or chiral nematic) liquid crystals have been used to produce panels that exhibit an optical memory effect (6). Applications of this concept to projection displays using photoconductors (7) and lasers (8) have also been realized. The use of smectic liquid crystals for a display technique using laser addressing has also been reported recently (9).

A novel family of cholesteric liquid crystals are currently being developed (10) for the decorative market. In contrast to most commercial liquid crystals, however, these materials exist as solid glasses at ambient temperatures. By application of pressure or shearing action, it is possible to produce textured, iridescent patterns in the films. It is anticipated that these materials will be used to produce decorative patterns on metals, fabrics, leather, ceramics, and paper.

References:

(1) G.H. Brown, Ed., Liquid Crystals 2, Part I, Gordon & Breach, New York, 1969.
(2) J.L. Fergason: work conducted at Kent State University.
(3) G.H. Heilmeier, L.A. Barton and L.A. Zanoni, Proc. IEEE, $\underline{56}$, 1162 (1968).
(4) W. Helfrich and M. Schadt, Appl. Phys. Lett., $\underline{18}$, 127 (1971)
(5) G.H. Heilmeier, J.A. Castellano and L.A. Zanoni, Mol. Cryst & Liq Cryst., $\underline{8}$, 293 (1969)
(6) G.H. Heilmeier and J.E. Goldmacher, Proc. IEEE, $\underline{57}$, 34 (1973)
(7) J.D. Margerum, J. Nimoy and S.Y. Wong, Appl. Phys. Lett. $\underline{21}$, 392 (1973)
(8) F.J. Kahn, Appl. Phys. Lett. $\underline{21}$, 392 (1973)
(9) F.J. Kahn, Appl. Phys. Lett. $\underline{22}$, 111 (1973)
(10) Organic Chemicals Dept., duPont de Nemours & Co.

APPLICATIONS OF LIQUID CRYSTALS TO INFORMATION DISPLAY,
FAULT DETECTION, AND MEDICAL THERMOGRAPHY

John M. Washick
Physicist
INTERNATIONAL LIQUID CRYSTAL CO.
26101 Miles Road
Cleveland, Ohio 44128

Recent advances in liquid crystal technology have resulted in an electric field operated nematic cell with unusual electro-optic properties. Additional developments have given rise to a method of determining casting wall thickness using a highly sensitive and responsive cholesteric liquid crystal. Cholesteric liquid crystals have also been used for generating thermograms in medical applications.

A Field Effect Device:

It has long been known that a thin film of nematic material placed between two glass plates whose surfaces have been treated and oriented exhibits an alignment of the long axis of the molecules near the surface as parallel to the direction of orientation. If the glass plates are positioned such that their orientation directions are parallel, the molecules have their long axes parallel throughout the material. If the glass plates are positioned with the orientation directions perpendicular, the liquid crystal exhibits a twisted structure.

Due to the optical properties of the nematic material, linearly polarized light whose polarization direction is parallel to the orientation direction will have the polarization direction turned by 90°. For an optically positive liquid crystal, the application of an electric field disturbs the twisted orientation causing the molecules to line parallel with the electric field. With the field turned on, linearly polarized light is not twisted in passing through the liquid crystal.

Polarizers attached to each glass plate such that the polarization directions are perpendicular results in a cell which passes light with the field off and blocks light with the field on. If the polarizers are oriented with parallel polarization directions the opposite effect occurs. By etching the metal coating to form a pattern, this construction can be used to display alphanumeric information.

The twisted nematic or field effect liquid crystal display described above can be operated in either the transmissive or reflective mode. In the former, the polarizers are oriented parallel to one another, and light from behind the display passes through those segments to which voltage is applied. The reflective display has the polarizers oriented in a perpendicular fashion with a diffuse reflector behind the rear polarizer.

The distinguishing features of the field effect device are: (a) Extremely low current of the order of 10 nA or less per square centimeter of active area; (b) Extended cell life due to the near absence of displacement current;

(c) Improved viewing angle; (d) Relatively voltage independent contrast ratio as this is a true bilevel device; (e) Low operating voltage of the order of eight volts for maximum contrast ratio; (f) Wide operating temperature range; (g) Contrast ratios of 25:1 to 100:1 depending on the operating conditions.

The field effect digital display has been used in several unique applications such as wrist watches, where low current, high contrast ratio, and wide viewing angle are used to an advantage. Other applications include digital panel meters, panel displays for home appliances and metering pumps.

Figure 1 - A four digit field effect liquid crystal display operating in the transmissive mode.

Cholesteric Liquid Crystals:

The temperature sensitivity of the cholesteric phase has been used in many applications such as nondestructive testing and biomedicine. One use of cholesterics has been the detection of casting faults in turbine blades by measuring the wall thickness. A system has been developed wherein the temperature pattern on the surface of the turbine blade resulting from a step change in the temperature of a fluid passing through the blade is detected by a cholesteric liquid crystal. Unlike other similar methods, this system measures variations in the wall thickness. The time interval associated with the temperature change reaching the cholesteric liquid crystal detector is affected by casting wall thickness, and therefore, casting voids and cracks show up via color indications. The temperature pattern is photographed at a specific interval after the forementioned step change using a synchronized strobe light, providing a permanent record of the structure of the blade. This concept is also applicable to other areas of fault detection where nondestructive testing and a permanent record is required.

Another promising application of cholesteric liquid crystals is medical

Figure 2 - A battery powered crystal oscillator wrist watch using a field effect liquid crystal display. The time is displayed continuously and battery life is one year or more.

thermography. Calibrated cholesterics are capable of producing a color thermogram with a sensitivity of 0.1°C and resolution of 1000 lines per inch over a large skin area. Although infrared thermography has been demonstrated effective in measuring the surface temperature and viewing the associated vascular pattern, the cost for such a system has limited its application. Liquid crystal thermography has been demonstrated as a lower cost and a more sensitive alternative. The technique involves applying a black skin coating to the area to be examined and covering the light absorbing coating with a calibrated cholesteric liquid crystal. The liquid crystal consists of a mixture designed to respond to a 4°C color temperature range with the coldest temperature indicated by red or orange and the warmest blue or violet. With direct illumination the patient is examined by first hand observation and the colors related to temperature using a calibrated chart. A color photograph provides a permanent record for future reference or further interpretation at a later date.

"""""""""""

PERFORMANCE AND CHARACTERISTICS OF SMECTIC LIQUID CRYSTAL STORAGE DISPLAYS

by

Frederic J. Kahn, D. Maydan, and G. N. Taylor
Bell Laboratories
Murray Hill, New Jersey

High optical quality displays have been developed using infrared laser addressed smectic liquid crystal light valves.[1] The light valves are capable of storing and displaying high resolution information - 3500×3500 elements on a 3.5×3.5 cm active liquid crystal area. The information is recorded by an intensity modulated infrared laser beam which is XY deflected and focused on the liquid crystal cell.

The focused beam locally heats visibly transparent but infrared absorbing electrodes adjacent to the liquid crystal layer. The resultant heating and subsequent cooling of the liquid crystal creates light scattering centers which are semipermanent and remain stored until erased. The recorded scattering-type image is projected on a screen with a simple projection system. High contrast, high luminance images are obtained without the use of Schlieren optics. The images may be selectively erased.

The display system is described in Fig. 1 and consists of a low power (<50 mW) YAℓG laser, a TeO_2 acoustooptic modulator and pattern generator, X and Y galvanometer mirror deflectors, a liquid crystal cell and a projection system.[2] The aperture shown in Fig. 1 is required in order to get good contrast with cholesteric light valves[3] but is not required with the smectic light valves due to the latters' wide angle scattering characteristics. The system is capable of recording both computer and/or graphic tablet generated images at a rate of about 10^5 resolvable elements per second. Writing speed has been found to scale linearly with laser power for laser powers in the 1 to 50 mW range. Laser powers on the order of 500 mW should permit writing at rates of 10^6 elements per second.

The liquid crystal cell described in Fig. 2 consists of a layer of smectic liquid crystal, nominally 14 μm thick, sandwiched between two transparent $In_{2-x}Sn_xO_{3-y}$ electrodes deposited on fused silica substrates. The unique electrodes[4] facilitate thermal writing with low power IR laser beams. They are sputter deposited in such a way that they strongly absorb the laser writing light in the wavelength range between 0.8 and 1.5 μm (about 35% absorption by each layer at 1.06 μm) and have little absorption of the projection light in the visible between 0.4 and 0.7 μm. Infrared HeNe or GaAs lasers could be used instead of YAℓG lasers as the writing source because of the broad band absorption of the electrodes.

The liquid crystal used in this device has a stable smectic phase extending from slightly above room temperature to about 70°C. The light valve is held at an operating temperature within several degrees of the isotropic transition temperature in order to minimize the laser power required for writing.

In the unwritten state, the smectic liquid crystal is ordered with its molecular axes normal to the glass, and appears fully clear and transparent. Heating the liquid crystal above its smectic-isotropic phase transition temperature disorders the material. On rapidly cooling back to the smectic phase, scattering centers are formed due to disorder which has been frozen in by the cooling. These scattering centers appear in the projected image as dark areas on a bright background. Positive dielectric anisotropy liquid crystals are used, and if the material cools back to the smectic phase in the presence of an ac electric field (obtained by applying a voltage between the transparent electrodes) the initially uniform, nonscattering texture is recovered. Application of the same ac electric field in the absence of heating and subsequent cooling causes no change in the texture of the liquid crystal.

Selective erasure is obtained by locally heating the liquid crystal with the laser beam in the presence of an ac electric field obtained by applying a voltage, typically

14 V-rms at 10 kHz, between the transparent electrodes. The selective erase speed and writing speed are comparable.

The X and Y mirror galvanometers are used to generate images in either a random access manner or with a raster scan at a maximum speed of 300 cycles per second. The acoustooptic TeO_2 Y-deflector serves as the intensity modulator. The same deflector is also capable of rapidly generating a linear array of up to 30 resolvable beam positions. When combined with the galvanometer scan, this permits the generation of characters and symbols at rates up to a few thousand characters per second.

A photograph of a projected image about 80 cm wide is shown in Fig. 3. The image was recorded on about a 2 cm wide area of a smectic liquid crystal light valve. The information in this case was generated by both a graphic tablet and a keyboard. The alphanumerics are of the 5×7 dot matrix type and were generated with the acoustooptic deflector and the X mirror galvanometer. A 500 W tungsten filament lamp and an f/2.5 projection lens provide black on white images with 55 fL highlight luminance and contrasts of 9:1 on a 20 ft^2 unity gain screen. Insertion loss of the light valve is 21%. With an f/5.6 lens and a 250 W tungsten lamp projected images have contrasts of 18:1. Typical modulation transfer functions measured for the smectic light valves are 80% at 40 line pairs/mm.[5]

The authors thank W. Q. McKnight, S. R. Williamson, and W. Nowotarski for their experimental assistance and D. B. Fraser for technical discussion.

REFERENCES

1. F. J. Kahn, Appl. Phys. Lett. $\underline{22}$, 111 (1973).

2. H. Melchior, F. J. Kahn, D. Maydan, and D. B. Fraser, Appl. Phys. Lett. $\underline{21}$, 392 (1972).

3. D. Maydan, H. Melchior, and F. J. Kahn, IEEE Conf. on Display Devices, New York, October 11-12, 1972.

4. D. B. Fraser and H. D. Cook, J. Electrochem. Soc. $\underline{119}$, 1368 (1972).

5. R. A. Heinz, unpublished.

FIGURE CAPTIONS

Fig. 1   Writing and projection system.

Fig. 2   Smectic light valve showing:

    a) erased state,
    b) writing state, and
    c) written state.

Fig. 3   Image from 2 cm wide liquid crystal area projected to about 80 cm wide on front projection screen.

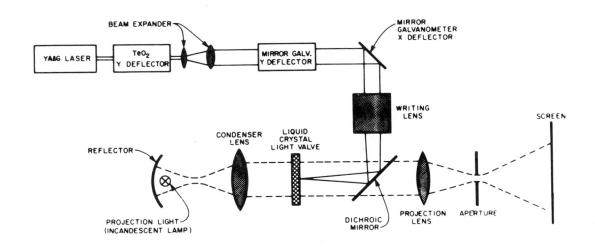

Figure 1.

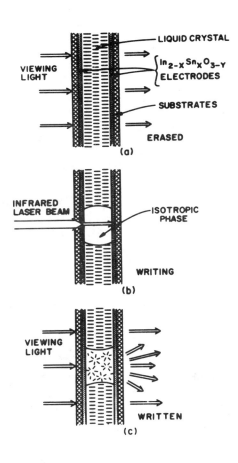

Figure 2.

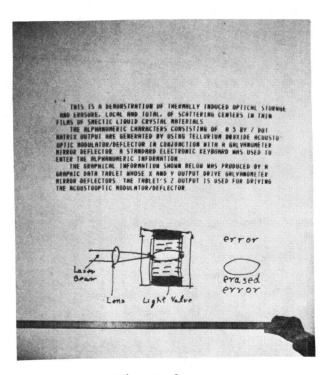

Figure 3.

MULTIPLEX OPERATION OF DYNAMIC SCATTERING LIQUID CRYSTAL DISPLAYS

C. R. Stein
Electronics Engineer
General Electric Corporate Research and Development
Schenectady, New York

The application of dynamic scattering liquid crystals to multi-element matrix displays is complicated by the absence of a sharp threshold in the scattering versus voltage characteristic. Consider, for example, a conventional voltage half-select system. A single "off" element in a group of "on" elements will experience half excitation continuously. It must, however, exhibit less response than the single "on" element in a group of "off" elements which receives excitation only during the short addressing time. The lack of a sharp threshold prohibits such simple half select systems because this condition cannot be met.

Recent papers (1, 2) described a novel and successful electronic technique which obviated this difficulty in some applications. In this system, the temporal coincidence of signals whose frequency components are above and below the space-charge cutoff frequency of the material is used to achieve the multiplexing function. In order to avoid the electrolytic degradation caused by dc operation, these signals are exclusively ac, having equal time integrals above and below the zero line.

Sample Applications

Calculators -- Numeric Displays

The application of this addressing technique to an 8 or 10 position calculator is a good one. Previous papers have described in some detail the wave form requirements of the devices and the schematic configuration of a cell laid out to operate in this manner (see Figure 1). Laboratory-made interfaces with a few of the popular calculator chip sets have used presently available, stock-item integrated circuit elements as the cell drivers. While repackaging of these demonstrators has not been done, inspection shows that the amount of electronics necessary to drive the liquid crystal cell is less than that used in at least one of the commercial units to drive its vacuum fluorescent display. Power requirements remain modest, although somewhat increased over the direct drive mode on an equivalent area basis.

Computer Terminals -- Alphanumeric Displays

The application of a multiplexed liquid crystal display to an alphanumeric computer terminal (3) has resulted in the first truly portable computer terminal. This model (Figure 2) was built to demonstrate the capability of the two-frequency multiplexing scheme in operating a cell having more information content than could conceivably be controlled without multiplexing. With a total display capability of 32 characters in the 16 segment starburst pattern, there are 512 addressable sites in the single cell. These are controlled by only 72 leads to the cell.

Housed in a 10 by 10 by 3 inch package, the terminal weighs 7 pounds and consumes about 10 watts of power. It has been built of off-the-shelf parts covering the range from discrete cell crivers to an LSI telephone line controller.

Limits of the Dynamic Scattering Mode of Liquid Crystal Operation

Typically the multiplexing process achieves its selectivity by use of a non-linearity in the response of the display to applied voltage. In this two-frequency scheme the interaction of high and low frequency signals is such as to enhance selectivity to the level of at least one part in 10, or to as much as one part in 16, with presently available dynamic scattering materials. Inasmuch as this scheme relies on straddling the space-charge cutoff frequency of the material, and that frequency changes with temperature, there is an additional trade-off between power consumption and the temperature range over which successful operation can be achieved.

Other Systems

Dynamic scattering is, of course, only one of the many possible operating modes of liquid crystal devices. Little has been published thus far concerning the possibilities of multiplexing the field effect modes, but work is progressing in several laboratories.

References
1. C. R. Stein and R. A. Kashnow, Applied Physics Letters, Vol. 19, page 343 (1971).
2. C. R. Stein and R. A. Kashnow, 1972 Digest of Technical Papers, Society for Information Display, June 1972, page 64.
3. C. R. Stein, 1973 Digest of Technical Papers, Society for Information Display, May 1973.

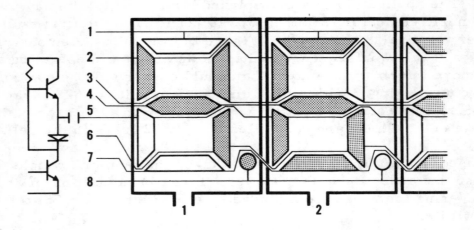

FIGURE 1. Schematic Cell Construction

FIGURE 2. Liquid Crystal Computer Terminal

PATENT DESCRIPTIONS

3,499,112 (U.S.) *Patented Mar. 3, 1970*

ELECTRO-OPTICAL DEVICE.

<u>George H. Heilmeier</u> and <u>Louis A. Zanoni</u>, *assignors to RCA Corporation.*
Application Mar. 31, 1967.

An electro-optical display device having a layer of a nematic liquid crystal composition of a type that scatters light due to turbulence in the layer created by the application of a voltage across the layer, which voltage is accompanied by a current in the layer, means for supporting the layer, and means for applying a voltage across the layer of a magnitude sufficient to cause turbulence of the layer in the region of the applied voltage.

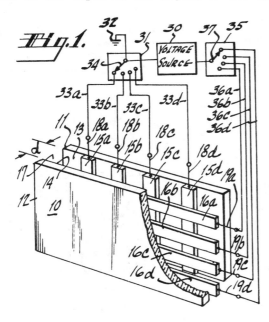

Figure 1 illustrates a novel crossed grid optical display device 10. The novel device 10 is comprised of back and front transparent glass support plates 11 and 12 respectively. The two plates 11 and 12 have essentially parallel inner faces 13 and 14 respectively that are separated by a distance d, which is generally in the range of about 5 to 30 microns. The back plate 11 supports, on its inner face 13, an array of parallel spaced transparent conductive back electrode strips (15a, 15b, 15c and 15d) are shown; but a much larger number of electrode strips may be used. The front plate 12 supports, on its inner face 14, an array of parallel, spaced transparent conductive front electrode strips 16a, 16b, 16c and 16d. The front strips 16 are positioned so that their longitudinal directions are substantially perpendicular to the longitudinal direction of the conductive back strips 15. Again, only four front strips (16a, 16b, 16c and 16d) are shown, but a much larger number may be used.

The space between the back and front plate 11 and 12 is filled with a medium so as to form a film 17. The film 17 is comprised of a nematic liquid crystal composition of the type that exhibits turbulent motion upon the application of a voltage which produces an electric current in the film, for example, a film comprised of anisylidene-p-aminophenylacetate, having an excess of mobile ions therein. The film 17 is a weak electrolyte.

The device 10 includes connection means 18a to 18d and 19a to 19d for applying a voltage to the conductive back electrodes 15a to 15d and to the conductive front electrodes 16a to 16d, respectively. As used herein, said connecting means and/or said conductive strips are included in the means for applying the electric field or voltage to the liquid crystal layer.

Figure 1 also includes a schematic representation of a circuit for operating the display device 10.

In a transmissive mode of operation, as shown in Fig. 2, a light source 21 is positioned on one side of the device 10 so that light is directed through the device in a direction substantially normal to the major faces of the plates 11 and 12. The observer 22 is on the opposite side of the device 10 from the light source 21. At less than a threshold field, the observer 22 sees the entire plate area as uniformly bright. When a voltage of sufficient magnitude is applied between a back electrode strip, and a front electrode strip, such as, for example, between electrode strips 15a and 16d via the connecting means 18a and 19d respectively, the film 17 in the volume defined by the intersection of the energized electrode strips is affected by the voltage and current caused to flow in it due to the applied voltage. This voltage and current flow causes turbulence in the film in this volume and gives rise to scattering of light incident on that portion of the device. The observer sees this region of his field of view become darker than the remaining plate area due to the light scattering.

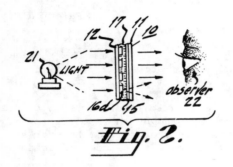

Fig. 2.

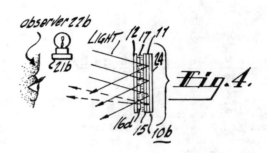

Fig. 4.

The preferred mode of operation is the reflective mode and uses a device 10b as illustrated in Fig. 4. Here, one support plate of the device 10b, e.g., the back plate 11, is made reflective rather than absorbing, for example by a specularly reflective coating 24 on the outer face of the plate 11. In the reflective mode of operation, a light source 21b and a viewer 22b are both positioned on the transparent plate side of the device 10b. The light source 21b, preferably but not necessarily, produces a collimated beam at such an angle that the light reflected from the reflective coating 24 does not strike the viewer 22b. When a voltage is applied across the electrodes of the device as previously described, light will be scattered in the region of the electrode intersection causing some of the light to be observed by the viewer.

A variety of transparent solids may be utilized for a transparent support plate, including the various types of glass, fused quartz, transparent corondum, and transparent plastics and resins. A non-transparent support plate may be made from the same materials as the transparent plate coated with a material, such as a black organic dye for absorption or a metallic film for specular reflection. The transparent conductive strips may be made, for example, by depositing thin layers of indium oxide or stannic oxide on the desired region of the plates 11 and 13. The spacing between the support plates may be maintained in any one of several ways. For example, by means of shims, clamps, or a suitable frame-like holder. In operation, the device is maintained at a temperature at which the nematic-liquid crystal composition is in its mesomorphic state.

A wide variety of liquid crystal compositions are useful in the novel electro-optical devices which operate by the dynamic scattering mode. Preferably, the nematic liquid crystal compositions have resistivities of between about 10^8 ohm-centimeters and 10^{11} ohm-centimeters. The preferred film 17 is comprised of anisylidene-p-aminophenylacetate (APAPA), having trace amounts of p,n-butoxybenzoic acid (BBA), therein. Although APAPA can be used without the BBA, it is somewhat milky and causes unwanted zero field reflection which thereby reduces the contrast ratio of the device. In operation, APAPA films require voltages in the order of 10 to 100 volts across a 1/2 mil film with operating temperatures in the range of 65°C. to 95°C. The light scattering typically has a two millisecond buildup and a 15 to 30 millisecond decay time in response to changes in the electric field.

The novel devices are operable under either A.C. and D.C. or pulsed D.C. voltages.

" " " " " " " " " "

3,499,702 (U.S.)   Patented Mar. 10, 1970

NEMATIC LIQUID CRYSTAL MIXTURES FOR USE IN A LIGHT VALVE.

<u>Joel E. Goldmacher</u> and <u>George H. Heilmeier</u>, assignors to RCA Corporation.
Application Dec. 5, 1967.

    The electro-optical effect due to alignment of domains of the nematic liquid crystal molecules in an electric field may be employed in transmissive, reflective, or absorptive type flat panel displays, in light shutters and other applications. Prior art liquid crystals of this type have relatively higy crystal-nematic transition temperatures. That is the temperature at which the material enters its nematic mesomorphic state is relatively high. Since the nematic liquid crystal composition must be operated while in its nematic mesomorphic state, it is therefore desirable to use compositions which have a low crystal- nematic transition temperature. It is also desirable to use materials that are highly polar so that the time required to align the domains in an electric field will be relatively short.

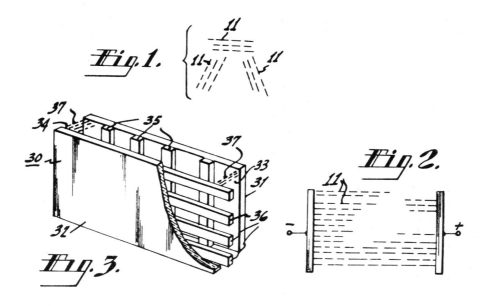

    In the absence of an electric field, a nematic liquid crystal composition of the type described herein is believed to have its molecules 11 arranged in small randomly oriented domains, as shown in Fig. 1, when in its mesomorphic state. Upon the application of a sufficiently high electric field to the liquid crystal composition it is believed that the domains tend to align themselves, as shown in Fig. 2, so that a substantial proportion of the domains and hence a substantial proportion of liquid crystal molecules are aligned essentially parallel to the direction of the applied field. The degree of alignment is a function of the strength of the field. One specific embodiment of this type of light valve is shown with reference to Fig.3.

    In the novel device the space between the front and back plates 31 and 32 is filled with a liquid crystal mixture 37 comprising equal weight proportions of p-n-ethoxybenzylidene-p'-aminobenzonitrile and p-n-butoxybenzylidene-p'-eminobenzonitrile. The mixture may be sealed in the device 30 by using an epoxy cement around the edges of the device.

    In order to increase the contrast ratio in the preferred transmissive mode of the device 30, it is preferred to use the device 30 between crossed polarizers 41 and 42, as shown in Fig. 4. When used in this manner, the device takes advantage of the birefringent properties of the nematic liquid crystal composition which causes rotation of the plane of polarization of polarized light incident thereon. In operation light 43 from the light source 44 is polarized when passed through the first polarizer 41. The polarized light is transmitted through the device 30 and when no field is applied to the device 30 a substantial portion of this light passes through the second polarizer 42. Hence with no field applied, the device appears uniformly bright to an observer 45. This is due to the random orientation of the domains which causes rotation of some of the polarized light which rotated light is then able to pass through the second polarizer 42. When an electric field is applied across the device 30, the domains of the novel liquid crystal composition align such that the liquid crystal molecules are parallel to the direction of the incident light. When this occurs, the plane of polarization is not rotated and hence the second polarizer 42 which is crossed with relation to the first polarizer 41 impedes the passage of light therethrough and the device in the region of the applied field appears dark to the observer 45.

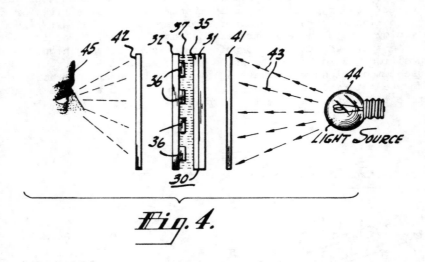

Fig. 4.

Using the same principle, the device can be operated in a reflective mode with just one polarizer. Another mode of operation of the novel device is in combination with pleochroic dyes. In this configuration a pleochroic dye is mixed with the nematic liquid crystal composition and rotation and alignment of the liquid crystal molecules causes rotation and alignment of the dye molecules. This effect thereby changes the absorption characteristics of the device when viewed in the transmission mode with polarized light. An example of a useful dye material is methyl red.

The liquid crystal compounds useful in the disclosed device are p-n-ethoxybenzylidene-p'-aminobenzonitrile; p-n-butoxybenzylidene-p'-eminobenzonitrile; and p-n-octoxybenzylidene-p'-aminobenzonitrile. In order to substantially lower the crystal-nematic transition temperature, mixtures of from 10 to 90 weight percent of p-n-butoxybenzylidene-p'-aminobenzonitrile with p-n-ethoxybenzylidene-p'-eminobenzonitrile or mixtures conating of from 15 to 70 weight percent of each of the three compounds are preferred.

The mixtures may be inserted in the liquid crystal device by having the planar supporting elements either separated by means of a shim or by sealing three sides together with an epoxy cement leaving a space between the elements. The liquid crystal composition is then heated to a temperature greater that its crystal-nematic transition temperature and injected into the space between the planar elements, for example, by means of a syringe.

Threshold voltages for a one-half mil thick cell containing the novel mixtures are approximately 10 volts A.C. or D.C. Switching time for switching of a one-half mill cell at 100 volts D.C. is in the order of 10 milliseconds for the rise time and about 30 to 100 milliseconds for the decay time. By rise time, it is meant the time necessary to align the domains. By decay time, it is meant the time it takes after the removal of the applied field for the liquid crystals to return to its random state.

" " " " " " " " " "

3,503,672 (U.S.)   Patented Mar. 31, 1970

REDUCTION OF TURN-ON DELAY IN LIQUID CRYSTAL CELL.

<u>Frank J. Marlowe</u>, assignor to RCA Corporation.   Application date: Sep. 14, 1967.

Turn-on response time of liquid crystal cell is decreased by applying pulses thereto at a level lower than the voltage threshold for dynamic scattering of the cell.

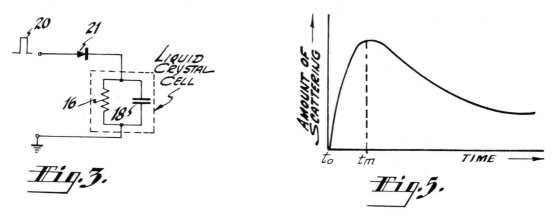

Fig. 3.

Fig. 5.

A simplified equivalent circuit for a liquid crystal cell is a resistor such as shown at 16, in shunt with a capacitor, such as shown at 18 (Fig.3). The crystal is excited by applying a short duration pulse such as 20 to the cell. In television applications, this pulse may have a duration of 0.06 millisecond which is the equivalent of one television line time. This implies that all of the elements of one television line are "addressed" at the same time. Operation in this way rather than an element of a line at a time, is preferred because it permits a greater length of time for capacitor 18 of the liquid crystal cell to charge. It is also important that the capacitor 18 retain its charge for a reasonable time interval to permit the dynamic scattering to take place. The function of diode 21 is to permit such storage. It prevents discharge of the capacitor through the source (not shown) which produces pulse 20 so that the capacitor must discharge through the liquid crystal itself as represented by resistor 16.

The amount of light scattering which occurs in the liquid crystal cell of Fig. 3 is as shown in Fig. 5. It takes short interval of time t_o to t_m, approximately 1-10 milliseconds (depending upon the temperature, field strength and particular material used) for the maximum amount of light scattering to be achieved. At time t_1, when there is no longer any voltage across the liquid crystal cell, there is still a considerable amount of scattering present, as the mechanical time constant is relatively long.

In the actual operation of the circuit of Fig. 3, it is found that if the liquid crystal cell is dark initially (is in its unexcited state), it requires a relatively large number of excitation pulses 20 to be applied before the crystal exhibits the light scattering characteristic shown in Fig. 5. This is shown in the top two waveforms A and B of Fig. 6. The excitation or "write" pulses in Fig. 6, are shown on the upper line A. While these pulses are shown to be of fixed amplitude greater than the dynamic scattering threshold of the crystal, in televisionapplications they would be video pulses and their amplitudes above the dynamic scattering voltage threshold level would correspond to the video information it is desired to write into a particular element successive excitations of that element. These write pulses are each of a duration of approximately 0.06 milliseconds and are at a repetition frequency of approximately 30 pulses per second.

The present inventor has discovered that the discharge time constant for a liquid crystal element does not remain constant but increases with the application of successive pulses. It is believed that it is for this reason that the element initially exhibits a long turn-on delay. Initially, that is, in response to the first excitation pulses, the time constant is relatively low and the capacitor 18 discharges relatively rapidly as illustrated by the solid line curve 30 of Fig. 7. At the time t_m, which is the time required for the amount of scattering produced in the crystal to reach its maximum value, the voltage V_A across the cell is extremely low, - lower than the dynamic scattering voltage threshold of the cell. Accordingly, no scattering is produced and this is borne out by the region t_o-t_s of the curve B of Fig. 6.

The low time constant is believed to be due to a low value of resistance 16 of the cell of Fig. 3. While the reason for this is not completely understood, according to theory developed by others, there are current carriers initially present in the liquid crystal. These may be free ions or impurities or perhaps other conducting particles, the nature of which is not fully

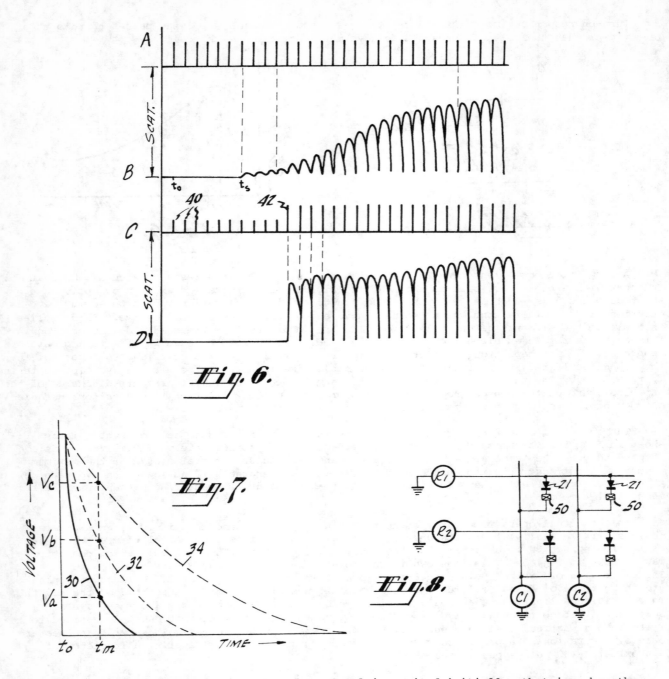

Fig. 6.

Fig. 7.

Fig. 8.

understood. The theory states that when the crystal is excited initially, that is, when the pulses are initially applied across the crystal, they cause these current carriers (negative and positive) to travel through the liquid crystal to the positive and negative conductors, respectively. This movement of current carriers through the liquid crystals corresponds to relatively low resistivity of the crystal.

According to the theory, as successive pulses continue to be applied to the liquid crystal, the free current carriers gradually are swept out of the liquid and reach the positive and negative conductors. During this period, the internal resistance, represented by resistor 16 of Fig. 3, gradually increases. As the value of the resistance increases, the liquid crystal cell discharge time constant increases correspondingly, and the shape of the exponential discharge curve also changes, as shown in Fig. 7. Here, curve 30 may represent the voltage across one particular liquid crystal cell in response to 1 pulse, the curve 32 to voltage is response to perhaps several successive pulses and the curve 34 the voltage in response to perhaps 15 pulses.

With increasing internal cell resistance, the voltage present across the cell at the time t_m increases. For example, this voltage increases from its initial value V_A through a value V_B to a final value of V_C. The dynamic scattering voltage threshold of the circuit is some value between V_A and V_C and, as soon as it is reached, the crystal begins to exhibit the dynamic scattering effect in the manner illustrated in the second waveform B of Fig. 6.

The solution of the present invention to the problem above is illustrated at C and D in Fig. 6. Again, the environment of commercial television and line-at-a-time excitation is assumed. Rather than allowing any element to be free of inputs, each element has continuously applied thereto voltage pulses of an amplitude somewhat lower than the voltage threshold for dynamic scattering of the element. This is illustrated in Fig. 6, row C, by the pulses 40. These pulses are of 0.06 millisecond duration and occur at a rate of approximately 30 pulses per second. If the threshold for dynamic scattering of a cell is 40 volts or so, the pulses 40 of Fig. 6 in row C may have an amplitude of 35 volts or so.

When it is desired to turn the liquid crystal cell "on", the amplitude of the pulses is increased as, for example, is shown at 42. The pulse 42 may, for example, have an amplitude of 100 volts. The result of the application of such a pulse is to turn on the cell "immediately" as illustrated in wave D. Note that in response to this first pulse 42, the amount of scattering produced is quite high, fairly close to the maximum amount of scattering which can be expected. The successive pulses after 42 slightly increase the scattering effect and, after a number of such pulses, the scattering effect reaches its maximum level.

A matrix of liquid crystal elements arranged according to the invention is shown in Fig. 8. While only two-by-two liquid crystal elements 50 are shown, in practice there are many more elements, than this in the matrix. The diodes 21 serve two functions, that of isolation between elements as well as that of permitting each cell to store charge. Every frame interval, that is, approximately every 30 milliseconds, a row of elements is "addressed" by raising the voltage output of a row voltage source such as R1 from -50 volts to +35 volts for 0.06 millisecond for one row interval. Each time a row of elements is addressed, all of the column pulse generators C1, C2 produce outputs of an amplitude representing the brightness of the video information to be written into the respective elements of a row. For example, if the particular cell is to remain "dark", the column generator for that cell produces an output of zero volts while the row generator for that cell produces an output of +35 volts. This voltage causes conduction through a diode and a voltage of 35 volts appears across the liquid crystal cell. The continuous application of such relatively low amplitude pulses to a cell maintains its internal resistance high; however, as the amplitude of the pulses is below the voltage threshold for dynamic scattering of the cell, the cell doesn't go on (does not produce light scattering).

When it is desired that a cell in a row light up, a column voltage pulse generator for that cell produces a negative-going voltage pulse at a level such that the difference between the column and row voltages exceeds the voltage threshold for dynamic scattering of the cell (assumed in the present example to be 40 volts).

3,503,673 (U.S.)   Patented Mar. 31, 1970

REDUCTION OF TURN-ON DELAY IN LIQUID CRYSTAL CELL.

<u>George H. Heilmeier</u> and <u>Louis A. Zanoni</u>, assignors to RCA Corporation.
Application Sep. 14, 1967.

The application to a liquid crystal cell of an electrical bias at a voltage level lower than the voltage threshold for dynamic scattering increases the effective internal resistance of the cell and substantially decreases the response time of the cell.

For Figs. 3, 5, 7 and explanatory text see U.S. Patent No. 3,503,672.

The inventors believe that the reason it requires so long for a liquid crystal element to turn on is that the resistor 16 is, in fact, non-linear. Initially, that is, in response to the first excitation pulse, this resistor exhibits a relatively low resistance value so that the capacitor 18 discharges relatively rapidly. The solution of the present invention to the problem above is shown in Figs 8 and 9. The liquid crystal cell 40 is biased by a bias source shown schematically at 42 to a level which may be substantially lower than its threshold level.

In practice, the bias which is found to give good performance may be of the order of 25% of the dynamic scattering voltage threshold. The diode 44 may be employed between the bias source 42 and the liquid crystal cell to prevent shorting the write voltage source 46 to ground.

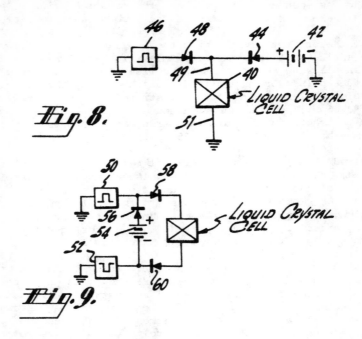

Fig. 8.

Fig. 9.

In the operation of the circuit of Fig. 8, the bias applied to the liquid crystal cell sweeps any current carriers which may be present in the liquid crystal to the relatively positively and negatively charged conductors in contact with the liquid crystal. Thus, the liquid crystal cell exhibits a high internal resistance at the time the source 46 applies an excitation pulse thereto and essentially immediately "lights up" at relatively high intensity.

Another form of the circuit of the invention is shown in Fig. 9. Here, the liquid crystal is excited by the concurrent removal of the back bias on diode 58 by the write source 50 and a negative pulse by the write source 52. (note that the source 50, during the periods between positive-going pulses, may produce an output which is negative relative to ground and source 52, during the periods between negative pulses may produce an output which is at ground.) The bias source, illustrated by battery 54, is connected to the liquid crystal cell through both its own diode 56 and through the diodes 58 and 60. As in the previous circuit, the bias voltage may be 25% or so of the threshold voltage for dynamic scattering of the liquid crystal cell.

""""""""""""

3,505,804 (U.S.)   Patented Apr. 14, 1970

SOLID STATE CLOCK.

Steven R. Hofstein, assignor to RCA Corporation.
Application Apr. 23, 1968.

The face plate of a solid state clock employs a liquid crystal of relatively limited life - not greatly different than that of the clock battery. The latter is integrated into the face plate structure so that both the face plate and battery are replaced at the same time.

The solid state clock of Fig. 1 includes an oscillator 10, which is preferably crystal-controlled, a counter 12 and a decoder 14. These three circuits are integrated onto a single chip. The outputs of the decoder 14 are applied to the display elements of the clock. For purposes of the present application, these are shown to consist of four, 7-segment display elements (see Fig. 4), two display elements for hours and two for minutes. One of these elements is shown in detail in Figs 2 and 3. This element includes a transparent face 16 formed of

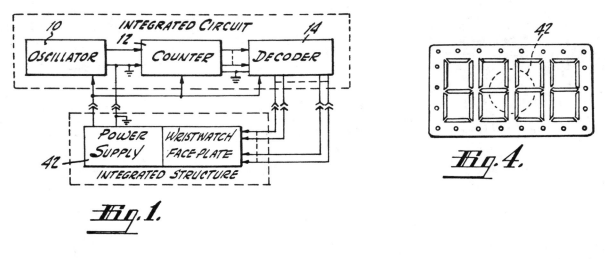

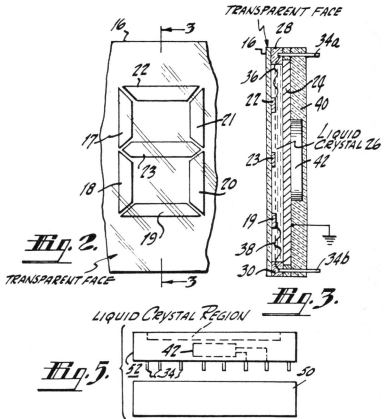

insulating material and having seven transparent conductive segments 17-23 secured to the underside thereof. Located between the transparent face and the opaque, conducting back plate 24 is the liquid crystal 26. The insulating gaskets 28 and 30 of Fig. 3 serve to seal the face plate structure and to space the back plate from the transparent face.

Each of the conducting segments 17-23 is connected to a different pin. Two such pins, 34a and 34b, are shown in Fig. 3. For purposes of illustration, short flexible leads 36 and 38 are shown connecting segments 22 and 19 to these two pins 34a and 34b. In practice, the connections instead may be made by means of very fine transparent conductors on the underside of the transparent face.

An insulation element 40 is located behind the back plate 24. A battery 42 is located within this insulator. One terminal of the battery may connect to the back plate 24 which, in turn, may provide the ground for the wrist-watch. This back plate may be connected via one of the terminals to integrated circuit 10, 12, 14. The other terminal of the battery 42 may be connected to another terminal which leads to the integrated circuit. The battery provides the necessary power to operate the circuit.

The life of the liquid crystal is relatively limited - very much less than that normally contemplated for a wrist-watch, for example. This problem is solved according to the invention by integrating the battery 42 into the wrist-watch face plate structure as shown in Figs 3, 4 and 5. The integrated circuit comprising oscillator 10, counter 12 and decoder 14 is located in the case (either plastic or metal) which forms the bottom portion 50 of the solid state clock. The remainder of the clock includes the face plate with the four alphanumeric characters as shown in Fig. 4 and with the battery 42 integrated into the same structure as the face plate.

"" "" "" "" "" ""

1,194,544 (British)   Patented June 10, 1970

LIQUID CRYSTAL DETECTOR.

<u>Byron B. Brenden</u> and <u>Howard R. Curtin</u>, assignors to The Battelle Development Corporation. Application Aug. 1, 1967; prior U.S. applications Aug. 3, 1966 and Feb. 8, 1967.

This invention relates to ultrasonic holography. According to the present invention, there is provided apparatus for detecting and displaying the interference maxima and minima of two interfering ultrasonic beams, comprising a thin film of liquid crystal adapted to be positioned at the area of interference of said two interfering ultrasonic beams, said liquid crystal having the property of responding to said two interfering ultrasonic beams by varying the rotation of plane polarized light impinging thereupon, wherein the thin film of liquid crystal is entrapped between two windows positioned and held by a frame. The changes in the properties of the liquid crystal may then be translated into variations in intensity of a beam transmitted through it. The apparatus of the invention is transformed, by two interfering beams of ultrasonic energy into a transient ultrasonic hologram which can be reconstructed with plane polarized light.

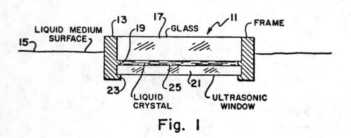

Fig. 1

Referring to Fig. 1, the detector 11 makes use of the unique properties of substances known as "liquid crystals." One property of cholesteric liquid crystals is that they are optically active, i.e., they rotate the plane of polarization of light which passes through them. The rotation may be as much as 18,000 degrees per millimeter of travel. In Fig. 1 the frame 13, when in use, is immersed in the liquid medium surface 15. A glass window 17 in the frame 13 is supported by a ridge 19 on the inner wall of the frame 13.

A second transparent plate 21 is supported by a second ridge 23 on the frame 13. The liquid crystal 25, such as cholesterol chloride, is disposed between the plates 17 and 21. The plates 17 and 21 may be two sheets of glass, two membranes, a sheet of glass and a membrane or some similar arrangement.

When the liquid crystal 25 is irradiated by two ultrasonic waves, variations in the thickness corresponding to variations in ultrasonic intensity are developed. Plane polarized light is transmitted through the irradiated thin layer of liquid crystal 25 and variations in the rotation of the plane of polarization are produced in accordance with the distribution of the interfering ultrasonic beams giving rise to the diffraction of light.

Figure 2 shows the detector 11 in use. Fig. 2 also shows a method of reconstructing a three-dimensional image of the object 43 from the ultrasonic hologram 47. The hologram 47 may be recorded on film when convenient.

(see Fig. 2 on the opposite page...)

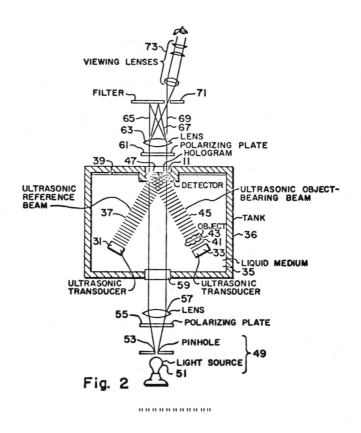

Fig. 2

3,519,330 (U.S.)   Patented July 7, 1970

TURNOFF METHOD AND CIRCUIT FOR LIQUID CRYSTAL DISPLAY ELEMENT.

George H. Heilmeier, assignor to RCA Corporation.
Application Sep. 14, 1967.

    For Figs. 3 and 5 and explanatory text see U.S. Patent No. 3,503,672.

    Liquid crystal element is turned off by applying a relatively short-duration, relatively high-amplitude voltage pulse thereto during a period in which the cell is not permitted to to store charge.

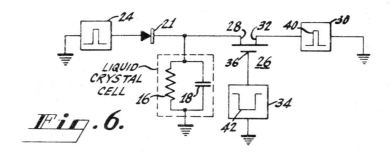

    A circuit according to the present invention for quickly turning off a liquid crystal cell is shown in Fig. 6 above. The excitation or "write" pulse source is shown at 24. The elements 16, 18 and 21 correspond to the like-numbered elements in the circuit of Fig. 3 (Pat. 3,503,672). In addition to these elements, the circuit of Fig. 6 includes a P-type insulated gate, field-effect transistor 26 connected at its drain electrode 28 to the liquid crystal cell. An erase pulse generator 30 is connected to the source electrode 32 of the transistor. A gate pulse generator 34 of low internal impedance is connected to the gate electrode 36 of the transistor.

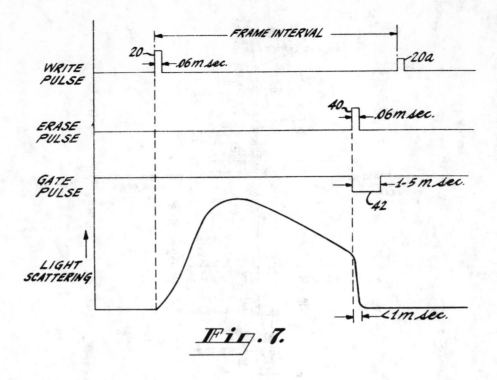

Fig. 7.

The operation of the circuit of Fig. 6 may be better understood by referring also to Fig. 7. Information is written into the cell by a pulse 20. A short interval (somewhat longer than the duration of gate pulse 42) before the following write pulse 20a occurs, an erase pulse 40 is applied to the source electrode of the transistor 26 concurrently with the application of a gate pulse 42 to the gate electrode 36. The erase pulse may be of relatively short duration - the same 0.06 millisecond duration, for example, as the write pulse, but should be of relatively high amplitude, greater than that of the domain alignment voltage threshold of the cell. The actual value of voltage which should be employed depends upon many parameters including the amount of quenching which is desired, the thickness of the cell, the temperature, and so on. As one example, for some cells approximately 1/2 mil (0.005 inch) thick, erase voltages form 75 to 150 volts of 0.06 millisecond duration have been employed.

The gate pulse 42 should be of somewhat greater duration than the erase pulse. For example if the erase pulse is of 0.06 millisecond duration, the gate pulse may have a duration of from 1 to 5 milliseconds. The gate pulse causes the impedance of the conduction path from the source-to-drain electrode of the transistor abruptly to change from a very high value of impedance to an extremely low value of impedance. The pulse 40 applied through the transistor to the liquid crystal cell causes a relatively high electric field to be produced across the cell. This causes the relatively large domains to align with the field in the manner shown in Fig. 8.

However, the pulse is of such short duration that there is insufficient time for ion current of flow through, that is, for oins to move a significant distance through the liquid crystal during the pulse interval.

After the erase pulse has terminated, the gate pulse is still present so that there is a low impedance path from the capacitor 18 of the liquid crystal cell through the low impedance of the conduction path of the transistor 26 and the low internal impedance of the erase pulse generator 30 to ground. Accordingly, any charge which momentarily accumulates on the capacitor 18 quickly discharges to ground. The time required for this discharge to occur is sufficiently short that ion transit induced turbulence is not created.

The light scattering characteristic obtained with the circuit of Fig. 6 is as illustrated in the last graph of the Fig. 7. The amount of light scattering produced drops off to an extremely low value within a period of less than a millisecond. During this period, the amount

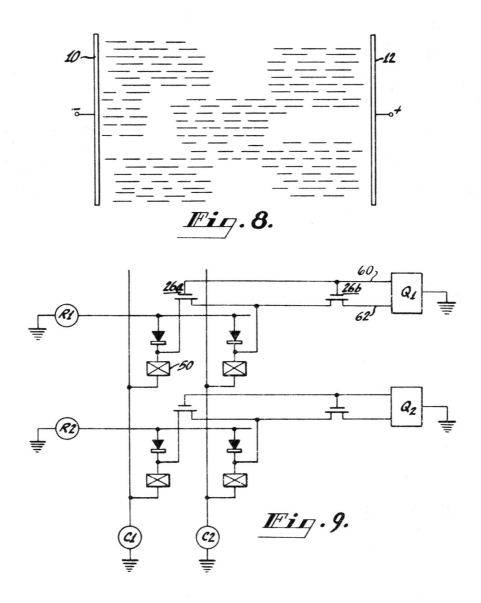

Fig. 8.

Fig. 9.

of light scattering does not drop to zero such as would be the case if the domains were completely unexcited but does drop to a very small value. The contrast ratio is typically between 1 and 1-1/2.

The liquid crystal cells of Fig. 6 may be arranged in a matrix in a manner shown in Fig. 9. While only a two-by-two matrix is illustrated, in practice there would be many more elements.

" " " " " " " " " " " "

1,201,230 (British)   Patented Aug. 5, 1970

CHROMATIC DISPLAY OR STORAGE DEVICE.

<u>The National Cash Register Company</u> (Inventor's name not given).
Application Jan 31, 1969.

For description see U.S. Pat. Nos. 3,578,844 and 3,600,060.
" " " " " " " " " " " "

3,524,726 (U.S.)   Patented Aug. 18, 1970

SMECTOGRAPHIC DISPLAY.

<u>Heinz A. de Koster,</u> assignor to General Time Corporation.
Application Apr. 4, 1968.

   Display device comprising a heat conductive substrate having a coating of crystalline liquid material which exhibits a colored smectic state when heated to a predetermined smectic temperature range. A temperature gradient in part encompassing the smectic temperature range is imposed on the crystalline liquid coating to produce therein a discrete, colored smectic band forming a display indicator. The temperature gradient may be produced by heating elements or thermoelectric elements spaced on the substrate; these are proportionally raised or lowered in temperature to cause the indicator to move. The display device may be linear or curvilinear in shape and means are disclosed for producing a multi-revolution indicator on a circular device.

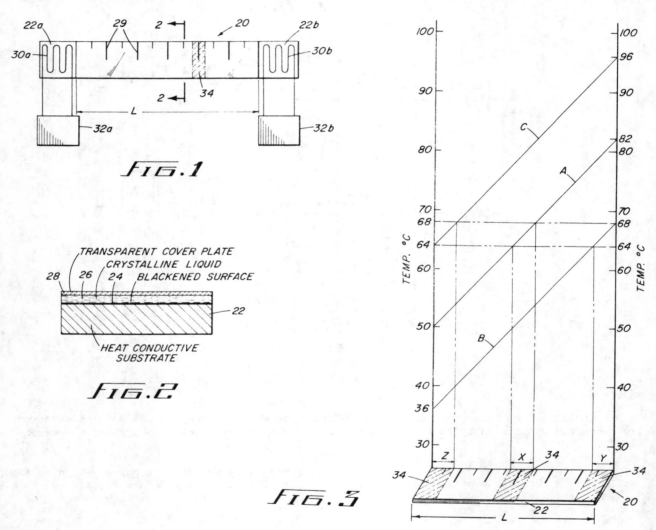

   Referring to Fig. 1, there is shown a linear smectographic display 20 such as might be used as the display dial of an electrical instrument. It comprises a substrate 22 (Fig. 2) formed of a good heat conducting material such as a metal like copper, silver or aluminum, or a highly thermal conductive glass or ceramics. Preferably, the visible surface 24 of substrate 22 is blackened by painting, anodizing or the like to provide visual contrast with the indicator of the display. Substrate 22 may be formed as a monolithic member; preferably, however, and particularly where size is a critical feature, it will be a layer which is applied by printed circuit, thick film, or thin film techniques to a backing member.

A thin coating 26 of crystalline liquid material is applied over the darkened surface 24 of substrate 22. Coating 26 in a typical application will be in the order of approximately 0.1 mm. or 0.004 inch thick. The crystalline liquids which are used for the purposes of the invention comprise a group or organic chemical compounds which exhibit a visually discernible, colored smectic state. Typical examples of such materials are thallium stereate, thallium oleate, cholesterine derivatives and certain unsaturated aliphatic carbonacides.

A transparent cover plate 28 of glass, plastic or the like is preferably placed over the coating 26 of crystalline liquid to protect it from contamination, or damage from contact; the cover plate may further be extended to surround the display to prevent heat dissipation and conserve power. It has been found that the use of a cover plate also improves the visibility of the smectic band. Cover plate 28 may also conveniently be provided with indicia 29 appropriate to the particular application for which display 20 is intended.

The temperature gradient producing the colored smectic band which acts as the indicator of display 20 is produced by heating one end of substrate 22 relative to the other. This can be accomplished by contacting one or both of the ends with heated or cooled fluids, with exothermic or endothermic materials, or with an open flame or the like. However, when the display forms a part of an electrical instrument and accuracy is required, the heating or cooling is preferably accomplished electrically.

Referring to Fig. 1 again, heat conductive tabs 22a, 22b are provided at each end of substrate 22. A heating element in the shape of a coil 30a, 30b is placed on or about each tab 22a, 22b, and each coil is connected to a variable current source 32a, 32b. Coils 30a and 30b may be formed of high resistance electrical wire, or may be applied on the tabs 22a and 22b by printed circuit or thin film techniques; where thin film is used, a strip may be substituted for the coil shaped elements shown in Fig. 1. The variable current sources 32a and 32b will, in the case of a display for an electrical instrument, be a part of or be controlled by the output circuitry of the instrument.

The operation of display 20 is best understood with reference to Figs. 1, 2 and 3. Assume that the crystalline liquid coating 26 is a material having a smectic temperature range of 64 to 68°C. If heating coil 30a is heated to 50°C. and heating coil 30b is heated to 82°C., the temperature gradient shown graphically by line A in Fig. 3 will be imposed over the length L of substrate 22. Under these temperature conditions, the smectic range of 64 to 68°C. will exist as a narrow temperature band X at the center of display 20, as shown by projection in Fig. 3. Accordingly, the crystalline liquid coating 26 within band X will be in the smectic state and form a highly visible colored indicator 34 on display 20.

By proportionally raising or lowering the temperatures of both coils 30a and 30b, indicator 34 can also be moved over the length L of display 20 as shown in Fig. 3. For example, by gradually lowering the temperatures of both coils 30a and 30b by 14°C. to respectively 36°C. and 68°C., the temperature gradient becomes that shown by line B, and the smectic temperature range and thus indicator 34 is gradually shifted to band Y at the right end of display 20. Conversely, by raising the temperatures of each coil by 14°C. respectively, to 64°C. and 96°C. (gradient line C), indicator 34 can be shifted to band Z at the left end of the display 20. Indicator 34 can, of course, be halted at any intermediate point between bands X, Y and Z by controlling the temperatures of coils 30a and 30b, and this control may be readily accomplished through connection with the output of an electrical instrument.

It can also be seen from Fig. 3 that the width of the smectic zone and therefore indicator 34 can be varied by varying the steepness of the temperature gradient; a steeper gradient produces a narrower indicator while a flatter gradient makes the indicator broader.

The smectographic display is not, however, limited to the linear type displayed above, and may be curvilinear in shape. Referring to Fig. 4 (next page), there is shown a circular smectographic display 36 which, like the linear display, can be used as the display dial for an electrical instrument, and which is particularly suitable for use as the face of a clock.

Due to the circular shape of display 36, a somewhat more complex means is preferably used to produce the temperature gradient required to generate an indicator which revolves continuously over the entire surface 40. Referring to Fig. 4 (next page), a plurality of thermoelectric elements 46, 48, 50, 52 and 54 are provided under substrate 38 (Fig. 5, next page) and divide up

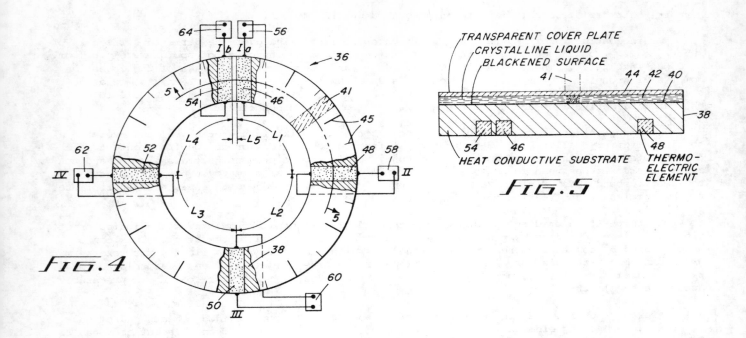

the substrate essentially into quadrants L1, L2, L3, and L4. Each quadrant functions in much the same manner as the linear display described above.

" " " " " " " " " " "

3,527,945 (U.S.) Patented Sept. 8, 1970

MOUNTING STRUCTURE FOR A LIQUID CRYSTAL THERMAL IMAGING DEVICE.

Gerald Jankowitz, assignor to Barnes Engineering Company.
Application Sept. 24, 1968.

A thermal image system is provided utilizing temperature sensitive liquid crystal detectors which detect thermal patterns in a field of view imaged thereon. In carrying out this invention in one illustrative embodiment thereof, the membrane containing the liquid crystal detector is supported through ribs or spokes of the same material as the membrane, to provide an extremely high thermal impedance between the body of the membrane containing the liquid crystal material and its external support.

Gradients due to differential radiation coupling to the liquid crystal are eliminated by using a diffusing reflective cone in front of the membrane, which forces all points on the membrane to optically see the same solid angle of temperature controlled chamber and environment.

Referring now to Fig. 1 (next page), and evacuated housing 30 is provided having an evacuation tube 34 which may be connected to a vacuum pump (not shown). The temperature of the housing is controlled by a temperature control means 32 which is shown schematically and which performs the function of controlling the temperature of the evacuated housing 30. The enclosure has a first window 24 which is transparent to infrared radiation. At least a second window 36 on the opposite side of the housing 30 is provided which is transparent to visual rediation but reflects infrared radiation by the provision of a dichroic coating. It will be apparent that the window 36 may be in the form of a plurality of windows, if desired.

Mounted in the housing 30 is a liquid crystal detector referred to generally with the reference character 10. The liquid crystal detector is comprised of a very thin membrane 12 on the order of 0.00025" thick made of a suitable material such as polyethylene terephthalate, commercially sold as "Mylar," which is blackened on one side 14 and carries the liquid crystal metarial 16 on the other side thereof. The present invention is not directed to the type of

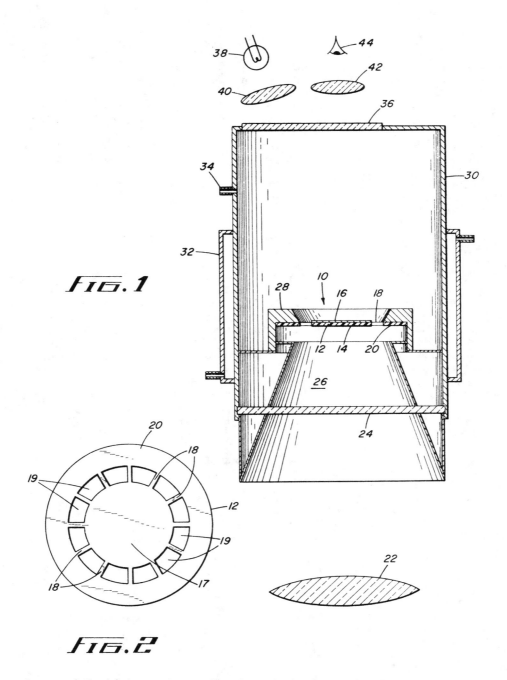

Fig. 1

Fig. 2

liquid crystal material which may be employed, and the invention is not considered limited to such a material, and includes all commercially available type liquid crystals.

As will best be seen in Fig. 2, the membrane 12 has a central body portion 17 on which the liquid crystal is mounted. An annular outer extremity 20 is eparated by the body portion 17 of the membrane 12 by ribs or spokes 18 which connect the body portion 17 to the rim 20. The spokes 18 of the membrane 12 are separated by open or cut-out areas 19. The membrane 12 may be constructed by cutting away the Mylar in areas 18 and leaving the spokes, or by melting away the undesired Mylar near its outer extremities.

Liquid crystal detector 10 is mounted on its outer periphery 20 to a support member 28 which is separated from the body of the membrane 12 by the spokes 18. The support 28 is mounted on a split diffusing reflective cone 26. On the opposite side of the chamber a source of illumination is applied via a collimating lens 40 through the window 36 onto the liquid crystal detector 10. The field of view of the imaging system is imaged by an objective lens 22 and the diffusing reflecting cone 26 onto the liquid crystal detector 10 through the first window 24. The visual image of the thermal image of the field of view thus applied is viewed by the eye 44 through a lens 42 through the second window 36. It will be apparent that separate windows may be utilized for the window 36, one for the source 38 and one for viewing by the eye 44.

The liquid crystal material 16 is deposited on the thin membrane 12 blackened at 14 to enhance the absorption of the thermal energy. The liquid crystal detector is mounted in an evacuated housing to eliminate thermal effects due to convection of the air. The two windows 24 and 36 in the evacuated housing 30 are provided to allow the infrared image from the field of view to be focused upon the detector and to permit the viewing of the visible image produced by the temperature sensitive material of the liquid crystal detector 10. The housing 30 is heated or cooled by the temperature controller 32 to bring the membrane 12 to the approximate temperature of operation which will depend on the operating characteristic of the liquid crystal material used for the liquid crystal detector 10.

In providing a thermal imaging system where the radiant energy to be measured is at approx. the same temperature as the environment, the temperature of the membrane on which the liquid crystal is mounted must be tightly controlled, for any irregularities or non-uniformities will impair the operation of the system. One of the problems is temperature non-uniformity across the membrane. This is caused by the fact that under usual conditions the percentage of the housing and the outside world seen by the emenents in the center and the edge of the membrane will be different, causing a compilation of differential radiation which produces variations in membrane temperature. On the visual side such non-uniformities will render a distorted version of the thermal image applied to the liquid crystal detector 10. This problem is treated in the present invention by optically forcing each element of the membrane 12 to see the same solid angle of temperature-controlled chamber and the outside world which is in the form of infrared energy, since the window 24 is transparent to infrared. Radiance due to differential radiation coupling can be eliminated. This optical technique is achieved by using a reflective cone 26 in front of the membrane 12. The cone 26 is so arranged that one side of the membrane sees over 2π steradians out of the chamber. This occurs either directly or after a reflection off of the cone. The other side of the membrane sees the inside of the thermal chamber 30. As an addition, the window 36 through which the membrane is viewed is coated with an interference filter which reflects the infrared but transmits the visible. Thus each element of the membrane 12 sees the same thermal radiation.

The other problem treated by the present invention deals with the problem of the absolute temperature of the membrane carrying the liquid crystal. This problem arises in supporting the liquid crystal detector 10 by some means. Since the support will normally have a much greater mass, the membrane will tend to assume the temperature of the support, thus greatly reducing the usable membrane area, since the support is massive thermally when contrasted to the membrane. This problem is treated, and greatly minimized, in the present invention by means of an extremely high thermal impedance between the body of the membrane 17 and the support 28. Materials to provide such a high thermal impedance would be difficult and would make the detector structure difficult if not impossible to manufacture in view of the thinness of the membrane. However, by using the membrane itself and reducing the area to the bare minimum by providing suppprting ribs as needed to hold the membrane, the thermal gradient can be held to the ribbed area, thus providing a maximum usable uniform membrane area. Summarizing, the thermal gradient between the membrane and the support is primarily due to conduction from the membrane to the holder and to eliminate this conduction gradient, portions of the membrane are cut away from the periphery of the membrane, providing spoked connections to its outer extremity, raising the conduction impedance to a sufficiently high level that the thermal gradient between the center of the membrane and the walls of the housing or the support occur close to the rim of the membrane and not to the center. Since the temperature difference between the support 28 and the membrane 12 determines the useful membrane area, by supporting the membrane through the ribs or spokes 18, it is possible to use almost the entire body 17 of the membrane even though several degrees of temperature difference will exist between the membrane 12 and the support 28.

" " " " " " " " " " " "

3,529,156 (U.S.) Patented Sept. 15, 1970

HYSTERETIC CHOLESTERIC LIQUID CRYSTALLINE COMPOSITIONS AND RECORDING DEVICES UTILIZING SUCH COMPOSITIONS.

James L. Fergason and Newton N. Goldberg, assignors to Westinghouse Electric Corporation.

Application June 13, 1966.

It has now been discovered, and the present invention is in large part based on this discovery, that certain liquid crystal cholesteric compositions may be formulated to provide a significant color change time lag after the change from the liquid phase to the cholesteric phase. Such hysteretic compositions may be employed, for example, to determine not only whether a given temperature has been exceeded but also to provide a record, for a significant time period of whether the temperature has been exceeded. The lag in color change will also occur if the compositions have been exposed to high intensity electric fields in the order of 20 volts per micron or to gas vapors, such as vapors of chloroform in concentrations in the order of 100 ppm in air.

It is apparent, of course, that not all cholesterogenic materials are hysteretic in their color transformations. Indeed, most of the heretofore known materials rapidly change color in response to environmental changes. The mixtures or combination of compounds of this invention return to their colored state, from a colorless liquid phase, only after a time delay in the order of 3 to 30 minutes. Moreover, the mixtures will have bright visible colors in the aligned cholesteric phase and will not crystallize in the range of normal temperatures, i.e. 0°F. to 300°F. The particular substrate employed, whether glass, polyester or some other clear plastic film, will have some effect on the actual color change delay time for a particular composition but the time delay phenomenon will be observed on all substrates.

The hysteretic liquid crystalline cholesteric mixtures of this invention contain (A) from about 15 to 40 percent, by weight, of at least one halide selected from the group consisting of cholesteryl chloride, cholesteryl bromide, cholesteryl iodide, cholestanyl chloride, cholestanyl bromide and cholestanyl iodide and (B) at least about 30 percent, by weight, of at least one compound selected from the group consisting of cholesteryl erucyl carbonate, cholestanyl erucyl carbonate, cholesteryl oleyl carbonate, cholestanyl oleyl carbonate, cholesteryl erucate, cholestanyl erucate, cholesteryl oleate and cholestanyl oleate.

Referring now to Fig. 1, there is illustrated as one example of the invention, a simplified temperature recording or information storing device 10 that includes a thin clear sheet of polyethylene terephthalate 11 having a thickness of about 6 microns. A pigmented film 12 containing carbon black dispersed in a thin layer of polyvinyl alcohol and having a thickness of about 12 microns is deposited on one side of the terephthalate film 11. The film 12 is applied as an aqueous solution or suspension and dried on the film 11. A film of a hysteretic liquid crystalline cholesteric mixture 13 having a thickness of about 10 microns is deposited on the other side of the terephthalate base 11. The film 13 is deposited by applying a 10-percent solution of the hysteretic mixture in petroleum ether onto the film 11 and evaporating the solvent.

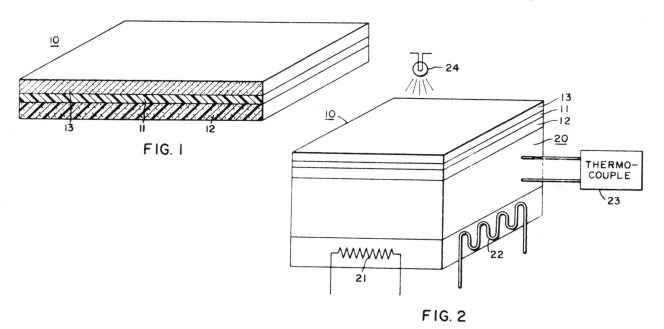

FIG. 1

FIG. 2

As an illustration of the practice of the invention, there is shown in Fig. 2, the recording device 10 placed so that film 12 is in intimate contact with an aluminum plate or block 20. The block 20 is adapted to be heated, as for example, by a resistance heater 21, and is adapted to be cooled either by lower ambient temperatures after the heating has ceased or by cooling means such as the coil 22 containing circulating cold water. A thermocouple 23 is embedded in the plate 20 so that an accurate measurement of the temperatures of the block can be made. A source of white light 24 is provided so that the color of the layer of hysteretic film in the device will be visible to the naked eye. The visible color is, of course, a known phenomenon associated with the light scattering properties of the cholesteric phase.

A series of simple recording devices was prepared, in the manner described hereinabove in conjunction with Fig. 1, using for film 13 a variety of simple admixtures or combinations of compounds in accordance with the constituent mixtures and composition ranges set forth above for the preparation of suitable hysteretic compositions.

TABLE I.—PROPERTIES OF HYSTERETIC MIXTURES

Example No.	Composition	Color of cholesteric phase	Clearing point, °C.
1	30% cholesteryl chloride, 70% cholesteryl erucate.	Red	37.5
2	20% cholesteryl chloride, 80% cholesteryl erucate.	Green	38.2
3	15% cholesteryl chloride, 85% cholesteryl erucate.	Blue	40
4	24% cholesteryl chloride, 76% cholesteryl oleyl carbonate.	Green	45
5	20% cholesteryl chloride, 80% cholesteryl oleyl carbonate.	Blue	45.4
6	30% cholesteryl chloride, 70% cholesteryl oleyl carbonate.	Red	46
7	26% cholesteryl bromide, 74% cholesteryl oleyl carbonate.	Green	43.8
8	26% cholesteryl bromide, 74% cholesteryl oleate.	do	41.1
9	26% cholesteryl chloride, 74% cholesteryl erucyl carbonate.	do	38.6

The compositions presented in Table I were found to be hysteretic and their room temperature colors, i.e. the color of the cholesteric phase, and their clearing points, i.e. the melt point or liquid transition point, were as indicated. In each case, at least a three minute time lag was observed in the color change, after the layer had been cooled to a temperature where the material was in the cholesteric phase.

The cholesteric layer may be deposited from other volatile organic solvents such as chloroform, halogenated hydrocarbons and volatile hydrocarbon solvents. The cholesteric layer need only be a few microns in thickness but may be in the order of 50 microns and more. It should also be understood that cholestanyl compounds may be substituted for the foregoing cholesteryl compounds to provide similar results.

As is apparent from the data in Table I, only a small variation in the clearing point occurs with varying amounts of compounds in a given combination of compounds. It has also been discovered that significant changes in clearing points may be made by the addition of certain clearing point modulating agents. Up to 35 percent, by weight, of (c) a clearing point elevating agent selected from the group of esters consisting of cholesteryl nonanoate, cholestanyl nonanoate, cholesteryl decanoate and cholestanyl decanoate may be included in the admixture of compounds (A) and (B) to raise the clearing point temperature of the deposited cholesteric layer. The hysteretic effect is present in the layers even though the ester has been added to raise the clearing point. However, the total amount of materials or compounds selected from (A) and (C) should not exceed 60 percent of the weight of the total mixture. As little as 1.0 percent of the esters in the group (C) will have noticeable and useful effect in elevating the clearing point temperature of a basic mixture of (A) and (B). In general, the addition of about 5 percent of any of the esters (C) will raise the clearing point about one degree centigrade. (See Table II)

TABLE II.—ELEVATION OF CLEARING POINT

Example No.	Composition	Color of cholesteric phase	Clearing point, °C.
10	28% cholesteryl chloride, 26% cholesteryl nonanoate, 46% cholesteryl oleyl carbonate.	Red	51
11	28% cholesteryl chloride, 10% cholesteryl nonanoate, 62% cholesteryl oleyl carbonate.	Red	44
12	28% cholesteryl chloride, 26% cholesteryl decanoate, 46% cholesteryl oleyl carbonate.	Red	51
13	28% cholesteryl chloride, 10% cholesteryl decanoate, 62% cholesteryl oleyl carbonate.	Red	44

In order to depress the clearing point of a given mixture of (A) and (B), up to about 10 percent of another class of modulating agents (D), in this instance known as depressing agents, may be included in the mixture. Fatty compounds which dissolve the admixture (A) and (B) without crystallizing will function as clearing point depressing agents. Suitable fatty compounds for this purpose include the fatty acids, fatty esters and fatty alcohols, both saturated and unsaturated. Syitable fatty acids are, saturated and unsaturated hydrocarbon acids with preferably over 8 carbon atoms, for example, oleic, stearic, palmitic, lauric, erucic, myristic and behenic acids. Suitable fatty esters are derived by reacting monohydric and polyhydric hydrocarbon alcohols, such as methyl and ethyl alcohol, ethylene glyceryl and glycerol with the fatty acids, for example, methyl oleate, ethyl oleate, cetyl oleate, diolein, triolein, tributrin, tripalmitin, tristearin and mixtures thereof, particularly mixtures of triglyceryl esters of fatty acids such as the naturally occurring olive, sperm, soybean, corn and tall oils. Suitable fatty alcohols are for example, oleyl, stearyl, cetyl and lauryl alcohols. The liquid fatty compounds are preferred but the solid compounds may be employed if the ultimate admixture with the combination of (A) and (B) does not crystallize. Crystallization will, of course, destroy the liquid crystal cholesteric phase. (See Table III)

TABLE III.—DEPRESSION OF CLEARING POINT

Example No.	Composition	Color of cholesteric phase	Clearing point, °C.
14	26% cholesteryl chloride, 73% cholesteryl oleyl carbonate, 1% oleic acid.	Green	36.4
15	26% cholesteryl chloride, 72% cholesteryl oleyl carbonate, 2% oleic acid.	...do...	35.3
16	26% cholesteryl chloride, 71% cholesteryl oleyl carbonate, 3% oleic acid.	...do...	33.4
17	26% cholesteryl chloride, 73% cholesteryl oleyl carbonate, 1% methyl oleate.	...do...	36
18	26% cholesteryl chloride, 72% cholesteryl oleyl carbonate, 2% methyl oleate.	...do...	35
19	26% cholesteryl chloride, 71% cholesteryl oleyl carbonate, 3% methyl oleate.	...do...	33
20	26% cholesteryl chloride, 73% cholesteryl oleyl carbonate, 1% triolein.	...do...	36.5
21	26% cholesteryl chloride, 72% cholesteryl oleyl carbonate, 2% triolein.	...do...	35.5
22	26% cholesteryl chloride, 71% cholesteryl oleyl carbonate, 3% triolein.	...do...	33.5
23	26% cholesteryl chloride, 73% cholesteryl oleyl carbonate, 1% oleyl alcohol.	...do...	36
24	26% cholesteryl chloride, 72% cholesteryl oleyl carbonate, 2% oleyl alcohol.	...do...	35
25	26% cholesteryl chloride, 71% cholesteryl oleyl carbonate, 3% oleyl alcohol.	...do...	33
26	26% cholesteryl chloride, 73% cholesteryl oleyl carbonate, 1% olive oil.	...do...	36.5
27	26% cholesteryl chloride, 72% cholesteryl oleyl carbonate, 2% olive oil.	...do...	35.5
28	26% cholesteryl chloride, 71% cholesteryl oleyl carbonate, 3% olive oil.	...do...	33.5
29	26% cholesteryl chloride, 73% cholesteryl oleyl carbonate, 1% sperm oil.	...do...	36
30	26% cholesteryl chloride, 72% cholesteryl oleyl carbonate, 2% sperm oil.	...do...	35
31	26% cholesteryl chloride, 71% cholesteryl olyel carbonate, 3% sperm oil.	...do...	33
32	26% cholesteryl chloride, 73% cholesteryl oleyl carbonate, 1% tributrin.	...do...	35.5
33	26% cholesteryl chloride, 72% cholesteryl oleyl carbonate, 2% tributrin.	...do...	34.5
34	26% cholesteryl chloride, 71% cholesteryl oleyl carbonate, 3% tributrin.	...do...	32.5

As noted heretofore, the device of Fig. 1 so long as it has a layer of hysteretic cholesteric liquid crystalline material, will function as a temperature recording device for the environment contacting the device. The device may be calibrated by the selection of a proper hysteretic film composition, to determine and record whether a particular temperature has been exceeded.

All of the heretofore described hysteretic compositions may be employed as thin films in devices which will record or store information on environmental conditions other than temperature. Again referring to Fig. 1, a 10 micron layer of a composition containing 28 percent cholesteryl chloride and 72 percent cholesteryl oleyl carbonate is deposited as the layer or film 13, the substrate sheet 11 and the pigmented film or layer 12 being as described. Such a device 10 will have an easily visible pronounced color at room temperature when exposed to white light. If a 200 volt field is applied across the 10 micron hysteretic cholesteric layer 13, the color will disappear and the layer will become colorless. The layer 13 will remain clear for about 3 weeks if it is not heated through its clearing point.

The same device may, instead, be employed to record the presence of contaminating vapors, for example. When exposed to air containing from 70 to 100 ppm of either chloroform or trichloroethylene, the color will disappear. The hysteretic layer will remain colorless for about ten minutes after it is removed from the contaminating atmosphere unless heated through its clearing point.

" " " " " " " " " " " "

3,532,813 (U.S.)   Patented Oct. 6, 1970

DISPLAY CIRCUIT INCLUDING CHARGING CIRCUIT AND FAST RESET CIRCUIT.

Bernard J. Lechner, assignor to RCA Corporation.
Application Sept. 25, 1967.

Matrix of display means each such means including a display element such as a nematic liquid crystal cell and a storage element. A circuit which serves both as a low impedance path to permit charge to be delivered to the two elements in parallel and as a fast reset circuit for the display element connects these two elements.

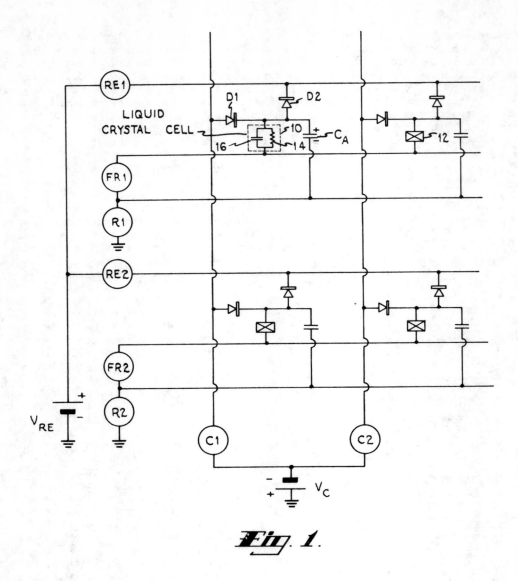

Fig. 1.

For purposes of illustration, a two-by-two matrix of storage elements is illustrated in Fig. 1. However, it is to be understood that, in practice, the matrix may have many more elements than this. Each display means includes a liquid crystal cell such as shown at 10 in equivalent circuit form and as shown at 12 in block form. Each such cell may be considered to consist of an internal resistance 14 shunted by a capacitance 16. Each display means also includes a capacitor C_A connected in shunt with the liquid crystal cell through the relatively low impedance of a fast reset pulse generator such as FR1.

The addressing means for each location comprises two diodes D1 and D2. One diode D1, employed to write (to charge the liquid crystal cell), is connected at its cathode to the display means and to the anode of diode D2 and at its anode to a column of the matrix. The other diodes, such as D2, which are employed to erase the respective cells, are connected at their cathodes to the rows of the matrix. The diodes such as D2 are normally reverse biased by a positive voltage supplied by a bias source shown as a battery V_{RE}. The diodes such as D1 are back biased by a negative voltage supplied by a bias source V_C.

There is a row pulse generator connected to each row of the matrix. Two such generators R1 and R2 are shown. There is a column pulse generator connected to each column of the matrix and two such column pulse generators are shown at C1 and C2. There is a reset pulse generator connected to each row of the matrix. There is also a fast reset pulse generator associated with each row of the matrix.

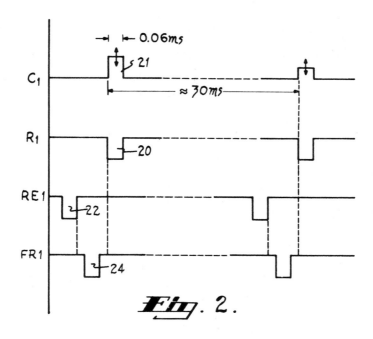

Fig. 2.

In the discussion which follows of the operation of the circuit of Fig. 1 both Fig. 1 and 2 should be referred to. The matrix of Fig. 1 may be addressed a row at a time. In other words, each time a row generator is actuated, all of the column generators are concurrently actuated so that, for example, one row interval, 0.06 millisecond in commercial television applications, is available for charging the liquid crystal cells of the row being addressed. (Shorter time durations than this such as the 0.01 millisecond retract interval, may be used instead.) The column generators C1, C2 and so on, provide pulses corresponding in amplitude to the amplitude of the signal to be displayed at each point along the row being addressed and the amplitudes of these pulses determine the extent to which the liquid crystal cells to which these pulses are applied, will "light up," that is, the extent to which these liquid crystal cells will scatter incident light.

To turn on a liquid crystal cell, a row generator such as R1 applies a negative pulse such as 20 of Fig. 2 to a row of liquid crystal cells at the same time that a column pulse generator such as C1 applies a positive pulse such as 21 to a column of elements. The row and column pulses in coincidence are of sufficient amplitude to overcome the bias on a diode such as D1 and charge flows through this diode and through the two paths consisting of, on one hand, the capacitor C_A and on the other hand, the fast reset pulse generator FR1 and the liquid crystal cell 10. The fast reset pulse generator has a relatively low internal impedance and therefore the two paths appear to the column and row generators to be simply the liquid crystal cell in shunt with the capacitor C_A.

In the event that it is desired that the liquid crystal cell light up, the amplitude of the column pulse must be such that when added to that of the row pulse, it exceeds the threshold of diode D1 and reverse biasd by source V_C by an amount sufficient to cause dynamic scattering to occur in the liquid crystal cell. The capacitor C_A, which normally has a capacitance of ten or more times that of the internal capacitance 16 of the liquid crystal cell 10, becomes charged in response to the concurrent receipt of a row and column pulse. After these pulses have terminated, this capacitor C_A discharges into the liquid crystal cell maintaining it in its dynamic scattering condition. It should be mentioned here again that the fast reset pulse generator FR1 has a relatively low internal impedance compared to the internal resistance of the liquid crystal cell so that the capacitor C_A appears to be directly connected across the liquid crystal cell.

The diode D1 is poled in the reverse direction with respect to the discharge of capacitor C_A so that this capacitor cannot discharge through a column generator. The diode D2 is poled in the forward direction with respect to the discharge of capacitor C_A, however, the bias source V_{RE} back biases diode D2 to an extent such that it cannot conduct. (As an aside, diode D1 is also back biased but by the source V_C.)

Immediately prior to the time that it is desired again to write information into a liquid crystal cell, the reset pulse generator for the row containing that cell is turned on. For example, for the cell 10, the reset pulse generator RE1 is turned on and it applies a negative pulse, such as 22 of Fig. 2 to the diodes D2 of row 1. This pulse is of sufficient amplitude to overcome the back bias on the diodes D2 and the capacitors C_A and 16 discharge through these diodes. The pulse duration may be perhaps 0.010-0.015 millisecond, however, longer pulses may also be used. The liquid crystal cell may be turned off by applying thereto a short-duration, relatively high-amplitude pulse during a period in which the liquid crystal cell is not permitted to retain charge. Such a pulse causes the domains of the liquid crystal cell to align and this causes the scattering exhibited by the cell to reduce to a very low value.

Fast reset circuits for producing such pulses are shown in Fig. 1 at FR1 and FR2. After a pulse produced by a reset generator such as RE1 terminates, a fast reset pulse generator such as FR1 applies a pulse across the capacitor C_A and the liquid crystal cell. The duration of this pulse, shown at 24 in Fig. 2, is sufficiently short that turbulence is not created in the liquid crystal. The actual duration depends upon the cell thickness, temperature, material and other parameters.

The fast reset pulse occurs after the pulse produced by the reset pulse generator and before the row and column pulses. As the fast reset pulse generator has relatively low internal impedance, the charge which accumulates on capacitors C_A or 16 discharges so rapidly that no turbulence can be created in the liquid crystal cell.

In the arrangement of Fig. 1, only one fast reset pulse generator is required per row. This fast reset pulse generator serves two functions as should be clear from the explanation above. It serves as a relatively low impedance connection between the capacitors C_A and the liquid crystal elements and it also serves to turn off the liquid crystal cells.

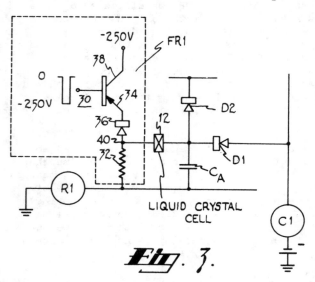

Fig. 3.

An embodiment of a fast reset pulse generator which is suitable for use in the present invention is shown in Fig. 3. It includes a PNP transistor 30 connected as an emitter-follower. The resistor 32 of the transistor circuit is connected to the emitter 34 through a diode 36. The resistor 32 may have a value of say 10,000 ohms and this is only a small fraction of the internal resistance of the liquid crystal cell 12. The latter may have an internal resistance in the range of 10^8-10^9 ohms or so.

In the operation of the circuit of Fig. 3, the resistor 32 normally acts as a relatively low value of impedance which connects the capacitor C_A across the liquid crystal cell 12. The base and emitter of the transistor are normally at ground and the transistor normally does not conduct.

When it is desired to apply a fast reset pulse to the circuit, a -250 volt pulse is applied to the base of the transistor and this drives the transistor into heavy conduction. Approximately -250 volts now appears at point 40 in the circuit and a fraction of this 250 volts develops across resistor 32. In the circuit shown, the portion of the 250 volts which appears across this resistor depends upon the value of the internal resistance of the low pulse generator R1. Preferably, the internal impedance is relatively low so that the major portion of the 250 volt pulse appears across resistor 32 and this fast reset voltage pulse is applied across the circuit consisting of the liquid crystal cell 12 and the capacitor C_A. Since the internal capacitance of the liquid crystal cell 12 is only a small fraction of the value of the internal capacitance of the capacitor C_A, (the capacitive reactance of cell 12 is much much higher than that of capacitor C_A) the major portion of the voltage pulse across resistor 32 appears across the liquid crystal cell.

The diode 36 is in the circuit to protect the emitter-to-base diode of transistor 30. In the absence of this diode, the row pulse, which may have an amplitude of -30 to -40 volts, would be applied directly across the emitter-to-base diode, in the reverse direction, and this would damage the particular transistor employed.

" " " " " " " " " " " "

3,540,796 (U.S.)   Patented Nov. 17, 1970

ELECTRO-OPTICAL COMPOSITIONS AND DEVICES.

<u>Joel E. Goldmacher</u> and <u>Joseph A. Castellano</u>, assignors to RCA Corporation.
Application March 31, 1967.

A novel class of nematic liquid crystal compositions and a novel display device employing the compositions wherein the compositions include a compound represented by the general formula

$$X-\bigcirc-CH=N-\bigcirc-Y$$

wherein X and Y are either alkoxy radicals or acyloxy radicals such that when X is an alkoxy radical Y is an acyloxy radical and vice versa. The novel display device consists of a film of a novel liquid crystal composition held between two supports plates, and parallel electrode strips on each of the support plates. The longitudinal axes of the parallel strips on one support plate is orthogonal to the longitudinal axes of the parallel stripes on the second support plate. The electrodes are in contact with the film. The device includes means for applying an electric field to the film so as to cause turbulence of the liquid crystal composition in the region of the applied field.

One of the features characterizing the novel electro-optical devices employing the novel family of compositions is the relatively low minimum operating temperature due to the low crystal-mesomorphic transition temperatures of members included in the family of compositions. Mixtures have been found which have a crystal-mesomorphic transition temperature below room temperature. Another feature is the wide temperature range in which the novel devices can be used.

In order for the novel compounds to produce best results in the novel devices they are preferably first purified so as to remove particulate impurities as well as impurities within the crystal structure. One measure of the purity is the resistivity of the compounds. The pure compounds as prepared herein generally have resistivities in the range of about 10^9-10^{10} ohm-centimeters. Resistivity measurements were performed using a cell 1 cm by 1 cm containing about a 1/4 mil thick layer of liquid crystal composition and applying a voltage of about 15 volts.

Examples of compounds included in the novel nematic liquid crystal composition are given in Table I along with their transition points as measured after repeated recrystallization of the raw material to a point of constant and reversible nematic-isotropic liquid transition.

Table II gives examples of mixtures included in the novel compositions. The numbers representing the components given in Table II refer to the numbers designating the novel compounds in Table I except that 11 is a known compound, namely p-(anisalamino)-phenylacetate. This compound has a crystal-mesomorphic transition temperature of 81°C. and a mesomorphic-isotropic liquid transition temperature of 110°C. Percent designation refers to weight percent based on the total weight of material in the composition.

Homogeneous mixtures included in the novel composition can be made over a wide variety of concentrations as the components tend to form homogeneous solid solutions over a wide range of concentrations. Mixtures can be prepared so that the melting point of the solid solution is less

TABLE I
[Compounds included in the novel liquid crystal compositions

$$X-\phi-CH=N-\phi-Y$$

Compound Example	X	Y	Crystal-mesomorphic transition temp., °C.	Mesomorphic-isotropic liq. transition temp., °C.
1	$CH_3CH_2\overset{O}{\overset{\|}{C}}$	$-OCH_3$	86	118
2	$CH_3(CH_2)_2\overset{O}{\overset{\|}{C}}O-$	$-OCH_3$	86	119
3	$CH_3(CH_2)_2\overset{O}{\overset{\|}{C}}O-$	$-OC_6H_{13}$	86	120
4	C_4H_9O-	$-O\overset{O}{\overset{\|}{C}}CH_3$	82	113
5	iso $C_5H_{11}O-$	$-O\overset{O}{\overset{\|}{C}}CH_3$	74	82
6	$C_6H_{13}O-$	$-O\overset{O}{\overset{\|}{C}}CH_3$	88	109
7	$C_8H_{17}O-$	$-O\overset{O}{\overset{\|}{C}}CH_3$	80	105.5
8	$C_9H_{19}O-$	$-O\overset{O}{\overset{\|}{C}}CH_3$	86	100
9	CH_3O-	$-O\overset{O}{\overset{\|}{C}}(CH_2)_2CH_3$	49-50	113
10	CH_3O	$-O\overset{O}{\overset{\|}{C}}(CH_2)_2CH_3$	55	100

TABLE II.—MIXTURES INCLUDED IN THE NOVEL LIQUID CRYSTAL COMPOSITIONS

Mixture Example	Components	Crystal-mesomorphic transition temp., °C.	Mesomorphic-isotropic liquid transition temp., °C.
A	50% 11, 50% 4	47	108
B	35.1% 11, 32.6% 4, 33.2% 7	40	103
C	25.5% 11, 24.5% 4, 50% 7	45	103
D	25% each of 11, 4, 6 and 7	39	104
E	50.1% 3, 49.9% 2	48	118
F	34.8% 3, 34.6% 2, 30.6% 1	53	117
G	⅓ wt. ratio each of 9, 4 and 11	22	105
H	⅓ wt. ratio each of 9, 10, 11	25	105
I	50% 9, 50% 10	45	106

than the melting point of the individual solid liquid crystal compounds comprising the solid solution. The preferred composition for a mixture is one in which a eutectic is formed, the crystal-mesophase melting point being at a minimum for a eutectic mixture. The mole percent of compounds used in forming a eutectic depends on the particular compounds in the mixture.

Mixtures can be prepared, for example, by weighing the pure crystalline components in a beaker so as to get the desired proportions. The components are then heated with stirring above their mesomorphic-isotropic transition temperatures. The homogeneous liquid thus formed is allowed to cool to 0°C. The resulting solid is generally a waxy homogeneous mass.

In general, the compounds included in the novel composition can be prepared, for example, by the condensation of a p-acyloxy-phenylamine with the appropriate p-alkoxybenzaldehydes. The reaction is carried out in a refluxing benzene solution with benzenesulfonic acid or acetic acid as a catalyst and facility for azeotropic removal of water. The compounds can be purified by recrystallization from hexane solution. The recrystallization is repeated until a constant mesophase-isotropic transition temperature is reached, followed by one additional recrystallization during which the hexane solution is filtered through a micro-pore filter. (Six detailed examples are given in the patent.)

"""""""""""

3,551,026 (U.S.)   Patented Dec. 29, 1970

CONTROL OF OPTICAL PROPERTIES OF MATERIALS WITH LIQUID CRYSTALS.

George H. Heilmeier, assignor to RCA Corporation.
Application Apr. 26, 1965.

The optical characteristics of a mixture of a thermotropic nematic liquid crystal and another material are controlled by varying the molecular orientation of the nematic liquid crystal with a modulating signal. The other material can consist of either a pleochroic dye or particulate matter. Controlling the molecular orientation of the nematic liquid crystal enables control of the polarization of light passing through the mixture; while local variations in the absorption spectra of the dye can be obtained with the modulating signal. A display panel with a video modulating signal is provided.

The inventor discovered that by controlling the molecular orientation of a nematic "host" material with a suitable means, e.g. an electric or a magnetic field, properties of "guest" materials mixed with the nematic host may be controlled. When certain dyes for example are mixed with a nematic substance, their absorption spectra are influenced by the molecular ordering of the nematic substance and can be controlled. Moreover, small particles of a guest material tend to assume an ordered orientation similar to the nematic host. If flake-like particles of aluminum, for example, are mixed with a nematic host the long dimensions of the flake-like particles tend to align themselves parallel to the molecular axes of the molecules of the nematic substance. When an electric field is applied to the nematic host which includes a guest material the resultin orientation of the host molecules results in a corresponding ordering of the guest material. This effect enables a control of the optical properties of the guest material.

The absorption spectrum of a guest pleochroic dye mixed with a host nematic substance may be controlled with an applied field. The absorption spectrum of a pleochroic dye is a function of the direction of polarization, with respect to the molecular axes of the dye molecules, of the light incident upon it. Virtually all dyes exhibit pleochroism to some extent. However, the degree of pleochroicism is greater in some dyes than in others. Dyes in which the pleochroic effect is pronounced are well known in the art. Two examples are methyl-red and indolphenol-blue. The color exhibited by a methyl-red guest mixed with a p-n butoxy benzoic acid host varies from orange to yellow depending upon the direction of polarization of the incident light with respect to the axes of the dye molecules. The color of indolphenol-blue in the same host varies between a very deep blue and a pale blue.

When a pleochroic dye is mixed with a nematic substance, the orientation of the dye molecules may be controlled by controlling the orientation of the nematic host molecules. Thus the color exhibited by the dye in plane polarized light may be controlled.

The absorption spectra of many materials are functions of the local electric field in the vicinity of the molecules of such materials. Therefore, if the local electric field in the vicinity of the molecules is controlled, the absorption spectrum may be controlled. It was found that a host nematic substance under the influence of an external electric or magnetic field may be used to control the local fields in the vicinity of guest molecules mixed with the nematic substance. When a material whose adsorption spectrum varies with local electric field is mixed with a nematic substance control of the orientation of the molecules of the host nematic substance results in a control in the electric field in the vicinity of the guest molecules and thus a control in the absorption spectrum of the material.

An example of a material whose absorption spectra varies with local molecular field is methyl-red. As noted above methyl-red exhibits pleochroism to a significant degree. It can be shown however that the absorption spectrum of methyl-red also varies as a function of local molecular electric field.

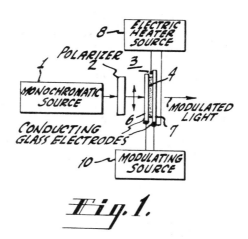

Fig. 1.

Any of the above mentioned effects may be employed to intensity modulate a light beam. Fig. 1 shows a light modulator constructed according to the present invention and employing a pleochroic guest in a nematic host.

The preparation of the mixture 4 of nematic substance and pleochroic dye depends in part upon the type of nematic substance used. Where the molecular axes of the nematic substances are parallel to the dipole moments of the molecules, a condition exhibited by butyl anysilidene amino cinnamate for example, no special preparation is required and the mixture may merely be placed between the two electrodes 6 and 7. However, where the molecular axes of the nematic substance are perpendicular to the dipole moments as in p-n butoxy benzoic acid, then it is desirable to establish an initial orientation of the molecular axes by the following procedure:

One of the transparent electrodes, e.g. the electrode 6, is treated with an acid such as HCl to form a roughened surface which provides many points at which the liquid crystal molecules may attach themselves. The mixture of nematic substance and dye is then applied to the roughened glass surface in a manner such that a preferential direction of the molecules is obtained. This may be accomplished by wiping the electrode in one direction with a cotton swab soaked with the mixture of nematic material and dye. The second electrode 7 is then placed over the mixture deposited on the first electrode 6. This procedure establishes a preferential orientation of the molecules of the mixture. The molecular axes tend to align in a direction parallel to the direction of wiping. Further orientation of the molecules is established when the field is applied across the mixture from the modulating source 10 establishing an alignment of the dipole moments.

The electrodes 6 and 7 are of conventional construction, an example being tin oxide coated glass. One of the electrodes serves as a heating element to maintain the temperature of the nematic substance at a proper value. Where the host is nematic at ambient temperatures, heat is not required. A source of modulating signals 10 is connected to the conducting portions of the electrodes 6 and 7 to establish an electric modulating field across the mixture 4.

The light beam from the polarizer 2 which enters the modulator 3 is polarized in a direction parallel to the paper as indicated. When no field is applied across the mixture 4 its color in plane polarized white light may be described as orange. The orange color is due to the lack of ordering of the dye molecules in the host. As a voltage ia applied across the mixture 4, its color in white light polarized in a direction parallel to the molecular axes changes from orange to yellow as the electric field changes from zero volts to approximately 10^4 volts per centimeter. The change is caused by the alignment of the dipole moments of the molecules parallel to the applied field. By varying the field applied across the mixture 4 with the source 10 the amount of ordering of the dye molecules is varied and therefore the amount of light absorbed by the material at any one frequency in the absorption band is varied. Since the source 1 is essentially monochromatic, the beam passing from the source 1 through the mixture 4 is intensity modulated by the modulating signals from the modulating source 10.

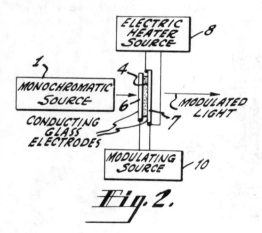

Fig. 2.

Figure 2 is a diagram of an intensity modulator where a dye whose absorption spectrum is a function of the electric field in the vicinity of the molecules of the dye is mixed with a nematic host. The construction of the modulator of Fig. 2 is similar to that of Fig. 1 except that no polarizer is required.

A modulator employing perticles dispersed in a nematic substance may be constructed along the same lines as the modulator of Fig. 2.

Figure 3 shows a third embodiment of the present invention. In Fig. 3 a mixture of a nematic host and a suitable pleochroic guest dye is employed as a display device. The source 56 generates light whose emission spectrum includes the absorption spectrum of the pleochroic dye of the mixture 50. For example, the source 56 may generate white light. The polarization modulator 57 may be of the conventional electro-optic type, for example, a KDP crystal. The polarization of the light beam from the polarization modulator 57 depends upon the value of the video signal applied across the modulator 57 from the source 58. The beam from the output of the polarization modu-

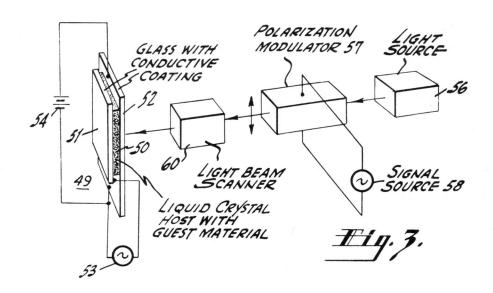

Fig. 3.

lator 57 is supplied to a scanning device 60 which causes the beam of light to scan electrode 52. The scanning device 60 may take any suitable form, for example, a rotating mirror scanner.

In the operation of the display device of Fig. 3 the electric field produced by the source 53 across the mixture 50 causes an ordering of the dye molecules mixed with the nematic liquid crystal in addition to the ordering established when the mixture was wiped onto one electrode. The electric field provided by the source 53 aligns the molecular dipole moments of the nematic host. This alignment causes a corresponding alignment of the dye molecules. The color exhibited by the panel 49 in plane polarized white light will therefore depend upon the direction of polarization of the light passing through the mixture 50 due to the pleochroicism of the dye molecules.

The scanning device 60 causes a light beam to scan the transparent electrode 52 and thus pass through the mixture 50. The polarization of the light beam scanning the electrode 52 is controlled by the polarization modulator 57 in accordance with the video signal applied to the modulator from the source 58. Thus, as the signal applied to the polarization modulator 57 varies, the color of the beam passing through the mixture 50 will vary. Where a mixture of methyl-red and p-n butoxy benzoic acid is employed as the mixture 50, the color may be controlled from orange when the incident light beam is polarized in a direction parallel to the molecular axes of the dye molecules to yellow for a polarization at right angles to this direction. Synchronization between the video signal applied to the polarization modulator 57 and the sanning of the light beam across the electrode 52 may be accomplished by conventional television techniques.

""""""""""""

1,219,840 (British)   Patented Jan. 20, 1971

REDUCTION OF TURN-ON DELAY IN LIQUID CRYSTAL CELL.

<u>RCA Corporation</u>. Application Sept. 13, 1968; prior U.S. application Sept. 14, 1967.

For description of this patent see U.S. Patents 3,503,672 and 3,503,673.

""""""""""""

1,220,169 (British)   Patented Jan. 20, 1971

TURN OFF METHOD AND CIRCUIT FOR LIQUID CRYSTAL DISPLAY ELEMENT.

<u>RCA Corporation</u>. Application Sept. 13, 1968; prior U.S. application 14 & 25 Sept. 1967.

For description of this invention see U.S. Pat. No. 3,519,330.

....

3,569,614 U.S.)   Patented Mar. 9, 1971

LIQUID CRYSTAL COLOR MODULATOR FOR ELECTRONIC IMAGING SYSTEMS.

<u>Thomas F. Hanlon.</u>
Application Apr. 10, 1969.

    This invention utilizes the principle that a nematic liquid crystal, which is nromally clear and transparent, is darkened upon application of a voltage thereto and can be darkened sufficiently to prevent light from passing therethrough. In a three color system, the instant color modulator blocks out two unselected solors by applying voltage to the liquid crystal behind the unselected color filters thus permitting the selected color to pass through the transparent liquid crystal behind its filter.

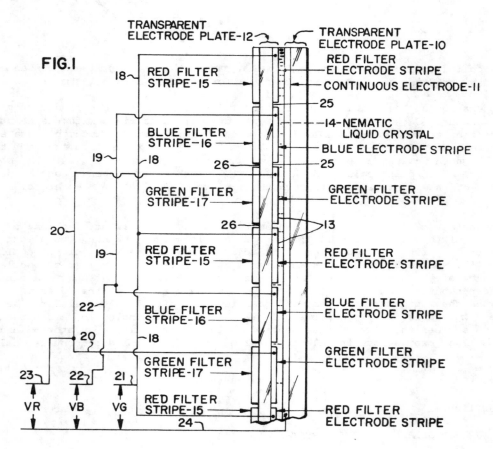

    Referring in detail to the drawing, the instant modulator has the form of a sandwichlike structure which includes two transparent electrode plates 10 and 12. The plate 10 has on one side thereof a multiplicity of closely spaced horizontal electrode stripes 13, for example, 800 to the inch, the spaces 25 between the electrode stripes being nonconducting. These two plates are positioned relatively close together with their alectrode bearing sides facing toward each other. The space between the two plates is filled with a nematic liquid crystal 14.

    In repeated sequences downward, Fig. 1, three color filters red 15, blue 16 and green 17 are shown applied or coated on the blank outer (left-hand) surface of the plate 12 in the form of stripes having the same vertical dimensions as the electrode stripes 13. Each such filter stripe lies precisely in a common horizontal plane with one of the electrode stripes. The filters

30

could instead be applied on the outer (right-hand) surface of the plate 10, not shown. The spaces between the electrode stripes, shown at 25, or those shown at 26 between the filter stripes, are darkened or made opaque in any suitable manner or by any suitable means such as, for example, by applying a black die, to provide light obstructing means to prevent light from a source from passing through the aligned spaces 25, 26. Preferably the spaces 25 are so treated.

For the sake only of this discussion, and assuming that a light source, not shown, is at the left of the modulator, Fig. 1, the electrode stripes behind the red filters will be referred to as the red electrode stripes, those behind the blue filters as the blue electrode stripes, and those behind the green filters as the green electrode stripes; they are so identified by legends in Fig. 1. Leads 18 tie all the red electrode stripes to a common lead 21, leads 19 tie all blue electrode stripes to a common lead 22, and leads 20 tie all the green electrode stripes to a common lead 23. An additional lead 24 extends from the continuous electrode 11.

When no voltage is applied between the lead 24 and any of the common leads 21, 22, 23, the modulator when seen through is white. When red is to be seen by a viewer positioned to the right of the modulator, Fig. 1, then voltages VB and VG are simultaneously applied, thus blocking out the blue and green. When blue is called for, voltages VR and VG are simultaneously applied, thus blocking out the red and green. And when green is called for, voltages VR and VB are applied, thus blocking out red and blue.

In the application of the instant modulator to produce a color television image in a black-and-white receiver, the standard NTSC color signal is converted to a field sequential color signal; then applying the voltage from this signal to the kinescope of the receiver and to the instant modulator converts this voltage to color images when the modulator is positioned between the kinescope and the viewer.

""""""""""""

3,569,709 (U.S.)　　　　　　　　　　　　　　　　　　　　　　　　　　　　　Patented Mar. 9, 1971

THERMAL IMAGING SYSTEM UTILIZING LIQUID CRYSTAL MATERIAL.

Martin R. Wank, assignor to Optical Coating Laboratory, Inc.
Application March 27, 1968.

This invention relates to a new and improved thermal imaging system useful in detecting flaws in mechanical equipment and defects in electronic circuitry and which is generally useful in determining the thermal characteristics of a monitored field or object. The invention also has application in the fields of medical diagnosis and thermographic printing.

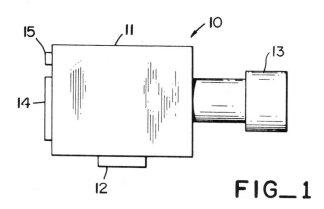

FIG_1

In the embodiment of the present invention illustrated in Fig. 1 there is generally provided a thermal imaging camera 10 comprised of a housing 11 and a tripod mount 12. An infrared transmitting objective lens 13, which may be, for example, a zoom lens, is provided for forming an image of an object or monitored field under study. At the back of the camera there is provided a screen 14 for viewing a thermally evoked image of the monitored field or object under study. Means may be provided for attaching a conventional camera to the viewing screen 14 for providing photographic pictures of the thermally evoked image projected on screen 14. A visual sight or eyepiece 15 is also provided for viewing directly the object or monitored field under study.

The optical system housed within camera 10 is illustrated in Fig. 2. The objective lens of the camera comprises an infrared transmitting lens 20 which may also be an infrared transmitting lens system such as a zoom lens. The infrared transmitting objective lens is adapted to form an image from infrared rays emitted by an object or monitored field 21. Positioned in the image plane of the infrared transmitting lens 20 is a field of liquid crystal material 22 having a

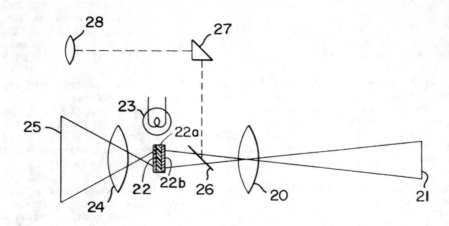

FIG_2

thermally responsive color spectrum. The film of liquid crystal material may be formed on a film base 22a such as Mylar. Preferably, the liquid crystal material is coated on the film base on the side opposite the infrared transmitting lens 20. The side of the film base facing the infrared transmitting lens is then provided with a dark coating 22b such as black paint for optimum thermal absorption. Positioned adjacent the film of liquid crystal material is a heater 23 such as a lamp for heat biasing the liquid crystal materials to the characteristic temperature at which the material exhibits thermally sensitive optical scattering properties. A conventional objective lens 24 is also provided to form an image of the film of liquid crystal material on a viewing screen 25 such as a frosted glass screen spaced from the lens 24.

A beam splitter 26 may also be provided interposed between the infrared transmitting objective lens 20 and the film of liquid crystal material 22 for directing a portion of the light via a reflector 27. The reflector may be a prism to an eyepiece lens 28 which permits an observer to view the object or monitored field under study directly.

In using the thermal imaging system disclosed above, the film of liquid crystal material 22 may be heat biased by heater 23 to a temperature just below the characteristic temperature range or band across which the material exhibits a thermally responsive color spectrum. An infrared image from infrared rays emitted by the object or monitored field under study may then be projected onto the film. A thermal image is evoked on the film of liquid crystal material having a coloration dependent upon the thermal characteristics of the object or monitored field.

According to the preferred method of thermal imaging, however, an image of the object or monitored field formed by infrared transmitting lens 20 is first projected on the film of liquid crystal material 22. The temperature of the film of liquid crystal materials is then raised by heater 23 to the characteristic temperature at which the liquid crystal material exhibits thermally sensitive optical scattering properties. The thermally evoked image of the object or monitored field then appears from the background of the film.

" " " " " " " " " " " "

3,572,907 (U.S.)   Patented Mar. 30, 1971

OPTICAL CELL FOR ATTENUATING, SCATTERING AND POLARIZING ELECTROMAGNETIC RADIATION.

<u>Ivan Cindrich</u>, assignor to Chain Lakes Research Corporation.
Application Apr. 28, 1969.

A optical cell for attenuating, scattering or phase shifting electromagnetic radiation such

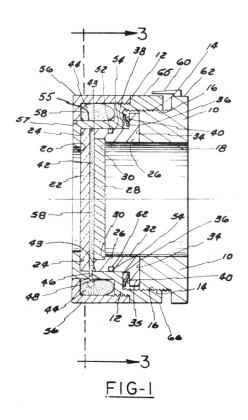

FIG-1

as light comprising an annular member telescopically engaged with a mounting ring to thereby define a chamber. The ends of the chamber are formed by windows supported in the mounting ring and in a cylindrical frame in contiguous relationship with the annular member. A fluid having light affecting properties fills the chamber and is transferable between the chamber and a reservoir in fluid communication with the chamber. Telescopic operation of the optical cell varies the length of the fluid light path between the windows, to thereby cause variable attenuating and the like effect upon traversing light.

Liquid crystal material may also be used and may be additionally affected, for example, by an electrical field applied between the inner surfaces of windows 28 and 22 which, may be suitably coated to act as transparent electrode surfaces.

" " " " " " " " " " "

1,227,616 (British) Patented Apr. 7, 1971

LIQUID CRYSTAL ELECTRO-OPTIC DEVICE.

<u>RCA Corporation</u>.
Application Nov. 15, 1968; prior U.S. application Dec. 5, 1967.

For description of this invention see U.S. Pat. No. 3,650,603.

" " " " " " " " " " "

1,228,606 (British) Patented Apr. 15, 1971

ELECTRO-OPTIC LIGHT VALVE.

<u>RCA Corporation</u>.
Application Nov. 15, 1968; prior U.S. application Dec. 5, 1967.

For description of this invention see U.S. Pat. No. 3,499,702.

" " " " " " " " " " "

3,575,491 (U.S.) Patented Apr. 20, 1971

DECREASING RESPONSE TIME OF LIQUID CRYSTALS.

<u>George H. Heilmeier</u>, RCA Corporation.
Application Oct. 16, 1968.

The speed of response of a liquid crystal is increased without degrading the crystal, by applying the crystal an electric field which is substantially greater value than the minimum value of field required to achieve substantially maximum light scattering from the crystal and, after a short interval, reducing the magnitude of the applied field to a value close to said minimum value.

Figure 1 illustrates the opacity of a 1/4-mil thick liquid crystal cell as a function of the voltage applied to the cell. It may be observed that substantially maximum opacity is obtained at about 60 volts and that higher voltages produce no significant increase in opacity.

To produce dynamic scattering in a liquid crystal at a rate greater than that obtainable with the minimum electric field E_o necessary to give substantially maximum scattering and yet to avoid degradation of the crystal by high fields, a field substantially greater than E_o is applied for a fraction of a second (10 ms.) and then decreased to E_o, e.g. as indicated by either the full (D.C.) or dashed (60 Hz.) lines in Fig. 2. The time between the leading edge of the turn on pulse and the state of maximum scattering may thus be reduced, e.g. from 10 milliseconds to less than 500 micro-seconds. In Fig. 3 wherein the crystal may be used as a light shutter, e.g. in a car or aircraft windscreen, when a photocell detects oncoming headlights or a flare or bomb burst, of above a predetermined light intensity a trigger circuit is actuated to produce a pulse 50. The pulse is applied to a turn-on circuit, such as shown in Fig. 6, whose output is as shown in Fig. 2, to close the shutter. The trigger may reset manually or automatically when the light intensity drops below a fixed level. It is also stated that the trigger may be omitted. In Fig. 6, the pulse 50 is differentiated at 52, 54, 56 to give a waveform similar to that of Fig. 2 which controls the conduction of a transistor 84 of an A.C. voltage from a source 92. The voltage developed across a resistor 90 is applied to a liquid crystal and capacitor in series. In Fig. 4 the crystal is placed directly in parallel with the resistor 56 of the differentiating circuit, while in Fig. 5 the differentiated pulse controls an FET so as to vary a D.C. field across the crystal.

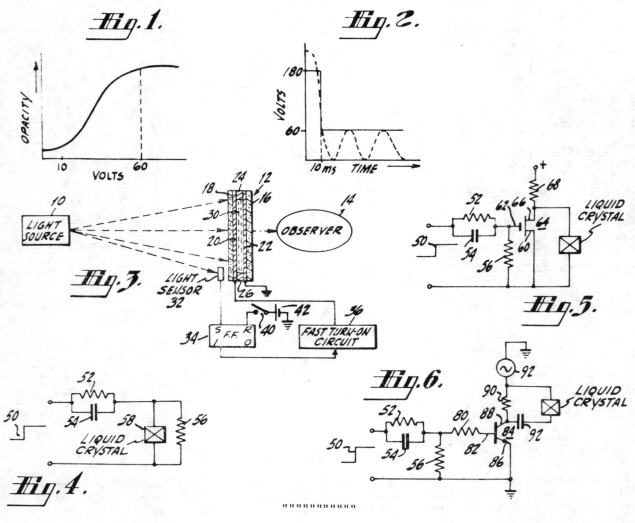

3,575,492 (U.S.)

Patented Apr. 20, 1971

TURNOFF METHOD AND CIRCUIT FOR LIQUID CRYSTAL DISPLAY ELEMENT.

Edward Oskar Nester and *Bernard Joseph Lechner*, assignors to RCA Corporation.
Application July 10, 1969.

The dynamic scattering state produced in a liquid crystal solely of the nematic type by the application of an electric field may be erased rapidly compared to the natural relaxation of the crystal following cessation of the electric field by applying an alternating voltage in the range 2-20 kHz, and preferably 7-8 kHz, to the crystal. It is believed that in this frequency range, the dipoles in the crystal are still able to align themselves with the electric field, whereas the mechanisms producing turbulence, i.e. injection of ions into the crystal have time constants which are too large.

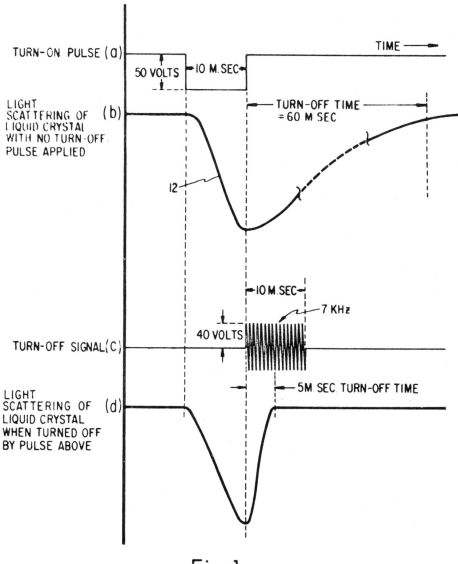

Fig. 1.

Figure 1 illustrates the effect of a 40 volt, 10 millisecond, 7 kHz pulse immediately following a 50 volt, 10 millisecond D.C. "turn-on" pulse, the turn-off time being reduced from 60 to 5 milliseconds. The A.C. voltage may be left on at all times except when it is desired to produce the scattering state, and the contrast ratio may be improved by up to a factor of 2.

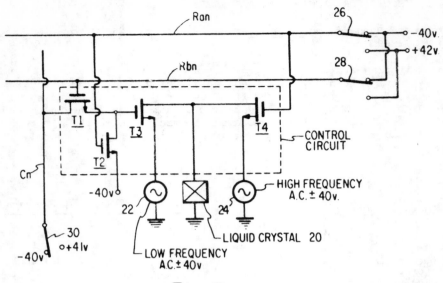

Fig. 3.

Figure 3 shows a suitable circuit for one cell of an array of liquid crystal cells using transistors which are "on" only when the gate voltage is one or more volts positive relative to the source voltage, and which are all "off" in the state shown. On switching row line R_{bn} to +42 volts, T1 is turned on but T3 remains off. If column lead C_n is simultaneously raised to +41 volts, T3 is turned on and a low frequency source 22 applies a field to the crystal 20 to produce the scattering effect. If row line R_{bn} is now turned to -40 volts, and subsequently column C_n is also returned to -40 volts, the charge stored in the capacitance of the gate electrode of transistor T3 keeps that transistor "on." To turn the crystal off, the voltage on row lead R_{an} is switched from -40 volts to +42 volts for about 10 milliseconds, thus turning trabsistors T2 and T4 on, consequently discharging the gate of T3 and removing the low frequency source 22 while admitting high frequency excitation from a source 22 to the crystal.

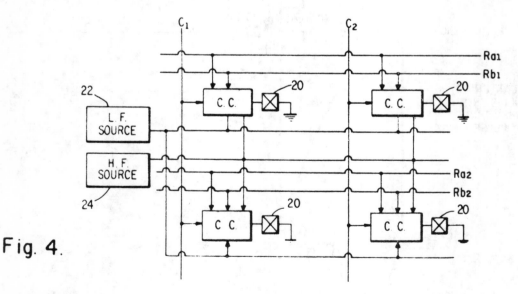

Fig. 4.

A matrix of such elements is illustrated in Fig. 4 wherein a row lead and a column lead must be simultaneously energized to switch a particular crystal on and the erase operation occurs a row at a time. Instead of an array of cells, a single crystal may be used with a matrix of individual electrodes.

3,575,493 (U.S.)

Patented Apr. 20, 1971

FAST SELF-QUENCHING OF DYNAMIC SCATTERING IN LIQUID CRYSTAL DEVICES.

George H. Heilmeier, assignor to RCA Corporation.
Application Aug. 5, 1969.

A liquid crystal element is operated within a critical frequency range such that upon removal of the exciting electric field, the dynamic scattering exhibited by the crystal decays within a matter of milliseconds - an interval which is substantially shorter than the usual natural decay time. The critical frequency range is related to the conductivity (which is temperature and voltage dependent) and dielectric constant of the liquid crystal and is precisely defined by equations in the detailed description below.

The alternating voltage employed to cause self-quenching of dynamic scattering to occur in a liquid crystal is in the frequency, the lower limit of which is about $1.4\sigma/2\pi D$ and the upper limit of which is within the range in which dynamic scattering occurs, where:

σ = conductivity in (ohm-centimeters)^{11},
$D = \varepsilon\varepsilon_o$,
ε is the dielectric constant of the liquid crystal material, and
$\varepsilon_o = 8.85 \times 10^{114}$ farads/centimeter and is the dielectric constant of free space.

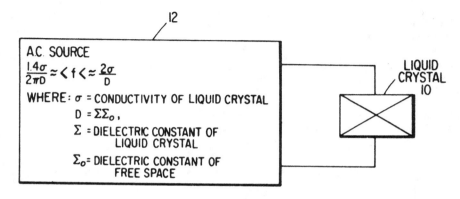

Fig. 1.

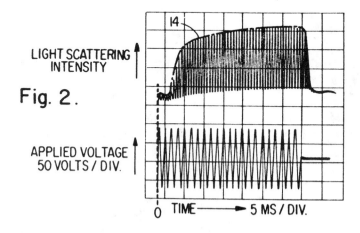

Fig. 2.

The liquid crystal element shown schematically at 10 in Fig. 1 is basically a parallel plate capacitor with a liquid crystal dielectric. The method of fabricating the device has been described previously in the literature. The plates are conductors at least one of which is transparent and they are usually formed on glass. The thickness of the liquid crystal layer between the plates may be of the order of 25 microns or less, that is, 1 mil or less and is held in place between the plates by capillary action.

In operating a liquid crystal element such as this, the present inventor has discovered that by controlling certain of the liquid crystal cell parameters and the operating frequency and amplitude of the alternating voltage employed to produce the dynamic scattering, that upon the removal of the voltage, the liquid crystal self-quenches in a very short interval of time. For example, in the circuit shown in Fig. 1 in which the liquid crystal has a material dielectric constant ϵ of about 3.5 and a conductivity of about 10^{110} (ohm-centimeters)^{11} and in which the frequency applied by source 12 is a gated 600 hertz sine wave, the performance shown in Fig. 2 is obtained.

The lower waveform of Fig. 2 is the voltage applied to the liquid crystal. The upper waveform of Fig. 2 consists of two parts. The oscillations within envelope 14 are obtained with the aid of a photomultiplier positioned to receive the light reflected from the liquid crystal 12 when it is excited by the lower wave of Fig. 2. The photomultiplier has a relatively short time constant and is able to follow the variations in intensity of the actual light scattering which is produced. Each time the applied voltage causes maximum ion motion, and this accord twice each cycle, the light scattering is maximum, and between these intervals the light scattering is minimum. This explains why the frequency of the light scattering wave is double that of the applied voltage wave. The heavy line 14, which the envelope of the light scattering curve, is what is observed by the eye. The eye integrates the amount of light scattered by the light crystal, as it is not able to follow the rapid change from maximum to minimum light scattering condition of the liquid crystal. This curve 14, for the major portion of its extent, is similar to what would be obtained by applying a direct voltage pulse to the liquid crystal or by applying a conventional alternating voltage turn-on pulse to the crystal, both as described in the introductory portion of this patent. However, with the system parameters adjusted in the manner discussed in detail below, when the alternating voltage produced by source 12 is terminated, the light scattering intensity very abruptly decays. The decay time, that is the turnoff time, is roughly 3 milliseconds. The same liquid crystal under direct voltage excitation has a turnoff time of approx. 100 milliseconds!

The present inventor has found experimentally that a liquid crystal which rapidly self-quenches after alternating voltage excitation exhibits the following properties:

1) For a given amplitude of the applied alternating voltage, there is a critical frequency range over which fast self-quenching is observed. Below this range, the decay time increases markedly while above it no dynamic scattering is observed.

2) Raising the amplitude of the applied alternating voltage while maintaining the frequency constant flat the decay time. Fast self-quenching can be restored if the frequency natural also raised. According to the theory of operation which has been formulated, this effect can be explained in terms of the conductivity of the liquid crystal, the conductivity being voltage dependent.

3) Raising the temperature of the liquid crystal while maintaining the amplitude of the applied voltage constant requires an increase in the frequency of the voltage to retain the fast self-quenching characteristic of the liquid crystal. Again, raising the temperature raises the conductivity and it is the conductivity which is one of the factors upon which the fast self-quenching effect depends.

4) As the conductivity of the material is increased, the frequency range of the applied voltage necessary to permit fast self-quenching is shifted to a higher range. If the conductivity of the liquid crystal is made too high, however, fast self-quenching is no longer observed.

It has been observed that when operating in the fast self-quenching mode, essentially 100 percent modulation of the scattered light is obtained. In other words, the liquid crystal produces substantially the maximum amount of light scattering which is possible. This behavior is obtained, however, only when the current through the liquid crystal cell leads the voltage by 55° to 60° or more. As the lead angle between current and voltage decreases to less than 55°, modulation decreases sharply from 100 percent. The voltage range over which the effect has been observed extends from roughly 40 to over 200 volts.

" " " " " " " " " " "

3,576,364 (U.S.) Patented Apr. 27, 1971

COLOR ADVERTISING DISPLAY EMPLOYING LIQUID CRYSTAL.
<u>Louis A. Zanoni</u>, assignor to RCA Corporation.
Application May 20, 1969.

The object of the present invention is to provide a relatively low-cost color picture display panel suitable for point-of-sale advertising and also for larger displays, such as billboards, employing liquid crystals.

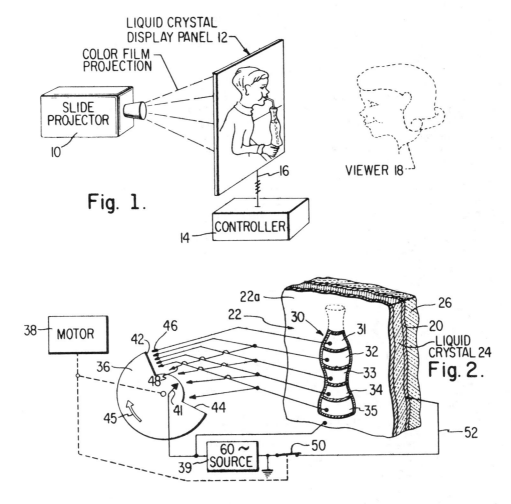

Fig. 1.

Fig. 2.

The system shown in Fig. 1 includes a slide projector 10 suitable for color transparencies and a liquid crystal display panel 12 serving as the screen for the slide projector. The operation of the display panel is under the control of controller 14. The picture appearing on the panel may be observed by a viewer 18 preferably located in front of the panel as shown. In the operation of the system of Fig. 1, a color film is projected onto the back surface of the liquid crystal panel. If the liquid crystal is in its inactive state, no picture is visible on the panel. However, when a voltage is applied to the electrodes of the panel of sufficient magnitude to cause dynamic scattering to occur, the color picture becomes visible, in color.

For advertising purposes, it is often desirable to simulate a "moving" picture. This is accomplished in the manner shown in Fig. 2. (Display: boy drinking soda pop through a straw).

Referring now to Fig. 2, which shows only the portion containing the bottle of the panel, the front and back electrodes are shown at 20 and 22 respectively, although their locations could be reversed. Each electrode is located between a glass sheet and the liquid crystal 24. To simplify the drawing, only one of the glass sheets, namely 26, is shown. The drawing is not to scale - the liquid crystal layer being very thin - perhaps a fraction of a mil (a typical value may be 0.0005 inch) in thickness.

The portion of the transparent back electrode 22 onto which the picture of the bottle is projected is illustrated at 30. It consists of five separate transparent electrode sections 31-35 which are insulated from one another and from the main portion 22a of the back electrode. The spacing between electrode section 31-35 and between these electrodes and the main electrode section 22a is relatively small - for example, one one-hundredth of an inch.

The electrode section 31-35 and 22a are connected to brushes, shown in the figure as arrow heads. These brushes are arranged to be contacted by the rotatable cam-shaped, switch element 36. The latter continuously may be driven by motor 38. A 60 cycle source 39 is connected at one terminal to a brush 41 which continuously engages the cam-shaped switch element 36 and is connected at its other terminal to ground and to the transparent front electrode 20.

In the operation of the system of Fig. 2, in the position of a cam 36 shown, the 60 cycle source is connected at one terminal only to the back electrode section 22a and is connected at its other terminal to the front electrode 20. It may be assumed that the slide projector 10 of Fig. 1 is projecting a color film onto the screen and that the color picture of the soda pop bottle superimposes over the corresponding region of Fig. 2. Since the entire backplate 22 is energized except for the electrodes 31 through 35, the picture becomes visible on the display panel except that the bottle of soda pop appears to be empty.

The motor 38 continuously drives the cam 36 clockwise, as indicated by arrow 45. When the edge 42 reaches the five radially aligned brushes immediately adjacent to that edge, the electrodes 31 through 35 are energized simultaneously, the liquid crystal between these five sections and the front electrode 20 becomes excited, and the soda in the bottle becomes visible in the color in which it is projected by the projector 10. The bottle appears to be full of soda, see t_1 in Fig. 3, and the remainder of the picture, as shown in Fig. 1, is also visible to the viewer.

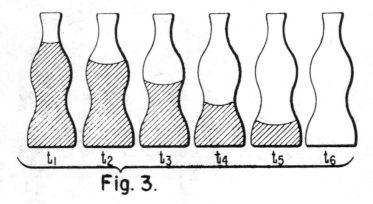

Fig. 3.

When, about two-thirds of the cam 36 cycle later, the other edge 44 of cam 36 passes beyond the brush 46, power is removed from electrode 31. The liquid crystal between this electrode and electrode 20 returns to its unexcited state, and the picture of the bottle changes from that shown at t_1 to that shown at t_2, both in Fig. 3. A short time later at t_3, edge 44 passes beyond the brush 48, disconnecting source 39 from electrode 32. Now power is absent both from electrodes 31 and 32, the liquid crystal between these two electrodes and the front electrode 20 is in its unexcited state, and the level of liquid appears to move down in the bottle as shown at t_3 in Fig. 3. This sequence of steps continues until the bottle appears empty.

If desired, by appropriate switching at some later time in the cycle, the entire picture may be erased. For example, there may be a switch 50 controlled by motor 38 and in series with the ground lead. It may be opened to disconnect the lead 52 for the transparent electrode 20 from power supply 39 and thereby to remove the electric field from across the liquid crystal. This causes the entire picture to be erased. One point worth mentioning is that the technique of the present invention may be employed either for selectively erasing or selectively displaying a color picture.

" " " " " " " " " " " "

3,576,761 (U.S.) *Patented Apr. 27, 1971*

THERMOMETRIC COMPOSITIONS COMPRISING ONE MESOMORPHIC SUBSTANCE, ONE CHOLESTERYL HALIDE, AND AN OIL SOLUBLE DYE SELECTED FROM THE GROUP CONSISTING OF DISAZO, INDULENE, AND NIGROSINE DYES.

<u>Frederick Davis,</u> assignor to Liquid Crystal Industries, Inc.
Application March 18, 1969.

Novel thermometric compositions capable of retaining an indication that they have exceeded a given temperature, which compositions comprise a mixture of at least one mesomorphic substance and at least one cholesteryl halide, the mixture exhibiting color in the cholesteric state at a first temperature and changing from that state at a second temperature, and an amount of oil-soluble dye sufficient to prevent the mixture from reverting to the color of the cholesteric state when the compositions are returned to the first temperature; thermometric elements comprising such compositions; and thermometric articles comprising said compositions.

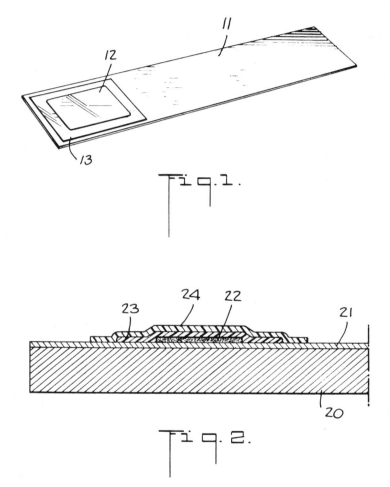

Figure 1 shows a simple thermometric article comprising a substrate or base 11 which is non-absorbent with respect to the thermometric composition. On substrate 11 is coated an area 12 of a cholesteric composition according to this invention. The cholesteric composition is overcoated with a thin film of material 13 to prevent contamination of the composition by dust, fluids, vapors, and the like. While the cholesteric composition 12 is shown as a rectangular area in Fig. 1, it will be appreciated that the configuration of the area can be any plane geometric figure.

Figure 2 shows a second embodiment of the invention which provides relatively inexpensive disposable clinical thermometers. Here, the base or substrate 20 is a piece of black paperboard stock which is coated with a thin layer of polyethylene 21 to protects the paper from the thermometric composition 22 coated thereon. The thermometric composition 22 is then overcoated with a layer of immiscible material 23 such as casein glue, polyvinyl alcohol, and the like. Where the thermometric article is to be utilized in an aqueous or humid environment it may be desirable to provide a second overcoating 24 of water-insoluble material such as polyethylene, methacrylate, and the like.

Electronic devices, printed circuit boards, light fixtures, electrical wiring, and other artifacts which are either temperature sensitive, or the failure of which produces a temperature rise, can similarly be monitored.

The following examples are given to illustrate embodiments of the invention as it is presently preferred to practice it.

Example I

A thermometric composition is prepared by admixing 30 parts of cholesteryl oleyl carbonate showing a color play at 5-6°C., 39 parts of cholesteryl oleyl carbonate showing a color play at 20-22°C., 25 parts of cholesteryl chloride having a melting point of 94-95°C., six parts of Solvent Black 7 NJD nigrosine dye, made by Allied Chemical. At room temperature this mixture displays a green color in thin films.

When the temperature of a thin film of this material is heated to 37°C., the color disappears and does not reappear after cooling below 37°C., even two weeks after exposure to the higher temperature. Bulk quantities of the mixture behave in the same manner. Agitating the composition after it has been heated above 37°C. and returned to room temperature (about 23 °C.), restores the original color.

Example II

A 10-mil (0.25 mm.) thick sheet of black polyvinyl chloride is coated by a silk-screen technique with a pattern of squares of the composition prepared in Example I. The thickness of the composition coated on the sheeting is about 2 mils. The sheeting and the thermosensing composition are then overcoated with casein white liquid glue and dried rapidly. After curing to dry the glue fully, a protective coating of impregnated acrylic polymer is sprayed onto the glue overcoat. The sheeting is then cut into thin strips such as shown in Fig. 1. These strips containing the area of thermosensing material are used to determine the oral temperatures of patients. Those with temperatures higher than the normal 37°C. cause the composition to become colorless and to remain in that condition. The colorless thermometers can be reused by merely rubbing the surface of the acrylic overcoat to restore the original color.

Further compositions according to the present invention appear below, wherein all amounts are in parts, "ChCl" is cholesteryl chloride melting at 94-95°C.; "High CHOlC" os cholesteryl oleyl carbonate showing a color play at 20-22°C.; "Low ChOlC" is cholesteryl oleyl carbonate showing a color play at 5-6°C.; "ChNo" is cholesteryl nonanoate; and the temperature is that at which the thermosensing compositions became colorless and remain in that condition:

Example	ChCl	High ChOlC	Low ChOlC	ChNo	Dye	Temp., °C.
III	27	73			6(1)	42
IV	27	58		15	6(1)	45.5
V	27	50	23		6(1)	38
VI	27	59	14		6(1)	35
VII	25	75			6(2)	48

(1) Solvent Black 7 NJD as in Example I.
(2) Spirit Nigrosine SSB, made by Allied Chemical.

It will be appreciated from the present description that the compositions of Examples III-VII could be used to prepare thermometers as set forth in Example II. Moreover the cholesteryl bromide or iodide or mixtures of these halides with each other and/or with the cholesteryl chloride can be used. Similarly, an oil-soluble disazo dye can be used in lieu of the nigrosine azine dyes. Further, other alkanoic esters of cholesterol or alkyl carbonate esters of cholesterol can be used in the foregoing examples to provide a broad variety of temperatures and temperature ranges for the thermometric compositions. Likewise, other cholesteric materials such as corresponding derivatives of β-sitosterol, stigmasterol, ergosterol, and the like can be substituted with comparable results.

" " " " " " " " " " "

3,578,844 (U.S.) Patented May 18, 1971

RADIATION SENSITIVE DISPLAY DEVICE CONTAINING ENCAPSULATED CHOLESTERIC LIQUID CRYSTALS.

Donald Churchill and James V. Cartmell, assignors to The National Cash Register Company.
Application Feb. 23, 1968.

This disclosure is directed to articles of manufacture, chiefly display devices containing encapsulated cholesteric liquid crystals which change color or shade of color not only upon application of an electric potential but also upon removal of the field. The image produced has a comparable outline to that of the path of the electric field. Three chromatic states are evident, the normal color (before the electric potential is applied), the color given off when the electric field is applied, and the color observed when the electric field is removed. All three chromatic states are readily discernible from one another. The encapsulation of the cholesteric liquid crystal provides an unusual advantage regarding electric field behavior because the third chromatic state (electric potential removed) has a much greater longevity with the encapsulated material versus unencapsulated material of identical composition.

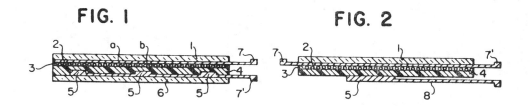

As shown in Fig. 1, transparent insulating protective layer 1 is directly in contact with transparent first electrode 2 which in turn is in intimate contact with the encapsulated cholesteric liquid crystalline member 3. The encapsulated cholesteric liquid crystalline member 3 is composed of an array or profusion of minute capsules, a containing cholesteric liquid crystalline material incorporated in and/or on polymeric film b. In addition to enhancing the optical properties of the encapsulated liquid crystalline member and improving the adhesion between the capsule cell walls and the first and second electrodes, the polymeric film b serves as an effective insulator between both of said electrodes. Next to encapsulated cholesteric liquid crystalline member 3 is located a black lacquer insulating film 4 to enhance the observation of the chromatic changes in the liquid crystalline member. Conductive elements 5 are located intermediate between lower protective layer 6 (usually glass or plastic) and black insulating film 4. While Fig. 1 shows three such conductors, it should be clearly understood that any desired number can be used and the configuration thereof can be arranged in any desired display pattern, sequence or shape. Upper and lower electrodes (leads) 7 and 7', respectively, allow the passage of an electric current from a suitable potential source (not shown) into electroconductive contact with both the first transparent electrode and the conductive elements 5. Upon subjecting the cholesteric liquid crystalline member to the electric field by completing the circuit; the chromatic representation viewed through the transparent layers of 1 and 2 will conform closely with the configuration defined by the field established between conductive elements 5 and the top electrode. That is to say that the confirguration established by the elements 5 will be reproduced, but in different color or intensity of reflectance from the surrounding film areas when looking down through the transparent protective layer 1 and transparent first electrode 2. Upon opening the circuit, a further chromatic change takes place in the configuration areas wherein the encapsulated cholesteric liquid crystalline material rapidly changes from the second chromatic state to the third chromatic state characterized by the ability of the encapsulated cholesteric liquid crystalline member to retain a substantially permanent and different chromatic state for extended periods of time under normal atmospheric conditions of temperature and humidity. If desired, a sealer tape or potting compound (not shown) can be applied to the outer peripheral edge(s) to aid in sealing the various layers against lateral and interfacial exposure to the atmosphere at the peripheral edge surfaces.

Figure 2 illustrates an alternative display device. A conductor lead wire 8 is attached to or made integral with conductor film 5. Leads 7 and 7', are attached to bus bars in contact with the transparent tin oxide electrocinductive film 2 at the left- and right-hand sides, respectively, oxide electrode. On passing a current through leads 7 and 7' through the upper tin oxide electrode the upper electrode can be heated to the isotropic transition temperature of the cholesteric liquid crystals thereby erasing the image produced during the third chromatic state. Upon cooling to ambient temperatures (room temperatures), the liquid crystals then return to their original (first) chromatic state.

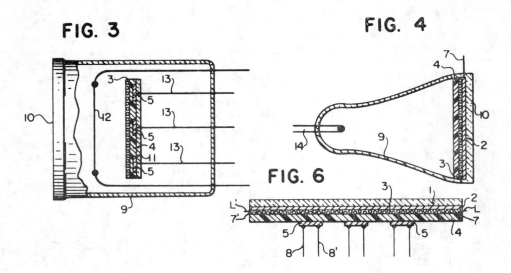

Figure 3 illustrates an article of the present invention wherein the field is established by electrons attracted to the surface of the encapsulated cholesteric liquid crystal film. A matrix of conductive elements, 5, is supported on insulator, e.g. fiberglass substrate in a vacuum tube 9 with a transparent viewing port or faceplate 10. The conductive elements are held at a positive potential by means of connecting the leads 13 to a potential source (not shown). A current is passed through thin, nontransparent resistive wire 12, so that it is heated and emits electrons. The electrons are drawn to the surface of the liquid crystal film 3, directly over the positively charged conductive elements. The field formed between the surface electrons and the conductive elements causes the liquid crystal to change color. After the potential is removed from the elements, 5, the pattern will remain until erased by heating to the isotropic transition temperature. This can be accomplished by means of any suitable energy source, e.g., heat lamp, heat wires or passing a current through the conductive elements.

The device of Fig. 4 is a conventional cathode-ray tube envelope 9 (CRT) with a transparent conductor 2 coated on the inner surface of the transparent faceplate 10. A film of encapsulated cholesteric liquid crystal (in polyneric binder) 3 is coated in direct contact with the transparent electrode 2 on the inner surface of the faceplate, and a black insulator film 4 is coated on the liquid crystal film. The transparent electrode 2 is held at a positive potential via lead 7 and electrons from the electron gun 14 are caused by suitable deflection techniques to impinge on the black insulator in the desired pattern. The local fields set up between the electrons at the insulator and the front electrode cause the liquid crystal to change color in those areas formaing a visual pattern when viewed from the front of the tube. The pattern is semipermanent and one sweep of the electron beam will suffice to establish a permanent image. This image may be erased by heating the liquid crystal above its isotropic transition temperature.

In the device of Fig. 5 a conventional CRT is prepared with a multitude of conductive wires, 15 extending through the faceplate 10 and flush with the outer surface on the faceplate.

In the device of Fig. 6, transparent electroconductive coating 2, e.g., tin oxide, is deposited on transparent protective element 1, which also serves as a substrate or support for the coating. Bus bars 7 and 7' are attached along the edges of coating 2 and have conductive leads L, L', respectively, attached thereto. Opaque, e.g., black, nonreflective insulator film 4 has a layer or film of encapsulated cholesteric liquid crystals (constituting an encapsulated cholesteric liquid crystal member 3) deposited thereon. Resistive conductor elements 5 can be deposited directly on insulator 4, or on a lower supportive and protective layer (not shown), such as layer 6 of Fig. 1. The encapsulated cholesteric liquid crystal member 3 is positioned in contiguous or closely spaced relationship to coating 2. Each resistive element 5 has a pair of leads 8, 8' attached thereto, thus enabling each resistive element 5 to be operable separately or in groups, e.g. by appropriate conventional switching. Resistive elements 5 can be formed, or deposited, in any desired configuration or design. The device of Fig. 6 has two primary modes of operation:

A) Field effect mode of operation with image storage and erasure;
B) Thermal mode of operation with storage option.

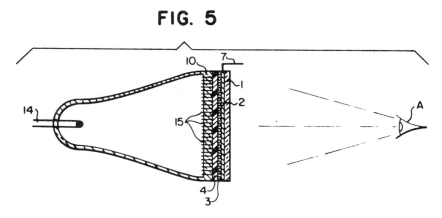

FIG. 5

Six examples are described in the patent specification regarding the encapsulation of cholesteric liquid crystals and formation of encapsulated cholesteric liquid crystalline members.

" " " " " " " " " " "

3,580,864 (U.S.) Patented May 25, 1971

CHOLESTERIC-PHASE LIQUID-CRYSTAL COMPOSITIONS STABILIZED AGAINST TRUE-SOLID FORMATION, USING CHOLESTERYL ERUCYL CARBONATE.

Newton N. Goldberg and James L. Fergason, assignors to Westinghouse Electric Corporation. Application Apr. 30, 1969.

Stability of cholesteric-phase liquid-crystal compositions is improved by incorporating therein an amount, up to 100%, of cholesteryl erucyl carbonate, a novel compound, made by reaction of cholesteryl chloroformate with erucyl alcohol and pyridine. The novel carbonate does not crystallize to true solid in several months, whereas most other liquid-crystal compositions transform to true solid within several hours to several days. Moreover, the novel carbonate does itself exhibit a color play, so that it may be used in very large amounts in cholesteric phase liquid-crystal compositions, without decreasing the vividness of the color play exhibited thereby.

Description of the preferred embodiments making cholesteryl erucyl carbonate:

A three-necked glass flask having standard-taper ground-glass joints was fitted with a reflux condenser, a stirrer and an addition funnel. The flask is charged with 0.75 mol (336.9 grams) of cholesteryl chloroformate and 450 milliliters of benzene. Heat was applied until a complete solution was effected. In a separate flask, 0.76 mol (246.7 grams) of erucyl alcohol was added to 300 milliliters of benzene, and heat was applied until complete solution was effected. Through the addition funnel, the erucyl alcohol solution was then added to the cholesteryl chloroformate solution. then, a solution of 0.75 mol (60.7 milliliters) of pyridine dissolved in 75 milliliters of benzene was added dropwise, through the addition funnel, to the contents of the flask. An exothermic reaction resulted, and a white solid was formed. After complete addition of pyridine, the contents of the flask were heated to and maintained at a reflux temperature for 1.0 hour.

The solid formed by the reaction, namely, pyridine hydrochloride, was vacuum-filtered by means of a Buchner funnel. The filtrate (a benzene solution containing the desired product) was treated with methanol, and an oily product came out of the solution. By means of a separatory funnel the oily (lower) layer was separated from the benzene-methanol layer. As prepared, the cholesteryl erucyl carbonate exhibited a color-play temperature range of about 25.5°C. to 26.5°C., being green at about 26°C. After repeated washings with water, the color-play temperature range of the material was raised to 33.5°C-34.5°C. The product oil exhibited no tendency to crystallize to a true solid, even after prolonged periods of storage in a refrigerator. Under refrigeration, the material became grease-like in consistency. When used by itself, the compound exhibits a color play within a narrow temperature range.

The novel carbonate is used by incorporating it, in amounts up to 100%, in cholesteric-phase liquid-crystal compositions. It may be mixed readily with any of a great number of compounds that are known to exhibit cholesteric-phase liquid crystals at temperatures about 20°C. The novel carbonate is effective when used in amounts as small as about 5 weight percent of the stabilized composition, and it may be used in greater amounts, up to 100% b-weight. In most instances, it will be desirable to use the compound by itself, or with relatively small amounts, up to about 30%, of other materials, but the possibility of using it in small amounts, such as 5-20 weight percent, simply to stabilize a composition whose color-play temperature range is in large part determined by the other materials present, is not to be overlooked.

Although the foregoing teachings and examples relate to cholesteryl erucyl carbonate, it will be possible to prepare and obtain results substantially similar with other compounds closely related in structure and properties. For example, it is apparent that the 5,6-position of cholesteryl may be saturated with hydrogen and/or halogen to yield cholestanol and halocholestanol, and the corresponding carbonates made from the compound may be expected to exhibit similar stabilizing effects.

" " " " " " " " " " " "

3,585,381 (U.S.)   Patented June 15, 1971

ENCAPSULATED CHOLESTERIC LIQUID CRYSTAL DISPLAY DEVICE.

Theodore L. Hodson, James V. Cartmell, Donald Churchill and Joe W. Jones, assignors to
The National Cash Register Company.
Application Apr. 14, 1969.

The present disclosure is directed to articles of manufacture, e.g., display devices, having an encapsulated liquid-crystal member of enhanced color purity, color contrast and visual resolution due to an overlying essentially transparent top layer having a substantially smooth exterior surface. This essentially transparent layer has an index of refraction which approximates that of the capsule cell wall material and any polymeric or other binder used in conjuction with the ancapsulated liquid crystals in the encapsulated liquid crystal layer or member.

In Figure 1 there is shown an opaque, e.g., black, background layer, 1, which can be a preformed opaque film or a substrate film having a black coating, e.g., ink, on its upper or lower surface. The opaque film 1 has on its upper surface an overlying layer formed of a plurality and more accurately a profusion of individual encapsulated liquid crystals 2 (or clusters of such individual encapsulated liquid crystals) located in association with a polymeric binder 3, which assists in retaining the capsules in substantially uniform distribution on the opaque substrate. The essentially transparent top layer 4, overlies and is in direct contact with the encapsulated liquid crystal layer throughout substantially its entire extent and has a smooth outer viewing (upper) surface, S, as illustrated in Fig. 1. The article of Fig. 1 can be made conveniently by applying an emulsion coating of encapsulated liquid crystals and binder to a paper or plastic substrate, e.g., Mylar, previously painted black. When the encapsulated liquid crystal layer has fully dried, a transparent polymer topcoat 4, having an essentially smooth upper surface can then be applied thereto. While Figs. 1 and 3 show outermost surface S of top layer 4 to be quite flat, it will be realized that the top coating procedure will result in some undulations as the top coating tends to assume in part the contours of the underlying capsules. These surfaces are "smooth" however, as defined above. When the top layer is formed by topcoating, the underlying cubstrate can be paper, wood or any nontransparent material, such as metalfoil, e.g., aluminum foil, plastic, e.g., poly(tetrafluoroethylene) "Teflon" etc.

In the article shown in Fig. 2 the encapsulated liquid crystal binder is applied as a slurry or emulsion by coating onto the sheet or 4, of transparent material, e.g., organic plastic, e.g., polyacrylate or silicone polymer, or inorganic material, e.g., glass. Top layer 4, is smooth on both its upper (exterior) and lower (interior) surfaces, S and S' respectively, and is of substantially the same thickness throughout. This structure can be secured readily by use of preformed sheets, e.g., of polished plate glass or plastic having substantially smooth surfaces. In the case of the Fig. 2 device, the transparent smooth top layer 4 is inverse coated (on surface S') with the encapsulated liquid crystal binder composition and the coating is allowed to dry thoroughly. Surface S', then becomes the interface between glass or plastic and encapsulated liquid crystals. Then this encapsulated liquid crystal layer is overcoated with a black or other opaque coating, 1, to aid in viewing the color change by incident light upon thermal and/or electric field stimulus. The thus formed device (article) is then inverted for viewing through transparent layer 4 as shown in Fig. 2. The exterior (lower) surface of background layer 1 can be either smooth or uneven as the case may be since it has nothing to do with the visual contrast and optical purity of the image observed through the transparent smooth surface layer 4.

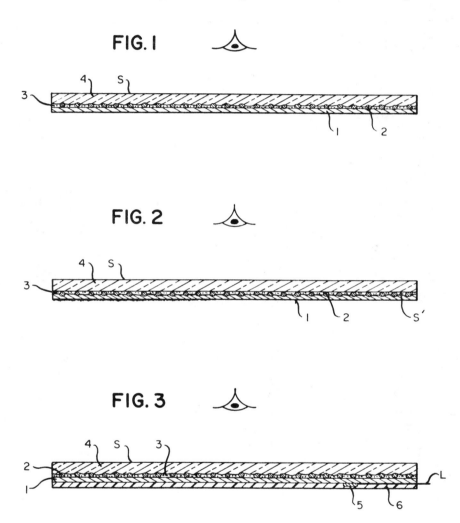

The article of Fig. 3 is formed by topcoating and hence is like that of Fig. 1, but it additionally includes one or more electrically conductive resistor heater element(s), 5, deposited on substrate 1 or at least positioned in thermally responsive communication with encapsulated liquid crystals 2, e.g., by conduction, convection or radiation. Electric lead(s), L, can be connected to a source of current, not shown. Optionally, there can also be included a base or insulating substrate 6.

Various natural and synthetic polymeric materials can be employed to constitute the polymeric binder matrix of the encapsulated cholesteric liquid crystal layer. Any transparent or substantially transparent polymeric material having an index of refraction from about 1.48 to 1.59 can be used. The encapsulated cholesteric liquid crystals can be associated intimately with the polymer matrix in a variety of ways. For example, the capsules can be deposited onto a polymer film, e.g., as a coating simply by spraying from a dispersion or emulsion of the encapsulated liquid crystal in a binder.

3,588,225 (U.S.)   Patented June 28, 1971

ELECTRO-OPTIC DEVICES FOR PORTRAYING CLOSED IMAGES.

Lawrence J. Nicastro, assignor to RCA Corporation.
Application Jan. 27, 1970

An electro-optical device for portraying closed images comprises at least two electro-optical cells in tandem. Each cell has a medium which is capable of modulating light in response to an applied voltage thereacross and electrodes on opposite sides of the medium for applying a voltage thereacross. The electrodes on at least one side of the medium of each cell consists of a plurality of separate electrodes so as to define an open electrode pattern. This pattern corresponds to a portion of the desired closed image to be portrayed. The individual cells of the device are arranged with respect to one another in a manner such that in operation the desired closed pattern is portrayed by the composite of the open patterns of the individual cells. Typical electro-optical medium is a nematic liquid crystal composition.

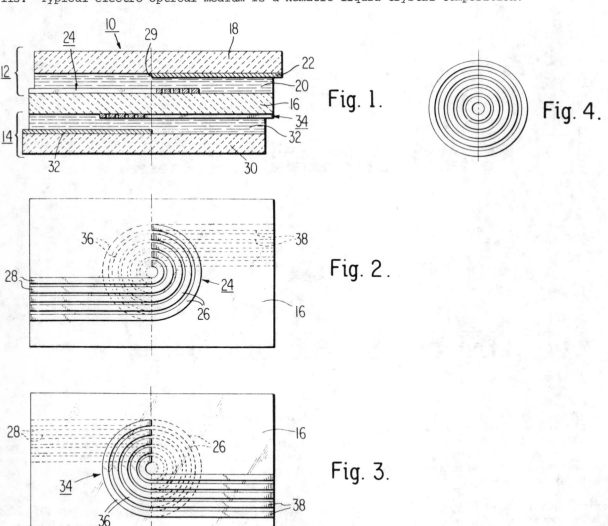

Referring to Fig. 1, there is shown a liquid crystal device 10 useful for portraying patterns consisting of concentric circles such as that shown in Fig. 4.

The device 10 comprises two electrooptical cells 12 ans 14 arranged in tandem and having a transparent common support plate 16. The first cell 12 comprises a first transparent support plate 18, spaced from the common support plate 16. The space between these support plates 16 and 18, which is typically from one-forth to one-half mil, is filled with a first nematic liquid crystal composition 20.

A first continuous transparent conductive electrode 22 is provided on the inner surface of the first support plate 18. This conductive electrode 22 is in contact with the liquid crystal composition 20 and completely covers the right half of the first support plate 18. The surface of the common support plate 16 which faces the support plate 18 is provided with a patterned transparent electrode array 24 which is in contact with the first liquid crystal composition 20. The pattern of the electrode array 24 corresponds to only a portion of the desired image to be portrayed. The electrodes of the particular electrode array 24, as shown in Fig. 2, consists of a concentric semicurcular portion 26 having parallel linear extansions 28 extending from one end of each of the semicircles. The concentric semicircles are positioned such that the diameter defining the semicircles 26 lies directly under the inner edge 29 of the first continuous electrode 22 and the semicircular portion 26 lies wholly under the first continuous electrode 22. The linear extensions 28 extend from the semicircles 26 to the left end of the common support plate 14. In this way, the linear extensions 28 do not overlap the first continuous electrode 22.

The first cell 12 is operated by applying a voltage across the liquid crystal composition between the first continuous electrode 22 and any one or combination of individual electrodes which comprise the electrode pattern 24. The liquid crystal composition will be activated only in the region of electrode overlap. Consequently, only the activated semicircular regions 26 of the electrode pattern 24 will cause an image to be seen by one watching either light transmitted through the device or light reflected from the device.

The second cell 14 is similar to the first cell 12. This cell 14 comprises, a second support plate 30 spaced from the common support plate 16; a second continuous transparent electrode 32 on the inner surface of the second support plate 30; a liquid crystal composition 32 filling the space between the second support plate 30 and the common support plate 16; and a second transparent patterned electrode array 34 as shown in Fig. 3 which corresponds to and provides the remaining portion of the desired image and consists of concentric semicircles 36 and parallel linear extensions 38 from an end thereof. The linear extensions 38 of the electrodes of the second electrode array 34 extend to the right edge of the common support plate 16. The second continuous electrode 32 covers the entire left half of the second support plate 30. In the same manner as described with reference to the first cell 12, the entire semicircular portion 36 of the second electrode array 34 overlaps the continuous electrode 32 while the linear extensions 38 do not overlap the continuous electrode 32.

The first and second cells are arranged such that the image formed from the composite of these cells appears as complete concentric circles. This is achieved by having the diameters of each of the two semicircular patterns lie in a common plane which plane is perpendicular to the surface of the device. However, to negate any error due to parallax one might slightly overlap the edges of the concentric circular patterns.

By applying a voltage between each set of electrodes, that is, between any selected one or more of the electrodes in each electrode array and the respective continuous electrode of each cell, one can form images of different combinations of circles.

" " " " " " " " " " "

3,590,371 (U.S.)　　　　　　　　　　　　　　　　　　　　　　　　　　Patented June 29, 1971

METHOD UTILIZING THE COLOR CHANGE WITH TEMPERATURE OF A MATERIAL FOR DETECTING DISCONTINUITIES IN A CONDUCTOR MEMBER EMBEDDED WITHIN A WINDSHIELD.

<u>Hugh E. Shaw, Jr.</u>, *assignor to PPG Industries, Inc.*
Application Dec. 31, 1969.

Circuit discontinuities in conductor members embedded in pieces of glass, such as windshields, are detected by placing in operative association with the glass a stratum of cholesteric phase liquid crystal material having appropriate color-change temperature-range characteristics. When current is passed through the conductors, color changes are observed in the vicinities of the operating conductors. By using for the cholesteric-phase liquid-crystal material a mixture of at least first and second ingredients, with the first and second ingredients spanning respectively different portions of the temperature range from about 20°C. to about 32°C. desirable and useful indications are rapidly obtained without the use of particular testing-environment conditions or particular glass-storage conditions.

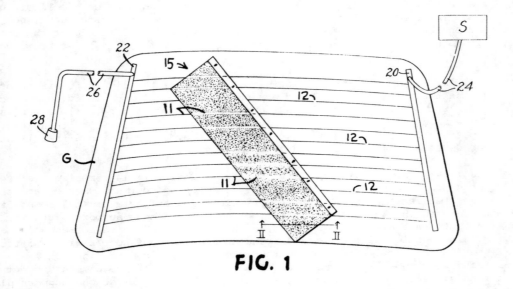

FIG. 1

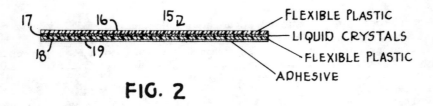

FIG. 2

Referring to the drawing, there is shown a substrate piece of glass or the like, the piece G having embedded therein a plurality of conductor members 11, some of which contain breaks as indicated at 12.

Positioned on the glass is a device 15 that comprises a layer of suitable cholesteric-phase liquid-crystal material 17 sandwiched between a stratum 16 of flexible plastic and a second stratum 18 of flexible plastic. To secure the piece 15 to the piece G, there may also be provided a strip 19 of double-faced adhesive material. The strata 16 and 18 may be made, foe example, of polyethylene terephthalate or other suitable transparent material.

The stratum 17 of cholesteric phase liquid crystal material may comprise, for example, a mixture of (in weight per cent) 45 percent cholesteryl nonanoate, 50 percent cholesteryl linolineate and 5 percent cholesteryl chloride. If this particular liquid crystal material is used, it has a color-change temperature of about 1° to about 100°C. This implies that it is necessary that the piece G and the testing environment be at about 0°C. or lower when the testing is commenced. As is indicated in Fig. 1, when electrical power is provided to the conductors 11 by means, for example, of suitable bus bars such as 20 and 22, the ones of the conductors 11 that contain the discontinuities or breaks 12 do not become warm, and consequently the portions of the test piece 15 that overlie them do not exhibit a color change, whereas the ones of the conductors 11 that overlie operative ones of the conductors 11 do.

It is most convenient to use a mixture of two or three suitable ingredients, with each of those ingredients being a compound or a mixture that responds in a different portion of the temperature that extends from about 20°C. to about 32°C. This usually makes it possible to obtain adequate indication, despite the lack of any exercise of control over the testing-environment temperature or the pretesting storage temperature of the pieces G.

" " " " " " " " " " "

3,592,526 (U.S.)   Patented July 13, 1971

MEANS FOR ROTATING THE POLARIZATION PLANE OF LIGHT AND FOR CONVERTING POLARIZED LIGHT TO NONPOLARIZED LIGHT.

John F. Dreyer, assignor to Polacoat Inc.
Application July 15, 1969.

A light transmissive device which will rotate the plane of polarized light and will also depolarize plane polarized light passed through the device, said device comprising a pair of directionally oriented supporting surfaces, such as unidirectionally rubbed glass or plastic plates, rotatable relative to each other and separated by a thin film of a nonoptically active nematic liquid crystal material, the device acting to rotate the plane of the polarized light when the directionally oriented surface first contacted by the entering light is either parallel or perpendicular to the plane of polarization of the entering light and the remaining directionally oriented surface is rotated so as to bring its oriented surface to the desired plane of polarization for the exiting light, the device also acting to convert plane polarized light to nonpolarized light when the oriented surface first contacted by the entering light lies at an angle of 45° with respect to the plane of polarization of the entering light and the directionally oriented surfaces are parallel to each other.

Referring to Figs. 1 and 2, a polarizing filter is indicated at 1, which is adapted to receive light from a light source 2. Upon passage of the light through the polarizing filter 1, the light will be completely polarized, its plane of polarization being indicated by the solid arrows 3, as seen in Fig. 1. A first supporting surface 4 overlies the polarizing filter, the supporting surface comprising a glass or plastic plate the uppermost surface of which has been unidirectionally oriented by rubbing, the direction of orientation of the first supporting surface being indicated by the dotted arrows 5. The film of nematic liquid crystal material is indicated at 6 and, in the embodiment illustrated, the liquid crystal film is enclosed within a gasket or mask 7 the thickness of which determines the thickness of the liquid crystal film. Such mask may be conveniently formed from a transparent material, such as cellulose acetate. A second supporting surface 8 overlies the liquid crystal film and its undersurface is also unidirectionally oriented, again by rubbing, as indicated by the dotted arrows 9.

With the supporting surfaces oriented in the manner illustrated in Fig. 1, the plane of polarization of the exiting light on the viewing side of the device will be in the direction of the solid arrows 10, the plane of polarization being parallel to the direction or orientation of the second supporting surface, as indicated by the dotted arrows 9. Thus, the angle between the direction of orientation of the two supporting surfaces, i.e., the angle defined between the dotted arrows 5 and 9, determines the degree to which the plane of polarization of the light is rotated. In the embodiment illustrated, it is shown rotated through an angle of 90°.

It will be understood, however, that the angle of rotation may lie anywhere between 0° and 360°, depending upon the extent to which the supporting surface 8 is rotated relative to the supporting surface 5. Thus, by rotating the second surface 8, the plane of polarization of the exiting light can be rotated at will to any desired angle relative to the plane of polarization of the entering light. The degree of rotation is uniform for all visible wave length and no color effects are observed if a proper thickness of the nematic liquid crystal film is maintained and the film is chosen so as to have no optical activity.

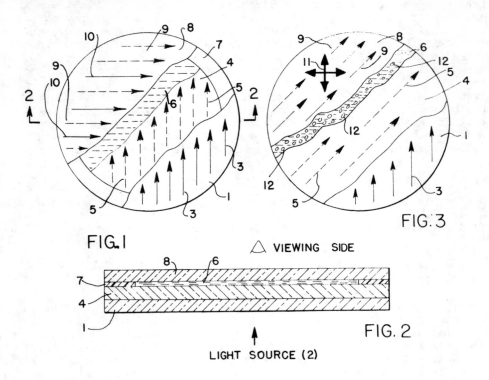

The same rotational effect can be achieved by rotating the first supporting surface 4 so that its direction of orientation lies at right angles with respect to the plane of polarization of the entering light. In other words, the orientation of the supporting surface on the side of the device where the light enters may be either parallel or perpendicular to the plane of polarization of the entering light. Otherwise some of the entering polarized light will be transformed into nonpolarized light.

Where it is desired to depolarize the light passing through the device, the parts will be oriented in the position illustrated in Fig. 3, wherein it will be seen that the first supporting surface 4 has been rotated so that its direction of orientation, as exemplified by the dotted arrows 5, lies at an angle of 45° with respect to the plane of polarization of the entering light, as exemplified by the solid arrows 3. In addition, the second supporting surface 8 has been rotated so that its direction of orientation, as exemplified by the dotted arrows 9 is parallel to the direction of orientation of the first supporting surface. Where this condition exists, the exiting light will be nonpolarized, as indicated by the crossed arrows 11. In other words, when the orientation of both supporting surfaces are parallel and lie at an angle of 45° with respect to the plane of polarization of the entering light, the exiting light will be completely depolarized. Under these circumstances, the device becomes a means for converting plane polarized light into nonpolarized light without inducing elliptical or circular polariation of the light; and such depolarization takes place uniformly throughout the spectrum.

If the direction of orientation of either or both of the supporting surfaces is rotated so as to lie at any angle less than or greater than, 45° with respect to the plane of polarization of the entering light (other than 90° or 0°), a combination of the effects of rotating the plane of polarization and depolarizing of the light will take place. In other words, as the direction of orientation of the supporting surfaces moves away from an angle of 45° with respect to the plane of polarization of the entering light, the degree of polarization will progressively decrease, but at the same time the plane of polarization of the polarized light which passes through the device on the viewing side will be rotated, the degree of rotation depending upon the extent to which the polar oriented surfaces are rotated.

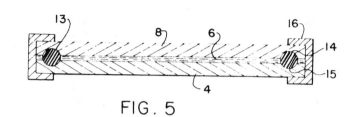

FIG. 5

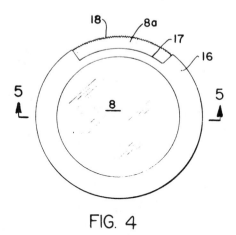

FIG. 4

Figures 4 and 5 illustrate an examplary form of device in accordance with the instant invention wherein the liquid crystal film 6 is retained between the supporting surfaces 4 and 8 by means of an O-ring 13 which may be conveniently received in mating annular grooves 14 and 15 formed in the film contacting surfaces of the supporting plates. The assembly may be conveniently mounted in a frame 16 having a cutaway portion 17 which acts to expose marginal edge portions of the supporting plates, one such edge portion being indicated at 8a in Fig. 4. For ease of rotation, the edges of the supporting plates may be milled, as indicated at 18, or otherwise configured to permit ready engagement for rotational movement.

It should be evident that diverse types of frames and mounting means may be employed, depending upon the size and intended mode of operation of the device. For example, assemblies may be made up with the direction of orientation of the two supporting surfaces in fixed relationship to each other, their directions of orientation being parallel if the device is to be used to depolarize the light, or their directions of orientation may be at any desired angle where it is desired to rotate the plane of polarization by a fixed angle.

Similarly, by employing liquid crystal materials which are no longer fluid, fixed filters or cells can be made. For example, it has been found that isobutyl methacrylate resin can be added to a number of liquid crystal materials, depending upon their chemical constitution, up to 50 percent by volume. The combination may be made by using a suitable mutual solvent, such as toluol. After coating one of the oriented supporting surfaces with the liquid crystal-resin mixture, the solvent is evaporated and the second oriented surface is placed in contact with the film so formed.

""""""""""

3,592,527 (U.S.)　　　　　　　　　　　　　　　　　　　　　　　　　　　　Patented July 13, 1971

IMAGE DISPLAY DEVICE.

Gary H. Conners and Paul B. Mauer.
Application Nov. 12, 1969.

This invention relates to a display device comprising a layered or sandwich structure which includes in order, a first transparent electrode, a photoconductive layer, a specular conductive mosaic layer, and liquid crystal material 22 which scatters incident light when subjected to an electric field, so that when an image, e.g. of a film 34, is projected on to layer 18 (the photoconductor), and a D.C. or low frequency A.C. potential is applied to electrodes 10, 12, ambient light such as from a source 40 is either reflected at 20, away from an observer 42 (dark image) or is scattered at least partly toward the observer and to an extent determined by the image intensity. The contrast of the viewed image is increased by placing a black area 45 so that it is visible after reflection at the mosaic 20. The mosaic 20 may be formed by vapour deposition of the dots within holes etched in an insulating photoresist 21.

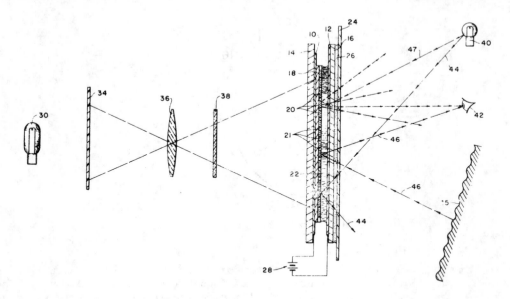

If the ambient light falls within the sensitive range of the photoconductor 18 it must be blocked, e.g. by using an opaque photoresist 21 or as shown by a filter 24. Thus if layer 18 is activated only by U.V. light, a U.V. transmitting filter and a U.V. blocking filter 24 are provided. The liquid crystal 22 may be such as to retain its light scattering condition after cessation of the electrical field, and in a modification of the device, Fig. 3 (not shown), electrodes 10 and 12 are provided with prongs, for connection to a D.C. or low frequency A.C. source for recording an image and for connection to a high frequency A.C. source for erasing the image. In another modification, Fig. 2 (not shown), the surface 45 is omitted and a quarter wave plate and a polarizer are successively placed over filter 24 to render the light from source 40 circularly polarized. After reflection from the mosaic such light will be blocked by the quarter-wave plate and polarizer combination. If the liquid crystal is light scattering, the light becomes depolarized and may be transmitted back through the quarter-wave plate and polarizer. Also suggested are suitable mixtures of nematic compounds, and of nematic and cholesteric materials, for a layer 22 without or with a memory function, and materials for the photoconductor 18, which may be responsive to infra-red, visible or ultra-violet light.

" " " " " " " " " " " " " "

3,597,043 (U.S.)   Patented Aug. 3, 1971

NEMATIC LIQUID CRYSTAL OPTICAL ELEMENTS.

<u>John F. Dreyer</u>, assignor to Polacoat, Inc.
Application May 2, 1969.

The production of visible motion patterns in a thin film of a nematic liquid crystal compound by inducing physical movement of the film, the film being formed between support plates at least one of which is transparent or translucent, means being provided to effect movement within the liquid crystal film.

Optical elements in accordance with the invention may be used as a means to detect the presence of any of the forms of energy which may be converted into mechanical energy, inclusive of electric, magnetic and acoustic energy, or to detect changes in any of the forms of energy which may be converted into mechanical energy.

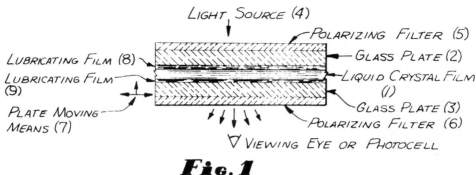

Fig. 1

Figure 1 of the drawing illustrates an optical element in which the film 1 of nematic liquid crystals lies between transparent or translucent supporting plates 2 and 3, which may comprise glass or a plastic material. Since light from the light source 4 will pass through the element, polarizing filters 5 and 6 will be provided on opposite sides of the liquid crystal film 1. These light polarizing filters do not have to be in a crossed position, the plane of polarization of one can be at any desired angle with respect to the plane of polarization of the other, depending upon the amount of light to be transmitted. In instances where the liquid crystal compound is dichroic, only one filter, usually the filter 6, is required to show the motion pattern generated when one or both of the plates is moved. A plate moving means is indicated at 7, it being understood that such means will comprise a vibrator or other means capable of effecting movement of the plates relative to each other and to the interposed film of liquid crystals.

The thickness of the liquid crystal film may vary depending upon the effect desired. If the film has a thickness of about one-half mil or less, color patterns will be produced ranging through the spectrum from red to blue. Above about 1 mil, the patterns will tend to be essentially monochromatic, depending usually on the basic color of the polarizing filters. For example, if the polarizers are of neutral gray color, gray and white patterns will be produced. Similarly, if the polarizers are of a rose tint, the patterns will vary from near white to dark rose. For pracyical purposes, film thicknesses up to about 10 mils produce excellent results and the applied films are normally colorless, although as previously indicated, the films may be colored depending upon the visual effects desired.

It has been found, however, that it is essential to the production of visible motion patterns that at least one, and preferably both, of the film contacting surfaces of the supporting plates be treated with a lubricating material which will remove the effects of surface orientation of the plates. For example, if the plates are rubbed or polished, or even cleaned, it has been found that a surface orientation results in the direction or directions in which the plates were rubbed, and that such surface orientation, in addition to being visible when the motion patterns are generated, will interfere with the generation of the motion patterns and, in some instances, will preclude their formation. Some form of such surface orientation appears to inherently exist in all forms of supporting plates, whether they are of glass, plastic or metal. It has been found that the effects of such surface orientation can be eliminated by providing thin films 8 and 9 of a lubricant at the interfaces of the plates and the liquid crystal film. A preferred lubricating material comprises a cationic wetting agent which will be adherent to the surfaces of the plates and at the same time nonabsorbent with respect to the liquid crystal film. Excellent results have been obtained using Intercol R A, which is a long chain fatty acid amide containing multiple amine groups, manufactured by Synthetic Chemicals, Inc. Other satisfactory wetting agents are Sotex 45 A, also manucatured by the above firm; Quaternary 0, which is a quaternary amine manufactured by Geigy Chemical Corporation; and Perma-Par-K, which is a substituted long chain secondary amine, produced by Refined Products, Inc. While not considered a cationic wetting agent, lecithin has also been found to be suitable as a lubricant for the liquid crystal film. Essentially, the function of the wetting agent is the formation of a molecular layer which will be attracted to the surface of the plates, which will not be dissolved or absorbed by the liquid crystal film, and which will act as a lubricant at the interface of the liquid crystal film and the supporting plates.

While in many aplications the film contacting surfaces of both supporting plates will be treated with the wetting agent, particularly where it is desired to prevent surface orientation of either plate from being imposed on the motion pattern, additional optical effects can be produced by intentionally forming rubbed patterns on the surface of one of the plates. For example, the plate may be rubbed, scratched or etched to provide fixed pattern or design which will become visible when physical movement is imparted to the liquid crystal film and it is viewed using two light polarizers in the case of colorless liquid crystal compounds and one polarizer in the case where a dichroic liquid crystal compound is employed, or where one of the supports comprises a reflective surface.

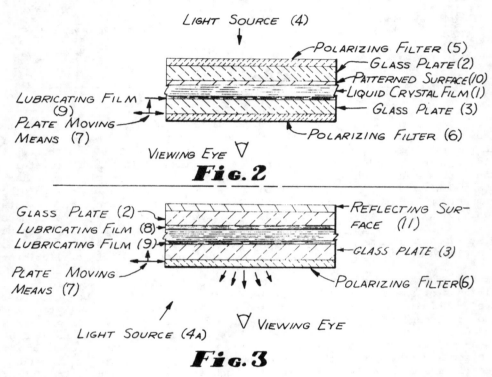

Figure 2 of the drawing illustrates a display device of the character just described wherein the surface 10 of plate 2 is rubbed, as by means of a slurry of rouge in water, to produce the desired rubbed pattern which may comprise all or a portion of the surface of the plate. Where this is done, the lubricant will be omitted at least in the areas on which the rubbed pattern has been impressed.

Another combination is illustrated in Fig. 3 wherein one of the supporting plates 3 is transparent or translucent and the second supporting plate 2 has a reflecting surface 11, which may comprise a mirror or a reflective paint. Alternatively, the supporting plate 2 may comprise a plated or polished metallic member. While, as before, it is preferred to employ lubricating films 8 and 9, in this instance only one polarizing filter 6 is required, the filter being on the side of the liquid crystal film 1 from which the element is viewed.

Still another combination is illustrated in Fig. 4 wherein the supporting surface on the viewing side of the element comprises a thin plastic sheet 12 which may incorporate a polarizing filter, the plastic sheet acting as a diaphragm sensitive to acoustic vibrations. The acoustic vibrations will be transmitted by the plastic sheet to the liquid crystal film as mechanical vibrations effective to induce motion patterns in the film. Readily identifiable motion patterns can be produced by varying the amplitude of the acoustic vibrations, the skilled observer thereby being able to effectively read sound.

It is also within the spirit and purpose of the invention to incorporate various particles in the liquid crystal film to cause it to respond with greater sensitivity to outside influences.

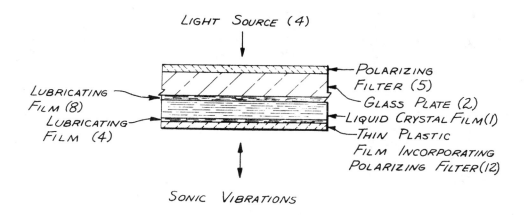

Fig. 4

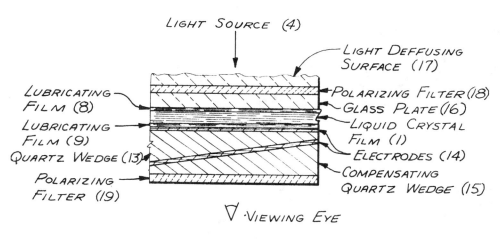

Fig. 5

For example, iron particles or iron oxide particles incorporated in the liquid crystal compound will produce a response from a magnetic field; and either dielectric or metallic particles of correct size (resonance frequency) will respond to a corresponding frequency.

In the embodiment of the invention illustrated in Fig. 5, a piezoelectric material in the form of quartz wedge is utilized as one of the supporting plates for the liquid crystal film. As seen therein a quartz wedge 13 takes the place of the transparent plate on the viewing side of the liquid crystal film 1, with transparent electrodes indicated at 14, formed from a material such as tin oxide, juxtaposed between the upper surface of wedge 13 and lubricating film 9 and between the inclined surface of the wedge 13 and a compensating quartz wedge 15 which acts to straighten thru, i.e., the compensating wedge acts to remove the wedge character of the assembly and hence maintain the light rays in a straight path without being bent as they pass through the device. In the illustrated embodiment, the opposing support plate 16 has a light diffusing surface 17 which may be conveniently formed by sandblasting the exposed surface of the plate, and in this instance the polarizing filter 18 is integrally incorporated in the plate 16. As will be understood, a second polarizing filter 19 is provided on the viewing side of the element unless, of course, the liquid crystal film 1 is dichroic, in which event the second filter may be eliminated. With such arrangement, variable voltages may be applied through the electrodes 14 to produce variable motion patterns in the liquid crystal film.

It will be understood, of course, that in each of the embodiments described, the supporting plates will be mounted in a frame or other support which will permit them to move under the influence of the applied mechanical energy. Frames formed from rubber or other resilient material serve nicely for such purpose, or the frames or other plate supporting structure may be rigid with the interposition of pads of a yieldable material supporting the plates from the frames.

3,597,044 (U.S.)   Patented Aug. 3, 1971

ELECTRO-OPTIC MODULATOR.

<u>Joseph A. Castellano</u>, assignor to RCA Corporation.
Application Sept. 3, 1968.

    A light valve comprises an acyloxybenzylidene anil nematic liquid crystal and means for applying an electric field thereto so that the light scattering and birefringent properties of the liquid crystal are altered due to the tendency of the liquid crystal domains to align preferred liquid crystal compositions for use in the valve are mixtures of p-n-alkoxy substituted (particularly butoxy to heptoxy) and p-n-acyloxy substituted (particularly acetyl to capryl, $C_7H_{15}CO$) benzylidene-p'-aminobenzonitriles, three such compositions being specifically described. The mixtures exhibit a depressed crystal-nematic transition temperature, e.g. 26°C. for one ternary mixture, they may be supercooled to below room temperature in the nematic state, and they are prepared by mixing the isotropic liquids, cooling to 0°C. to crystallize and then slowly heating the crystals until above the nematic transition temperature.

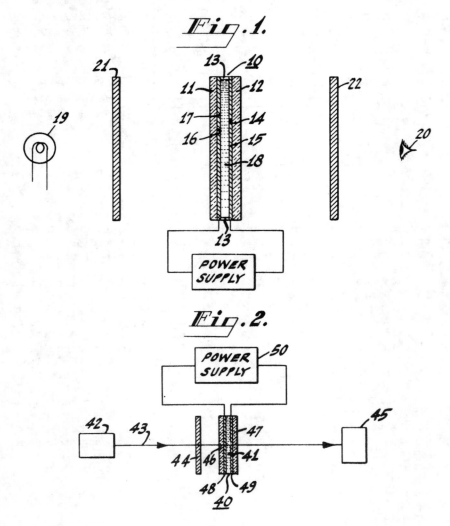

A first embodiment of light valve may include means for polarizing the light reaching the liquid crystal, or crossed polarizers. The liquid crystal may contain 1/2-5% of a pleochroic dye, e.g. indophenol, which co-operatively orientates with the crystal when an electric field is applied, resulting in a change from a coloured to a colourless or differently coloured state. Switching speeds of up to 15,000 cycles/sec. are possible e.g. for television displays.

A second embodiment, Fig. 2 for modulating a helium-neon laser beam, includes a polarizer, a light valve including 1% indophenol blue and a CdS photodetector. The polarizer is oriented so that transmission factors of 3 and 50% are obtained with no field, and a field of 400 kV/cm. respectively. Instead of varying an applied D.C. voltage to modulate the beam, it is possible to apply an A.C. voltage and vary the frequency thereof.

"""""""""""

3,600,060 (U.S.)   Patented Aug. 17, 1971

DISPLAY DEVICE CONTAINING MINUTE DROPLETS OF CHOLESTERIC LIQUID CRYSTALS IN A SUBSTANTIALLY CONTINUOUS POLYMERIC MATRIX.

<u>Donald Churchill and James V. Cartmell</u>, assignors to The National Cash Register Company. Application Feb. 23, 1968.

This disclosure is directed to articles of manufacture, chiefly display devices, containing minute "naked" droplets or inclusions of cholesteric liquid crystal material in a substantially continuous polymeric matrix, said liquid crystal material changing color or shade of color not only upon application of an electric potential but also upon removal of the field. The image produced has a comparable outline to that of the path of the electric field. Three chromatic states are evident, the normal color (before the electric potential is applied), the color given off when the electric field is applied, and the color observed when the electric field is removed. All three chromatic states are readily discernible from one another. The polymer matrix protects the cholesteric liquid crystal droplets from aging and enhances electric field behavior because the third chromatic state (electric potential removed) has a greater longevity with the matrix-bound material versus unprotected material of identical composition but no polymeric matrix.

Figure 1 shows the presence of cholesteric liquid crystal member 3 comprised of a plurality of minute, individual, "naked" droplets or inclusions, a, of cholesteric liquid crystal material confined within a substantially continuous matrix, b, of polymeric material. The droplets have a random shape and are reasonably uniformly distributed (dispersed) within the polymer matrix. An optional thin, protective barrier film, e.g., polyethylene glycol terephthalate "Mylar" polyester can be deposited on one or both major surfaces of liquid crystal member 3 to prevent deleterious solvation of surface located liquid crystal material.

<u>Formation of cholesteric liquid crystal member</u>:

The cholesteric liquid crystal members employed in accordance with this invention are comprised of a profusion of randomly shaped inclusions of cholesteric liquid crystal material dispersed in a substantially continuous solid polymeric matrix. These members can be prepared readily by dispersing (emulsifying) minute droplets of cholesteric liquid crystal material in a dryable liquid solution of film-forming polymeric material and then coating, casting or otherwise depositiong the solution upon the desired surface. Also films, sheets, layers, etc. can be preformed, e.g., by casting, and then assembled into the composite structure at the desired time. Coating and casting solutions can be prepared readily by adding the cholesteric liquid crystal compound or mixture to a solution, e.g., aqueous solution of film-forming polymer matrix material, e.g., polyvinyl alcohol, using a stirrer, mixer, blender or equivalent agitation device until a liquid crystal droplet size range of from about 5 to about 20 microns is obt-ined. This emulsion can be coated on the substrate directly, e.g., by means of a draw-down applicator, onto a blackened substrate, e.g., of "Mylar" to a wet film thickness of about 10 mils (0.0254 cm.) and air dried, e.g., at about 25°C. Film thickness can be increased by repeated sequences of coating and drying. The dried emulsion film can be stripped from its substrate and utilized as a preformed film which can be optionally opacified or blackened for use in the articles of this invention. Various mixtures of cholesteric liquid crystals can be used with various polymeric film-forming matrix materials. In accordance with this invention films can be formed which contain from about 30 to about 95 weight percent of cholesteric liquid crystal material, in droplet form, with the remainder being polymer matrix material. Usually, however, the liquid crystal droplets represent from about 50 to about 90 weight percent of the total film weight (droplets plus polymer matrix).

Layers cast from an emulsion of the liquid crystal material and then dried are dry to the touch (although containing liquid inclusions; are relatively unaffected by brushing contact with foreign bodies; are substantially immune to solute contamination, e.g., absorption of extraneous vapor; and are not subject to rapid deterioration by selective nucleation crystallization; and, in cases where crystallization does begin, it is stopped from further areawise development by a boundary of the polymer matrix.

Another feature of the incorporation of polymer matrix bound droplets of the cholesteric liquid crystalline materials into a system to provide an electric potential-activated sensing or display device is the utilization of various mixtures of liquid crystals as to droplet size and chromatic response for indicating and/or displaying a wide range of specific levels of electric potential. Such a system, in one case, can comprise a plurality of layers, each layer comrpising one, two or more types of droplets having different mixtures of chromatically responsive cholesteric liquid crystalline materials. These devices can be tailor made to accomplish the desired task by variation of characteristics imparted thereto by any one of the following adjustments: (a) electric field response range; (b) size of the liquid crystal droplets; (c) type and thickness of the polymer matrix material; (d) specific composition of the cholesteric liquid crystalline material(s), and the like, all to the purpose of choosing a response suitable for a given proposed use.

"""""""""""

3,600,061 (U.S.)   *Patented Aug. 17, 1971*

ELECTRO-OPTIC DEVICE HAVING GROOVES IN THE SUPPORT PLATES TO CONFINE A LIQUID CRYSTAL BY MEANS OF SURFACE TENSION.

<u>*George H. Heilmeier and Louis A. Zanoni*</u>*, assignors to RCA Corporation.*
Application March 21, 1969.

A device comprises a layer of a liquid crystal sandwiched between support plates. The device includes means for confining the liquid to a particular region of the device.

It has been discovered that by providing means such as grooved support plates, the liquid of the device may be confined to a region separated from the edges of the device. In addition, the grooves also act to provide the liquid with space in which to expand upon heating of the device.

An embodiment of a novel device 10 may be described with reference to Fig. 1. In the figure, there is shown a first transparent support plate 11 and a second transparent support plate 12 between which is sandwiched a liquid crystal layer 13 which exhibit dynamic scattering. The first and second plates 11 and 12 each have a pair of parallel grooves 14 and 15 respectively on the inner surfaces 16 and 17 thereof. The grooves, as shown, extend across opposite edges of each support plate. However, the grooves may be made to terminate at a point prior to the edges. The first and second support plates 11 and 12 are arranged such that the grooves 14 on the first support plate 11 are orthogonal to the grooves 15 on the second support plate 12. The liquid crystal layer 13 is confined to the area defined by the pairs of grooves 14 and 15 and indicated by ABCD on the figure. The surface tension along the edges of the grooves 14 and 15 is what causes the liquid crystal layer 13 to be so confined.

The first and second support plates 11 and 12 each have a transparent conductive coating 18 and 19 on the inner surfaces thereof so as to be in contact with the liquid crystal layer 13. The device which modulates light due to a scattering effect, may be operated by applying a voltage source (not shown) connected to the transparent conductive coatings 18 and 19. The configuration

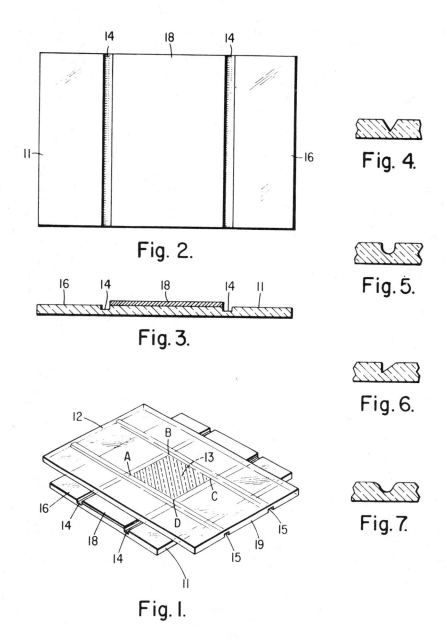

of grooves as shown in Fig. 1 is preferred due to the ease in making contact with the conductive coatings on each support plate.

The grooves of one support plate in the device 10 can be seen more clearly with respect to Figs. 2 and 3. The cross-sectional shape of the grooves is not limited to that shown in Fig. 3 and many cross-sectional configurations for the grooves are feasible.

Figures 4-7 indicate several alternate cross-sectional configurations for the grooves.

" " " " " " " " " " "

3,604,930 (U.S.)   Patented Sept. 14, 1971

METHOD AND APPARATUS FOR DISPLAYING VISUAL IMAGES OF INFRARED BEAMS.

<u>Phillips J. Allen</u>, assignor to The United States of America as represented by the Secretary of the Navy.   Application March 5, 1970.

A method and device for displaying a visual pattern of an incident beam of infrared radiation includes a film of minute droplets of a liquid crystal composition encapsulated in gelatin. The nature of the composition is such that within a certain temperature range its optical properties change with a change of temperature, so that it appears to change color. The film of liquid crystal is thermally biased to its critical temperature so that when incident infrared radiation is absorbed by the composition, the temperature thereof is raised which causes a color change in the zone of the film on which the radiation is incident.

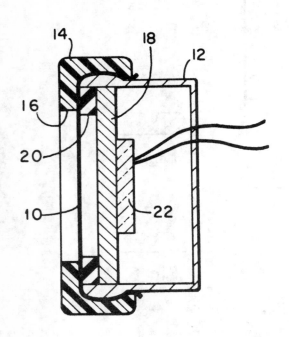

Laser imaging techniques in the near infrared-sensitive films, electronic image converters, infrared-sensitive Vidicons, and infrared phosphor screens. However, none of these has useful sensitivity beyond about 2 microns.

Liquid crystal screens offer a number of attractive advantages for infrared laser imaging, such as high thermal sensitivity and real-time viewing in bright light, but normal exposure to the atmosphere and ultraviolet light often results in total loss of sensitivity, sometimes in a manner of hours.

Rapid deterioration of performance does not seem to be a problem with encapsulated liquid crystals, in which the liquid crystal material is sealed in minute gelating capsules some 10-30 microns in diameter. These minute gelatin capsules are then suspended in a water-latex vehicle to form a slurry which, when spread over a surface and allowed to dry, leaves a randomly packed film or coating of the tiny capsules of liquid crystal which adhere because of the latex. The heat sensitive screen thus formed is ready for use as soon as drying is complete.

A 4-inch diameter "tunacan" infrared-imaging detector (shown in Fig. 2) employs a 1-mil Mylar diaphragm sprayed on one side with ultraflat black Krylon lacquer as absorber, and coated on the outside surface with Type R-33 encapsulated liquid crystal (manufactured by the National Cash Register Company, Dayton, Ohio). The screen 10 thus prepared is stretched over the mouth of a can 12 and held taut using a plastic can cover 14 with the center portion removed at 16.

In this detector, the necessary thermal bias is provided by a heated baffle plate 18 air spaced an eight of an inch behind screen 10 by means of spacer ring 20. Less than 2 watts of electrical power to a 40-ohm resistor 22 epoxied to plate 18 are required to maintain screen 10 at proper operating temperature (about 33.5°C. for the type encapsulated liquid crystal used). A regulated DC supply (not shown) may be used to stabilize heater input power, but an automatic temperature control is preferred.

This detector will give a good visual indication with a 10-micron flux of 10 milliwatts per square centimeter incident on the screen, and a narrow one milliwatt beam at 1.15 microns has been detected with this device.

" " " " " " " " " " " " "

1,246,847 (British)   Patented Sept. 22, 1971

LIQUID CRYSTAL DISPLAY ELEMENT HAVING STORAGE.

RCA Corporation (Inventor's name not given)
Application Nov. 18, 1968; prior U.S. application Dec. 13, 1967.

A light valve or display device comprises a mixture of at least one nematic liquid crystal of the type that exhibits non-destructive turbulent motion when an electric current of sufficient magnitude is passed therethrough, i.e., upon the application of a D,C, or low frequency electric field the single phase homogeneous molecular solution separates into two phases causeing incident light to be scattered, with at least one member of the group consisting of cholesterol, cholesterol derivatives and cholesteric liquid crystals. The mixture preferably comprises cholesteryl oleate and cholesteryl chloride, and at least one member of the group consisting of anisylidene-p-aminophenyl acetate, p-n-butoxybenzylidene-p-aminophenylacetate and anisylidene-p-aminophenylbutyrate. The mixture exhibits storage of its light scattering emulsion-like state after removal of the electric field initiating the state. To return the mixture to its transparent state a relatively high frequency A.C. field is applied to the mixture. In the display device a thin film of the mixture is disposed between two planar elements, one of the elements being transparent, the other being transparent, reflective or absorptive depending upon the descired mode of operation of the device. The planar elements have transparent row and column conductors for applying electric fields to selected areas of the liquid crystal film. The Patent Specification includes a table of alternative nematic liquid crystals, cholesterol derivatives and cholesteric liquid crystals, and describes various mixtures by weight.

"""""""""""""

3,612,654 (U.S.)   Patented Oct. 27, 1971

LIQUID CRYSTAL DISPLAY DEVICE.

<u>Richard I. Klein and Sandor Caplan</u>, assignors to RCA Corporation.
Application May 27, 1970.

The device comprises a front transparent substrate having a transparent electrode on the inner surface thereof, a rear substrate, and a liquid crystal material disposed between the two substrates. The inside surface of the rear substrate is covered with a light-reflecting material, e.g., silver, covered in turn with a thin layer of dielectric material, e.g., silicon dioxide. Disposed on the dielectric layer is a patterned electrode of either a transparent material, e.g., tin oxide, or a light-reflecting material, e.g. silver.

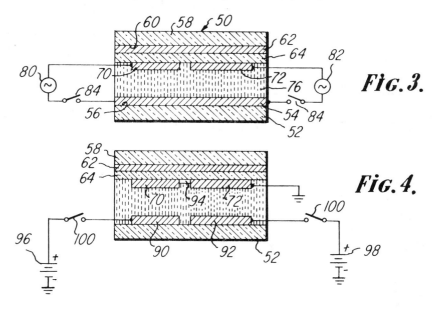

A display device 50 in accordance with the instant invention is shown in Fig. 3. The device 50 comprises a front transparent substrate 52 of, for example, glass, having a transparent electrode 54, e.g., of tin oxide, covering the inside surface 56 thereof. The device further comprises a rear substrate 58 of, for example, glass, which need not be transparent. Covering the inside surface 60 of the rear substrate 58 is a light reflective layer 62 of, e.g., aluminum or silver. Covering the layer 62 is a thin filmlike layer 64 of a transparent insulating material, e.g., silicon oxide, silicon dioxide, silicon nitride, chrome oxide, or the like. The layer 64 is preferably as thin as practical, being in the order of 2,000 A. Disposed on the layer 64 are a pair of electrodes 70 and 72 having the shape of a circle and a square, respectively.

The electrodes 70 and 72 can be light reflecting in the direction towards the front substrate, being, e.g., of aluminum or silver, or can be transparent, e.g., of tin oxide. Disposed between the two substrates 52 and 58 is a thin film 76 of a liquid crystal material.

Two separate alternating voltage sources 80 and 82, and switch means 84 are provided, whereby a voltage can be selectively applied between the electrode 54 on the front substrate 52 and either or both of the electrodes 70 and 72.

In use, the device 50 is both illuminated and viewed from the front substrate 52 side of the device. By applying a voltage between the electrode 70 and the electrode 54, a luminous circle is displayed on a mirrorlike background. By applying a voltage between the electrode 72 and the electrode 54, a luminous square is displayed.

The close spacing between the reflective layer 62 and the electrodes 70 and 72 substantially completely eliminates the problems of the prior art devices. Thusm for example, oqing to the close spacing, the shadows cast by the electrodes 70 and 72, if opaque, are substantially invisible to the viewer regardless of the angle of view of the devices, i.e., short of viewing the device edge-on, the shadows are substantially hidden by the electrodes. Likewise, if the electrodes 70 and 72 are transparent, the reflection of the luminous images in front of the electrodes 70 and 72 from the reflective layer 62 appears substantially directly behind the electrodes 70 and 72 regardless of the angle of view of the device. Thus, the light reflection contributes to the image brightness without subtracting from the clarity thereof. Further, the edges of the electrodes 70 and 72, owing to the close spacing thereof to the background layer 62, are likewise less noticeable to the viewer.

It is noted that, with the voltage switching means shown, a problem with the device 50 is that if the electrodes 70 and 72 are shorted to one another via the reflective layer 62 and pin holes through the insulator layer 64, the application of a voltage to either electrode will result in the display of both images, i.e., neither image could be individually displayed. In another embodiment, shown in Fig. 4, this problem is avoided. In this embodiment, the transparent electrode on the front substrate 52 is patterned into two segments 90 and 92 each corresponding in shape and being aligned with a different electrode 70 and 72, respectively. The electrodes 70 and 72 are electrically connected by means of a narrow conductive strip 94 on top of the insulating layer 64. Both electrodes 70 and 72 are connected to ground potential. Two D.C. voltage sources 96 and 98, each having a ground terminal, are each connected to a different electrode segment 90 and 92, respectively, through switches 100.

In operation, either a square or circular image, or both, is displayed upon the application of a voltage to the appropriate segment 90 and 92, or both. The presence of the strip 94 has no effect on the liquid crystal film because there is no electrode on the substrate 52 opposite thereto, i.e., a projection of the strip 94 onto the substrate 52 does not intercept any portion of the electrode thereon.

"""""""""""

3,613,351 (U.S.)   *Patented Oct. 19, 1971*

WRISTWATCH WITH LIQUID CRYSTAL DISPLAY.

<u>Richard S. Walton</u>, *assignor to Hamilton Watch Company.*
Application May 13, 1969.

Disclosed is a liquid crystal display and power supply system particularly suited for a wristwatch. A bar segment digital display is subjected to an electric field to energize selected portions of the display in accordance with the output of a timekeeping source within the watch. The display is separately powered either from solar cells or from a piezoelectric transducer actuated by watch movement.

Referring to the drawings, Fig. 1 shows at 10 a wristwatch having a watchband 12 and a time display face generally indicated at 14. Fig. 2 is a partial cross-section showing one of the bar segments such as the uppermost segment station 18. Fig. 3 shows a block diagram of the overall drive system for the watch 10 of Fig. 1. In that figure, a suitable power source such as an electronic oscillator is labeled S and this may take the form of any suitable pulse generator the output of which is fed to a series of dividers 40, 42 and 44. The dividers convert the timed output from source S (which by way of example only may be a highly accurate crystal controlled solid state oscillator) into binary decimal signals which in turn are converted by decoders D into the energizing code for the bar segment display.

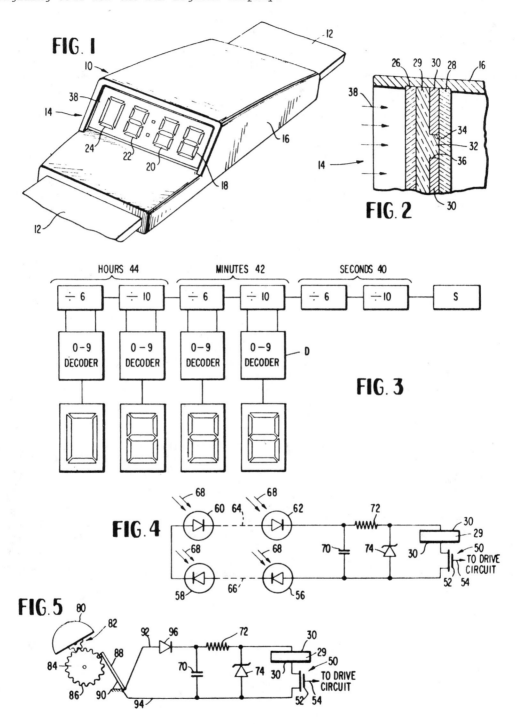

Of the three states of mesomorphic behavior, the nematic state exhibits the electromagneto-optic effect utilized in the present digital time display. It is understood that when the portions of the liquid crystal 29 disposed between the electrodes of the bar segments remain in an unenergized state, the dark background of the rear plate 28 is seen through the front transparent plate 26 and the transparent liquid crystal. Accordingly, in viewing the display's face

14, the dark background afforded by rear plate 28 is seen through the transparent electrode material 30 (essentially the entire display face 14) and through the portions of the liquid crystal lying between the electrodes of the unenergized bar segments. However, the electrodes which are energized through the logic circuitry of Fig. 3 energize the liquid crystals between them such that incident light is scattered. This incident light which is reflected appears white against the dark surrounding background.

Figure 4 shows the manner in which the liquid crystal forming one of the bar segments is energized. In Fig. 4, the liquid crystal 29 is illustrated as connected to the spaced electrodes 30 through a transistor switch 50, preferably in the form of an insulated gate field effect transistor. Gate electrode 52 of the switch is connected by lead 54 to the output of one of the decoders D of Fig. 3. When switch 50 is closed, the circuit is completed to electrodes 30 and liquid crystal 29 is energized to become reflective.

It is characteristic of liquid crystals that they require for energization an electric field of from 10 to 40 volts, but with a power consumption that is only 100 microwatts per square inch. This relatively high voltage/low current for energization is provided in the watch 10 by incorporating in the watch a plurality of photovoltaic cells or solar cells such as those indicated at 56, 58, 60 and 62. The photovoltaic cells are connected in a series to provide the relatively high voltage and low current necessary to energize liquid crystal 29. The connection is illustrated by dash lines at 64 and 66 to indicate that more photovoltaic cells can be used if desired. Since only about 30% of the watch face illustrated in Fig. 1 is used by the display, the solar cells 56, 58, 60 and 62 may be mounted on face 14 about the display to collect ambient light as indicated by the arrows 68. Alternatively, the solar cells may be mounted on the top of the watch case if so desired, where more room is available for the incorporation of a greater number of cells.

Connected to the output of the solar cells is an accumulator in the form of a capacitor 70. The accumulator is provided for energizing the liquid crystal 29 during those times when the ambient light 68 is too low to provide sufficient power to drive the crystal. The energy from the ambient light produced by the solar cells is generated during light periods and stored in the accumulator 70 for later use. The accumulator 70 which can take the form of a rechargeable battery rather than a capacitor is connected to electrodes 30 through a current limiting resistor 72. A Zener diode 74 can be provided if desired to control the voltage level, but diode 74 is optional.

Figure 5 shows a modified version wherein the liquid crystal 29 is energized from a piezoelectric transducer instead of the solar cells of Fig. 4. In the embodiment of Fig. 5 the watch is provided with an eccentric or oscillating weight 80 of the type conventionally used in automatic or self-winding watch movements, which weight drives a gear 84 through a gear train generally indicated at 82. The teeth 86 of gear 84 succsessively engage and oscillate or flex the end of a piezoelectric rod or bar 88 rigidly held at its other end at 90. Bar 88 is preferably formed of a suitable piezoelectric ceramic such as barium titanate or the like. The electrical signal is developed by piezoelectric bar 88 is conducted by leads 92, 94 through a rectifier diode 96 to the accumulator 70 in the form of a capacitor. The output from the accumulator passes to current limiting resistor 72 and the optional Zener diode 74 in Fig. 5 to the electrodes 30 energizing crystal 29. Again the accumulator can be a rechargeable battery rather than a capacitor.

" " " " " " " " " " " "

3,614,210 (U.S.)   *Patented Oct. 19, 1971*

LIQUID CRYSTAL DAY/NIGHT MIRROR.

<u>Sandor Caplan</u>, assignor to RCA Corporation.
Application Nov. 6, 1969.

With reference to the drawing, a liquid crystal mirror 10 is shown comprising a support member 12 on which is mounted, by means of a swivel knuckle 14, a housing 16. Rigidly mounted within the housing 16, as by being clamped between front and rear portions thereof, is a liquid crystal cell comprising a pair of front and rear rectangular glass substrates 20 and 22, respectively, the substrates having a thickness, for example, of 125 mils. The front substrate 20 is transparent, and comprises a soda lime glass having a light transparency in excess of 95 percent.

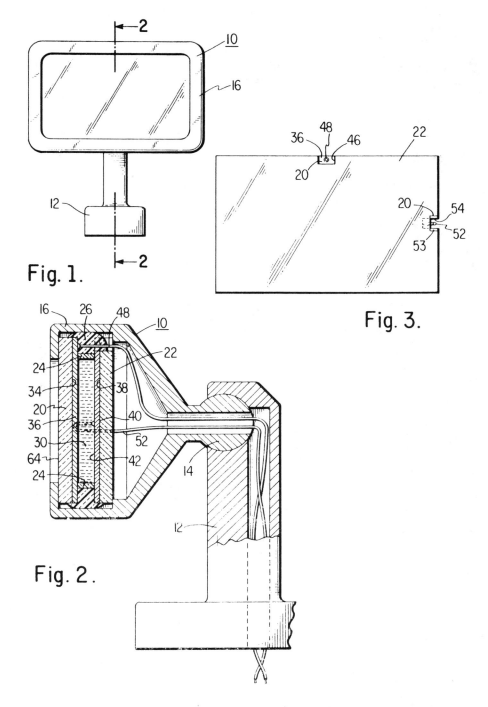

Fig. 1.

Fig. 3.

Fig. 2.

The two substrates 20 and 22 are maintained in spaced apart relation by means of a shim 24 of an insulating material, e.g., mica.

Disposed between the substrates 20 and 22, and maintained therebetween by the shim 24 and a seal 26 is a thin, e.g., 0.0005 inch thick nematic liquid crystal material that exhibits turbulent motion upon the passage of current therethrough.

The film 30 of liquid crystal material is normally substantially transparent to light. However, when an electric field of sufficient magnitude is applied across the film the current flow causes turbulence in the film and places the liquid crystal material in a light-scattering state, i.e., it forward scatters light incident thereon. The scattering of light completely diffuses the light incident on the film 30, thus preventing the transmission of light images therethrough. When the electric field is removed, the liquid crystal returns to its transparent state. The inside surface of 34 of the front substrate 20 is coated with in thin layer 36 of a transparent conductive material, e.g., tin oxide. The inside surface 38 of the rear substrate 22 is provided with a layer 40 of an electrical conductive material, e.g., nickel, having a specular surface 42.

For the purpose of providing access to the conductive layer 36, for making an electrical connection thereto, a notch 46 (Fig. 3) is provided through the substrate 22 at the periphery thereof exposing a portion of the substrate 20 and the layer 36 thereon. A lead member 48 is electrically connected to the exposed portion of the layer 36 by means of, for example, an electrically conductive silver epoxy cement. The lead 48 passes outwardly through the housing 16 and through the supporting member 12.

In use, the mirror 10 is mounted on a vehicle, e.g., an automobile, in such manner that both the front surface and the rear surface, 64 and 42 respectively, of the mirror are visible to a viewer in thevehicle. For example, the mirror 10 can be mounted in identical fashion as standard mirrors are mounted either inside or outside of automobiles, and the exact adjustment in the positioning of the mirror is made via the swivel knuckle 14.

The mirror 10 is operated as follows... When a reduced brightness or attenuated image is desired, as in nighttime use of the mirror, a voltage, either AC, DC, or pulsed DC, is applied between the conductive layers 36 and 40 via the terminal means 48 and 52 to place the liquid crystal material of the film 30 in its light-scattering mode. The liquid crystal film thus becomes blocking of transmission of light images therethrough, whereby reflection of images from the specular surface 42 is prevented. Images are thus reflected only from the outside surface 64 of the substrate 20 and from the interfaces between the substrate 20 and the layer 36, and the film 30. Owing to the poor reflectivities of the outside surface and the material interfaces, attenuated brightness images are reflected to the viewer. When maximum brightness reflected images are desired, as in daytime use of the mirror, the liquid crystal film is not energized.

The images reflected from the mirror during reduced reflectivity operation of the mirror appear on a luminous white background. To reduce the undesirable background luminescence, the diffuse reflectance of the overall mirror is preferably lower than 45% of the diffuse reflectance of standard white, and preferably is about 30% of standard white. One means to provide the desired low diffuse reflectance is to use, as in the instant embodiment, a specular surface 42 having a reflectivity significantly less than the reflectivity of standard mirrors. In the instant embodiment, using a nickel layer 40 having a specular surface 42 of 65 % reflectivity, the diffuse reflectance of the overall mirror 10 is 30 percent of standard white. Generally, using a specular surface 42 having a reflectivity of less than 75% will provide the desired low mirror diffuse reflectance of less than 45%.

" " " " " " " " " " " "

1,251,790 (British)   Patented Oct. 27, 1971

IMAGE CONVERTER.

Thompson-CSF (Inventor's name not given)
Application June 10, 1970; prior application in France June 11, 1969.

An image converter comprises a layer of photo-conductive material 1, Fig. 1, and a layer of nematic liquid crystal 2 between a pair of transparent electrodes 3 across which a D.C. voltage is applied. In use when radiation 6 to which the photo-conductor is sensitive is incident thereon, the voltage across the liquid crystal layer is increased and incident radiation 5 which is not absorbed by the photo-conductor is diffused 7, i.e., scattered. Thus, if a light image is projected on the photo-conductive layer an image is reproduced by the liquid crystal layer. With no incident radiation 6 the converter acts as a mirror since the face of the photo-conductive layer is smooth. With zinc sulphide as the photo-conductor an incident ultraviolet image is converted into a visible image, and since the photo-conductor is sensitive even to low illumination levels, the converter also acts as an image intensifier. If a photo-conductor which is sensitive to visible radiation is used then the photo-conductor layer is isolated from the radiation 5 by a thin layer of opaque carbon black disposed between the photo-conductive and liquid crystal layers (See Fig. 3). The image from a CRT may be projected on to the converter (Fig. 4) for large-screen

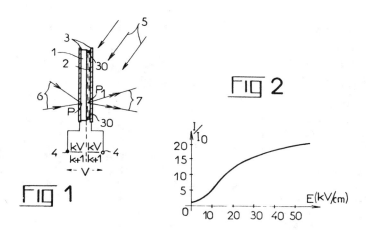

projection, and instead of reflecting the auxiliary light beam 5 from the converter, it may be transmitted therethrough, e.g., by using a semi-transparent mirror between the CRT and the converter (Fig. 5). Infra-red images instead of ultra-violet images may be converted by an appropriate choice of the photo-conductor material.

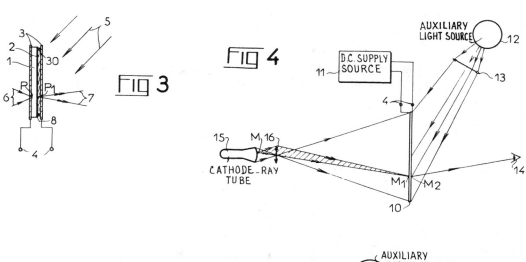

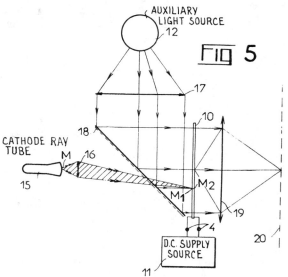

3,617,374 (U.S.)   Patented Nov. 2, 1971

DISPLAY DEVICE.

<u>Theodore L. Hodson, John G. Whitaker and Joe W. Jones</u>, assignors to The National Cash Register Co.
Application Apr. 14, 1969.

   The present disclosure is directed to articles of manufacture, e.g., display devices,
having self-contained controlled resistor means for generation of heat upon electrical activation,
which means can be formed in any desired configuration and thickness, and an encapsulated liquid
crystal layer responsive to variations in heat to present a display, which can be polychromatic.
The resistor is comprised of conductive ink deposited on one major surface (usually the lower
surface) of an opaque, substantially electrically nonconductive layer having a layer of encapsulated cholesteric liquid crystals in direct contact with at least a portion of the other major
(e.g., upper) surface of the opaque layer. Heating due to the resistor produces reversible
color changes in the encapsulated liquid crystals resulting in the display.

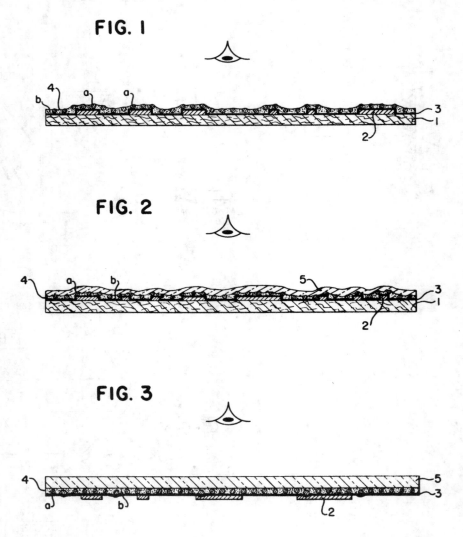

   The invention will be discussed in greater detail in conjunction with the drawings. All
six figures of the drawings are cross-sectional views illustrating the various component layers
contained in these articles of manufacture. The articles of Figs. 2, 3, 4, and 6 have a smooth,
essentially transparent brightness-enhancing top layer contiguous with the encapsulated liquid
crystal layer and of similar index of refraction with respect to the capsular wall material and
binder therein. The articles of Figs. 3 and 4 are formed by inverse coating, viz. coating of
the various layers on a transparent substrate which is then inverted for viewing of the display.

The articles of Figs. 2 and 6 are formed by top coating procedures, and the substrate need not be transparent. The articles of Figs. 1 and 5 have no such top layer and likewise can use non-transparent substrates.

As shown in Fig. 1, base or substrate 1, which need not be transparent, has deposited thereon various conductive ink elements 2 to define a pattern or configuration. An opaque substantially nonconductive layer 3 is located on and in direct contact with electroconductive ink elements 2 and overlying encapsulated liquid crystal layer 4 which, in turn, is comprised of encapsulated liquid crystals a and binder b. Fig. 2 is like Fig. 1 with the addition of a top layer 5 deposited by a topcoating procedure. Fig. 5 is like Fig. 1 except that in the article of Fig. 5 the configuration of the display is determined by the configuration of the various elements 4 of the encapsulated liquid crystal layer, the conductive ink layer 2 being coated over substantially the entire other major surface of opaque layer 3. In Fig. 1, the configuration of the display is determined largely by the configuration of the conductive ink elements 2 defining the conductive ink layer with the encapsulated liquid crystal layer 4 being deposited over a substantially larger portion of opaque layer 3. Fig. 6 is like Fig. 5 but it contains a smooth transparent top layer (deposited by top coating) over each portion 4 of the encapsulated liquid crystal layer pattern or configuration.

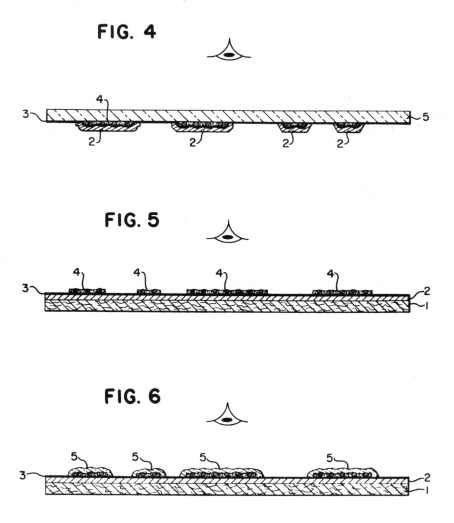

In the article of Fig. 3, the transparent top layer 5 also serves as a forming support for depositing the encapsulated liquid crystal layer 4 followed by the opaque nonconductive layer 3, and the configuration of the conductive ink layer is defined by conductive ink portions 2. Upon deposition of all the aforementioned layers, the article is inverted so that the display can be viewed by the observer through the transparent, brightness-improving layer 5. Such articles are referred to herein an "inverse coated" due to the aforementioned inversion prior to use. Fig. 4

is prepared by inverse coating as in Fig. 3 but the encapsulated liquid crystal layer 4 is placed thereon in selected areas thereof to define a pattern instead of covering substantially the entire surface of transparent top layer 5, as in Fig. 3.

The article of the present invention offers great flexibility in that the color response in the liquid crystal capsules can be varied by varying the resistance of the conductive ink elements.

Four examples are included to illustrate the invention in great detail.

"""""""""""

3,619,254 (U.S.)   Patented Nov. 9, 1971

THERMOMETRIC ARTICLES AND METHODS FOR PREPARING SAME.

<u>Frederick Davis</u>, *assignor to Liquid Crystal Industries, Inc.*
Application March 18, 1969.

The present invention provides thermosensing articles which can be prepared and then maintained for an indefinite time, as well as articles which will continue to function in a hostile environment. Methods for preparing such articles are described below...

Example I.

A 10-mil black polyvinyl chloride sheet is silk-screened with a cholesteric composition to provide a one mil thickness of composition on the sheet in patterns of squares about one-sixteenth inch on a side with nine such squares per unit of the pattern. The cholesteric composition is a mixture of 25 parts of cholesteryl chloride having a melting point of 94-95°C., 30 parts of cholesteryl oleyl carbonate showing a color play at 5-6°C., 39 parts of cholesteryl oleyl carbonate showing color play between 2-=22°C. (the lower temperature range cholesteryl oleyl carbonate is less pure and consequently has a lower color play range), and 6 parts of Solvent Black 7 NJD, made by Allied Chemical. This material changes from a green color to a colorless condition at a temperature of 37°C.

The silk-screened sheet is then fed into a coating machine having a vacuum backing to hold the sheet to a cylinder as it is passed through an aperture 15 mils in height. A white casein glue is flowed onto the silk-screened sheet to provide an overall coating thickness of about 5 mils on the sheet. As the coated sheet is withdrawn from the aperture, it is heated on the surface by a hot air blast and by infrared radiation rapidly to set the casein. The sheet is then dried for 5 minutes.

The dried sheet bearing the cholesteric composition and casein coating is sprayed with acrylic lacquer to provide a protective coating 2 mils thick. The sheet is then cut into strips so that each one contains a unit of 9 squares of the cholesteric composition.

These rectangular strips are then placed in a patient's mouth so that the cholesteric composition is subject to the oral temperature. The strip is withdrawn and if the squares on the strip show a green color, the patient's temperature is not above 37°C. If the squares are colorless, the patient's temperature is above 37°C.

Example II.

An 11-by-4 inch sheet of cellulose acetate butyrate is silkscreened with a cholesteric composition comprising 45 parts of cholesteryl oleyl carbonate, 45 parts of cholesteryl nonanoate, and 10 parts of cholesteryl benzoate (having a color play range of 26.5-30.5°C.) in the form of an array of 1/8-inch squares.

The silk-screened sheet is then fed into a coating machine as described in Example I so that it is coated with white casein glue blackened by the addition of 20 percent carbon black. The coating is dried rapidly with heat and infrared radiation. The dried coating is then overcoated with acrylic lacquer as in Example I.

The finished sheet has a color play temperature range of 26.5-30.5°C. so that when a hand is placed on either surface of the sheet a vivid change of colors takes place. At the lower temperatures, the cholesteric composition squares are bright red and as the temperature increases they pass through orange, yellow, green, and blue to a deep violet. At higher temperatures the colors entirely disappear.

Example III.

A 10-mil black polyvinyl chloride sheet is sprayed through a mask with the cholesteric composition set forth in Example I to form a pattern of various geometric figures. The sheet is then coated with casein glue, dried, and overcoated with acrylic lacquer as in Example I to provide a large panel which loses its green color when heated to 37°C. The color does not return spontaneously, but the green color can be restored by briefly rubbing the acrylic surface.

Example IV.

A 1-mil thick black nylon sheet is printed with a dot pattern of the cholesteric composition of Example II on a small printing press. The thickness of the cholesteric dots is one mil. The printed sheet is then coated with white casein glue as in Example I, dried, and overcoated with acrylic lacquer. The dots show a color play at 26.5-30.5°C.

Example V.

Example II is repeated using "Liquitex" acrylic polymer emulsion flatted with colloidal silica, and a similar thermosensing article is obtained.

Example VI.

Example II is repeated using a mixture of 25 percent "Liquitex" acrylic polymer emulsion and 75 percent casein glue, and a similar thermosensing article is obtained.

Example VII.

A strip of 5-mil black polyvinyl chloride is silk-screened with the cholesteric composition of Example II and dipped in white casein glue. The glue is rapidly dried by infrared radiation and held for an additional five minutes. It is then overcoated with acrylic lacquer spray as in Example I to provide a thermosensing article.

" " " " " " " " " " "

3,620,889 (U.S.)   Patented Nov. 16, 1971

LIQUID CRYSTAL SYSTEMS.

Donald H. Baltzer, assignor to Vari-Light Corporation.
Application June 11, 1968.

A principal object of the present invention is the provision of liquid crystal system incorporating cholesteryl esters, which systems are relatively permanent and highly resistant to physical damage, the liquid crystal materials being incorporated in a plastic resin either in the form of a film or in the form of a casting.

A liquid crystal compound normally has a somewhat milky appearance when in its solid or crystalline phase. As the compound is heated it enters its liquid crystal stage in which it begins to fuse and soften but is still not a true liquid. It is in this phase that the color display phenomenon takes place; but the compound is still essentially milky abd optically active. Upon further heating the liquid crystals enter the true liquid phase in which they become clear or transparent. There is a relatively sharp line of demarcation between the liquid crystal phase and the true liquid phase and the conversion from one to the other is reversible within a relatively narrow temperature range.

In contrast to the foregoing, the clouding and clearing phenomenon exhibited when the liquid crystals are dissolved in a heat-cured resin is wholly different. This can possibly be best explained by reference to Fig. 1 of the drawings, which is a graph illustrating the effect of temperature on a heat cured resin-liquid crystal system. As seen therein, the system

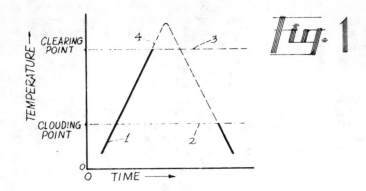
Fig. 1

is opaque or cloudy at low temperatures, as illustrated by the solid line 1. It is believed that at relatively low temperatures the liquid crystals will have been precipitated out of solution and exist in the plastic carrier as discrete or isolated entities entrapped in pockets as it were in the surrounding plastic. As the system is heated a clouding point 2 is reached at which the liquid crystals begin to soften although they do not truly dissolve until the clearing point 3 is reached, which may be at a substantially higher temperature. As the temperature is raised between the clouding and clearing points the system remains essentially cloudy and clearing does not noticeably take place until the temperature of the system closely approaches the clearing point 3. Upon passing the clearing point the liquid crystals dissolve and become a true liquid, as indicated at 4, and the system becomes clear, i.e., transparent. Upon subsequent cooling, however, the system will remain clear as its temperature passes downwardly through the clearing point 3 and willl remain essentially clear until the clouding point 2 is again reached and the liquid crystals precipitate out of solution.

An appreciable temperature span may exist between the clearing and clouding points in such systems, depending upon the particular liquid crystal compound being used and its solubility in the plastic carrier. It is this clearing and clouding phenomenon which is utilized to produce systems having numerous useful applications.

In accordance with the invention, the liquid crystal compound or mixture is dissolved in a plastic resin. In a typical example, the liquid crystals may be admixed with an acrylic resin, such as Rohm and Haas Acryloid B-66 acrylic resin, which is a 40 percent solid solution in toluene. While different liquid crystal compounds will function best at slightly different concentrations depending upon their compatibility with the resin, which in turn depends upon molecular structure and weight, as well as other characteristics of the compound, it has been found that 1 to 3 parts by weight of the liquid crystal compound to 14 parts by weight of the resin usually produces excellent results. The resultant liquid crystal system is clear and may be applied as such to a test plate or other surface. Upon application, the coating is allowed to dry at room temperature, thereby evaporating the solvent in the resin carrier. In order to insure complete solvent removal, the coated plate or the like may be heated in an oven for a short period of time at moderate temperature, i.e., 120°F.-150°F. to drive out residual solvents.

While acrylic resin has been found highly suited as a carrier for the liquid crystals, other plastic resins may be employed, including styrene, polyester, epoxy, polyvinyl butyral, plyvinyl chloride, polyvinyl acetate and polycarbonate.

Where two liquid crystal compounds are admixed, the resulting mixture is then dissolved with the resin in proportions given. For example, highly successful systems have been formulated by mixing one-third cholesteryl nonanoate and two-thirds cholesteryl oleyl carbonate with a 40% solids solution of acrylic resin in a ratio of 1 part by weight of the combined liquid crystal compounds to 14 parts by weight of the resin.

While liquid crystal containing plastic coatings may be applied to an object which has been previously painted black where the natural color of the object is such that it is difficult to perceive the color changes, the instant invention contemplates the addition of an insoluble black pigment to the coating mixture. It has been found that using Vabot's Sterling R carbon black pigment, a concentration between 0.05 percent to 1 percent of carbon black by weight to the weight of the liquid crystals provides excellent visibility of the color phenomenon. It is essential, that the pigment is insoluble when incorporated in the liquid crystals and/or in the liquid crystal, plastic resin system.

In systems in accordance with the invention the liquid crystal compounds may be dissolved in a heat-curing resin, such as an acrylic resin monomer, and the mixture poured into a mold, which may comprise an opposing pair of platens having an edge sealer, whereupon the material is cured at elevated temperatures to form a hard cast sheet of plastic having the liquid crystals dispersed therein. Sheets so formed display the characteristic of changing in a visually perceptible manner from a cloudy or opaque condition to a clear or transparent condition, or vice versa, depending upon variations in their temperature.

With reference to Fig. 3, which illustrates a situation wherein the curing temperature of the system is above the clearing point of the liquid crystals and the clouding point is below normal room temperature, the cured system will be clear as it is removed from the mold (as indicated by the solid line 1a) and will remain clear when cooled to room temperature. In fact, it will remain clear until additionally cooled to the clouding point of the system. As the clouding point is approached the cast sheet will begin to cloud and will become opaque as its temperature is lowered below the clouding point (as indicated by the dotted line 4a) and it will remain opaque until its temperature is elevated to at least the clearing point where it again begins to clear and becomes transparent (as indicated at 1b). It will be evident, however, that once the sheet has been rendered opaque, it will remain opaque at room temperature and hence the sheet in ordinary use may be made either clear or opaque depending upon its prior history.

Figure 4 illustrates a system wherein the clouding point of the system is above room temperature. In this instance the sheet will remain clear, as indicated by the solid line 1c, until cooled below the clouding point and hence will opaque, as indicated by the dotted line 4b, when it reaches room temperature. Upon subsequent reheating it will turn clear when heated above the clearing point, as indicated by the solid line 1d, and will remain clear until again cooled below the clouding point, as indicated by the dotted line 4c.

It will be understood that the temperatures at which the sheets will clear or cloud will vary, depending upon the particular liquid crystal compounds employed, and also the characteristics of the particular plastic resin in which they are dissolved.

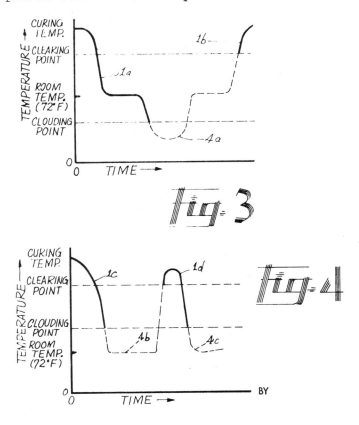

Such cured systems lend themselves to a variety of uses. Fig. 7 illustrates a cast sheet 11 having the properties of the system illustrated in Fig. 3, which may be used as a data storage device. The sheet, while in the opaque condition, is subjected to infrared heating in the limited areas 12 to cause the sheet to become transparent in those areas, thereby providing light transmitting apertures which may be utilized in conjunction with a scanning device incorporating a light detector which will be energized in accordance with the position of the apertures. When it is desired to erase the apertures, the sheet may be cooled, as by means of a thermoelectric device, to return the apertured areas to their initially cloudy or opaque condition, whereupon different apertured areas may be impressed in the sheet by subsequent reheating.

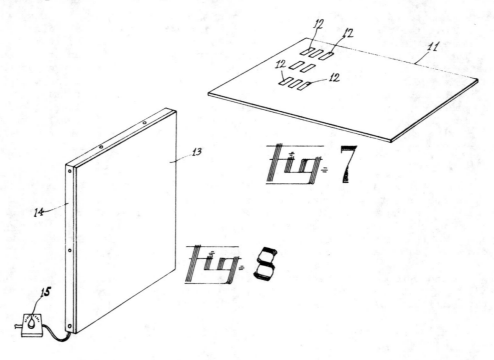

Figure 8 illustrates the use of a sheet having the properties of the system illustrated in Fig. 4 used as an adjustable windowpane which can be made either transparent or opaque by controlling its temperature. As seen therein, a plastic element 13 contains a liquid crystal compound which will normally become cloudy at room temperature and hence, under normal circumstances, the windowpane would be opaque for privacy. The element 13 is, however, combined with an electrically conducting sheet of glass 14 which acts as a heating element, the glass sheet being controlled by a rheostat 15 by means of which the plastic sheet may be heated to render it transparent, whereupon when the heating element is turned off the unit will cool to room temperature and the opacity will return to the plastic sheet 13.

" " " " " " " " " " " "

3,622,224 (U.S.)   Patented Nov. 23, 1971

LIQUID CRYSTAL ALPHA-NUMERIC ELECTRO-OPTIC IMAGING DEVICE.

<u>Joseph J. Wysocki and Robert W. Madrid,</u> assignors to Xerox Corporation.
Application Aug. 20, 1969.

An imaging system having a multiplane series of alphanumeric images in conjunction with an electro-optic imaging cell suitable for displaying one or more of the images.

In Figure 1 a typical imaging cell 10, sometimes referred to as an electroded imaging sandwich, is shown in cross section wherein a pair of transparent plates 11 having substantially transparent conductive coating 12 upon the contact surface, comprise a parallel pair of substantially transparent electrodes. An imaging member wherein both electrodes are transparent is preferred where the imaging member is to be viewed using transmitted light; however, such

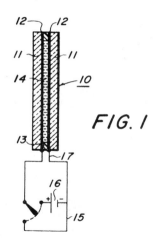

FIG. 1

imaging members may be viewed using reflected light thereby requiring only a single transparent electrode while the other may be opaque. The transparent electrodes are separated by spacing member 13 which contains voids which form one or more shallow cups or tanks which contain a suitable film or layer of electro-optic material which comprises the active element of the imaging member. In operation a field is created between the electrodes by an external circuit 15 which typically comprises a source of potential 16 which is connected across the two electrodes through leads 17. The potential source may be either DC, AC, or a combination thereof.

Cholesteric liquid crystals or a mixture of cholesteric liquid crystalline substances is used in an electrode sandwich such as that described in Fig. 1, and an electrical field across the liquid crystal film cause an electrical field-induced phase transition to occur wherein the optically negative cholesteric liquid crystalline substance transforms into an optically positive liquid crystalline state. This transition is believed to be the result of the cholesteric liquid crystal transforming into the nematic liquid crystalline mesophase structure. Cholesteric liquid crystals in the cholesteric state are typically translucent, for example, like a milky white, opalescent layer, when first placed in the unbiased electrode sandwich. When the electric field is placed across the liquid crystal film, the field-induced phase transition is observable because the liquid crystal film becomes transparent in areas where the field is present. Such transparent areas allow transmitted light to pass through them substantially unaffected. Mixtures of liquid crystals can be prepared in organic solvents such as chloroform, petroleum ether, methylethyl ketone and otjers, which are typically subsequently evaporated from the mixture thereby leaving the liquid crystalline mixture. Alternatively, the individual liquid crystals of the liquid crystalline mixture can be combined directly by heating the mixed components above the isotropic transition temperature.

In addition, nematic liquid crystalline materials are also preferred electro-optic imaging materials in various embodiments of the present invention. Nematic liquid crystals are particularly suited for the inventive system because many nematic materials are typically transparent in the absence of electrical fields, and become opaque or light diffusing, translucent when placed under the influence of electrical fields as in the present invention.

Smectic liquid crystalline materials are also suitable for use in the present invention.

In various embodiments of these electro-optic imaging cells, contrast enhancing devices such as polarizers or crossed polarizers may be advantageously used with the imaging cell.

In the imagign cell described in Fig. 1, the electrodes may be of any suitable transparent conductive material. The spacer gasket, 13, which separates the transparent electrodes and contains the imaging material between said electrodes, is typically chemically inert, transparent, preferably not birefringent, substantially insulating and has appropriate dielectric characteristics.

The basic electro-optic imaging cell described in Fig. 1 is further described in Fig. 2 wherein the desired image configuration is defined by the shape of the void areas in spacer gasket 13. In such embodiments, the desired image area may be thought of as the shape of these void areas, the shape of the spacer gasket, or more correctly the shape of the electro-optic material within the voids defined by the shape of the spacer gasket. In operation, there is an electric field across the entire area of the spacer 13, however, the image caused by the response of the electro-optic material occurs only in the area 19 where the electro-optic imaging material is present.

In Figure 3 another embodiment os the electro-optic imaging cell described in Fig. 1 is shown. In Fig. 3 the desired image is defined by the shape of at least one electrode, and therefore by the shape of a corresponding electrical field. In operation this embodiment will produce an electric field in areas where there are parallel electrodes, i.e., between the electrode in the desired image configuration, and the opposite electrode, whether or not the opposite electrode is also in the desired image configuration.

In Fig. 4 a preferred embodiment of the advantageous multiplanar electro-optic imaging system of the present invention is shown in a partially schematic, exploded isometric view. It will be understood that the alphanumeric units illustrated in Fig. 4 as being spaced apart, are typically stacked in contact one behind the other. The individual imaging cells of the multiplanar system illustrated in Fig. 4 are similar to the cell illustrated in Fig. 2.

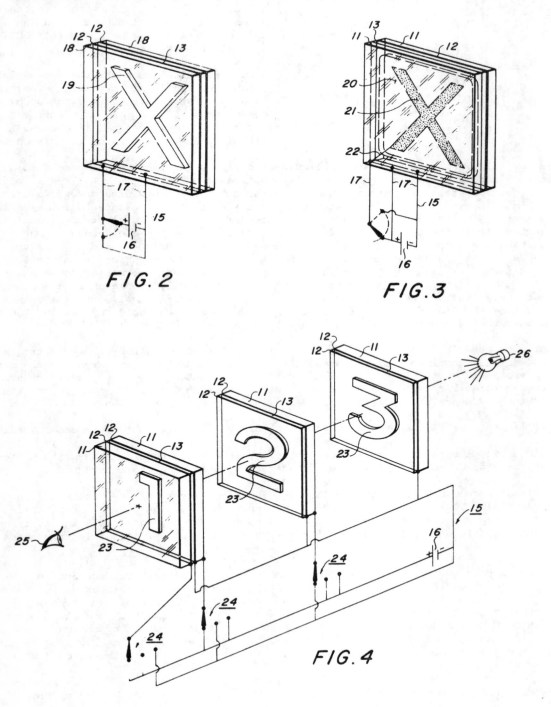

FIG. 2

FIG. 3

FIG. 4

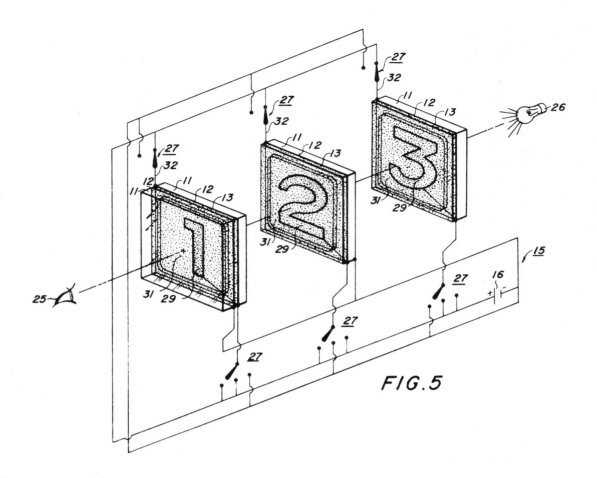

FIG. 5

When the device illustrated in Fig. 4 is used with an electro-optic imaging material in areas 23, and said material responds to the application of an electrical field across it by becoming transparent, i.e., as cholesteric liquid crystalline materials typically do, the device may be more advantageously used with image contrast enhancing devices such as polarizers or crossed polarizers. For the devices used with electro-optic imaging materials which respond to the electrical field across them by becoming translucent and light scattering, i.e., as nematic liquid crystalline substances typically do, it may often be just as advantageous to use the system without the image contrast enhancing devices.

Figure 5 illustrates another preferred embodiment of the multiplanar electro-optic imaging system of the present invention in a partially schematic, exploded isometric view. The individual imaging cells which comprise the system are similar to the cell described in Fig. 3, and the embodiment of Fig. 5 may be thought as of a series of electro-optic imaging cells like the one illustrated in Fig. 3 stacked one behind the other so that the images produced by each may be seen in a single imaging area.

When electro-optical material which is normally substantially transparent in the absence of an electrical field is used in the system described in Fig. 5, the desired image cell of the imaging system of Fig. 5 may be selectively imaged by selectively activating the cell through the use of external circuit means 15. The electro-optic material then responds by becoming light-scattering and translucent or even opaque in the desired image area having an electrical field across the electro-optic material. Nematic liquid crystalline materials are an example of electro-optic materials which are typically transparent in the absence of electrical fields and which may be imaged in the mode described here.

In Figure 6 yet another preferred embodiment of the multiplanar electro-optic imaging system of the present invention is shown in a partially schematic, exploded isometric view.

This embodiment comprises a single electro-optic imaging cell having a stacked, multiplanar series of electrodes in different imagewise configurations substantially on one side of the electro-optic imaging material. As illustrated in Fig. 6, the imaging cell comprises transparent support plate 11 having substantially transparent electrically conductive coating 12 thereon which acts as one electrode in the imaging cell. The electro-optic imaging material is contained within the void or shallow cuplike area enclosed by spacing gasket 13 sandwiched between transparent support plates 11. The desired image is defined by substantially transparent conductive coatings in imagewise configuration 29 and/or in background configuration 30 on any suitable transparent support member 11, and a multiplanar series of such shaped electrodes is stacked in front of or behind the electro-optic imaging material. Where the system having the multiplanar series of shaped electrodes also has full-area, substantially transparent electrodes 12 on both sides of the imaging material, then shaped image electrodes may be placed on both sides of the imaging cell and used in conjunction with the full-area electrode on the opposite side of the cell to produce the desired image.

Also, where both image and background areas on the same surface are substantially transparent electrodes and they may be operated as a full-area electrode, separate full-area electrodes are unnecessary in the embodiment where stacks of imagewise electrodes are placed on both sides of the imaging material. Such an arrangement helps minimize the distance between any image electrode and its corresponding full-area electrode on the opposite side of the imaging material, which tends to enhance the quality of the image. (Nine examples are described.)

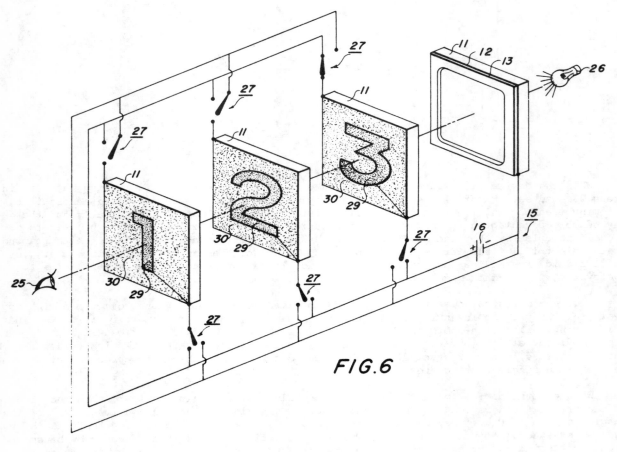

FIG.6

3,622,226 (U.S.)   Patented Nov. 23, 1971

LIQUID CRYSTAL CELLS IN A LINEAR ARRAY.

Dennis L. Matthies, assignor to RCA Corporation.
Application Nov. 19, 1969.

The novel array of liquid crystal cells is an improvement over prior-art arrays in that it provides for the contact exposure of light-sensitive recording elements, thereby utilizing all of the available light, increasing the speed of exposure, and eliminating the need for an expensive lens system.

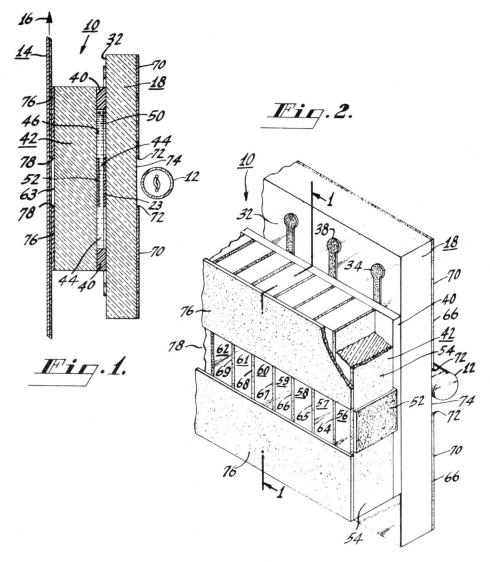

Fig.1.

Fig.2.

Referring to the drawins, there is shown a novel linear array 10, a light valve, of individually addressed liquid crystal cells disposed between a light source 12 and a light-sensitive recording element 14, such as an electrophotographic recording element 14, such as an electrophotographic recording element commonly used in the electrophotographic printing art. The recording element 14 is adapted to be moved vertically across the array 10 in the direction of an arrow 16 by any suitable means, not shown. In describing the array 10, the descriptive terms of direction, such as "vertical" and "horizontal", for example, are merely relative, and it is within the contemplation of the present invention to orient the array 10 in any desired position as necessitated by the applications and equipment in which it is incorporated as a part.

The array 10 comprises a backplate 18 of a light-transmitting electrically insulating material, such as glass or any suitable light-transmitting plastic material. A plurality of substantially rectangular electrodes, such as electrodes 20, 21, 22, 23, 24, and 25, for example, are disposed on the front surface 32 of the backplate 18 as shown in Fig. 3. The electrodes 20-25 are light-transparent and may comprise indium oxide or tin oxide coated on the backplate 18.

Adjacent electrodes on the front surface 32 are formed with oppositely disposed connecting leads, such as leads 34, 36, and 38 for the electrodes 20, 21, and 22, respectively. By having the leads of adjacent electrodes extend in opposite directions, electrical connections to the leads are facilitated and the possibility of shorting adjacent leads is greatly reduced.

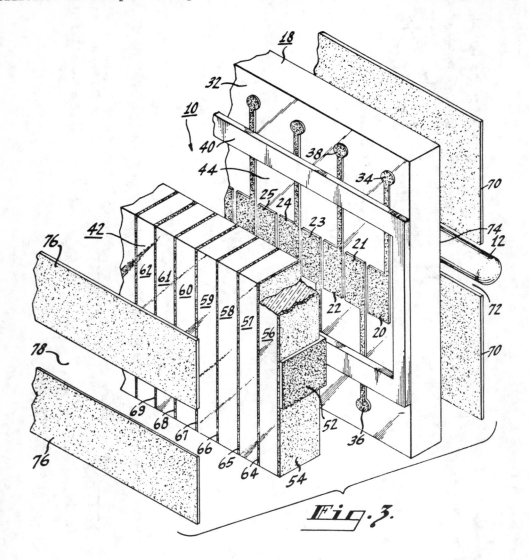

Fig. 3.

Each of the rectangular electrodes 20-25 is about 10 mils wide and 10 mils in hight, and spaced from its adjacent electrode by about 1 mil. Most of the dimensions of the array 10, except where specifically indicated, are not critical. Also, the drawings are not to scale, certain parts being exaggerated to illustrate their structures.

The backplate 18 is about 60 mils in thickness, about 2 inches in height, and may be as wide as desired, depending upon the application to which the array 10 is to be put. While only a few electrodes 20-25 are illustrated in the fragmentary view of Fig. 3, the array 10 may comprise as many electrodes as necessary to provide the number of cells desired.

A rectangular spacer 40, in the form of a square washer of an inert plastic material, such as "Teflon", for example, is disposed on the front surface 32 of the backplate 18 so that the horizontal portions of the spacer 40 lie across the leads of the electrodes 20-25 and its vertical portions are parallel to the vertical edges of the backplate 18, as shown in Figs. 2 and 3. The spacer 40 may have a thickness in the range between one-fourth and one-half mil. A front plate 42 whose height and width are of the same dimensions as the overall height and width of the spacer 40 is placed against the spacer 40 for forming, with plate 18, a reservoir or a chamber 44 for liquid crystals, as shown in Fig. 1 and 3.

The chamber 44, defined within the spacer 40 by the front surface 32 of the backplate 18 and the rear surface 46 of the front plate 42, is filled with a liquid crystal material 50. The liquid crystal material 50 may be a single, organic, thermotropic, nematic compound, or a mixture.

When the chamber 44 is filled with the liquid crystal material 50, the chamber may be sealed permanently by placing thin polyethylene gaskets (not shown) between the spacer 40 and the plates 18 and 42 and heating the sandwich consisting of the plates, the gaskets, and the spacer to a temperature of about 110°C. until the spacer 40 adheres to the plates 18 and 42, the fused gaskets acting as an adhesive. Any other suitable adhesive may also be used.

A common electrode 52 is disposed on the rear surface 46 of the front plate 42. The common electrode 52 is disposed directly opposite to the linearly aligned electrodes 20-25 on the front surface 32 of the backplate 18 and cooperates with each of the plurality of electrodes 20-25 to form a plurality of liquid cells. The common electrode 52 is brought out to a sidewall 54 of the front plate 42 to provide electrical contacting means for an electrical connection thereto. It is also within the contemplation of the invention for the array 10 to employ a plurality of separate electrodes instead of the electrode means in the form of the common electrode 52. Each liquid crystal cell of the array 10, however, is defined by a pair of oppositely disposed electrodes, that is, each of the electrodes 20-25 forms a separate liquid crystal cell with the common electrode 52. The width of each liquid crystal cell is the width of the smaller of the two electrodes in the pair that defines the cell.

The front plate 42 is a laminae of a plurality of light-transmitting bodies, such as similarly aligned light tunnels 56, 57, 58, 59, 60, 61, and 62 fixed to, or integral with, each other and disposed to transmit light from the rear surface 46 to the front surface 63 of the front plate 42. Each of the light tunnels 56-62 comprises a parallelepiped structure whose width is no greater than the width of the electrode defining a liquid crystal cell and whose front to back length, that is, the distance between the front surface 63 and the rear surface 46, is at least three times its width. The width of a light tunnel may be smaller than the width of an electrode defining a liquid crystal cell. The light tunnel 56, as shown in Figs. 2 and 3, is sealed to the spacer 40 and, therefore, is not used to direct light from a cell.

Opposite sidewalls of each of the light tunnels 56-62, that is, the adjacent sidewalls of the adjacent light tunnels 56-62, are coated or covered with, or separated by, layers 64-69 of suitable light-absorbing material, such as black paint, dark adhesive material, or vanes of black material. The light tunnels 56-62 may be a plurality of parallelepiped transparent glass structures laminated together with a black adhesive, such as black epoxy glue. The front plate 42 may also be a light-transmitting laminated structure of a plastic material, such as "Light Control Film" supplied by the 3-M Company (Minnesota Mining and Manufacturing C.). The "Light Control Film" consists of a plurality of parallelepiped structures separated by black light-absorbing plastic vanes in a structure of the dimensions described for the light tunnels 56-62 comprising the front plate 42.

The array 10 is provided with masking means to absorb light from the light source 12 that soes not pass directly through the liquid crystal cells. To this end, a rear mask 70 formed with a horizontal slit 72 is applied to the rear surface 74 of the backplate 18, and a front mask 76 formed with a horizontal slit 78 is applied to the front surface 63 of the front plate 42. The masks 70 and 76 comprise a light-absorbing material, such as black paint, for example, and the slits 72 and 78 are aligned with the electrodes 20-25. The height of each of the slits 72 and 78 of the masks 70 and 76, respectively, and the height of each of the electrodes 20-25 and the common electrode 52 are substantially equal and the slits and the electrodes are in alignment.

In operation, the liquid crystal cells are illuminated by light from the light source 12, the light being directed by the mask 70 to pass through the light tunnels 56-62. When a cell is in its transparent state, the light from the light source 12 goes directly through the cell and strikes the light-sensitive recording element 14 in substantial contact with the front mask 76. When the liquid crystal cell is in its light scattering state, light from the light source 12 is scattered so that it hits the black light-absorbing layers 64-69 of the tunnels in the front plate 42 and/or the light-absorbing mask 76. Hence, the light output is diminished to a point where the recording element 14 is not exposed.

Thus, any light from the light source 12 that strikes either the light-absorbing masks 70 or 76 or the light-absorbing layers 64-69 of the light tunnels 56-62 is absorbed, and only light passing directly through the light tunnels reaches the recording element 14 to expose it. By absorbing all but the direct rays of light from the light source 12, the novel array 10 provides means to expose the recording element 14 with rays of light (from each cell) of better contrast and resolution than possible with prior-art arrays.

3,623,795 (U.S.)   Patented Nov. 30, 1971

ELECTRO-OPTICAL SYSTEM.

<u>George W. Taylor and Arthur Miller</u>, assignors to RCA Corporation.
Application Apr. 24, 1970.

There is a need in, for example, the data processing field, for fast, efficient light deflectors, light valves and the like. The object of this invention is to provide new and improved systems for meeting this need.

A system embodying the present invention includes a material whose optical properties change sharply in a small temperature range. The material is heat biased to a temperature in or close to this range and a beam of light is directed at the material. In response to a signal applied to the material, its temperature is changed through a relatively small range such as from a value on one side to a value on the other side of a certain critical temperature and the change in optical properties of the material which results sharply changed characteristics such as deflection angle, polarization direction or other parameters of the light beam.

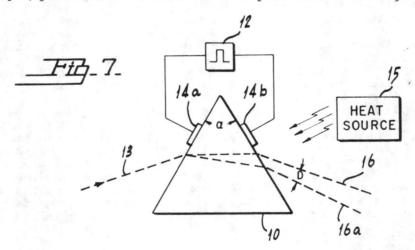

Figure 7 illustrates a light deflector. The system includes a prism shaped ferroelectric element 10 formed of a material which preferably exhibits a relatively large change in index of refraction at the Curie point. A heat source 15 applies heat to the material for maintaining it at a temperature below but close to the Curie temperature. The source 15 may be an oven, a source of infrared radiation, or other means for stably maintaining the crystal at the desired temperature. The system also includes a pulse source 12 connected to a pair of electrodes 14a and 14b secured to the surface of the prism. The prism 10 may be of cross section only slightly larger than that of the light beam.

In the operation of the system of Fig. 7, a beam of light, illustrated by dashed line 13, is directed at one face of the prism and is refracted in conventional manner by the prism. The emerging light beam is shown at 16, the amount of bending of the light being a function, of course, of the index of refraction of the ferroelectric material 10.

In order to deflect the light beam from the position indicated at 16 to a new position 16a, a short duration pulse is applied by source 12 to the electrodes 14a and 14b. The effect of the pulse is momentarily to heat a portion of the crystal, including at least the region through which the light beam is passing, to a value above the Curie point. The result is a sharp change in the index of refraction of the material 10 in the region thereof which is heated and this results in the deflection of the light beam through an angle D. For a given material, the amount of beam deflection is a function of the angle "alpha" of the apex angle of the prism. It is possible with a single prism and using a material such as SbSI to obtain a deflection angle of up to several degrees.

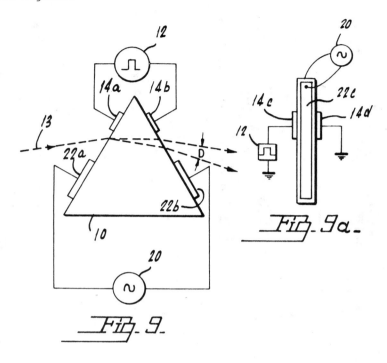

In the arrangement of Fig. 9, the same principle of operation is involved as is discussed above in connection with Fig. 7. The means employed for heating is the TANDEL effect. An AC voltage is applied by source 20 to the electrodes 22a and 22b. The amplitude of 22b is sufficient to switch the polarization between its two states during each cycle, that is, it is of sufficient amplitude to cause the entire hysteresis loop of the material to be traversed. In addition, the frequency is such that the switching domains in the material can follow the driving field. Tests have shown that good temperature stability is obtained by using a reasonably high frequency such as 100 kilohertz for the source 20 although other frequencies may be used.

The operation of the system of Fig. 9 is similar to that of the Fig. 7 system. By means of the TANDEL effect, the temperature of the prism 10 is stabilized at a value close to but slightly below the Curie point. To obtain beam deflection through an angle D, the pulse source 12 applies a short duration pulse to the crystal for heating a portion thereof through which the beam passes to value greater than the Curie point.

It should be mentioned that in the embodiments of the invention discussed above, the electrode configurations illustrated are merely examples. There are many other possibilities which are suitable. As one example, the electrodes 14a, and 14b of Figs. 7 and 9 may be transparent and the beam of light may be shined through these electrodes and through the ferroelectric material between them. As a second example the heating electrodes 22 may surround the signal electrodes 14 in a manner similar to that shown in Fig. 12. As a third example, either or both pairs of electrodes may be formed on the broad prism faces (those in the plane of the drawing in Fig. 9). This is illustrated schematically in Fig. 9a in which the signal electrodes 14c and 14d are on the parallel faces of the prism. In this embodiment the TANDEL electrodes are transparent and are on the angular sides of the prism, only one of these electrodes 22c being visible.

In all of the systems described thus far, the ferroelectric material has been a solid material. However, the invention is equally operative with a liquid which exhibits ferroelectric properties such as a liquid crystal. This material is considered to be ferroelectric because it exhibits the following characteristics:

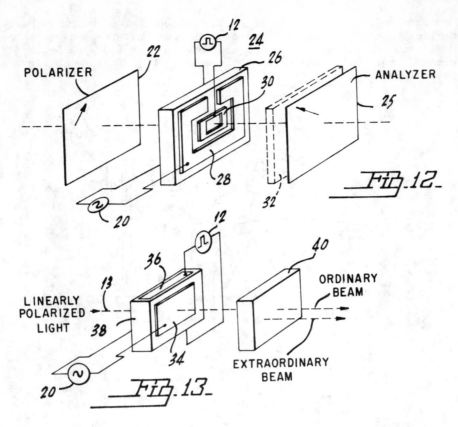

Fig. 12

Fig. 13

1. An order-disorder transition, that is, the crystal changes from an ordered phase (nematic) to a disordered phase (isotropic) at a particular temperature.
2. The transition is very sharp and generally can be considered to be a first order transition.
3. As in the case of solid ferroelectrics, there are large changes in the thermal, dielectric and optical properties at the transition.
4. Switching transients similar to those seen in solid ferroelectrics can be measured in the nematic phase. Also, low frequency hysteresis loops have been observed. Relaxation time of ions limit this frequency.

There are a number of important characteristics of liquid crystals which make them interesting for use in the systems described and illustrated here. For one, changes in birefringence of between 0.1 and 0.5 have been observed for liquid crystals at their transition temperature. This is large indeed and permits very large retardations to be imparted to the two components of light. Another interesting characteristic of liquid crystals is that the latent heat required for the phase transition from nematic to isotropic is relatively small. In addition, many of the liquid crystal materials have a transition temperature only a few degrees above room temperature. Thus, the thermal energy needed to produce the phase transition (which is supplied electrically) is quite modest.

Because the hysteresis loop for liquid crystals can only be traced at low frequencies and because the power dissipated in the liquid crystal during electric field switching in the nematic phase is often quite small, it is preferred to temperature stabilize the material below its nematic-isotropic transition temperature by dielectric heating rather than by the TANDEL effect. As an alternative, the liquid crystal material may be placed in intimate contact with a solid ferroelectric material whose Curie temperature closely matches the transition temperature of the liquid crystal material. Under such circumstances, the liquid crystal can be temperature stabilized just below its transition temperature by TANDEL stabilizing the temperature of the solid ferroelectric substrate.

An arrangement in which a liquid crystal is employed may be quite similar to that shown in Fig. 13. Here, the liquid crystal is in the position of element 38 and consists, for example, of a liquid crystal film. The electrodes 34 which may be formed of tin oxide, for example, are located on the inner surface of two glass plates and the electrodes 36 are not needed. These two glass plates (not shown in Fig. 13) enclose between them both the electrodes just described and the liquid crystal. Arrangements comparable to this may be employed for the other embodiments of the invention. This arrangement may be operated by employing a heating voltage of an amplitude and frequency to produce dielectric heating. The signal source may be in series or parallel with the heat source for modulating the heat source voltage.

"""""""""""

3,625,591 (U.S.)   Patented Dec. 7, 1971

LIQUID CRYSTAL DISPLAY ELEMENT.

<u>Marvin J. Freiser and Ivan Haller</u>, assignors to International Business Machines Corporation. Application Nov. 10, 1969.

An electro-optical display device employing a cell comprising a nematic liquid between two glass plates whose inner surfaces have been coated with transparent electrodes. Such coated electrodes are rubbed with a cloth or filter paper so that the liquid crystal film becomes oriented along the direction of rubbing. Such film will be homogeneous and strongly birefringent so that when an electric field is applied to the electrodes of the cell and the latter is viewed between crossed polarizers, a marked contrast in light between the quiescent and active states, respectively, of the cell is observed even in the presence of ambient illumination of the cell.

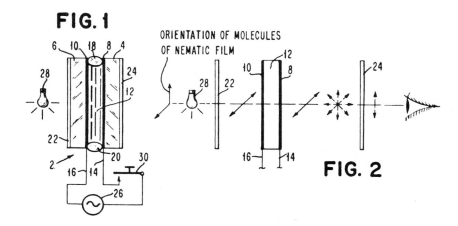

The present invention achieves high contrast in light intensity between the quiescent and active states of a nematic cell be relying upon two features, namely, (1) a preorientation of the nematic fluid on the walls of the cell and (2) the use of polarizers for observing the difference in light intensity between such two states. The employment of such two features has resulted in contrast ratios of 20 to 1, and in certain cases, ratios of 100 to 1 or higher can be attained.

The electro-optical device 2 of Fig. 1 comprises two transparent substrates 4 and 6, such as glass or other inert material having good light transmissive characteristics. The substrates 4 and 6 have plane opposed parallel faces on which are deposited electrically conductive light-transparent electrodes 8 and 10, of the order of 1000-3000 A. in thickness. Tin oxide is one of the many acceptable material well known in the art. Prior to inserting a suitable nematic liquid 12 between electrodes 8 and 10, the latter are rubbed with cloth or leather in the same direction so that the two electrode surfaces permit all the molecules of any liquid crystal insertedtherebetween to assume a uniform orientation.

The spacing between electroded glass plates 4 and 6 is between 5 to 300 microns , and the spacing may be maintained by shims, raised bevels on the opposed parallel plates 4 and 6, or by any other suitable means. Although capillary action will be sufficient to maintain the liquid crystal 12 in the thin space between the electrodes 8 and 10, after electrical leads 14 and 16 are connected to such electrodes 8 and 10, respectively, molten glass beads 18 and 20 can be dropped over the spacings to prevent loss of the fluid. Secured to the outer surfaces of each glass plate 6 and 4 are crossed polarizers 22 and 24. A low-frequency (up to several kilocycles)

A.C. source 26 is connected to lead 14 and 16.

With the aid of Fig. 2, operation in Fig. 1 will be described. Assume that a diffused light source 28 is observed through cell 2 at a location in front of the cell 2 as shown in the drawing. Such diffused light 28, when passing through polarizer 22, will become polarized in a direction parallel to the plane of the drawing. Upon passing through the nematic liquid 12 in its quiescent state (switch 30 is open and no voltage is across electrodes 8 and 10), the light remains polarized in the plane of the paper. However, since polarizer or analyzer 24 is crossed with polarizer 22, such polarized light does not pass through analyzer 24 and the observer sees a dark background despite the presence of ambient light in the neighborhood of the cell 2.

When switch 30 is closed so that the low-voltage A.C. source is applied across electrode 8 and 10, the liquid crystal 12 conducts current and such conduction produces a chaotic redistribution of the molecules of the liquid crystal film. Such redistribution of the molecules of the liquid crystal serves to depolarize the light (see dotted portions of Fig. 1), and analyzer 24 passes only that light which is polarized perpendicular to the plane of the drawing, which light is seen by the eye of the observer.

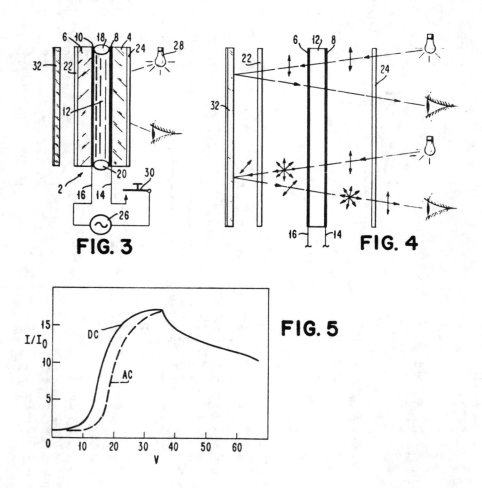

The embodiment of Fig. 3 is similar to the cell of Fig. 1 save that the polarized light that traverses the cell 2, impringes on a mirror 32, and then returns through the cell back to the eye of the observer. As seen in Fig. 4, the diffused light source 28 enters the cell 2 from the same side ad the observer. Such diffused light, on passing through analyzer 24, becomes polarized perpendicular to the plane of the drawing. With switch 30 opened, such polarized light traverses cell 2 unchanged, but is unable to pass through polarizer 22 that is crossed with polarizer 24. As a result, no light is reflected from mirror 32, so that the observer sees cell 2 as an opaque background. When switch 30 is closed so as to activate cell 2 (see bottom portion of Fig. 4) diffused light source 28 becomes polarized perpendicularly to the plane of the drawing as it passes through polarizer 24, but is depolarized after passing through the now activated cell 2. Such depolarized light, upon passing through polarizer 22, becomes polarized in the plane of the drawing and remains so polarized after reflection from mirror 32 and passage again through polarizer 22. The polarized light is again depolarized upon its passage through active cell 2 and becomes depolarized again, exiting from analyzer 24 as light that is polarized at right angles to the plane of the paper and observable by the viewer.

It is understood that many liquid crystals can be used in the present device. For example, APAPA and MBBA, the latter being particularly desirable in that it is a room temperature nematic.

Figure 5 is a plot of transmission of light through the cell 2 as a function of applied voltage, when the cell contained APAPA (nematic properties between 85°-90°C.) so that the cell was maintained at the nematic temperature while both AC and DC voltages were applied to the cell. The solid line shows a threshold of about 10 volts for DC operation and the dotted line shows a threshold of about 15-16 volts, with the intensity of light transmitted being given in arbitrary units. In general, the AC curve is similar to the DC curve at low frequencies, but at higher frequencies, the AC curve shifts to the right of the DC curve.

Tests on the cell 2 were made using DC voltages between 0 and 60 volts and AC voltages between 0 and 80 volts. Conductive coatings of tin oxide are of the order of 1000-2000 A. Prior to applying the nematic fluid to such glass supported conductive coatings, a piece of cheese cloth was spread onto a flat surface and each conductive coated glass was rubbed across the fixed cheese cloth about 25-30 times so that the molecules of the liquid crystal when spread across the conductive coatings prior to clamping or securing the coated surfaces in facing relationship, would all be aligned in a direction parallel to such coatings.

It has been observed that the combination of aligning the molecules of a liquid crystal in the nematic cell in combination with the use of crossed polarizers on the outer surfaces of the glass supports results in contrasts of at least 20 to 1 between light seen when the cell 2 is active and when the cell 2 is quiescent. Additionally, the cell 2, even in ambient light, is opaque when viewed by transmitted light or by reflected light. It is to be understood that multiple cells 2 can be employed in a matrix, utilizing grids to actuate selected cells in the matrix to provide an optical display device. Such matrix can also be used as substitutes for presently used light valves, optical switches, optical modulators, etc.

" "" "" "" "" "" ""

3,627,408 (U.S.) Patented Dec. 14, 1971

ELECTRIC FIELD DEVICE.

<u>James L. Fergason</u>, assignor to Westinghouse Electric Corporation.
Application June 29, 1965.

An electric field sensitive device for providing a visual indication in response to an electric field incorporating liquid crystalline materials of the type exhibiting the cholesteric phase.

Referring to Figure 1, there is illustrated a simplified voltage sensitive device including a layer 11 of a mixture of liquid crystalline materials of the cholesteric phase that exhibit a change in optical properties in response to an electric field impressed across the layer 10.

In Fig. 1, a suitable liquid crystal material of the cholesteric phase exhibiting the field effect such as 20 percent by weight of cholesteryl chloride with 80 percent by weight of cholesteryl oleyl carbonate. The percentage of Cholesteryl chloride may vary from 15 percent to 50 percent by weight of the mixture.

The layer 11 is sandwiched between two transparent electrical conductive layers 13 and 15. Electrical conductive leads 17 and 19 from the respective conductive layers 13 and 15 provide means os applying an electric potential illustrated as source 21.

The application of a voltage across the layer 11 by the source 21 results in a field across the layer 11. The electric field modifies the optical properties of the layer 11. These properties include optical rotation, selective scattering and selective transmission.

Referring to Figure 2, there is illustrated a voltage sensitive device including a liquid

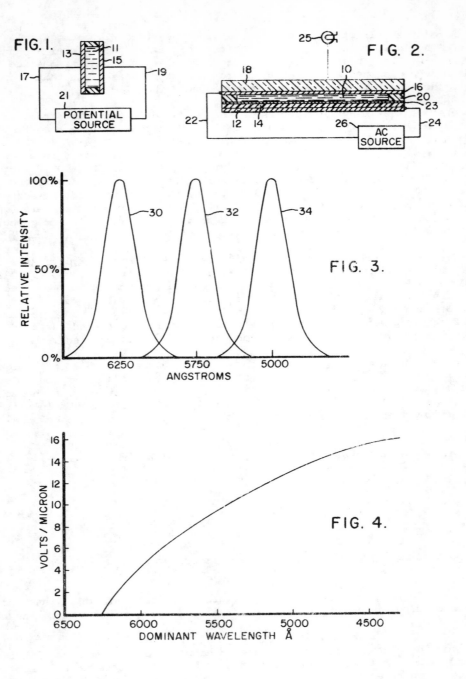

crystalline material of the cholesteric phase. This device utilized the selective scattering properties of the liquid crystalline layer. The device consists of a layer 10 of suitable liquid crystal materials, for example, a layer consisting of 45 percent by weight of cholesteryl bromide, 30 percent cholesteryl oleate and 25 percent cholesteryl nonyl-phenyl carbonate. This layer 10 when illuminated by white light, which provides uniform intensity over the spectrum, exhibits properties as best illustrated by the curves in Figs. 3 and 4.

The electric field sensitive device illustrated in Fig. 2 consists of a layer of a suitable electrically conductive layer of a material such as aluminum, a layer 14 of a suitable insulating material such as aluminum oxide, a suitable black coating 23 of a suitable dye, the layer 10 of a suitable liquid crystalline material, a layer 16 of a suitable electrical conductive material transmissive to visible light of a suitable material such as stannic oxide and a support layer 18 of a suitable material transmissive to visible light such as glass and supporting the layer 16. The function of the layer 14 is to provide a carrier for the light-absorbing layer 23 and to insure against electrical breakdown of the cell. In some applications, the layers 14 and 23 could be omitted.

Referring to Fig. 2, on application of potential across the layer 10 by means of a voltage source 26 the light from the source 25 will be selectively scattered back from the layer 10 in accordance with the voltages applied. For example, as illustrated in Fig. 3, with no voltage or a very small voltage from source 26 applied, light will be scattered from the liquid crystalline layer 10 to give a red color to an observer as illustrated by curve 30. With a field of about 6 volts per micron applied across the liquid crystalline layer 10 the light scattered from the liquid crystalline film 10 will be in the range as indicated by curve 32 so as to give off a green light. On application of a higher voltage of about 12 volts per micron light will be scattered from the liquid crystalline layer 10 so as to give a blue color and in the range indicated by curve 34. The bandwidth of scattering is about 200 angstroms. In Fig. 4, there is a plot of the dominant wavelength of scattering with respect to the electric field for the device shown in Fig. 2.

The molecular structure of the layer 10 is arranged in layers substantially with the molecules within each layer oriented in parallel alignment. The molecular layers are very thin with the long axis of the molecules parallel to the plane of the payers. Because of the peculiar shape of the cholesteryl molecules, the direction of the long axis of the molecules in each layer is displaced slightly from the corresponding direction in the adjacent layers. The overall displacement traces out a helical path. It is found that when selected liquid crystals are placed in an electric field parallel to the direction of alignment or optical axis, the wavelength of maximum scattering or color will be shifted from long wavelengths to short wavelengths that is from red toward blue.

It is found that such a device as illustrated in Fig. 2 is usable as a volt meter, having a very high impedance in excess of 500 megohms. It can also store charge for several seconds thus retaining its color once charged. By placing an AC field on the material, it was found that the liquid crystal could change its color at a rate in excess of 20 cycles per second indicating a response time as fast or faster than the human eye. Further it has been shown that only fields in the direction of the liquid crystal axis are effective, thus making very high definition possible. That is, if two areas of charge are very close together only that oart of the fields which are in the normal direction parallel to the optical axis will have an effect on the liquid crystal, thus imropving resolution. It is estimated that the resolution of a 20 micron thick film would be in excess of a thousand lines per inch. The properties of this device make this material ideally suited to a number of high resolution and fast display systems. The material is ideally suited for matching with an electron beam since both are of very high impedance. Additional embodiments relate to storage tubes; a radiation image converter; a projection system; an arrangement for displaying a multicolor image.

" " " " " " " " " "

3,627,699 (U.S.)　　　　　　　　　　　　　　　　　　　　　　　　　　　　　　　　　　　Patented Dec. 14, 1971

LIQUID CRYSTAL CHOLESTERIC MATERIAL AND SENSITIZING AGENT COMPOSITION AND METHOD FOR DETECTING ELECTROMAGNETIC RADIATION.

<u>Newton N. Goldberg</u> and <u>James L. Fergason</u>, *assignors to Westinghouse Electric Corporation. Application Apr. 30, 1969.*

Electromagnetic radiation in the frequency range of $10^{12.5}$ to 10^{17} cycles per second is detected, using a cholesteric liquid-crystal material to which there has been added, in the case of radiation of lower frequency, a suitable oil or oil-soluble dye, and in the case of radiation of higher frequency, a phototropic material such as beta-carotene or cholesteryl p-phenyl azophenyl carbonate, a novel compound

The invention is illustrated by the following specific examples...

<u>Example I</u>

To a piece of polyethylene terephthalate film about 0.00025 inch thick, there is applied on one side a coating of black spray enamel. Then, on the other side of the film there is applied a liquid made by mixing into 500 milliliters of petroleum ether the following materials:

5 grams of brilliant green dye, 10 grams of oleyl cholesteryl carbonate, 10 grams of cholesteryl benzoate, and 25 grams of cholesteryl nonanoate. This is permitted to dry, to leave a stratum about 20-30 microns thick, on one side of the film and at a temperature of about 50°C. this film appears green. Maintained at about such temperature, with electromagnetic radiation of infrared frequency of sufficient amplitude incident thereon, the film turned bluish-green.

Example II

Same as Example I but 5 grams of methyl orange dye replaces the 5 grams of brilliant green dye. At a temperature of about 20°C. this film appears orange. Maintained at about such temperature, with electromagnetic radiation of infrared frequency of sufficient amplitude incident thereon, the film turns yellow or brown.

Example III

Same as Example II. At a temperature of about 50°C. the film appears green. Maintained at about such temperature, with electromagnetic radiation of visible frequency of sufficient amplitude thereon, the film turns bluish-green or blue.

The patent specifications describe ten additional examples.

""""""""""""

3,628,268 (U.S.)   Patented Dec. 21, 1971

PURE FLUID DISPLAY.

Richard N. Johnson, assignor to The United States of America as represented by the Secretary of the Army.   *Application May 28, 1970.*

This invention relates to display devices, and more particularly to a device for providing a visual display of an array of fluid sources such as may be derived from the output of a plurality of pure fluid amplifiers.

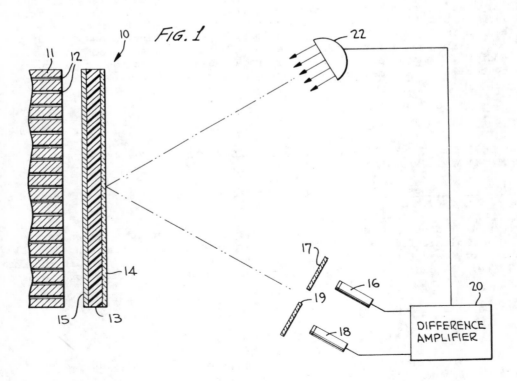

Referring to Figure 1, the visual display is indicated generally at 10. A thermally sensitive layer 13, whose optical characteristics are a function of temperature, is supported by a first coextensive supporting member 15 and a second coextensive supporting member 14. Thermally sensitive layer 13 comprises a liquid crystal layer whose color pattern is a function of the temperature pattern along its surface.

Supporting member 15 may comprise a thin support film of material such as aluminum oxide or that known as Mylar, and should be a very thin sheet consistent with providing suitable support layer 13. In addition, it should have a low thermal capacity and be provided with a thin coating of opaque material. Support member 14 provides support on the viewing side of the display. Generally it should be a transparent protective layer having a low thermal capacity. Thin Mylar, glass or other suitable material may be used for this purpose.

A temperature pattern is applied to thermally sensitive layer 13 by means of a plurality of fluid conduits or channels 12 which are embedded in supporting structure 11. The arrangement may take the form of a two-dimensional array as shown in Fig. 2. Each of channels 12 may conduct fluid flow from each of a plurality of pure fluid amplifiers (not shown).

FIG. 2

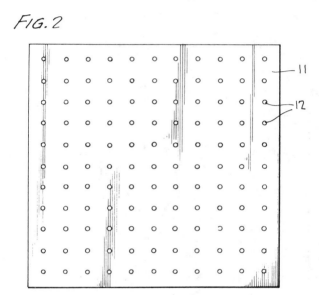

In operation, one or more fluid channels 12 direct a flow of cool fluid to impinge upon supporting member 15. This impingement upon surface 15 causes localized cooling which is transferred to thermally sensitive layer 13. Accordingly, a temperature pattern is applied to thermally sensitive layer 13 which responds by changing its color pattern to correspond to the applied temperature pattern. A visual indication of this temperature pattern is produced on the viewing side of display 10.

As is well known, liquid crystal layers only exhibit the unique property of color change in response to temperature while the layer is maintained within the predetermined temperature range. Accordingly, it is necessary to maintain layer 13 within the predetermined ambient temperature range at all times. This can be accomplished in a number of ways depending upon the type of sensitive layer which is employed and the degree of sensitivity desired. Referring again to Fig. 1, a radiant energy heat source 22, which may comprise a suitable heat lamp, is directed at thermally sensitive layer 13 to maintain its desired temperature. A control system is provided to shut off the radiant energy source 22 when the predetermined temperature is reached, and to turn on the source when the average temperature falls below the predetermined level. The temperature of layer 13 is sensed by a first photosensitive detector 16 which is provided with a filter 17 and a second photosensitive detector 18 provided with a filter 19. The filters are selected such that each photosensitive detector is responsive only to a given portion of the radiant energy emitted by layer 13, and a difference output is obtained by means of difference amplifier 20 which in turn is applied to radiant energy source 22 to turn it on and off when necessary.

"""""""""""

1,258,739 (British)   Patented Dec. 30, 1971

OPTICAL INFORMATION STORING SYSTEM.

Thomson-CSF (Inventor's name not given)
Application July 1, 1970; prior application in France July 2, 1969.

For description of this invention see U.S. Pat. No. 3,663,086.

3,637,291 (U.S.)   Patented Jan. 25, 1972

DISPLAY DEVICE WITH INHERENT MEMORY.

Clarence W. Kessler, Theodore T. Trzaska, assignors to The National Cash Register Company.
Application Feb. 11, 1970.

This invention relates to a visual display device which utilizes, for the display screen, a layer of encapsulated liquid crystals capable of exhibiting two stable chromatic states: i.e., a translucent state upon being momentarily subjected to an electrostatic field, and an opaque state upon being momentarily exposed to heat and the subsequent cooling thereof.

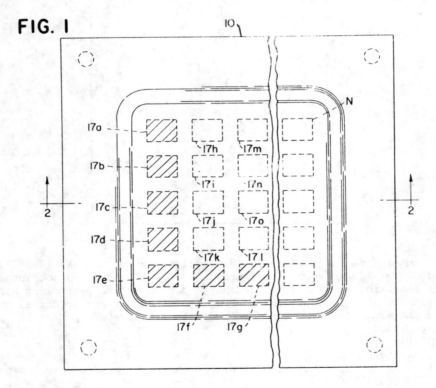

A plurality of thin-film thermal heating elements 17, 30, 31...N, formed of a high-resistance material such as tantalum, are positioned on a substrate 18, in any desired configuration, by conventional deposition techniques. The substrate 18 may be formed of a material having high electrical resistant qualities and good thermal insulating qualities, such as glass or alumina. The thermal elements in the described embodiment are positioned in an array resembling a matrix (i.e., rows and columns) although not necessarily restricted thereto. In the illustrated embodiment (Fig. 1), there are five thermal heating elements to a column, where three columns are required to form one character. The number of thermal heating elements in a row is determined by the number of characters desired, it being understood, of course, that the number of heating elements in each row and column can be increased or decreased, depending upon the requirements of the user. The thermal heating elements are each individually connected to an electrical source and are isolated from one another by a thin insulating layer 19. The layer 19 may be

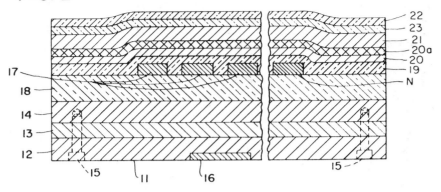

FIG. 2

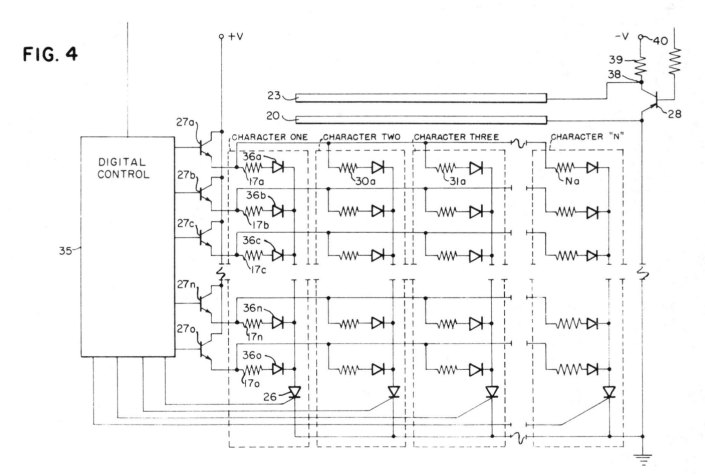

FIG. 4

formed of glass. A conductive layer of tantalum or other suitable material is uniformly deposited over the entire surface area of the glass layer 19 and functions as one field electrode 20. The surface 20a of the field electrode 20 not in contact with the glass layer 19 (second surface) is oxidized for the purpose of providing a dark background.

A thin layer of encapsulated liquid crystals 21 is conventionally applied to the oxidized surface of the tantalum layer 20 as a coating simply by spraying from an emulsion of the encapsulated liquid crystal material in a transparent polymeric binder. One form of encapsulated liquid crystals successfully utilized in the invention consisted of 70 percent cholesterol nonanoate, 25 percent cholesterol chloride, and 5 percent cholesterol cinnamate. It was found that liquid crystals having the above composition, when encapsulated, exhibited properties not found in unencapsulated liquid crystals of the same composition. It was further discovered that the above described liquid crystals, when encapsulated, shift to one stable chromatic state (translucent) when subjected to an electric field and remain in that state after removal of the field. It was also found that they shift from the translucent chromatic state to a second chromatic state (opaque) when heated above their clear point and then cooled below their clear point. The cooled encapsulated liquid crystal material remains in the opaque state until made translucent by being again subjected to an electric field. This stability (ability to remain either in an opaque state or in a translucent state) lends itself quite well for the display and storage of visual information.

The encapsulated liquid crystal layer 21 is covered with a flexible glass plate 22, which has a transparent conductor 23 applied thereon. This transparent conductor 23 functions as a field electrode and can be formed of any transparent conductive material, e.g., a metal deposit thin enough to be transparent and thick enough to be conductive. For staisfactory results, the conductor 23 should cover an area equal to that covered by the encapsulated liquid crystals 21. Representative materials that could be used for the transparent conductor 23 are gold, tin oxide, etc. all of which can be applied to the glass surface 22 by conventional deposition or coating techniques. The glass plate 22 is placed on the encapsulated liquid crystal layer in such a manner that the transparent conductor is in intimate contact with the encapsulated liquid crystal layer 21. However, the device can operate with the transparent conductor 23 separated from the liquid crystal layer 21 by the glass plate 22, if desired. The entire assembly is ealed by glass solder or glass cement, thereby forming a hermetic structure, which precludes the netry of any impurities into the liquid crystal layer 21.

The substrate 18 and its amalgamated structure are positioned on a heat stabilizer unit 11, which may be fabricated from a material capable of conducting heat, such as aluminum. The heat stabilizer unit 11, in this instance, comprises two layers of aluminum 12 and 14 and a resistive heating element 13. The layers are held together by conventional means such as fasteners 15. Bonded to the heat stabilizer unit 11 is a temperature-sensing element 16, which may be a conventional thermistor. The heat stabilizer unit serves to establish a reference temperature, so that the power required by each thermal heating element necessary to cause adjacent encapsulated liquid crystals to pass through their transition state into an isometric liquid remains substantially constant.

In operation, the areas of the encapsulated liquid crystal layer selected for the display are heated to their clear point by selected thermal elements 17, 30, 31...N. The heat necessary to raise the temperature of the encapsulated liquid crystals past their transition state into an opaque state is determined by the clear point of the liquid crystal material used, and the heat requirement for any given area of the display is the same as that required for any other area, since the entire encapsulated liquid crystal layer 21 is held at a constant temperature by the heat stabilizer unit 11.

While the temperature controller 24 may take any suitable form, it can be implemented quite satisfactorily with a typical temperature controlling circuit found in most SCR application manuals, more particularly such as found on page 118 of the G.E. SCR Manual, Second Edition. The constant temperature of the encapsulated liquid crystal layer facilitates the dynamic operation of the display in the event that such is desired, since the power requirements for each thermal heating element remain constant.

Figure 4 illustrates a typical electrical circuit arrangement utlized in operating the display device 10. It should be noted that it is not necessary for the electrical circuit arrangement to correspond to the physical arrangement shown in Fig. 1, since the thermal elements 17, 30, 31,...N can be connected and controlled in any manner desired. The illustrated circuit arrangement is designes to drive a display of "N" characters. It will be understood that the characters may represent alphabetic, numeric, or alphanumeric characters, dependent upon the selected thermal elements. In the illustrated embodiment, each character has a maximum of 15 associated thermal elements 17a-17o, 31a-31o, 31a-31o...Na-No. The thermal elements of each character are electrically paralleled with respect to each other. Corresponding thermal elements of each character are in parallel with each other; e.g., the thermal elements 17a, 30a, 31a...Na are all connected in parallel. For purposes of simplicity and brevity, only the operation or the character labeled "one" in Fig. 4 will be discussed. The character "one" is composed of 15 thermal elements 17a-17o arranged in three columns (Fig.1). The electrical connections can be better understood by referring to Fig. 4. The thermal element 17a is connected to the emitter of an element select driver 27a, which may be a transistor. The collector of the driver 27a is connected to a positive voltage source (+V), and the base is connected to a control device 35, which may be a conventional digital computer or other suitable output or enclding device. The other end of the thermal element 17a is connected to the anode of a diode 36a, for the prevention of any sneak paths. The cathode of the diode 36a is connected to the anode of a character select driver 26, represented by a silicon controlled rectifier. The gate of the character driver 26 is connected to the control device 35, and the cathode of said driver is grounded. The remaining thermal elements of character "one" and of the other characters of the display 10 are similarly connected.

The field electrode 23 is connected to a load resistor 39 and to the collector of a transistor 28 at the junction 38. The other end of the resistor 39 is connected to a negative voltage (-V) source 40. The field electrode 20 is connected to the grounded emitter of the transistor 28, and the base of the transistor 28 is connected to the control device 35.

Prior to writing the desired information on the display device, the entire encapsulated liquid crystal layer 21 is rendered translucent by the application of a potential between the field electrodes 20 and 23, which are momentarily energized upon the application of a pulse to the base of the transistor 28. In order to write information on the display medium, it is necessary to erase areas of translucency corresponding to the desired character from the display medium. For example, in displaying the letter "L," it is necessary to apply heat to the encapsulated liquid crystals immediately adjacent to the heating elements 17a-17g of the character 1. This is accomplished by pulsing the character select driver 26 into conduction and simultaneously pulsing the element select driver transistors 27a-27g into conduction, thereby establishing a current in each of the thermal heating elements 17a-17g of a magnitude sufficient to raise the temperature of the adjacent encapsulated liquid crystals above their point. The encapsulated liquid crystals, upon cooling below their clear point, will transform into a light-reflecting opaque area resembling the letter "L." The specific composition of the encapsulated liquid crystals determines the wavelength of the light that will be reflected by the opaque surface and therefore manifested as a color in the eyes of a viewer. In the instant embodiment, the characters displayed will appear green when subjected to normal incident light. The portion of the encapsulated liquid crystal layer not rendered opaque by the selected heating elements retains its translucency and permits the greater portion of incident light to pass therethrough to be absorbed by the dark surface 20a provided by the oxidized tantalum surface of the electrode 20. Therefore, the green characters would be observed on a dark, alsmost black background, providing a readily discernible display. The dark background is not shown in Fig. 1, so as not to obscure the drawings.

It can be seen (Fig.4) that the generation of the characters for the entire display can be written into the display by sequentially energizing one character select driver at a time along with the associated thermal heating elements forming the particular character desired.

"""""""""""

1,263,277 (British)   Patented Feb. 9, 1972

DISPLAY APPARATUS.

RCA Corporation (Inventor's name not given).
Application Apr. 29, 1969; prior U.S. application Apr. 23, 1968.

A watch has a replaceable face plate assembly 52, Fig. 5, which comprises a battery 42 located in an insulator 40 at the back of a liquid crystal digital display unit, so that the battery and display unit must be replaced together. The display consists of four alphanumeric characters each of which comprises seven transparent, conductive segments 17-23 attached to the inner surface of a transparent insulating face 16. A liquid crystal 26 is sandwiched between the face 16 and an opaque conducting back plate 24. Alternatively, the back may comprise conducting segments which are attached to the insulator 40 in alignment with those on the face 16.

The application of a voltage to appropriate segments of the display characters excites the adjacent portions of the crystal 26, which appear to light up. The watch is governed by a crystal-controlled oscillator 10, Fig. 1, a counter 12, and a decoder 14, all of which are integrated on a single circuit chip and contained in a back structure 50. Connections between the face plate assembly 52 and the integrated circuit are made through pins such as 34a, b when the assembly 52 is attached to the watch back structure 50.

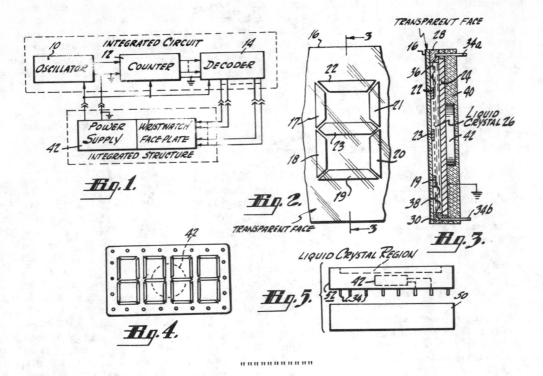

"""""""""""

1,263,278 (British) Patented Feb. 9, 1972

LIQUID CRYSTAL DISPLAY APPARATUS.

RCA Corporation (Inventor's name not given).
Application Apr. 23, 1969; prior U.S. application Apr. 23, 1968.

The Specification describes a watch similar to that of Specification No. 1,263,277 and the illustration is the same as shown above.

"""""""""""

3,642,348 (U.S.) Patented Feb. 15, 1972

IMAGING SYSTEM.

Joseph J. Wysocki, James E. Adams and Robert W. Madrid, assignors to Xerox Corporation.
Application Oct. 20, 1969.

A system which transforms a cholesteric liquid crystalline material from its Grandjean or "disturbed" texture state to its focal-conic or "undisturbed" texture state by an applied electrical field, and an imaging system wherein a cholesteric liquid crystalline member is imaged in a desired image configuration by the electric field-induced texture transition system.

In the advantageous system of the present invention it has been discovered that when cholesteric liquid crystals or a mixture of cholesteric liquid crystalline substances is used in an electrode sandwich such as shown in Fig. 1, that electrical fields across the liquid crystalline film cause an electrical field-induced texture transition to occur wherein a cholesteric liquid crystalline material initially in its Grandjean or "disturbed" texture is transformed to its focal-conic or "undisturbed" texture. The Grandjean texture is typically characterized by selective dispersion of incident light around a wavelength λ_o (where $\lambda_o = 2np$ where n=the index of refraction of the liquid crystalline film and p=the pitch of the liquid crystalline film) and optical activity for wavelengths of incident light away from λ_o. If λ_o is in the visible spectrum, the liquid crystalline film appears to have the color corresponding to λ_o, and if λ_o is outside the visible spectrum the film appears colorless and nonscattering. The Grandjean texture of cholesteric liquid crystals is sometimes referred to as the "disturbed" texture.

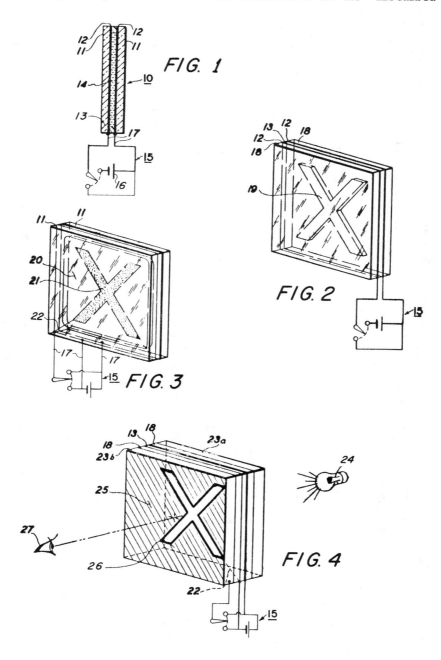

The focal-conic texture is also typically characterized by selective dispersion but in addition this texture also exhibits diffuse scattering in the visible spectrum, whether λ_o is in the visible spectrum or not. The appearance of the focal-conic texture state is typically milky-white when λ_o is outside the visible spectrum. The focal-conic texture of cholesteric liquid crystals is sometimes referred to as the "undisturbed" texture.

For example, in the inventive system when cholesteric liquid crystals are placed in the unbiased electrode sandwich, they initially appear colored, or colorless and transparent. If the electrode sandwich is observed between polarizers as illustrated in Fig. 4, the imaging sandwich appears colored or black. When the electrical field is placed across the liquid crystalline film, the field-induced texture change is observable because the liquid crystalline film becomes white in the imaged area when the imaging sandwich is observed in transmitted or reflected light. The described imaging system thereby produces a white image on a dark or colored background. However, it is clear that either field or nonfield areas in the liquid crystalline imaging sandwich may be used to create the desired image, with or without the use of polarizers or other image enhancing devices.

In Figure 2 the embodiment of the liquid crystal imaging member described in Fig. 1 is shown with the desired image defined by the shape of the void areas in the spacer gasket 13. As before, transparent electrodes 18 are separated by the spacer 13, but the entire desired image area 19 comprises the liquid crystal film or layer. In this embodiment the entire inner faces of the transparent electrodes comprise substantially transparent conductive coating 12 and the conductive coating 12 and the conductive coatings are electrically connected to external circuit 15. In operation there is an electrical field across the entire area of the spacer 13, however the image caused by the electrical field-induced texture change in the liquid crystal film, causes imaging to occur only in the area 19 where the liquid crystal film is present.

In Fig. 3 the desired image is defined by the shape of an electrode, and therefore by the shape of the corresponding electrical field. The imaging member here comprises transparent plates 11 separated by spacer gasket 13 having void area 20 filled with liquid crystals, said area comprising substantially the entire area of spacer layer 13. The desired image is defined by the shape of the substantially transparent conductive coating shown at 21, which is affixed to the inner surface of one or both of the transparent support plates 11, and is typically affixed only in the desired image configuration.

Example I

A liquid crystal imaging cell for observing the electrical field-induced texture change is prepared as follows: a mixture of liquid crystalline materials of about 59 percent cholesteryl chloride and about 41 percent cholesteryl nonanoate is prepared. A layer of the liquid crystalline mixture is placed between and in contact with a pair of substantially transparent electrodes of NESA coated glass available from PPG Co. The transparent electrodes are separated by about 1-1/2 mils by a Mylar spacer, made of Mylar polyester resin available from DuPont. The electrodes are connected in circuit with an electrical generator and a source of white light is placed behind the imaging sandwich.

The liquid crystalline layer initially appears substantially clear. The electrical field is then applied across the imaging layer. In the presence of the field, the imaged portion of the liquid crystal imaging layer assumes a white, light-scattering appearance. The threshold voltage for the about 1-1/2 mil sample is about 15 volts at about 25°C., or a field strength of about 4×10^3 volts per centimeter. At voltages slightly above this threshold, the Grandjean to focal-conic texture change is clearly observable in this system.

The cholesteric-nematic phase transition occurs for this mixture of liquid crystals at about 100 volts at about 25°C., or at field strengths of about 3×10^4 volts per centimeter.

Seven more examples are described in the patent specification.

" " " " " " " " " " "

3,645,604 (U.S.)   Patented Feb. 29, 1972

LIQUID CRYSTAL DISPLAY.

D.-T. Ngo, assignor to Bell Telephone Laboratories, Inc.
Application Aug. 10, 1970.

This invention relates to displays, and more particularly, to displays upon which images are generated by the selective energization of individual display cells or elements.

In Figure 1 of the drawing an illustrative liquid crystal display embodiment of the invention is shown comprising a stack of conventional liquid crystal display devices 101, 102, and 103 for generating mural images by the selective energization of individual ones of the cross-point display cells of each device. Illustratively, in Fig. 1, the three display devices 101, 102 and 103 each comprise a coordinate array of cross-point display cells, the display cells of

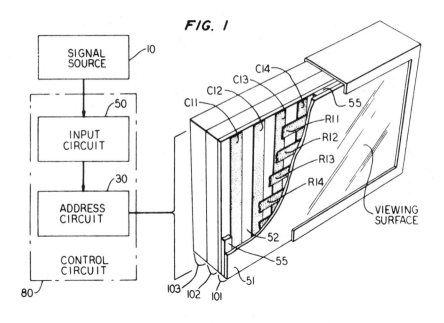

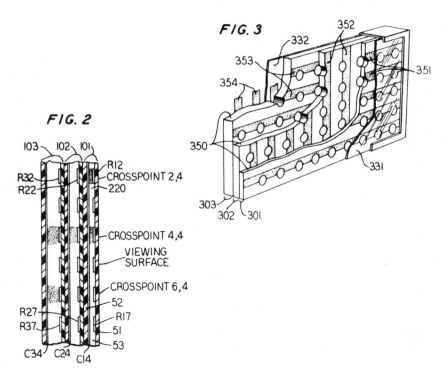

each device being substantially in registration with the corresponding display cells of the other display devices in the stack. However, it will be apparent from the description herein that the display cells may be employed in any form of array desired for particular display applications: for example, the display cells may be arranged in spiral rows or in concentric circles for radar display applications.

The cross-point display cells of each display device are defined by respective sets of row and column conductors, such as row conductors R11-R14 and column conductors C11-C14 of device 101 shown in Fig. 1, which are illustratively disposed on substrates, such as substrates 51 and 52. Substrates 51 and 52 are in turn spaced apart, such as by spacers 55, and a substantially uniform layer of liquid crystal display material 53 is disposed therebetween.

Display devices 101, 102 and 103 each utilize the dybamic scattering of light in the liquid crystal material regions at selected cross-point display cells for generating images. The light source (not shown in the drawing) may be located an back of or in front of the display, depending upon the mode of or in front of the display, depending upon the mode of operation chosen, whether transmissive, absorptive or reflective. The reflective mode of operation is generally preferred for display applications, often with ambient light serving as the light source, a reflective backplate typically being located at the back of the display facing the viewing surface. Thus, in Fig. 2 the reflective backplate would be located to the left of device 103 with the reflective surface thereof facing to the right.

A selected display cell of a particular display device in the stack is addressed via address circuit 30 under control of control circuit 80, by the application of coincident signals to the particular row and column conductors defining the selected display cell. The voltage thus extended across the selected display cell by the coincident row and column signals is above a threshold level sufficient to cause light scattering turbulence in the liquid crystalline material at the selected cell. At the same time, however, the voltage extended across the other display cells connected to the addressed row and column conductors is insufficient to cause significant light scattering at these other cells. If the magnitude of the signal applied across the display cell is increased, the turbulence and thus the light scattering caused thereby increases. In known liquid crystal display devices, this direct relationship between the applied signal magnitude and the degree of light scattering (and hence cell opqueness) is used to provide gray scale, typically via modulation of the signals applied to the individual display cell. However, in accordance with the present invention, gray scale is provided advantageously by applying signals of a fixed predetermined magnitude to the individual display cells, eliminating the heretofore requirement for modulation and digital-to-analog conversion circuitry. The liquid crystal material is of different thickness in the respective display devices 101, 102 and 103, such thatwith like magnitude signals applied thereto the cells of each device provide different degrees of light scattering. For example, assuming the reflective mode of operation, the different degrees of light scattering may be such that device 101 is provided with a light reflectance coefficient of 'k', device 102 with a light reflectance coefficient of 'k^2', and device 103 with a light reflectance coefficient of 'k^4'. Consequently, gray scale levels substantially corresponding to the standard television gray scales may be provided by selecting the respective display device thicknesses such that 'k' corresponds to the ratio between adjacent gray scales, illustratively on the order of 0.69. Addressing of the individual cross=point display cells in Fig. 1 may be effected via conventional addressing or scanning techniques, such as those known to the display and television art.

An alternative display embodiment is shown in Fig. 3, wherein each display device 301, 302 and 303 comprises a layer of dielectric material 350 having holes disposed therein filled with liquid crystal material to define an array of multilevel display cells. The display devices are disposed in housing 330 having a transparent viewing surface 331 and a backplate 332. Corresponding cells of the several devices in the stack are substantially in registration with one another, the holes of the devices thus being aligned and advantageously filled with the same liquid crystal material. As depicted illustratively in the embodiment of Fig. 3, the row and column conductors, are each shared by two adjacent display devices in the stack. Thus, column conductors 352 are shared by display devices 301 and 302, and row conductors 353 are shared by devices 302 and 303. An arrangement of 'n' display devices therefore would require only n+1 sets of conductors interleaved with the display device levels, alternating between row conductors and column conductors such as shown in Fig. 3. THis arrangement advantageously facilitates manufacture by significantly reducing the number of sets of conductors and, consequently, the number of connections which must be made to the display devices. Moreover, the embodiment of Fig. 3 permits closer spacing between display device levels in the stack than the embodiment of Fig. 1.

"" "" "" "" "" "" ""

3,647,279 (U.S.)   Patented Mar. 7, 1972

COLOR DISPLAY DEVICES.

Edward N. Sharpless and Frederick Davis, assignors to Liquid Crystal Industries, Inc. Application May 27, 1970.

A display device for exhibiting a color pattern, said device comprising container means having a light-transmitting section and a juxtaposed darker hued or opaque section, a quantity of of liquid crystalline material interposed between said container sections and encapsulated within said container means, said material having a characteristic of selective light scattering to exhibit color patterns within a range of temperatures at which said display device is normally utilized, and means for peripherally sealing one said container section to the other. Means can also be provided for applying deformational stress to the liquid crystal to vary its color pattern.

The novel display device or liquid crystal cell is provided firstly with a light transmitting wall to permit viewing of the contained liquid crystalline material. Secondly, the liquid crystalline cell is desirably associated with means for inducing deformational stresses within the contained liquid crystalline material.

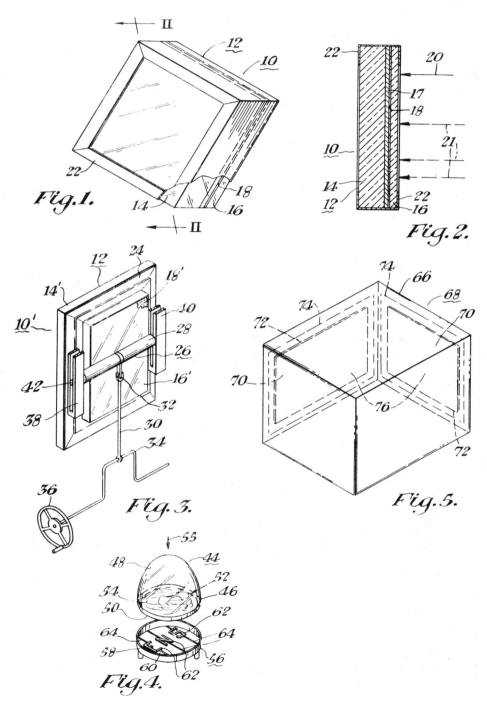

With reference to Fig. 1 of the drawings, a display device 10 in the form of a liquid crystalline support 12 is illustrated therein. In this arrangement, the support 12 in the form of a flat container having opposed container wall sections or structures 14, 16 of any suitable size and shape. In the arrangement shown, the wall sections 14, 16 are substantially coextensive although this is not an essential requirement. In point of fact, one of the wall structures 14, 16 can be significantly smaller than the other wall structure (Fig. 3), as long as one wall structure is joined about its periphery to the other wall structure, for example, in the manner described below. Likewise the wall structures 14, 16 need not be of flat configuration as illustrated but can be of some other configuration for example parallelopipedon or hemispherical as illustrated in Figs. 4 and 5. It is contemplated however, that any geometrical or nongeometrical, symmetrical or nonsymmetrical shape or form can be employed for either or both of the wall structures 14, 16.

In the arrangement as illustrated in Figs. 1 and 2, at least one of the wall structures 14, 16 is sufficiently thin or is made of a suitably plastic material as to lend a resilient or flexible character to the wall structure. Thus, the wall structure, such as the wall structure 16 can be bent or otherwise deformed toward the wall structure 14 when a force is applied more or less transversely thereto as denoted by arrow 20. Such force can be applied at various locations on the wall 16 as denoted by dashed arrows 21.

By thus bending one of the wall structures 14, 16 relative to the other, the liquid crystalline material 18, which is supported, in this example, between the wall structures 14, 16 in film-like form, is caused to flow generally away from the region of applied force (arrow 20, 21) to other regions of the volume confined within the liquid crystal container 12. The application of the force 20 and the resultant flow of the liquid crystal 18 develops shear and other deformational stresses within the liquid crystal 18. Such stresses modify the light scattering and attendant transmittance characteristics of the liquid crystal material 18 and result in an endless variety of color changes and patterns.

In order to observe these aesthetic color changes one of the container sections, for example, the section 14, is light transmitting, and desirably transparent, to permit the display device 10 to be observed from a side away from the application of deformational forces 20 or 21. The clear container section 14 can be fabricated from polacrylic, polycarbonate, polybutyrate, glass or other suitable material.

At least a portion of the other wall structure 16 can be made dark opaque or of a more or less transparent but darker hued material for optimum visual characteristics, which result from viewing only the light scattered from the display device 10, in particular from its liquid crystal layer 18. The darker hued container section 16 may be a buff gray or other neutral color although desirably a darker coloration will make the color patterns of a liquid crystal more obvious.

At least one of the container sections 14, 16 is joined about its periphery to a surface of the other wall structure. In the Fig. 1 arrangement, such joining means are further arranged to peripherally seal one wall structure to a surface of the other. In the display device 10, such joining and sealing means include a pressure-sensitive tape 22, which is compatible with the material of the wall structures 14, 16 and covers their coextensive peripheral edges. The liquid crystal 18 is thereby sealed in the context of film thicknesses within the space defined by the slightly separated wall structures 14, 16 and the peripheral tape 22.

It is desirable to provide the light transmitting section 14, particularly when transparent, with appreciable thickness to enhance the variable color patterns of the liquid crystal 18 and to create an illusion of depth. When the display device is substantially planar as in Fig. 1 and 2, the container section 14 should be in the neighborhood of about one-eighth inch or more in thickness although such thickness is not essential to the invention and can be varied depending upon a specific application of the display device. The liquid crystalline material 18 is selected from one or more of those materials which exhibit variation in light-scattering and attendant transmittance characteristics under deformational stresses. For the purposes of our invention, we employ a cholesteric liquid crystalline material which exhibits a relative optical phenomenon attendant to the selective scattering characteristic of this category of liquid crystal. The latter characteristic is the stress or shear sensitivity of certain cholesteric materials whereby the selective (light frequency) scattering characteristic is varied upon the application of deformational stresses.

Specific examples of liquid crystalline compositions useful for our present invention appear below, wherein all amounts are in parts. "ChCl" is cholesteryl chloride melting at 94-95°C; "High ChOlC" is cholesteryl oleyl carbonate showing a color play at 20°-22°C; "Low ChClC" is cholesteryl oleyl carbonate showing a color play at 5°-6°C; "ChNo" is cholesteryl nonanoate; and the upper temperatures are those at which the compositions become colorless.

Example	ChCl	High ChOlC	Low ChOlC	ChNo	Color	Color play range (C°)
I.....	27	73			Red	Below 0 to 48
II....	27	58		15	Red	Below 0 to 51.5
III...	27	50	23		Red	Below 0 to 38
IV....	27	59	14		Red	Below 0 to 35
V.....	25	75			Green	Below 0 to 48
VI....	22	78			Blue	Below 0 to 46

In the display device 10 of Figs. 1 and 2 it is contemplated that the forces 20, 21 can be applied manually, for example, by pressing or stroking the container section 16 with the fingers. A single force can be applied as designated by arrow 20 or alternatively multiple forces can be applied as desired as denoted by arrows 21. The edge-sealing tape 22 can be of the pressure-sensitive variety, desirably of the light-transimtting or transparent type.

The display device 44 can be utilized, foe example, as an entertaining and ornamental novelty for a table or desk top. A relatively slight pressure upon the rounded surface of the hemispheroidal container section 48 will apply compressional forces to the resilient or displaceable container section 50 resting, for example, directly upon the table or desk top. This in turn will cause various flow patterns within the liquid crystal 54 depending upon the magnitude and location of the applied forces. As a result an interesting and entirely unexpected variable color display is produced.

The bottom area 60 of the stand 56 may incorporate the owner's initials denoted in this example by reference numeral 62. The design, message item, or indicia 62 can be fabricated from any suitbale structural material, plastic or metallic, and desirably are arranged such that their undersurfaces seat flushly against the table or desk top. The upper surfaces of the design or message items 62 project sufficiently above the remainder of the bottom structure 60 and are supported in this example by connecting links 64. In consequence only the message items 62 are engaged by the conatiner section 50 when the display device 44 is seated in the stand 58. when so arranged the message items or indicia 62 depress the flexible container section 50 at their top surface areas with the result that the items appear as a discrete and contrasting coloration within the color pattern 54 of the liquid crystal.

Other geometric shapes can be utilized in addition to the hemispheroidal one of Fig. 4. For example, Fig. 5 illustrates another geometric, transparent member 66, exemplary in the form of a cube, forming part of display device 68. One or more faces of the cube 66 can be utilized as a component container section of a corresponding number of liquid crystal containers or cells.

Similar geometric shapes are illustrated in Fig. 6, 7, and 8 which respectively show parallelopiped, pyramidal, and prismatic shapes.

The aforementioned liquid crystal color patterns (which can be varied by the application of deformational stress as described above) can be employed as unexpectedly decorative and entertaining backgrounds for items such as coins, models, fossils, precious and semiprecious stones, specimens and the like embedded in the transparent member.

In Figure 10, however, the adaption of the novel display device to form relatively large surface areas is exemplified. The latter form of the display device 108 is incorporated in an article of furniture, in this example table 110. For maximum effect the display device 108 is applied to top structure 112 of the table 110.

Another arrangement of the novel display device is exemplified by display container 140 of Figs. 12 and 13. The display container 140 or aesthetic novelty includes a light-transmitting member 142, which can be fabricated froma polyacrylic resin in sufficient thickness to give the aesthetic novelty 142 sufficient rigidity or structural strength. For example, if the aesthetic novelty 140 is of the order of about 4" square, the light-transmitting member 142 can be of the order of about one-eight inch in thickness, although a greater or lesser thickness can be employed as evident from Figs. 16 and 16A described below. Desirably, the light-transmitting

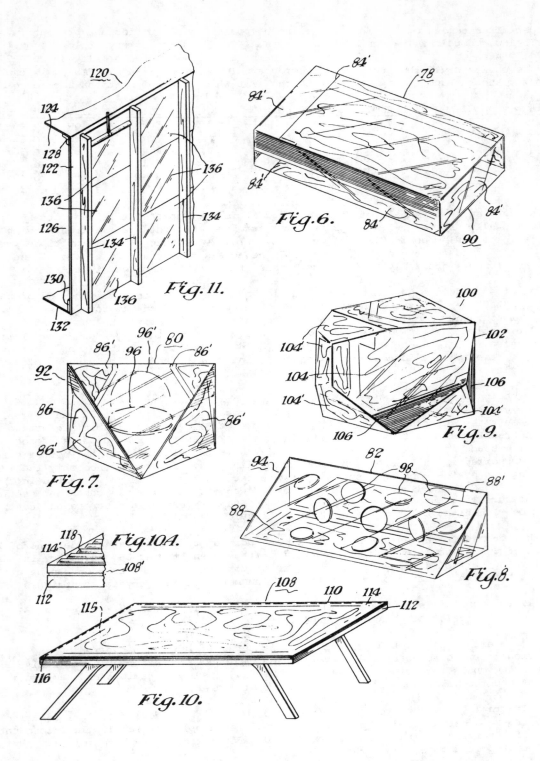

member 142 is fabricated from a fully transparent polyacrylic resin to enhance the color patterns of the liquid crystal material 144 encased between the light-transmitting member 142 and a desirably darker-hued or dark-opaque film or sheet 148 adhered to the upper surface (as viewed in Figs. 12 and 13) of the flexible film 146. Other light-absorbing means can be substituted such as described with reference to other figures. In this arrangement, the film 146 can be formed from a sheet of PVC plastic or the like to which the coating of pressure sensitive adhesive is applied entirely over one surface thereof. The PVC sheet or film 146 and the application of the adhesive thereto can be formed by conventional techniques.

The area occupied by the liquid crystalline material 144 can be demarcated by a sheet of heavy paper or cardboard or by a second plastic film or sheet 148, which can be pressed into adhesive engagement with the central area of the adhesive film 146. Use of the film layer 148 prevents the juxtaposed surfaces of the film 146 from adhering to the underside of the light-transmitting member 142 and thus delineates a shallow pocket for the liquid crystalline material 144.

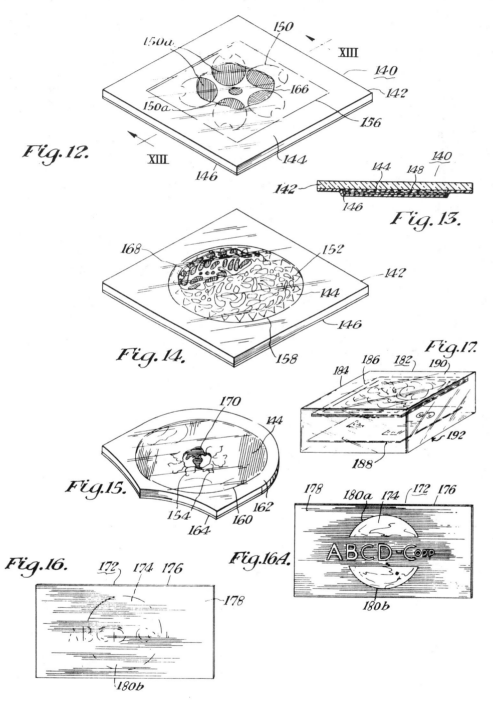

The film layer 148 can be coated or formed from a material having a dark or other contrasting color relative to the predominating color of the liquid crystalline material 144. Printed messages (not shown) or various designs, e.g., the design 150 (Fig. 14) or 154 (Fig. 15), can be applied to the film or sheet layer 148. Such designs, foe example, the designs 150, 152 can be printed in darker colors or shades upon a light background or alternatively as evident from the design 154 in Fig. 15, the design can be delineated in lighter colors against a darker background. Also, the designs can be more or less geometrical as shown in Fig. 12 or random as shown in Fig. 14 or pictorial as shown in Fig. 15. The unique cooperation of the contrasting colors of the film layer 148, when provided with a design of some sort such as those described above, is evident when the flexible film 146 is depressed in the area of the film layer 148 to apply deformational stresses to the liquid crystalline material 144. The alternate thinning and thickening of the liquid crystalline layer considerably enhances the varying color patterns resulting from deformational flows in the liquid crystal. Interest in the liquid crystalline patterns is heightened, with the variation in thickness or depth of the liquid crystalline material above the various contrasting colors or shades of the designs imparted to the film layer 148.

As a further enhancement of the color pattern variation and interest therein, an air bubble 166 (Fig. 12) can be introduced into the liquid crystal area 156, along with the liquid crystalline material. The air bubble 166 operates to thin the juxtaposed portions of the contained liquid crystalline material, and such thinning provides an interesting variation in the resulting color patterns. Also, interesting differences in reflection occur at the air bubble, depending on viewing angle. In addition, as the flexible backing member 146 is depressed or deformational stresses are otherwise applied thereto, the bubble 166 tends to break up into a number of smaller bubbles exhibiting variable patterns, depending on the amount and area of pressure application, to further increase the viewer's interest in the color patterns.

In Figs. 16 and 16A, another form of the novel color display device 172 is illustrated with optional commercial aspects. The display device 172 can be fabricated from relatively thin material, for example in the shape of a calling card or the like.

A further modification of the novel display device 182 is shown in Fig. 17 and is arranged in this example as a largely transparent novelty such as a paper weight or the like. The display device 182 includes in this example a solid block 184 of transparent material, such as one of the polyacrylic resins. The transparent block 184 is provided in accordance with this aspect of our invention with a first embodiment 186 of liquid crystalline material and a second embodiment 188 of a design or lettering such as a slogan, motto, the owner's name or initials or the like.

From the foregoing it will be apparent that the novel display device is capable of a wide variety of applications, some of which are illustrated in this application and others of which have been alluded to. The patent involves 20 Claims and 54 drawing figures.

" " "" "" "" "" "

3,647,280 (U.S.)   Patented Mar. 7, 1972

LIQUID CRYSTAL DISPLAY DEVICE.

<u>Richard I. Klein and Sandor Caplan</u>, assignors to RCA Corporation.
Application Nov. 6, 1969.

This invention relates to liquid crystal display devices, and particularly to display devices of the type presenting a plurality of fixed images.

With reference to Fig. 1, a liquid crystal display device 10 is shown capable of presenting three different images, as follows: a first image A of a circle; a second image B of a vertical bar; and a third image C of a horizontal bar. The three images overlap one another.

The display device 10, as shown in Figs. 1 and 2, comprises a pair of front and rear transparent glass substrates 12 and 14, respectively. The two substrates 12 and 14 are maintained in a spaced-apart relation by means of a shim 16 of an insulating material, e.g., mica. Disposed between the substrates 12 and 14, and maintained therebetween by the shim 16, as well as by a seal 18 of, e.g., glass frit, extending about the peripheries of the substrates 12 and 14, is a thin film 20, e.g., 0.0005 inch thick, of a liquid crystal material. In the instant embodiment, a nematic liquid crystal composition is used of the type that exhibits turbulent motion upon the passage of current therethrough.

The liquid crystal material, hence, the film 20, is normally substantially transparent to light. However, when an electrical voltage sufficiently in excess of a threshold voltage is applied across the film, a current is caused to flow through it. The current flow causes turbulence in the film and places the liquid crystal material in a light-scattering state, i.e., it forward scatters light incident thereon. The scattering of light gives the material a somewhat milky or cloudy appearance. As the applied voltage is increased, the amount of scattering increases until the film 20 becomes substantially opaque to the passage of images therethrough. With further increases of voltage, the amount of light scattering reaches a maximum of saturation level, which is not increased by still further increases in voltage. When the electric field is removed, the liquid crystal returns to its transparent state.

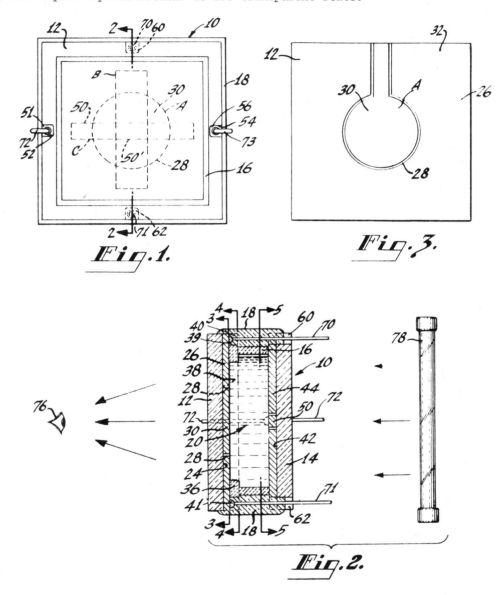

The inside surface 24 of the front substrate 12 is coated with a thin layer 26 of a transparent electrically conductive material, e.g., a 1000 A thick layer of tin-oxide. The first image A, a circle (Figs. 1 and 3), is defined in the layer 26 by means of a fine opening or line 28 through the layer 26 dividing the layer 26 into two separate spaced-apart segments 30 and 32. The line 28 is wide enough to provide electrical isolation of the segments 30 and 32, for the voltages used in the device, but is quite narrow in comparison with the size of the first image.

The segmented layer 26 can be provided by known techniques, e.g., sputtering of the conduc-

109

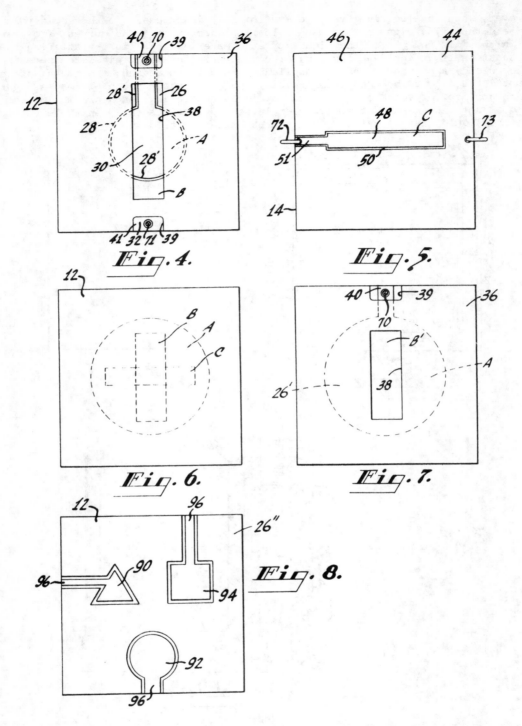

tive material onto the substrate 12, and selective etching of the layer 26 to provide the line 28.

Covering portions of the layer 26 is a thin layer 36 (Fig. 2) of a transparent dielectric material, e.g., 5,000 A thick layer of silicon dioxide or silicon nitride. An opening 38 (Fig.4) is provided through the layer 36 exposing a portion of the conductive layer 26. The opening 38 defines the second image B, a vertical bar.

The opening 38, it is noted, overlies and wxposes portions 28' of the line 28 through the conductive layer 26. This affects the appearance of the second image B, as described hereinafter. Also, openings 39 are provided through the layer 36 to expose peripheral portions 40 and

41 of the two segments 30 and 32, respectively, of the layer 26. This facilitates the making of electrical connections to the segments 30 and 32, as described hereinafter.

Covering the inner surface 42 (Fig.2) of the substrate 14 is a thin layer 44 of a transparent electrically conductive material, e.g., 1000 A thick layer of tin-oxide. The layer 44 is segmented (Fig.5) into two spaced-apart segments 46 and 48 by means of a thin opening or line 50 formed, e.g., by etching, through the layer 44. The segment 48 defines the third image C, a horizontal bar. A portion 51 of the segment 48 extends to the edge of the substrate 14.

In a different embodiment, not shown, the dielectric layer 36 covers the layer 44 on the substrate 14 rather than, as in the illustrative device 10, the layer 26 on the substrate 12.

Terminal means are provided for making separate electrical connections to each of the four segments 30, 32, 46, and 48. To this end, peripheral notches 52 and 54 (Fig. 1) are provided through the substrate 12 to expose the extending portion 51 of the segment 48 and a peripheral portion 56 of the segment 46 on the substrate 14. Likewise, notches 60 and 62 (Fig.2) are provided through the substrate 14 to expose the peripheral portions 40 and 41 of the segments 30 and 32, respectively, on the substrate 12. As previously noted, the peripheral portions 40 and 41 of the layer 26 are exposed by the openings 39 through the layer 36 (Fig.4). Fine leads 70, 71, 72, and 73, are connected to the exposed segment portions 40, 41, 51, and 56, respectively, by, e.g., a silver epoxy paste.

In operation of the device 10, the device is disposed between a viewer 76 (Fig.2) and a source of illumination, e.g., a fluorescent bulb 78.

With no voltage applied to the device 10, the liquid crystal film 20 is transparent, and the front viewing surface of the device appears uniformly illuminated by the light source 78.

To display the first image A, i.e., the circle, a fluctuating switching voltage, either AC or pulsed DC is applied between all of the conductive layer 44 on the substrate 14, via the leads 72 and 73, and the first image defining segment 30 of the conductive layer 26 on the substrate 12, via the lead 70. No voltage is applied to the segment 32 of the layer 26. The portion of the liquid crystal material film 20 between the segment 30 and the conductive layer 44, and only that portion, is thus switched into the light scattering state. Stated differently, a fluctuating current path is created through the film 20, the path having a cross-sectional shape corresponding to the shape of the segment 30. Owing to the scattering of the light by the switched portion of the film, this film portion, having the shape of the circle image A, appears less bright to the viewer than the surrounding transparent portions of the film 20. The circle is thus displayed as a dark image on a light background.

By proper selection of the magnitude of the energizing voltage, the presence of the dielectric layer 36 on the conductive layer 26 has little effect upon the appearance of the first image A. Where the opening 38 (Fig.4) through the layer 36 exposes portions of the image A defining segment 30, the liquid crystal film 20 is in direct contact with the segment 30 (Fig.2) and the voltage on the segment 30 is coupled directly to the liquid crystal film 20. Where the segment 30 is covered by the layer 36, and thus separated from the film 20 thereby, the fluctuating voltage on the segment 30 is coupled to the film 20 capacitively through the dielectric layer 36. By using a sufficiently high voltage, all portions of the film 20 opposite the segment 30 are driven into or close to the saturation level of light scattering regardless of the difference in coupling between the different portions of the film 20 and the segment 30. Thus, the degree of light scattering is substantially uniform over the entire extent of the switched film portion.

To the extent that the segment 30 defining the image A overlies portions 50' (Fig.1) of the line 50 defining the image C, these line portions 50' appear as bright lines in the dark circle image. This occurs because the portions of the liquid crystal film 20 disposed between these line portions 50' and the segment 30 are not exposed to a switching voltage and thus remain transparent. By the use of a segmenting line 50 of small width, in comparison with the size of the displayed image, the thin, bright light is not objectionable.

To display the second image B, i.e., the vertical bar, a DC switching voltage is applied between all of the conductive layer 26, via the leads 70 and 71, and all of the conductive layer 44, via the leads 72 and 73.

Since a DC voltage is used, only these portions of the liquid crystal film 20 directly engaged with the conductive layer 26 through the opening 38 in the dielectric layer 36 are switched to the light-scattering state. The other portions of the film separated from the conductive layer 26 by the dielectric layer 36 are electrically isolated from the DC voltage on the layer 26, and are not switched from the transparent state. Thus, a dark image of the vertical bar, corresponding to the shape of the opening 38 (Fig.4) through the dielectric layer 36, is provided on a light background.

Again, to the extent that the layer 36 opening 38 overlies portions 28' (Fig.4) of the line 28 in the conductive layer 26, or portions of the line 50 in the conductive layer 44 (Fig.1), these line portions appear as bright lines in the dark image.

With respect to the formation of the second image B, it is noted, it is immaterial whether the dielectric layer 36 covers the layer 26, as in the illustrative embodiment, or the layer 44.

To display the third image C, i.e., the horizontal bar, an AC voltage is applied between the segment 48 (Fig.5) of the layer 44, via the lead 72, and all of the conductive layer 26, via the leads 70 and 71. No voltage is applied to the segment 46 of the layer 44.

Again, using a sufficiently high exciting voltage to excite the film 20 into or close to its saturated light-scattering level, the presence of the dielectric layer 36 has little or no effect upon the appearance of the third image. Also, to the extent that the image C overlies portions of the line 28 of the image A (Fig.1), these line portions appear as bright lines in the dark third image.

In the device 10, it is noted, portions of projections of each image, A, B, and C fall outside the outlines of the other images. A defining segment 30 is necessary in this embodiment to serve as an electrode in the formation of the images B and C. Likewise, the segment 46 surrounding the image C defining segment 48 is necessary to serve as an electrode in the formation of the images A and B.

Another embodiment of the invention is shown in Figs. 6 and 7. In this embodiment, the second image B, the vertical bar, and the third image C, the horizontal bar, are both smaller, and lie entirely within the confines of the first image A, the circle. Owing to this dimensional relationship of the images, the conductive layer 26' (Fig.7) on the front substrate 12 comprises merely a single segment (plus a connector extension 40 therefor) having the shape of the circle image A. That is, since projections of both the second and third images B and C fall within the outline of the conductive layer 26' defining the image A, no additional conductive material on the substrate 12, surrounding the image A defining layer, is required for the formation of the images B and C.

To provide more than three images, although not overlapping, either or both of the conductive layers on the substrates 12 and 14 can be provided with two or more image defining segments. As shown in Fig. 8, for example, the conductive layer 26" on the substrate 12 is provided with three segments 90, 92, and 94, defining images of a triangle, a circle, and a square, respectively. Leads, not shown, are provided connected to an extending portion 96 of each segment, whereby individual ones, or combinations of images can be provided.

" " " " " " " " " " " "

3,648,280 (U.S.)   Patented Mar. 7, 1972

THERMOCHROMIC LIGHT-FLASHING SYSTEM.

James W. Jacobs, assignor to *General Motors Corporation*.
Application May. 15, 1970.

Referring to the drawings, in Fig. 1 a light-warning or flashing device 10 is illustrated that includes an opaque case or enclosure 12 having a top 14 and a base 16 which has a socket 18 depending therefrom. The socket 18 has a first and second electrical terminal 20, 22 directed therefrom which are adapted to be connected across a power source.

At the top of the socket 18 the terminal 20 is formed in a cylindrical shape 24 to receive the base 26 of an incandescent lamp 28. A contact 30 on the base 26 is held against the terminal 22 in spring biased relationship therewith. The lamp 28 more particularly includes an evacuated

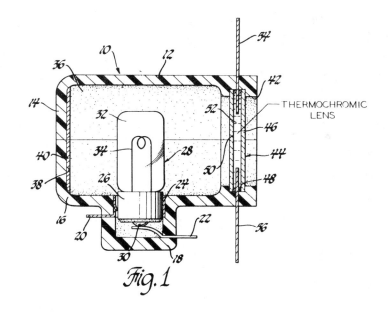

Fig. 1

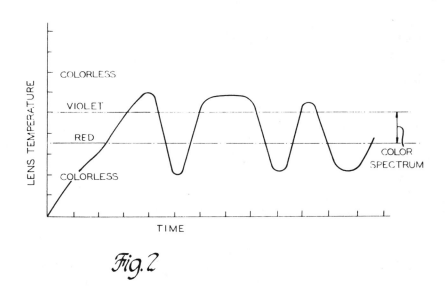

Fig. 2

glass bulb portion 32 which encloses an electrically energizable filament 34 whereby when the terminals 20, 22 are connected across a power source the filament 34 will be energized into an incandescent range to produce a source of visible white light.

The lamp 28 more particularly is located within an open interior 36 formed between the top 14 and the base 16 of the case 12. The rear wall 38 of the interior space 36 is covered by a surface 40 of light-reflective material. The opposite wall of the case 12 is opened at 42 whereby light from the bulb 28 and reflected from the surface 40 is directed from the interior 36 to a point exterior the enclosure 12.

In accordance with certain principles of the present invention the constantly energized single light source represented by the lamp 28 serves as a heat source for a thermochromic lens 44 which fits across the opening 42 to be exposed to all light energy passing either directly from the lamp 28 or the reflective surface 40.

In the illustrated arrangement the lens 44 includes a transparent outer surface or face member 46 which is seated within a groove 48 formed in the upper and lower parts 14, 16 of the enclosure 12 around the opening 42 therein.

It also includes a transparent face or surface member 50 that is spaced apart from the member 46. This member also is seated within the groove 48 and likewise extends across the full planar extent of the opening 42.

Between the lens surfaces 46, 50 is a thin layer of cholesteric liquid crystals 52 which are representative of materials which change color in response to a change in temperature. These materials are derivatives of cholesterol, they flow like a liquid, and simultaneously exhibit the optical properties of a crystal by selectively scattering light. Particular liquid crystal systems can be sensitive to temperature changes of 1°C. to produce color changes through the full visible spectrum.

The flasher 10 also includes a plurality of elongated metallic heat transfer fins 54, 56 that serve to reduce the temperature of the lens 44 in accordance with the changes in the temperature or convective flow pattern of ambient air being directed across the outside of the case 12.

In the illustrated arrangement the surfaces 46, 50 serve the purpose of an infrared heat sink so as to raise the temperature of the crystals 52 so as to cause it to change from a red color to a violet color within the color change temperature range.

Outside this color change the temperature range, as shown in Fig. 2, the lens will become colorless. This along with the change in the cooling action of airflow across the fins 54, 56 will cause the temperature of the lens to be cycled through thereby to return through the violet and red phases in a repetitive manner.

The combination of continued heating from the single light source 28 and the cooling action of the fins 54, 56 will produce a continual rise and fall of colors in the spectrum from red to ultraviolet and back to red as shown in Fig. 2 whereby the constantly energized single visible white light source will be able to produce a variable color changing flashing effect from the device 10.

""""""""""

3,650,603 (U.S.)   Patented Mar. 21, 1972

LIQUID CRYSTAL LIGHT VALVE CONTAINING A MIXTURE OF NEMATIC AND CHOLESTERIC MATERIALS IN WHICH THE LIGHT SCATTERING EFFECT IS REDUCED WHEN AN ELECTRIC FIELD IS APPLIED.

George H. Heilmeier and Joel E. Goldmacher, assignors to RCA Corporation. Application Dec. 5, 1967.

An electro-optic light valve comprises a mixture comprised of a nematic liquid crystal composition of the type whose molecules align in an electric field with at least one member of the group consisting of cholesterol, a cholesterol derivative and a cholesteric liquid crystal. The light valve includes means for applying an electric field to this mixture.

It has been found that by mixing cholesterol, and/or a cholesterol derivative, and/or a cholesteric liquid crystal with a nematic liquid crystal composition of the type whose molecules align in an electric field, one achieves a mixture which is highly light scattering. The mixture also results in depresssion of the crystal-nematic transition temperature of the nematic liquid crystals. When an electric field is applied to such a mixture, so as to align the nematic liquid crystal molecules of the mixture with their axes substantially parallel to the direction of the incident light, the mixture essentially no longer scatters the light as it would in its unexcited state and becomes transparent to the incident light. This discovery makes possible light valves and display devices having improved contrast ratios as compared to prior art liquid crystal devices of the type which depend upon alignment of nematic liquid crystal molecules in an electric field. A preferred embodiment of a novel liquid crystal element is shown in Fig. 3 in the form of a crossed grid reflective electro-optic display device 30.

In the novel device 30, the space between the front and back plates 31 and 32 is filled with a mixture 39 comprising 80 weight percent of a nematic liquid crystal composition and 20 weight percent of cholesterol derivatives. The nematic composition consists essentially of equal weights of p-n-ethoxybenzylidene-p'-aminobenzonitrile and p-n-butoxybenzylidene-p'-aminobenzonitrile. The cholesterol derivatives consist essentially of 23 weight percent of cholesteryl chloride and 77 weight percent of cholesteryl oleate. The mixture may be sealed in the device 30 by using epoxy cement around the edges of the device 30. The nematic composition by itself has a crystal-nematic transition temperature of about 41°C. However, this transition temperature is reduced to about room temperature when mixed with the cholesteryl compounds.

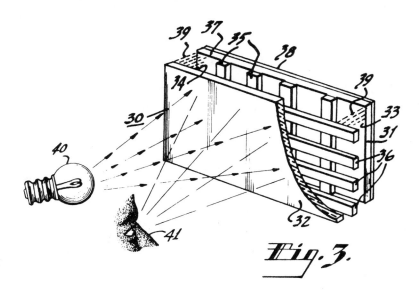

Fig. 3.

It is preferred that the nematic liquid crystal portion of the overall mixture comprise at least 50 weight percent of the total mixture, and preferably from about 75 to 95 weight percent of the overall mixture. However, it is possible to use mixtures containing as little as 5 percent nematic liquid crystals and up to 95 percent of cholesteric material.

The novel mixtures can be prepared by placing the desired proportions of the pure compounds in a vial and heating these compounds until a uniform isotropic liquid solution is formed. Generally, heating to a temperature of about 100°C. is sufficient. This isotropic liquid is then allowed to cool slowly to room temperature. During cooling the mixture enters its mesomorphic state and, depending upon the particular mixture, may solidify. The novel mixtures can be incorporated into the device by, for example, placing some of the mixture in its mesomorphic state on one support plate and carefully placing the other support plate over the first so that the mixture spreads to form a uniform film between them. Another technique is to have the spaced supports separated by shims and/or sealed on three sides and injecting the mixture into the space with a syringe.

"""""""""""

3,650,608 (U.S.)   Patented Mar. 21, 1972

METHOD AND APPARATUS FOR DISPLAYING COHERENT LIGHT IMAGES.

Charles E. Baker, assignor to Texas Instruments, Inc.
Application Dec. 23, 1969.

A method and apparatus for destroying the scintillation or speckling effect observed when a beam of coherent light is impinged or directed upon a stationary screen includes a relatively thin layer of an organic nematic mesomorphic compound which is impressed with a voltage above the particular threshold level of electric field which causes the compound to scatter light.

Referring now to Fig. 1, a coherent light source 10, illustrated schematically as a laser, directs a coherent light beam 12 through light modulator 14 in which the intensity or brightness of the light beam is varied in accordance with a predetermined input signal. From the

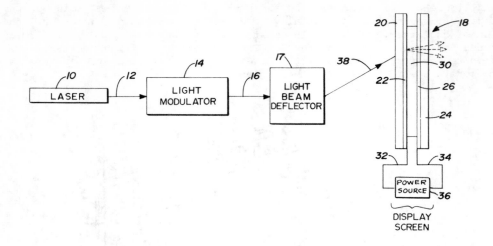

light modulator 14 the intensity modulated laser beam 16 is directed into a scanning mechanism or light beam deflector which causes the light beam to be deflected or scanned in a predetermined pattern. This pattern is projected or displayed upon a screen, generally designated 18. The rear projection display screen 18 includes a first transparent substrate 20, for example, a planar sheet of glass. The substrate is coated on one side with a layer or coating 22 of material which is optically transmissive and electrically conductive. The display screen also includes a second substrate 24 similar to substrate 20 which is also coated with a layer or coating 26 of material which is electrically conductive, and in this embodiment also optically transmissive. Sandwiched between the two layers 22 and 26 and in electrical contact therewith, is a relatively thin layer 30 of an organic nematic mesomorphic compound. The layer has an optimum thickness of less than 20 mils, preferably less than 10 mils, and most preferably has a thickness in the range of 1 to 4 mils. Leads 32 and 34 are connected respectively to the electrically conductive layers 22 and 26. The leads are also connected to a power source 36 which impresses a sufficient voltage when energized to cause the organic nematic mesomorphic compound to diffuse light which strikes it.

To the eye of the observer then, the beam of coherent light 38 striking the rear side of the projection screen 18 appears to be a bright stationary display point, not exhibiting the heretofore troublesome phenomenon of speckling or scintillation. The image created on the screen, of course, can be a variety of configurations, depending upon the particular manner in which the coherent light beam is modulated and/or deflected. An exemplary image can be produced by a raster scan of the screen, displaying, for example, a dynamic image such as a conventional line television image. As previously mentioned, this type of display would be of great importance in a large screen dynamic display such as that utilized in closed circuit television, conferences, and public information displays. Heretofore, this type of display has been subject to the scattering or scintillation effect which increases with the distance of the observer from the display screen. In addition the screen of the present invention can be utilized to improve other coherent and light image applications, such as transparency projection and holography.

Mesomorphic materials suitable for the practice of this invention are organic thermotropic nematic compounds. It will be understood that the screen 18 is maintained in the temperature range in which the compound exhibits the nematic mesophase.

The threshold value for the voltage gradient impressed across or electrical field applied to the particular organic nematic thermotropic compound varies with the compound itself and with the particular distance between the two electrodes. However, for most of the organic nematic thermotropic compounds, the applied voltage is of the order of magnitude of about 1,000 volts per centimeter.

"""""""""""

3,652,148 (U.S.)   Patented Mar. 28, 1972

IMAGING SYSTEM.

<u>Joseph J. Wysocki, James E. Adams, James H. Becker, Robert W. Madrid and Werner E.L. Haas</u>, assignors to Xerox Corporation.
Application May 5, 1969.

A system transforming an optically negative liquid crystalline substance to an optically positive liquid crystalline mesophase by an applied electrical field, and an imaging system wherein the electrical field-induced transition images a liquid crystalline member.

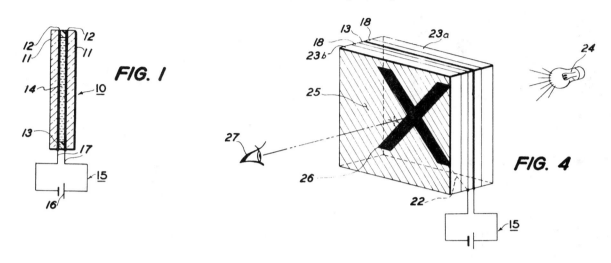

In the surprising and advantageous system of the present invention it has been discovered that when cholesteric liquid crystals or a mixture of cholesteric liquid crystalline substances is used in an electrode sandwich such as that shown in Fig. 1. that high electrical fields across the liquid crystal film cause an electrical field-induced phase transition to occur wherein the optically negative cholesteric liquid crystalline substance transforms into an optically positive liquid crystalline state. This transition is believed to be the result of the cholesteric liquid crystal transforming into the nematic liquid crystalline mesophase structure. Cholesteric liquid crystals in the cholesteric phase are typically translucent, for example, like a milky white, opalescent layer, when first placed in the unbiased electrode sandwich. When the high electric field is placed across the liquid crystal film, the field-induced phase change is observable because the liquid crystal film becomes transparent in areas where the field is present When viewed between polarizers with transmitted light, for example, as described in conjuction with Fig. 4, the areas in which the field-induced phase change has taken place appear dark, while the unchanged, translucent, light scattering and birefringent, cholesteric areas still retain their light appearance. Any other means suitable for enhancing the contrast of the imaged areas may also be used in place of the polarizers. Hence, it is seen that either field or non-field areas in a liquid crystalline imaging sandwich may be used to create the desired image, with or without the addition of means for image enhancement.

In Fig. 4 a liquid crystal imaging member comprising a pair of substantially transparent electrodes 18 sandwiching a spacer 13 containing a liquid crystal film is shown being observed between polarizers 23. Light from source 24 is planar polarized while passing through polarizer 23a, scattered by the translucent cholesteric liquid crystalline substance in non-image areas 25, and is transmitted by the field-induced nematic state areas 26. A viewer 27 then sees the planar-polarized light which passes through polarizer 23b which originated from source 24 and was scattered and passed through the non-image portion of spacer 13. Although the light was polarized by polarizer 23a in a plane crossed with the plane of polarizer 23b, the scattering effect of the cholesteric liquid crystal between thesubstantially transparent electrodes scattered sufficient amounts of the originally planar-polarized light to allow some of it to pass through polarizer 23b. However, in the image areas 26, the effect of polarizers 23, when said polarizers have their respective planes of polarization crossed, is to cut off the light transmitted through polarizer 23a and transformed image area 26 so that the image area 26 appears dark, as illustrated.

In Fig. 5 another preferred embodiment is illustrated wherein an electron beam address system is provided for the generation of an imagewise field across a liquid crystalline imaging member.

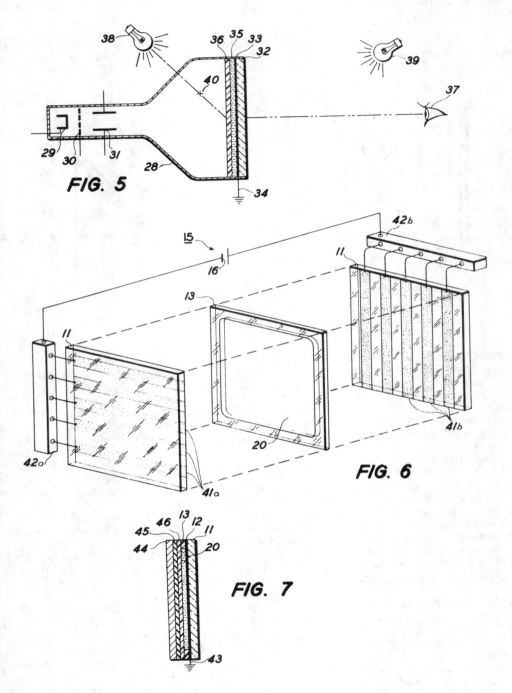

FIG. 5

FIG. 6

FIG. 7

The electron beam address system does not itself illuminate the image which it creates. However, external sources of light 38 and 39 are illustrated to show that the liquid crystalline imaging member comprising the face of vacuum tube 38 may be observed by light transmitted, as from light source 38, or reflected, as from light source 39. Alternatively, transmitted light may originate from a source placed inside vacuum tube 28 at a location indicated, for example, at point 40. However, the in-tube source of illumination should be so placed so as to not interfere with the electron beam which creates the image on the face of the tube.

In Figure 6 and X-Y address system suitable for imaging a liquid crystalline imaging member is illustrated in exploded isometric view. The liquid crystalline imaging film is placed in void area 20 within the transparent and substantially insulating spacer-gasket 13. The liquid crystalline film and spacer 13 are sandwiched between a pair of substantially transparent electrodes comprising transparent support plates 11 upon which strips of substantially transparent, conductive material 41 is coated. These electrodes are oriented so that conductive strips 41b and conductive strips 41a on the respective electrodes cross each other in an X-Y matrix or grid.

In Figure 7 the advantageous liquid crystal imaging member of the present invention is shown in a system where said member is imaged by an electrostatic latent image on an electrostatic latent image support surface such as a photoconductive layer. In Fig. 7 typically transparent, support substrate 11 is shown with typically substantially transparent conductive coating 12 which is grounded at 43. Adjacent substantially transparent conductive coating 12 spacer-gasket 13 contains liquid crystalline imaging layer 20 between the substantially transparent grounded electrode and an electrostatic latent image support surface. The electrostatic latent image support surface may be any substantially insulating surface which is capable of supporting an electrostatic latent image. The electrostatic latent image supported by said surface, when taken in conjunction with grounded substantially transparent conductive layer 12 creates an imagewise electrical field across liquid crystalline imaging layer 20, and said field causes the advantageous optical negative-positive, cholesteric-nematic phase transition of the advantageous system of the invention. A typical electrostatic latent image support surface is a photoconductive plate, foe example, as typically used in xerography. As illustrated, conductive supporting layer 44 supports a layer of photoconductive material 45, typically comprising selenium, mixtures of sulphur and selenium, phthalocyanine, or any other photoconductive insulating layer. Between the photoconductive insulating layer 45 and liquid crystalline imaging layer 20 there may be an insulating protective layer 46 which may retain the liquid crystalline substance in some embodiments, and at the same time layer 46 protects both the liquid crystalline layer and the photoconductive insulating layer 45.

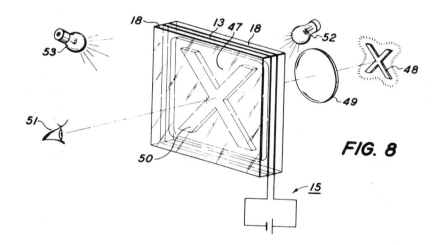

FIG. 8

In Figure 8 a liquid crystalline imaging member is shown imaged by a thermal image projection address system. Here the liquid crystalline imaging member comprises substantially transparent electrodes 18 separated by spacing gasket 13 which encloses a cholesteric liquid crystal whithin space 47 comprising substantially all of the area within spacer gasket 13. In this embodiment of the invention a source of a thermal image 48, here a heat source in the desired image configuration, is shown in position with conventional means 49 for focusing and projecting a thermal or optical image, and the thermal image 50 appears in the liquid crystalline film in the area where the liquid crystalline substance is heated into the temperature range required for the optical negative-positive, cholesteric-nematic phase transition, while at the same time, the liquid crystalline imaging member is biased by external circuit 15 so that the field across the cholesteric liquid crystalline film is sufficient to cause the phase transition when the imaged area of the film reaches the transition threshold temperature.

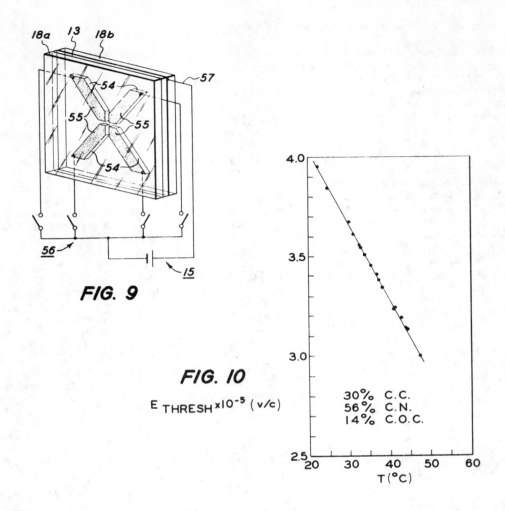

FIG. 9

FIG. 10

$E_{THRESH} \times 10^{-5}$ (v/c)

30% C.C.
56% C.N.
14% C.O.C.

Figure 9 illustrates a uni-planar, multiple cell, liquid crystalline imaging member suitable for use with the electrical field-induced phase transition imaging system of the present invention. In the imaging member of Fig. 9 the transparent electrodes 18 are separated by the typical spacer gasket 13 which contains voids 54 which contain the liquid crystal film. Corresponding areas 55 are the shaped, substantially transparent, conductive coating on the inner surface of electrode 18a. Each one of these cells 55 is capable of being selectively imaged either individually, or jointly, as desired, through the use of switching system 56 in external circuit 15 which is also connected to the substantially transparent conductive layer on the inner surface of substantially transparent electrode 18b, by electrical lead 57. It will be appreciated that uni-planar, multiple cell, imaging members such as the one in Fig. 9 may be designed so that various combinations of the desired image cells may be imaged to create any desired figure or character in any language or number system. It will also be appreciated, that either a character-shaped spacer or a character-shaped electrode system can be used in such a uni=planar image system

The experimental data for threshold field necessary for phase transition is expressed in volts per centimeter times 10^{-5}, and the temperature is expressed in degrees Centigrade. It is observed that the threshold field decreases rapidly with increasing temperature. The isotropic temperature of the mixture of about 30:56:14 cholesteryl chloride; nonanoate; oleyl carbonate, respectively, is about 57°C. This isotropic temperature is the point at which the liquid crystal mixture changes to the completely liquid physical state.

The Threshold Field versus Temperature relationship of Fig. 10 defines the operating range for the electrical field-induced cholesteric-nematic phase transition imaging member of the invention having a cholesteric liquid crystal imagign layer comprising the mixture of about 30%CC: 56%CN:14%COC. The preferred operating ranges for this mixture are: temperatures in the range of about 20° to about 57°C and electrical field strengths in the range of about 2.7×10^5 to about

4.0×10^5 volts/cm of liquid crystalline material.

Fourtyeight additional examples are described in this patent specification.

" " " " " " " " " "

3,653,745 (U.S.)   Patented Apr. 4, 1972

CIRCUITS FOR DRIVING LOADS SUCH AS LIQUID CRYSTAL DISPLAYS.

Roger A. Mao, assignor to RCA Corporation.
Application June 11, 1970.

The object of the present invention is to provide improved circuits for applying alternating voltage excitation to loads such as liquid crystals.

A pair of amplifying means each having a control electrode, an input electrode and an output electrode and a load such as a liquid crystal connected between said output electrodes. In response to one relationship of signals applied to said control electrodes, an alternating turn-on voltage is applied via at least one of said amplifying means to said load and in response to another relationship between said input signals, the light scattering produced by the load, when a liquid crystal, is reduced to a relatively low value.

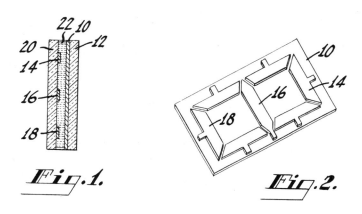

Fig. 1.   Fig. 2.

In the operation of the display of Figs. 1 and 2, when a direct voltage of sufficient amplitude or a relatively low frequency alternating voltage, such as one at 60 hertz, of appropriate amplitude, is applied between a segment such as 18 and the backplate 10, turbulence is created in the liquid crystal 22 in the region thereof between the segment 18 and the backplate 10 and light scattering occurs at this region. When the voltage is removed the liquid crystal relaxes to its original condition. The time required for this relaxation to take place is of the order of tens to hundreds of milliseconds. This relaxation (erasure) of the scattering effect can be speeded up by applying a relatively high-frequency A.C. voltage to the crystal, say between 2 to 20 kilohertz.

In the circuit of Fig. 7, the inverter amplifier 70 includes two NPN-transistors 72 and 74 and a PNP-transistor 76. Similarly, the inverter amplifier 78 includes two NPN-transistors 80 and 82 and a PNP-transistor 84. Transistor 72 is connected at its emitter to ground and at its collector to the base of transistor 76 via resistor 86. The emitter of transistor 76 is connected to a source of operating potential V and the collector of transistor 76 is connected to the collector of transistor 74. The emitter of transistor 74 is connected to ground. The amplifier 78 is similarly interconnected. The liquid crystal load 90 is connected between the collectors of transistors 74 and 80.

In the operation of the circuit of Fig. 7, an alternating voltage driving signal V_D is applied to input terminal 92 and a complementary driving signal $\overline{V_D}$ is applied to input terminal 94 via inverter 96. With switch 98 in the position shown, when V_D goes high, transistors 74 and 80 are forward biased. During this same period, $\overline{V_D}$ goes low turning off transistor 72 turning it off. Thus, transistor 76 and 84 are turned off so that no power is lost. When V_D goes low,

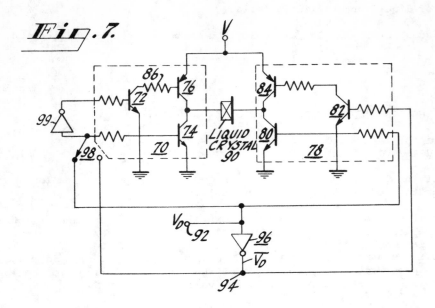

Fig. 7.

transistors 74 and 80 are cut off. During this period, $\overline{V}_D$ goes high, forward biasing transistors 72 and 82. Accordingly, a small amount of emitter to base current flows through the emitter-to-base diode of transistor 76, resistor 86 and the emitter to collector path of transistor 72 and a small amount of base current flows also through transistor 84 and the collector-to-emitter path of transistor 82 to ground, but these amounts of current are not very significant. Current does not flow through the emitter to collector paths of transistors 76 and 84, as transistors 74 and 80 are cut off.

When switch 98 *is* thrown to the other position, push-pull operation results. For example, when V_D is high and $\overline{V}_D$ is low, transistor 80 is forward biased and transistor 74 is cut off. Transistor 72 is forward biased and it turns on transistor 76. Accordingly, current flows through the emitter-to-collector path of transistor 76 and the liquid crystal element 90 and through the emitter to collector path of transistor 80. Thus, a voltage is applied across the liquid crystal in one direction for exciting the liquid crystal. During the next half cycle, transistors 74 and 84 are turned on permitting a voltage in the opposite direction to develop across the liquid crystal 90.

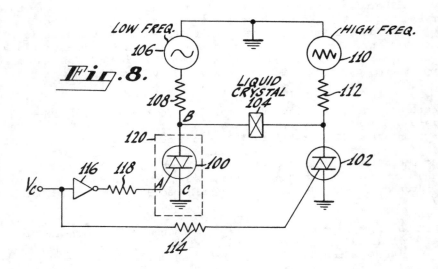

Fig. 8.

The embodiment of the invention shown in Fig. 8 employs unipolarity alternating voltage excitation just as the circuit already described. However, as contrasted to the previous circuits, The Fig. 8 circuit employs fast alternating voltage turn-off rather than permitting the liquid crystal naturally to relax to its erased condition. The circuit of Fig. 8 includes two triacs 100 and 102 connected at their first anode to ground and at their 2nd anode to liquid crystal element 104. A source 106 of low frequency alternating current is connected to triac 100 through resistor 108. A source 110 of high frequency alternating voltage is connected to triac 102 via resistor 112. A direct voltage V_C is applied via resistor 114 to the control or gate electrode of triac 102 and its complement is applied via inverter 116 and resistor 118 to the control electrode of triac 100. The low frequency may be some frequency of say up to 100 hertz or so and the high frequency may be in the tens of kilohertz range - a 20 kHz signal being suitable for example. In both cases sine- or square-waves are suitable.

When the voltage V_C is of a high value to turn-on triac 102, the inverter 116 applies a low signal to the control electrode of triac 100 to turn-off this triac. In this case, the low frequency alternating voltage produced by source 106 passes through resistor 108 and the liquid crystal element 104 and the triac 102. This voltage is of an amplitude sufficient to cause dynamic scattering to occur. The high frequency erase alternating voltage provided by source 110 passes through resistor 112 and triac 102 to ground and has essentially no effect on the liquid crystal 104.

When it is desired to erase the liquid crystal, the value of the voltage V_C is changed to a low value such that triac 102 is turned off and triac 100 is turned on. Now the high frequency turn-off signal from source 110 passes through resistor 112 and through the liquid crystal 104 to ground via the triac 100. This causes the liquid crystal quickly to be erased. The low frequency voltage from source 106 passes through resistor 108 and triac 100 to ground and has essentially no effect on the liquid crystal 104.

It is desirable in driving displays such as the one shown in Figs. 1 and 2 that the amplifying elements be integrated in one package. At the present state of the technology, triacs are not integratable. Accordingly, in these applications it is advantageous to substitute for the triacs 100 and 102 semiconductor devices which can be integrated. Fig. 9 shows a suitable circuit, that is, the circuit of Fig. 9 may be substituted for the triac within dashed block 120 and a similar circuit such as shown in Fig. 9 may be substituted for triac 102.

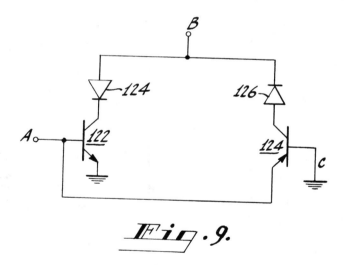

Fig. 9.

The circuit of Fig. 9 includes complementary symmetry transistors, that is, NPN-transistor 122 and PNP-transistor 124. Transistor 122 is connected through diode 124 to terminal B and transistor 124 is connected through diode 126 to terminal B. The diodes 124 and 126 are poled in the forward direction with respect to the collector-to-emitter paths of their respective transistors. The diodes 124 and 126 are used to circumvent the inherently low base-emitter avalanche breakdown, that is, about 6 volts, of transistors 122 and 124.

In the operation of the circuit of Fig. 9, when the signal applied to terminal A is high, transistor 122 conducts when the voltage at B goes relatively positive and transistor 124 conducts when the voltage at B goes relatively negative. The circuit, in other words, when so biased, operates as a bidirectional circuit. On the other hand, when A is made low so that zero base current flows, both transistors 122 and 124 are cut off.

It should be clear from the explanation above that the Fig. 9 circuit may be substituted into the Fig. 8 circuit and operation generally similar to that described for Fig. 8 obtained.

The circuit in Fig. 8 cannot be used with liquid crystal displays with a common backplate. In other words, with an arrangement such as shown in Fig. 8, rather than having a continuous backplate conductor such as 10 in Fig. 2, there would have to be discrete backplate electrodes corresponding in shape to the discrete frontplate electrodes shown, and each such pair of electrodes would have to be driven by a circuit such as shown in Fig. 8.

"""""""""

3,654,606 (U.S.)   Patented Apr. 4, 1972

ALTERNATING VOLTAGE EXCITATION OF LIQUID CRYSTAL DISPLAY MATRIX.

Frank J. Marlowe and Edward O. Nester, assignors to RCA Corporation.
Application Nov. 6, 1969.

The object of the present invention is to provide an improved means for operating a liquid crystal display matrix providing the advantages of alternating voltage excitation, but without appreciably increasing the display matrix complexity or cost.

The matrices of the present invention include a display element at each location and means for selectively exciting said elements during successive frames. The last named means includes means providing paths for charging the elements in one sense during alternate frames and means providing paths for charging said elements in the opposite sense during the remaining frames.

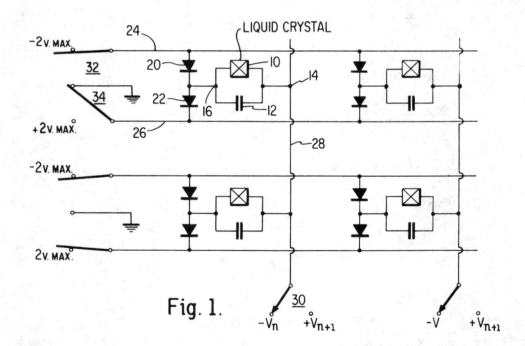

Fig. 1.

The matrix of Figure 1 includes, at leach location, a liquid crystal element 10 and a capacitor 12 in shunt with the liquid crystal element. The parallel connected circuit 10, 12 is connected at once terminal 14 to a column lead 28 and its other terminal 16 to the anode-to-cathode connection of two diodes 20 and 22. Diode 20 is connected at its anode to one row lead 24 and diode 22 is connected at its cathode to second row lead 26.

A column lead, such as 28, is placed at a voltage of $-v_n$ or $+v_{n+1}$ where $-v_n$ is the video level to which the element is to be charged during one frame and $+v_{n+1}$ is the video level to which that same element is to be charged during the next frame. The actual value of the voltage $-v_n$ may vary or not from column to column, depending on the brightness level to be represented at the respective column row intersections and is in the range from 0 to some maximum value $-v_{max}$ which may be -30 to -40 volts or so. Similarly, $+v_{n+1}$ is in the range from zero to +30 to +40 volts or so.

The means for connecting column 28 to the sources providing the respective voltages is shown as a mechanical switch 30, however, it is to be understood that, in practice, electronic means such as transistors or the like, are employed instead.

The upper row lead of each pair of row leads may be connected either to a $-2v_{max}$ source or to ground, that is, to 0 volts. The lower row lead of each pair may be connected either to 0 volts or to $+2v_{max}$. As in the case of the column conductors, the means for connecting the upper and lower rows are shown as mechanical switches 32 and 34, however, in practice, they are electronic switches such as transistors or the like.

Figures 2a to 2f are drawings to explain how the circuit of Fig. 1 operates (not shown).

While there are several additional forms of the invention which are possible which operate on the same principle as the circuit of Fig. 1, for purposes of the present application, only one other such circuit is shown (Fig.3) and is discussed.

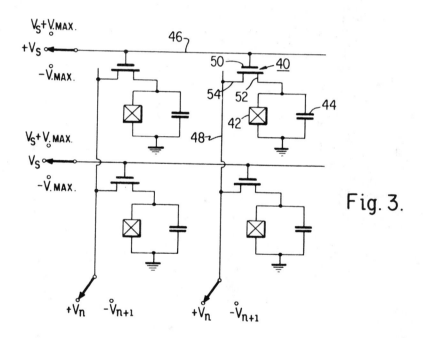

Fig. 3.

The circuit includes, at each matrix location, an N-type, field-effect transistor (FET) 40, such as one of the metal oxide semiconductor (MOS) type, and a shunt circuit connected between one end of the conduction path of the transistor and ground, the FET is of the enhancement type and does not conduct between source and drain unless the gate voltage is $+v_n$ relative to the source. The shunt connected circuit is the same as the one of Fig. 1 and consists of a liquid crystal element 42 and capacitor 44. In the matrix of Fig. 3 there is one row lead, such as 46, and one column lead, such as 48, per matrix location. Each row lead may be connected to one of three different voltages $-v_{max}$, $+v_s$ or $v_s + v_{max}$. Each column lead may be connected to a video voltage, $+v_n$ or $-v_{n+1}$, where v_n and $-v_{n+1}$ are the video voltage levels during the n'th and n+1th rows, respectively.

For the field-effect transistor matrix shown in Fig. 3, the system requirements are slightly more complicated than those for direct voltage driven matrices, as negative and positive column drivers are needed. Nevertheless, the number of elements in the matrix remains unchanged. As a matter of fact, the excitation method of the present invention can be substituted for any comparable direct voltage excitation circuit with reset capability with no increase in matrix complexity. This is an important advantage. The only change needed in the matrix is that the isolation elements such as the diodes of Fig. 1 or the field-effect transistors of Fig. 3 must withstand higher off state voltages for alternating voltage excitation than for direct voltage excitation.

""""""""""""

3,655,269 (U.S.)   Patented Apr. 11, 1972

LIQUID CRYSTAL DISPLAY ASSEMBLY HAVING INDEPENDENT CONTRAST AND SPEED OF RESPONSE CONTROLS.

<u>George H. Heilmeier</u>, assignor to RCA Corporation.
Application Jan. 25, 1971.

According to the teachings of the present invention, control of the speed of response, i.e., over the rise time of the exhibited electro-optical effect, can be had by varying the amplitude of the applied energizing signal, while control of the contrast can be independently had by varying the frequency of that signal.

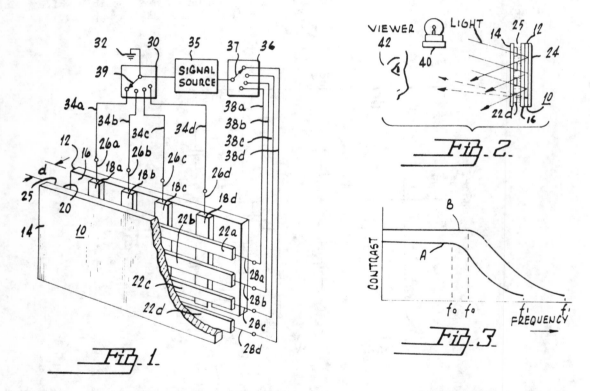

In Fig. 1, the liquid crystal display assembly includes a crossed grid optical display device 10 having parallel back and front transparent glass support plates 12 and 14 of between 5 to 30 micron separation. Interspersed between 12 and 14 is a thin layer of liquid crystal material 25, anisylidene-p-aminophenylacetate (APAPA) for example. In this display assembly, the back plate 12 is made reflective, in one manner by coating its outer face 24 with a material such as an evaporated film of nickel or aluminum. The assembly of Fig. 1 additionally shows apparatus for operating the display device 10. A back strip commutator switch 30 is included, having its common contact connected to one side of a signal source of variable amplitude and frequency 35 and to ground 32. A plurality of commutator contacts of the switch 30 are connected to the back strip connections 26a-26d through leads 34a-34d. Also, a front strip commutator switch 36 is provided, having a plurality of contacts connected to the front strip connections 28a-28d via leads 38a-38d and a common contact connected to the signal source 35.

As is known, a plurality of predetermined regions can be brightened by sequentially energizing more than one electrode strip from each back and front set. The percent of light scattered, and hence the degree of brightening for each element, can be modulated by controlling the amplitude of the applied signal voltage above the threshold, thereby affording a gray scale.

However, it has been found that when modulating this amplitude so as to provide a moderate-to-low amount of brightening (i.e., less than maximum contrast), the speed of response of the excited electro-optical effect may well be too slow for most video applications. It has also been found, though, that for a given amplitude signal above the threshold, the contrast will vary as a function of frequency in the low audio range. That is, given for example that amplitude signal which will produce maximum brightening, the contrast can be changed to afford a gray scale by varying the signal frequency. Since the speed of response remains essentially dependent only upon the amplitude of the energizing signal, an independent control of the exhibited contrast can be had.

It has been observed that the contrast varies as a function of signal frequency in the manner illustrated by the curves of Fig. 3. Curve A represents the contrast variation for a signal amplitude of 40 volts rms while Curve B represents the variation with a 60 volt rms signal. The frequency f_0 at which the contrast begins to fall off as a function of frequency is dependent upon the amplitude of the energizing signal, by the thickness of the liquid crystal cell and by the inverse of the crystal material resistivity. With the aforementioned APAPA mixture of 6.4×10^8 ohm-cm resistivity at 25°C for example, f_0 was measured for a 1/2 mil thick cell to be approximately 700 hertz at the 40 volt energizing level (Curve A) and approximately 1 kHz at the 60 volt rms level (Curve B). The frequency f_1 at which the contrast approached a minimum for the applied signal voltage was measured to be approximately 2 kHz for the 40 volt signal and 2.5 kHz for the 60 volt case.

These observations are employed in the liquid crystal display assembly of Fig. 1 by utilizing for the signal source 35, one of controllable frequency as well as one of variable amplitude. Selecting a signal amplitude of a value such that an optimum speed of electro-optical response will result will then permit independent contrast adjustment by proper frequency selection to provide an adequate gray scale for video applications. Such a signal source may comprise a frequency modulated oscillator, for example, wherein applied scanning signals may be employed to vary bias voltages so as to control the frequency of oscillation thereof.

In the reflective mode of operation illustrated in Fig. 2, a light source 40 and a viewer 42 are both positioned on the transparent plate side of the display assembly. Alternatively, operation can be achieved by modulation of light transmission or light absorption.

""" """ """ """ """ """

3,655,270 (U.S.)   Patented Apr. 11, 1972

ELECTRO-OPTICAL DISPLAY DEVICES USING NEMATIC MIXTURES WITH VERY WIDE TEMPERATURE RANGES.

Linda T. Creagh.
Application June 1, 1971.

It is an object of the present invention to provide a display device utilizing a nematic mesophase composition having a broad temperature range.

Ternary nematic mesomorphic compositions which may be utilized with the present invention include compositions consisting essentially of 4-ethoxybenzylidene-4'-n-butylaniline, 4-methoxybenzylidene-4'-aminophenyl butyrate and bis-(4'-n-octyloxybenzal)-2-chloro-1,4-phenylenediamine. These compositions are nematogenic, that is, in the nematic mesophase, through a broad range of temperatures, including room temperature. Specific ternary compositions were prepared by placing the appropriate weights of each of the materials in a sealed vial. The vial was warmed until the materials were isotropic, and was then placed in an ultrasonic bath at that temperature for about 2 to 3 hours. The vial was then allowed to cool slowly to room temperature in the ultrasonic bath. Exemplary compositions include the following, wherein the materials were purified, by conventional techniques, such as recrystallization and distillation, to a level such that the total amount of impurities comprised from 1/2 to 2 percent by weight of mixture; DC voltages were utilized to effect dynamic scattering.

Example I

A composition was prepared having 65 percent by weight of 4-ethoxybenzylidene-4'-n-butyl-aniline, 15 percent by weight of 4-methoxy benzylidene-4'-aminophenyl butyrate and 20 percent by weight of bis-(4'-n-octyloxybenzal)-2-chloro-1,4-phenylendiamine. The composition prepared as above described exhibited a crystal to mesomorphic transition temperature of -6°C and a mesomorphic to isotropic liquid transition temperature of 105°C. The composition was placed in an electro-optical display cell and the response characteristics were tested at 25°C. With 40 volts applied across the display cell the composition exhibited a rise time of 20 milliseconds and a decay time of 20 milliseconds. With 60 volts applied across the cell, the composition exhibited a rise time of 5 milliseconds and a decay time of 5 milliseconds. The threshold voltage for producing dynamic scattering was 6.5 volts. The maximum contrast ratio of the composition was 20:1.

Eleven examples are given in the patent. A summary of the compositions produced and the temperature ranges in which the nematic mesophase was exhibited is depicted in Table I.
Material A represents 4-ethoxybenzylidene-4'-n-butylaniline;
Material B represents 4-methoxybenzylidene-4'-aminophenylbutyrate;
Material C represents bis-(4'-n-octyloxybenzal)-2-chloro-1,4-phenylendiamine.

Example No.	Material, weight %			Crystal to mesophase temp.	Mesomorphic to isotropic temp.
	A	B	C		
1	65%	15%	20%	-6°C	105°C
2	50%	35%	15%	-9°C	105°C
3	50%	30%	20%	-8°C	110°C
4	50%	25%	25%	8°C	112°C
5	45%	35%	20%	-10°C	110°C
6	45%	30%	25%	10°C	112°C
7	40%	40%	20%	-12°C	110°C
8	40%	35%	25%	-15°C	115°C
9	40%	30%	30%	13°C	120°C
10	35%	40%	25%	0°C	113°C
11	31%	31%	38%	14°C	118°C

The response characteristics of the compositions tested in a 1-mil electro-optical display cell at 25°C are summarized in Table II below...

Example No.	Material, weight %			Applied Voltage	Rise Time	Decay Time	Threshold Voltage	Contrast Ratio, Max.
	A	B	C					
8	40%	35%	25%	60	5 msec	5 msec	7 volts	30:1
				40	7	10		
				26	10	10		
1	65	15	20	60	5	5	6.5	20:1
				40	20	20		
3	50	30	20	60	5	75		
				40	5	60		
				20	20	70		
11	31	31	38	60	18	70		
				40	20	70		
				20	30	100		
10	35	45	25	80	3	80	6.5	40:1
				60	4	70		
				40	7	80		
				20	30	90		
9	40	30	30	80	6	80		
				60	6	75		
				40	10	70		
				30	30	70		
				20	35	85		

It is to be appreciated that by varying the relative ratios of the specific materials, the upper and lower limits of the temperature range at which the composition is in the mesomorphic state may be varied while maintaining a broad overall mesomorphic temperature range.

" " " " " " " " " " "

3,655,971 (U.S.)   Patented Apr. 11, 1972

IMAGING SYSTEM.

Werner E.L. Haas, James E. Adams, James H. Becker and Joseph J. Wysocki, assignors to Xerox Corp.
Application Aug. 12, 1969.

A system wherein a film of liquid crystalline material is exposed to ultraviolet radiation thereby producing a visible image.

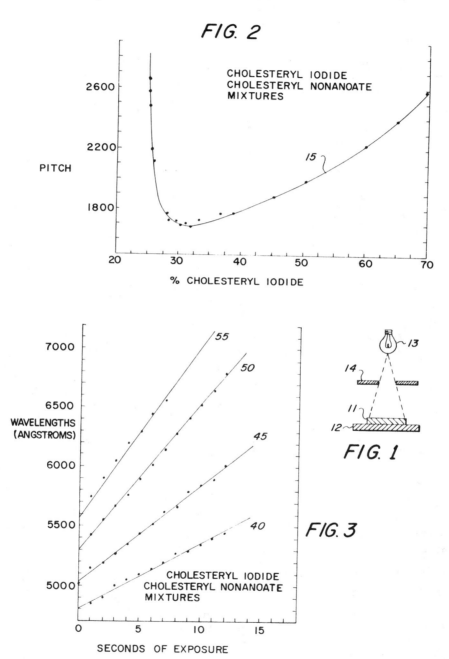

Figure 1 illustrates an exposure system wherein various samples of liquid crystals may be exposed to ultraviolet radiation by providing a film of the liquid crystalline material 11 upon a suitable substrate 12 and then exposing the material to ultraviolet radiation from a source 13

which may be any suitable source of ultraviolet radiation, for example a high pressure, short arc, mercury lamp.

A particularly preferred mixture of cholesteric liquid crystalline materials suitable for use in the present invention comprises a mixture a cholesteryl iodide and cholesteryl nonanoate. It has been shown that in the invention system the exposure of a liquid crystalline film containing cholesteryl iodide results in a pitch shift and an accompanying release of iodine. It is believed that the color or pitch shift in this material is in part due to a change of concentration of the cholesteryl iodide in the liquid crystalline mixture. It is believed that the exposure to ultraviolet light converts the cholesteryl iodide into one or more new compounds; however, the chemical composition of such compounds as well as the roles they may play in the inventive system are not clearly known.

In the process illustrated in Fig. 1, the liquid crystalline film 11 is shown being exposed through a mask 14. This mask may be in any desired image configuration, and it may be differentially transmissive to the radiation, for example like the photographic transparency.

In the inventive system the selective exposure of liquid crystalline materials to ultraviolet radiation surprisingly changes the color of the liquid crystalline material continuously over the entire spectrum in response to continuously varied total exposures. This exposure generally permanently changes the color of the exposed liquid crystalline material. However, diffusion or mechanical shearing or other mixing events can result in erasure rendering the film reusable.

Where liquid crystalline films suitable for use in the present invention are prepared by dragging a leveling slide across the surface of the sample, the film is mechanically aligned or disturbed and shows beautiful reflected colors. Where cholesteric liquid crystalline films are used and aligned in this manner, it is believed that the axes of the helicoidal region, which are characteristic of the cholesteric mesophase, are aligned approximately perpendicular to the substrate upon which the imaging film is coated. Where the liquid crystal is dissolved in a suitable solvent, applied to a suitable substrate and the solvent allowed to evaporate off without leveling alignment, the film is an unaligned or undisturbed film.

Exposure of an unaligned film through a suitable mask generally does not result in an immediately visible image. However, a latent UV exposure image is produced on the film, and that latent image can be developed by heating the film or by exposing the film to an organic solvent vapor, such as chloroform, methyl ethyl ketone, petroleum ether or others. Where mechanically aligned films are used, the image is generally instantaneously visible and no development step is required.

When observed at a given angle, a given liquid crystalline material exhibits a characteristic color which is associated with its pitch. Similarly, a specific mixture of liquid crystalline materials has a characteristic color for given observationconditions. Fig. 2 shows the dependence of the characteristic color of a mixture of liquid crystals, here, for example, mixtures of cholesteryl iodide and cholesteryl nonanoate, upon the composition of various mixtures having the same two components. In Fig. 2 the abscissa shows the composition of the liquid crystalline mixture, and the ordinate is a measure of the pitch of the liquid crystalline sample. The color of the liquid crystalline material is dependent upon this pitch.

In Figure 3 date showing the change in color corresponding to the quantity of exposure to ultraviolet radiation are shown for a number of various liquid crystalline mixtures. The mixtures used in the experiments from which the data of Fig. 3 were collected are mixtures of the cholesteryl iodide-cholesteryl nonanoate system whose characteristics are also described in the data of Fig. 2. In Fig. 3 the abscissa is expressed in seconds of total exposure, which is a measure of the period of time of which the liquid crystalline sample was exposed to ultraviolet radiation. It will be appreciated that the total quantity of ultraviolet exposure acting on the liquid crystalline materials is directly proprtional to the time of exposure provided the magnitude of the exposure, the distance from the source, and other factors remain substantially constant. The ordinate in Fig. 3 shows the color of the liquid crystalline material having been exposed to ultraviolet radiation, corresponding to the total time of exposure; the colors are expressed in Angstrom units.

It is clear that the advantageous system of the present invention is useful as an imaging system wherein color images are produced by selectively exposing a liquid crystalline film with various intensities of ultraviolet radiation.

It is a particularly advantageous feature of the present invention that the exposed liquid crystalline film, which may be thought of as a new photographic plate, is directly and instantaneously imaged upon exposure without any development step which is common in other photographic processes.

Example I.

A liquid crystalline film comprising a mixture of about 50% cholesteryl iodide and about 50% cholesteryl nonanoate is prepared by heating the liquid crystalline materials in a crucible to a temperature above the isotropic transition temperature of the mixture, mixing the heated materials thoroughly, and depositing the heated mixture onto a glass slide. Samples of approx. uniform thickness of about 2 mils are prepared in this manner by cutting a hole of the desired area in a Mylar spacer about 2 mils in thickness, placing the spacer with the heated liquid crystal mixture and dragging a glass slide across the spacer and mixture thereby filling the hole in the spacer and removing excess mixture while shearing and smoothing the sample. The shearing action of the leveling slide aligns the molecules in the sample. The pitch of the sample is determined using the observation apparatus described below.

The plate having the film of liquid crystalline material thereon is then exposed to ultraviolet radiation less than about 3,000 A in wavelengths, produced from a General Electric BH-6 high-pressure, short arc mercury lamp. Exposure time is controlled by a shutter. The film material is exposed while in its cholesteric liquid crystalline mesophase. The film material is in its aligned state and shows various reflected colors. During ultraviolet exposure, the color of the film material shifts toward the red, as observed in the fixed observation system. Using this exposure system, about 150 millijoules/cm^2 of ultraviolet radiation with wavelengths below about 3,000 A shifts the color peak about 1,000 A.

Samples prepared as shown above are exposed in incrementally increasing amounts of ultraviolet radiation, and incremental changes in colors of the samples changing from blue to red are observed with increasing total exposure.

The ultraviolet exposed liquid crystalline film is observed with a fixed observation apparatus having a source of monochromatic incident light which transmits the incident light beam through a collimating lens system and then to the exposed liquid crystal film which is displayed on a dark background on a spectrometer stage. Since the color of light scattered from a liquid crystal film is a function of the angle of incidence and the angle of observation, the light source and exposed sample are oriented so that the angle of incidence is fixed at about 60° from normal to the plane of the sample. Similarly, the angle of observed, reflected light is fixed at about 30° from normal to the plane of the sample where the reflected light is detected by a photodiode detector in conjuction with an oscilloscope read-out system. A mechanical chopper located in the line of the incident beam provides a reference signal for use with a lock-in amplifier.

Shifts in bright colors of the liquid crystalline material are observed with the eye. Data illustrating the color shifts corresponding to amounts of ultraviolet radiation are shown on the "50" line in Fig. 3. This data shows that the effect of UV radiation is to shift the color of the liquid crystalline sample toward the red.

Example V.

A liquid crystalline film comprising a mixture of about 20% cholesteryl iodode and about 80% cholesteryl nonanoate is prepared by dissolving the liquid crystals in petroleum ether in a concentration of about 1g/10cc and coating the solvent mixture on a polished copper plate, about 3 cm x 4 cm in size. The ether is then allowed to evaporate. The film has a thickness of about 5 microns and is an unaligned liquid crystalline film. The imaging plate is exposed to ultraviolet radiation through a steel mask using a xenon flashtube, Novatron-186 available from Xenon Corporation in the exposure process of Example I. The liquid crystalline material in its originally unaligned state is almost colorless. No image pattern is immediately visible. The exposed film is slowly heated to about 40°C and the image becomes clearly visible.

A total of twelve examples are described in the patent specification.

" " " " " " " " " "

ADDITIVE FOR LIQUID CRYSTAL MATERIAL.

Ivan Haller and *Harold A. Huggins*, assignors to International Business Machines Corporation. Application Dec. 9, 1970.

It is often desirable for device applications and in the measuring of certain physical properties to provide a nematic layer in a homeotropic texture, i.e., large areas in which the optic axis is oriented perpendicularly to the boundary. Heretofore, homeatropic texture layers have been prepared by pretreating the surfaces which subsequently enclose the nematic layer with a strongly oxidizing acid such as nitric acid or chromesulfuric acid. The technique has been found to be undesirable in many situations because it cannot be used with oxidation-sensitive (transparent conductive coated) surfaces. Particular care is required to avoid the touching of a surface so treated, and homeotropic nematic layers produced by this technique tend to be unstable, i.e., their homeotropic property tends to decrease in an undesirable short period.

In accordance with the invention, a nematic liquid crystal material which assumes a homeotropic texture is provided by dissolving an additive material having the formula

$$RR_3'N^+X^-$$

wherein R is an alkyl radical having 10 to 24 carbon atoms, R' is a methyl or ethyl radical, and X- is an anion derived from a simple acid, in a nematic material. A typical example of the additive is hexadecyltrimethylammonium bromide. The homeotropc texture provided by this composition, when placed between two conducting transparent plates, is readily deformed by a voltage applied to the plates and can thereby be utilized to control the transmission of light.

In the drawings, Fig. 1 is a schematic depiction of an improved nematic cell incorporating the material according to the invention.

Fig. 2 is a diagram which illustrates changes in polarization of light passing through the cell for two states of the cell.

Fig. 3 is a modification of the cell of Fig. 1 for use with reflected light, and

Fig. 4 illustrates the changes in polarization of light for the two states of the cell of Fig. 3.

There follows a detailed description of the making and testing of the inventive homeotropic nematic material...

An 0.49 percent solution of hexadecyltrimethyl ammonium bromide in p-methoxybenzylidene-p-butylaniline was prepared by mixing 0.0024 and 0.495 grams, respectively, of the materials in a vial, heating to a temperature where the nematic liquid first turned isotropic (approximately 46°C) and shaking until the last specks of the solid mass dissolved. A drop os the solution at room temperature was placed between untreated glass plates whose separation was effected by a 2-mil thick teflon gasket. The sample was then examined between crossed polarizers on a polarizing microscope over its entire nematic temperature range and extinction was observed independently of the direction of the sample relative to the polarization direction. Upon the shearing of the liquid crystal by selective motion of the two plates, light was observed to be transmitted. These observations indicated that the optic axis of the nematic fluid, at rest, is perpendicular to the glass plates, i.e., because of the additive, it assumed a homeotropic texture.

In a second experiment, the same procedure as described above was followed with the exception that the glass slides which were employed were first repeatedly rubbed on a cheese cloth in a given direction. Such rubbing treatment is known to normally cause pure p-methoxybenzylidene-p-butylaniline, which is subsequently enclosed, to assume a texture in which the optic axis lies in the plane of the glass and in the rubbing direction. The inventive solution, as described hereinabove, however, was again observed to assume the homeotropic texture.

In a third experiment, the same solution was enclosed between electrically conducting tin-oxide-coated glass plates wherein it again exhibited a homeotropic texture. A DC voltage was

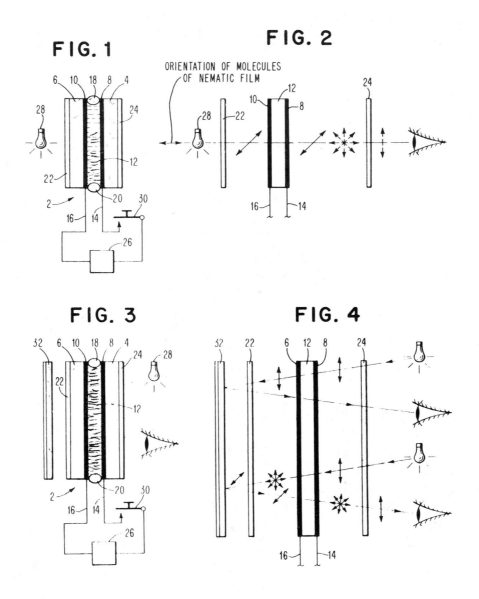

then applied to the plates, while the appearance of the sample was monitored on the polarizing microscope and the intensity of the transmitted light (between crossed polarizers) was measured with a CdS photocell replacing one of the eyepieces of the binocular microscope. With an applied voltage of 0 to 5 volts, the structure appeared uniformly dark. At 7.5 volts however, the liquid crystal material became transparent and, under magnification, it appeared stationary.

The contrast ratio, i.e., transmitted light intensity at 7.5 volts relative to that at zero volts was found to be 16. With applied voltages of 10 volts or higher, flow patterns and turbulence appeared similar to those patterns and turbulence observed in nematic materials of any texture. The maximum contrast which was achieved was 31 at an applied voltage of 24 volts. In a comparison, the maximum contrast ratio measured under similar conditions, of elements containing nematic liquid crystals with no additive, and aligned in the plane of the boundary surfaces in the polarization direction was found to be only 11 to 17.

The horizontal lines in nematic material 12 are intended to indicate that the nematic material being used in devices (Figs 1 & 3) is the material prepared according to the invention described above. The material 12 has a threshold electric field which, when exceeded, causes the deformation of the homeotropic texture and thereby the depolarization of polarized light passing therethrough. When an electric field is applied across the nematic crystal material which exceeds that threshold, depolarization of polarized light passing therethrough takes place. A total of four examples are described in the patent.

"""""""""""

3,661,444 (U.S.)   Patented May 9, 1972

COMPOUNDED LIQUID CRYSTAL CELLS.

<u>Dennis L. Matthies</u>, assignor to RCA Corporation.
Application Dec. 2, 1969.

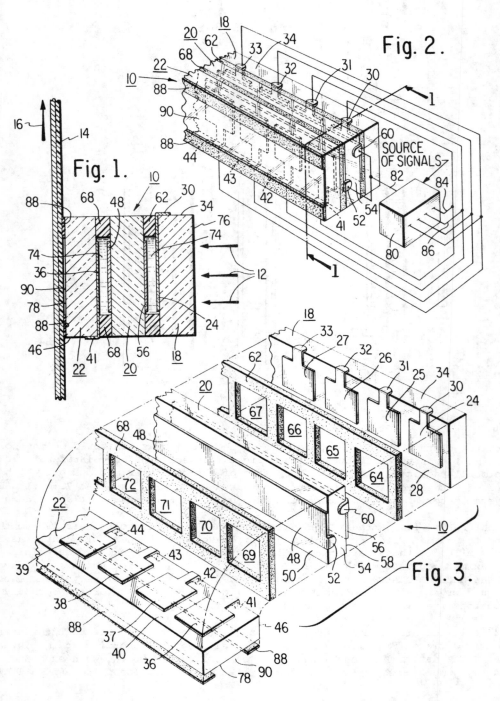

Compounded liquid crystal cells comprise a set of at least two cells, one cell being disposed behind the other, that is, in tandem alignment, so that a ray of light directed through one of the cells in the set must pass through the other cell of the set. Novel arrays of compounded liquid crystal cells may comprise first and second pluralities of liquid crystal cells. Each plurality of cells has a similar alignment and is parallel to the other plurality of cells so that a ray of light directed through one of the cells in the first plurality must also pass through a cell in the second plurality.

Figure 1 is a vertical cross-section of an improved array of compounded liquid crystal cells, taken along the plane indicated by the dashed line 1-1 in Fig. 2, showing the array disposed between a light source and a fragment of a light-sensitive recording element;
Figure 2 is a fragmentary perspective view of the novel array of compounded liquid crystal cells, viewed from the front, top and right side, showing schematically a source of signals electrically connected to sets of compounded liquid crystal cells; and
Figure 3 is a fragmentary exploded perspective view of the novel array of compounded liquid crystal cells shown in Fig. 2.

Referring now particularly to Fig. 2, there is shown schematically a circuit for operating the novel array 10 of compounded liquid crystal cells. A source 80 of signals, such as from the output of a computer or a facsimile system, foe example, is connected to the array 10 to energize the liquid crystal cells in each set of compounded liquid crystal cells simultaneously. Thus, to operate one set of compounded liquid crystal cells, one output terminal 82 of the signal source 80 is connected in parallel with the front and back common electrodes 48 and 56, via terminal leads 52 and 60, respectively; and another output terminal 84 of the signal source 80 is connected in parallel to the leads 41 and 30 of the electrodes 36 and 24 on the plates 22 and 18, respectively. To operate the next set of liquid crystal cells, another output terminal 86 of the signal source 80 is connected in parallel with the leads 42 and 31 of the electrodes 37 and 25 on the plates 22 and 18, respectively. In a similar manner the remaining sets of compounded liquid crystal cells are connected so that both of the liquid crystal cells in each set can be either energized and deenergized simultaneously.

Masking means are provided to absorb light that does not pass directly between oppositely disposed electrodes in a set of compounded liquid crystal cells. To this end, a mask 88 is disposed on the front surface 78 of the front plate 22. The mask 88 comprises a pair os spaced apart parallel strips of a light-absorbing material, such as black paint, foe example, separated by a slit 90 through which light rays directed through the array 10 pass.

In operation, the array 10 is illuminated by the light source 12 so that light rays are directed, e.g., perpendicularly, to the rear surface 76 (Fig. 1) of the back plate 18. When both of the compounded liquid crystal cells in a set are in their transparent state, light from the light source 12 passes directly through them, as shown in Fig. 1, and strikes the light sensitive recording element 14 which is in substantial contact with the front surface 78 of the front plate 22. When both of the compounded liquid crystal cells in a set are in a light-scattering state, as when each pair of oppositely disposed electrodes in each of the compounded liquid crystal cells is energized from the signal source 80, light from the light source 12 is scattered, and the light output is diminished to a point where the recording element 14 is not exposed.

By providing the novel array 10 with a plurality of sets of compounded liquid crystal cells, a direct ray of light passing through any set is not only scattered by a rear liquid crystal cell of the set, but is also scattered by a front (compounding) liquid crystal cell in the set. Hence, the contrast efficiency between transmitted light and scattered light through the array 10 is greatly improved without decreasing the writing speed and without increasing the voltage to energize a liquid crystal cell.

" " " " " " " " " " " "

1,273,779 (British)

Patented May 10, 1972

ELECTRO-OPTIC LIGHT MODULATOR.

RCA Corporation (Inventor's name not given)
Application May 28, 1969; prior U.S. application Sept. 3, 1968.

For description of this patent see U.S. Pat. No. 3,597,044.

3,663,086 (U.S.)   Patented May 16, 1972

OPTICAL INFORMATION STORING SYSTEM.

George Assouline, Eugene Leiba, and Erich Spitz, assignors to Thomson-CSF. Application June 23, 1970; prior application in France July 2, 1969.

The invention relates to an optical data storage system.

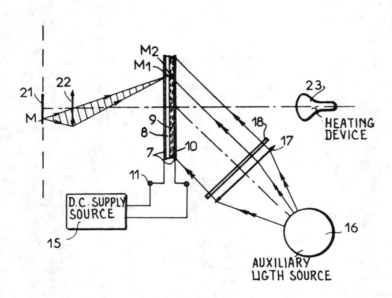

A data store has a first layer formed from a mixture of nematic liquid crystal and of a dichroic substance and a second, adjacent, layer of photoconductive material, the two layers being held between transparent electrodes. The first layer has one or more absorption lines if its molecules are aligned to the direction of polarization of incident light and which disappear if the molecules are perpendicular to the light polarization.

Light polarized parallel to the alignment of the molecules and to the electrode plane is incident on one of the electrodes and has a frequency corresponding to an absorption line. Data to be stored is projected on to the other electrode. A voltage is supplied to the electrodes such that where the photoconductive material is rendered conductive by the light the voltage across the first layer re-aligns the molecules so that light from source 10 is no longer absorbed. This light then maintains the photoconductive material conductive when the projected data is removed. A heater may be provided to maintain the layers at a desired temperature. The store may hold digital data or may store the display from a cathode-ray tube.

The first layer may comprise p-n butoxybenzoic acid and methyl red. An insulating layer may be provided between the electrodes and first layer. Data is erased by removing the voltage or extinguishing source 16.

3,663,390 (U.S.) *Patented May 16, 1973*

METHOD OF CHANGING COLOR PLAY RANGE OF LIQUID CRYSTAL MATERIALS.

<u>James L. Fergason</u> and <u>Newton N. Goldberg</u>, assignors to Western Electric Corporation. Application Sept. 24, 1970.

Increasing dosages of X-ray radiation progressively lower the color-play range of cholesteric-phase liquid-crystal materials. The effect is enhanced when an effective amount of an iodine-containing compound is used in the liquid-crystal material. Novel iodine-containing compounds are described, and articles are disclosed that give direct reading indication of the dosage of X-ray radiation that they have received, without need for a separate development operation.

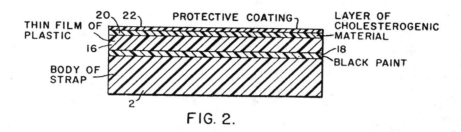

FIG. 2.

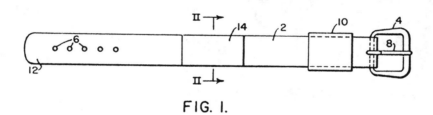

FIG. 1.

A dosimeter in accordance with the invention may comprise simply a short piece of plastic film, painted black on one side and coated on the other side with the film about 5-50 microns thick, made by applying a suitable cholesterogenic material solution to the film and permitting the solvent to evaporate therefrom, and then preferably applying another transparent protective film thereover to exclude contaminants and the like. The piece of plastic film is strapped to the wearer's wrist or taped to his forehead or otherwise placed in close contact with a part of his body, in order that variations in temperature to which the dosimeter is exposed may be minimized. Thus, observed color changes may be attributed to the quantity of X-ray radiation received by the dosimeter.

Accordingly, in accordance with one embodiment of the invention, reference being made to Figs. 1 and 2, there is provided an X-ray dosimeter comprising a strap 2 having at one end a buckle 4 and in its body near the other end a plurality of holes 6, into which the finger 8 of the buckle 4 may be fitted. There is also provided a sleeve 10 to receive the portion 12 of the strap after it has been fitted, foe example, to the wearer's wrist. There is provided, moreover, a region 14, that serves as the sensitive element of the dosimeter and contains cholesterogenic material.

With particular reference to Fig. 2, it will be seen that in the region 14, there is secured to the body of the strap 2, by adhesive or other suitable means (not shown), a thin film of plastic 16, preferably of polyethylene terephthalate or the like, upon which there has been provided on one side a coating 18 of black paint or the like (a black spray paint may be used, although carbon black or other black material is also suitable) and on the other side of which there is provided a very thin (5 to 50 microns) layer 20 of cholesterogenic material, preferably iodine-containing cholesterogenic material. Above the layer 20 is a layer 22 comprising a plastic protective coating to keep contaminants and the like from entering the layer 20. The layer 22 is, of course, suitably light-transmitting, being translucent or transparent. For the layer 22, transparent nail polish or the like may be used.

As an alternative, the cholesterogenic material is microencapsulated in transparent protective material.

A dosimeter of the kind indicated above is used by being strapped to the wrist of the wearer or otherwise placed into contact with a portion of the wearer's body (or otherwise being provided with an environment of substantially constant temperature), so that, when X-ray radiation becomes incident thereon, there will be a change in observed color that is indicative of the dosage of X-ray radiation incident upon the dosimeter.

It is to be understood, of course, that the extent to which a color change is observed will depend upon the nature of the cholesterogenic material used, the temperature environment of the dosimeter, and the dosage of X-ray radiation. Consider, for example, a cholesterogenic material that has a color-play temperature range, when unirradiated, of from 23 °C (red) to 31°C (blue). The material exhibits green at 26°C. If, for example, the color-play temperature range is shifted downwardly, using X-ray radiation of a given degree of intensity, by 1°C after 10 minutes of exposure, 3°C after 1 hour of exposure, and 10°C after 5 hours of exposure, it will in many instances be possible to say, if the temperature environment of the dosimeter is known, the dosage of X-ray radiation that it has received, simply from an observation of the color that it exhibits. For example, when a dosimeter containing a cholesterogenic material having the above-indicated response to X-ray radiation of given intensity is maintained at 20°C, the observed color is black until about one hour of exposure, at which time the observed color is red. With about 2-1/2 hours total exposure, the observed color is green.

Eight specific examples relating to the performance of specific cholesteric-phase liquid-crystal materials in the presence of X-ray radiation are described in the patent specifications.

""""""""""""

3,666,881 (U.S.)　　　　　　　　　　　　　　　　　　　　　　　　　　　　Patented May 30, 1972

ELECTRO-OPTICAL DISPLAY DEVICE EMPLOYING LIQUID CRYSTALS.

Richard J. Stein.
Application Dec. 28, 1970.

An electro-optical display system comprises a display screen including one or more juxtaposed sets of nematic liquid crystal and photoconductive layers bounded by transparent electrodes. The screen is scanned by scanning light while modulated bias voltages are applied across the liquid crystal and photoconductive layers. Images appearing on the screen are viewable by reflected ambient light to which the photoconductive layers are transparent. Colored dyes mixed with the nematic liquid crystal result in colored images on the screen. Cholesteric liquid crystal mixed with the nematic liquid crystal retains the images on the screen for predetermined periods of time. The images are viewable at either side of the screen. The screen can be internally illuminated.

The principles of the invention can be applied to other applications such as image conversion, amplification, color image storage, color densitometry, X-ray imaging, and so on.

Referring to Figure 1, there is shown a display system 50 useful for black and white TV or monochrome oscilloscope applications. The display screen is a single cell 10 having one photoconductive layer 25 and one liquid crystal layer 20 enclosed by spacer 22. A scanning light spot which may be polarized emitted by projector or scanner 52 is used to scan a raster onto the plane of the photoconductor. This may be a raster of 525 interlaced lines as conventionally used in current TV broadcast practice. The photoconductor 25 will be chosen so that the range of its photoresponse does not significantly overlay the desired range of light wavelengths which are to be viewed. This allows a reflective display which is not sensitive to ambient light. An optical filter 53 may be added to further restrict the possibility of the display being sensitive to ambient light.

3,666,948 (U.S.) *Patented May 30, 1972*

LIQUID CRYSTAL THERMAL IMAGING SYSTEM HAVING AN UNDISTURBED IMAGE ON A DISTURBED BACKGROUND.
Bela Mechlowitz, James E. Adams, Werner E. L. Haas, assignors to Xerox Corporation.
Application Jan. 6, 1971.

This invention relates to imaging systems, and more specifically, to an imaging system wherein the imaging material having cholesteric liquid crystalline characteristics. Furthermore, this invention more specifically relates to a novel system of thermally imaging such a liquid crystalline imaging member.

In Figure 1 an imaging member 10 suitable for use in the advantageous system of the present invention is illustrated wherein substrate 11 supports a layer of imaging composition comprising material which exhibits a cholesteric liquid crystalline mesophase.

FIG. 1

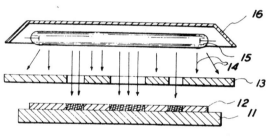

FIG. 2

An imaging member 10 which is prepared so that the composition layer comprising material in the cholesteric liquid crystalline mesophase is in its Grandjean or "disturbed" texture state is imaged in the system of the present invention by the imagewise application of thermal energy or energy which is capable of producing an imagewise thermal effect in the layer of imaging composition. One embodiment of the advantageous system of the present invention is illustrated, for example, in Fig. 2 wherein the imaging composition layer 12 is shown being imagewise exposed through a stencil-like mask 13 to thermal radiation 14 which is emitted by a source 15 under shield 16. Any source of thermal energy, for example such as lasers, gas discharge lamps, and others, may be used as the source of energy in the imagewise exposure step. Furthermore, in addition to the imagewise mask or stencil exposure system, any suitable means of providing an imagewise exposure may be used. Even light pencils or thermal styli are suitable for use in various embodiments.

In the inventive system the thermal energy which is applied in imagewise configuration to the layer of imaging composition is applied in sufficient amounts so that the imaging composition comprising material having a cholesteric liquid crystalline mesophase is, in the imagewise exposed areas, heated to a temperature at least about the liquid crystalline-liquid isotropic transition temperature of the imaging material having the cholesteric liquid crystalline mesophase characteristics. After the imagewise application of sufficient thermal energy to raise the temperature of the imaging composition in the imagewise exposed areas to a temperature at least about the isotropic transition temperature, the source of thermal energy is removed, and the imaging composition is allowed to cool into the cholesteric liquid crystalline mesophase temperature of the composition.

Upon cooling into the cholesteric liquid crystalline mesophase temperature range, the imagewise exposed areas of the imaging composition typically exhibit the focal-conic or "undisturbed" texture state, thereby providing an imaged member having image areas of the imaging composition comprising a material having cholesteric liquid crystalline mesophase in the focal-conic texture state with background areas of the imaging composition comprising material having a cholesteric liquid crystalline mesophase in the Grandjean type texture state.

In various embodiments of the imaging member of the present invention it may be advantageous to use substrate materials which are translucent or substantially transparent to visible light. In such embodiments, the imaging member may be suitable for use as an image transparency. The effect of thermally induced texture change in the imaged areas is to destroy polarization of transmitted light in the image areas whereas the background areas have little effect on polarization except in some small wavelength region.

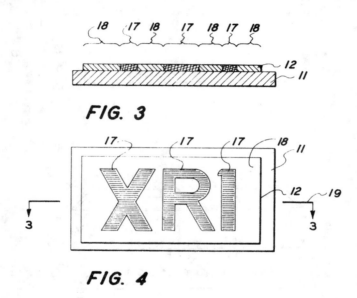

Figure 3 illustrates in partially schematic, cross-sectional view, an imaged member wherein the layer of imaging composition 12 exhibits imaged areas 17 in the focal-conic or "undisturbed" cholesteric liquid crystalline texture state, while the background areas 18 exhibit the Grandjean or "disturbed" cholesteric liquid crystalline texture state. The thermally induced texture transition imaging effect of the present invention is typically a bulk effect. The bulk effect is illustrated in Fig. 3 where imaged areas 17 of the layer of imaging composition 12 are schematically illustrated as being transformed throughout the entire cross-section of the composition.

Figure 4 is a top view of the imaging member of Fig. 3 (wherein the view of Fig. 3 is a cross-section along line 19) schematically showing imaged areas 17, in the focal-conic texture state on background areas 18, in the Grandjean texture state. Although the imaged areas 17 in Fig. 4 are here illustrated as being the darker areas, the contrast and density of an imaged member produced by the advantageous system of the present invention, may vary from one embodiment to another. Fig. 4 is intended to be a representation of an imaged member wherein the imaged areas 17 are clearly optically distinguishable from the background areas 18.

Example I

An imaging member is provided by dissolving a solute mixture comprising about 57 percent cholesteryl formate and about 43 percent cholesteryl nonanoate in petroleum ether solvent, mixing the solution thoroughly, and depositing a layer of the solution on a thick, stable black plastic substrate. The solvent is allowed to evaporate, leaving a layer of imaging composition comprising cholesteric liquid crystalline materials in a thickness of about 12 microns on the substrate. After solvent ecaporation, the film of imaging composition is uniformly provided in the Grandjean texture state by shearing the film by dragging the edge of a glass slide across the surface of the film. The film exhibits a green color. The imaging member is then imaged by the texture transition imaging system by placing the member, imaging composition side up, spaced a few microns under an imagewise metal mask, and placing the masked member about 25 cm from a BH6-1 mercury arc lamp. The lamp is activated thereby exposing the member through the mask for an exposure time of about 10 seconds. The imagewise exposure heats the image areas to temperatures about or above the cholesteric liquid crystalline-liquid isotropic transition temperature of the imaging composition, and the imaged areas exhibit a texture transition. A clearly optically distinguisable image, corresponding to the mask, of almost colorless imaged areas on a green colored background, appears almost immediately upon exposure. The composition is allowed to cool into its cholesteric liquid crystalline mesophase temperature range.

The imaged areas exhibit the focal-conic or "undisturbed" texture state while the background areas of the cholesteric liquid crystalline imaging composition exhibit the Grandjean or "disturbed" texture state. Eighty-nine additional examples are given in the patent.

" " " " " " " " " " " "

3,667,039 (U.S) Patented May 30, 1972

ELECTRICITY MEASUREMENT DEVICES EMPLOYING LIQUID CRYSTALLINE MATERIALS.

Andre Garfein, Wilhelm Rindner, David C. Rubin, assignors to The United States of America as represented by the Administrator of the National Aeronautics and Space Administration. Application June 17, 1970.

Disclosed are measuring instruments utilizing liquid crystalline elements that exhibit visible change in response to input signals above given threshold levels. By applying the input signals non-uniformly so that the threshold value is exceeded in only certain portions of the liquid crystalline element, visible discontinuities are created therein. The locations of these discontinuities are made dependent on the magnitudes of the input signals, direct readouts of which are provided by suitably calibrated indicia.

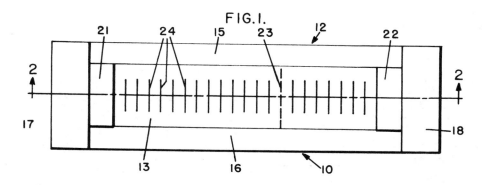

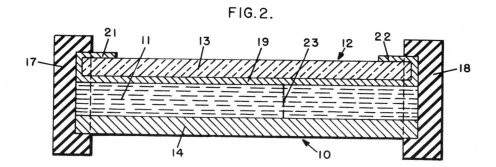

Referring to Fig. 1 and 2 there is shown a voltage indicating device 10 including a body of liquid crystalline material 11. Retaining the liquid crystalline material 11 is a housing 12

including an upper wall plate 13 formed of a transparent electrical insulator such as glass. Disposed parallel to the top wall 13 is a bottom wall formed by an electrically conductive electrode plate 14. A pair of side walls 15 and 16 extend between the top and bottom walls 13 and 14 on opposite sides of the liquid crytslline body 11. Closing the ends of the housing 12 are cup-shaped end walls 17 and 18 composed of a suitable electrical insulator material. Deposited on the inner surface of the top plate 13 is a transparent thin film resistor electrode 19 made of, for example, tin oxide. The deposited thin film resistor material extends around the ends of the top plate 13 and over limited portions of its top surface forming electrical input contacts 21 and 22.

During the use of the voltage indicating device 10, the bottom electrode plate 14 is connected to one end of the thin film resistor electrode 19. The contacts 21 and 22 are then connected to an input voltage being measured creating an electric field across the liquid crystalline body 11 between the electrodes 19 and 14. Because of the potential gradient existing in the electrode 19, the electric field is of nonuniform strength that increases or decreases along the crystalline body's longitudinal axis between the contacts 21 and 22.

Assuming that the applied voltage is within the dynamic range of the instrument 10, the electric field induced over a given longitudinal portion of the crystalline body 12 will exceed its threshold transition level resulting in a visible change in its optical properties. For example, in a preferred embodiment using the crystalline material 11, cholesteryl chloride 30 percent, cholesteryl oleyl carbonate 70 percent and with electrode 14 blackened, the liquid crystal appears to experience a color change from red to black in response to the application of an electric field above a given threshold strength of approximately 1 kV/cm. The color change extends along the longitudinal axis of the crystalline body 11 between the positive input contact 22 and a transition line 23 shown in Fig. 1. The exact position of the transition line 23 created by the color discontinuity corresponds, of course, to that position at which the applied field strength equals the threshold value. Obviously, the relative position of the transition line along the longitudinal axis of the crystal is directly dependent upon the magnitude of the voltage applied between the contacts 21 and 22. Thus, a direct readout of the measured voltage is obtained by noting the position of the discontinuity 23 with respect to suitably calibrated indicia 24 engraved on the top plate 13 and extending along the longitudinal readout axis of the crystalline body 11.

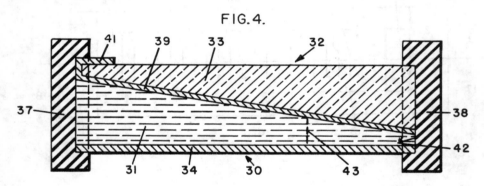

FIG. 4.

As shown in Figure 4, the adjacent surfaces of the upper and lower walls 33 and 34 are non-parallel so as to establish a non-uniform thickness for the crystalline body 31. During use of the voltage indicating instrument 30 the bottom electrode plate 34 and the input contact 41 are connected to receive an input voltage being measured. This creates an electric field across the liquid crystalline body 31 between the electrodes 34 and 39. Because of the crystalline body's non-uniform thickness, the strength of the applied field increases in a direction from input contact 41 toward the body's narrow end 42.

" " " " " " " " " " "

1,276,523 (British)   Patented June 1, 1972

LIQUID CRYSTAL DISPLAY ASSEMBLY HAVING INDEPENDENT CONTRAST AND SPEED OF RESPONSE CONTROLS.

<u>RCA Corporation</u> *(Inventor's name not given).*

Application July 23, 1969; prior U.S. application July 31, 1968.

For description of this patent see U.S. Pat. No. 3,655,269.

" " " " " " " " " " "

3,668,861 (U.S.)   *Patented June 13, 1972*

SOLID STATE ELECTRONIC WATCH.

<u>Hiromitsu Mitsui</u>, *assignor to Kabushiki Kaisha Suwa Seikosha.*
Application Nov. 10, 1971; prior application in Japan Nov. 17, 1970.

    A watch is provided with a liquid crystal display system operable at high levels of ambient illumination and a solid state display system operable at low levels of ambient illumination. A photo-transistor selects the display system appropriate to the ambient light level.

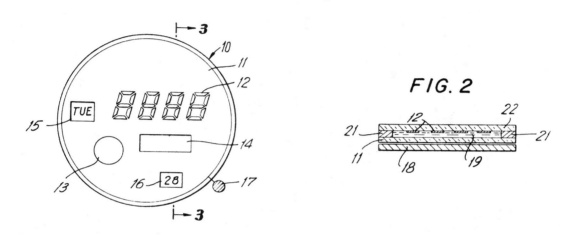

    A solid state electronic watch in accordance with the present invention is generally indicated by the reference numeral 10. The watch has a transparent base plate 11. The display portion of the watch is made up of segments 12, in the embodiment shown in Fig. 1 each of the indicia has seven segments. Photo-transistor 13 which is responsive to incident light controls a divider, oscillation and driving circuit and crystal vibrator portion 14, using an MOS (metal oxide semiconductor) transistor. Day and date display elements 15 and 16 are preferably mechanically activated. The time display consisting of the segments 12 is activated only on pressing a switch 17. The selection of the display mode is, of course, controlled by the phototransistor 13, but the level incident light at which the choice changes can be set manually. This adjustment is not shown. Similarly, the current fed to self-luminous element 18 (Fig. 2) may also be adjusted by means not shown.

    A cell for holding liquid crystal material is formed by spacers 21 between cover 22 and transparent base plate 11. Liquid crystal material 19 normally is transparent, but when a sufficiently high voltage transverse to the cell is impressed between appropriate segments 12 and transparent base plate 11, a desired numeral becomes visible by reason of dispersion of incident light from the surroundings. It should be noted that the brighter the incident light, the more light is dispersed, and consequently the brighter the indicia appear. Consequently, indication of the time by this method is eminently satisfactory where the level of illumination is high. Where the level of illumination is low, or where illumination is completely absent, photo-transistor 13 passes current through appropriate display elements of gallium-arsenide crystal 18. As is evident, the lower the level of the ambient illumination, the more strongly will the self-luminous display elements appear.

    Another ambodiment is shown in Fig. 3 wherein a gallium-arsenide display element, an MOS

FIG. 3

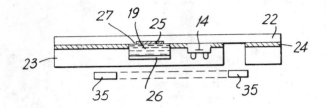

FIG. 4

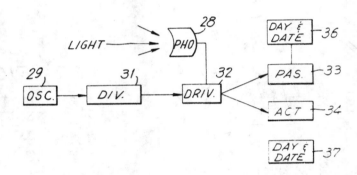

transistor and a liquid crystal cell are mounted monolithically on one semi-conductor base plate 23. Base plate 23 consisting of a sapphire or a spinel separated from cover 22 by sealing layer 24. Electrode 25 is provided for driving liquid crystal 19 into the dispersion state. Self-luminous element 26 of gallium-arsenide is mounted on semi-conductor base plate 23. Display cell 27 is so arranged that each desciRed segment is illuminated from the rear by infra-red radiation emitted by electroluminscent element 26. By using the liquid crystal material in combination with the infra-red radiating self-luminous material, visible light is produced and the power requirement is very low.

The arrangement of the electrical components of the system is shown in the electrical block diagram of Fig. 4 wherein the signal of crystal oscillator 29 which constitutes the time standard for the system is divided by divider 31 which employs an MOS transistor. The output from divider circuit 31 is supplied to driving circuit 32 from which energy is then transferred either to liquid crystal system 33 or self-luminous system 34. The selection is made by the photo-transistor circuit 28. Since the energy expended in operating liquid crystal display system 33 is in the range of microwatts, this display system is termed "passive." The self-luminous display system 34 consumes energy in the range of milliwatts and is therefore termed "active." Where the base plate 23 is of semi-conductor material as shown in Fig. 3, then the photo-transistor 13 may be mounted in the base plate.

" " " " " " " " " " " "

3,669,525 (U.S.)   Patented June 13, 1972

LIQUID CRYSTAL COLOR FILTER.

James E. Adams, Werner E.L. Haas, assignors to Xerox Corporation.
Application Jan. 6, 1971.

An optical filter system capable of transmitting a single wavelength band or a plurality of wavelength bands of incident radiation while simultaneously rejecting substantially all other wavelengths of incident radiation is described. The optical filter system utilizes liquid crystal films having optically negative properties. Typically the optical filters of the invention will be used at or near room temperature. Thus, it is preferred to employ liquid crystalline substances which have a liquid crystal state at or near room temperature. Generally speaking, the liquid crystal substance will preferably have a liquid crystal state at the desired operational temperature.

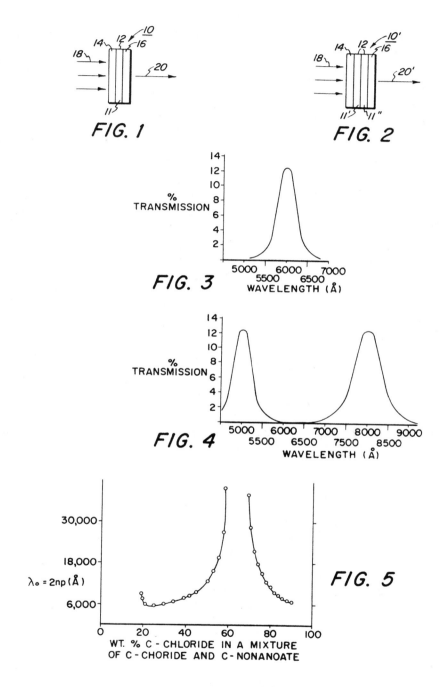

The liquid crystal films employed in the optical filter system of the invention will typically have a thickness of from 0.5 to about 20 microns. The liquid crystal film 11 is typically tacky, soft, viscous, glassy or liquid and thus is preferably encased in a protective outer casing 12 to protect the film from foreign matter such as dust, insects or the like. The purpose of protective outer casing 12 is to keep the liquid crystal film 11 in place and free of any contamination. Linear polarizer 14 and linear analyzer 16 may be selected from any of many various materials. For optimum results the linear polarizer and the linear analyzer are arranged with a pre-determined angular relationship between their axes of polarization of about 90° since, at this condition, the cooperative action of the two members is effective to prevent the transmission of incident unpolarized light through the linear analyzer.

For optimum results, the optical filter 10 is preferably arranged in the path of the ligth beam in a manner such that the incident radiation, represented by arrows 18, reaches the filter at normal incidence with liquid crystal film 11 being preferably disposed so that the helical axis of the liquid crystal substance is in the direction of the light propagation. Should the incident radiation not be at normal incidence to the filter device so that the helical axis of the liquid crystal substance is not exactly along the direction of light propagation, the optical filter will function according to the invention however with some deterioration in the bandwidth of the emergent beam. The characteristics of the emergent beam, represented by arrow 20 are shown in Fig. 3. In this exemplary instance the liquid crystal film is comprised of a composition of 20% cholesteryl chloride by weight in cholesteryl nonanoate having a λ_o value of about 6000 A. Thus, the optical filter conveniently substantially completely extinguishes all the wavelengths of light radiation in the incident light beam except those within the region centered around about 6000 A.

Figure 2 illustrates another embodiment of the optical filter system of the invention. The optical filter 10' is constructed in the same manner as described previously with the exception that it has two liquid crystal films 11' and 11" placed in series between the linear polarizer 14 and the linear analyzer 16. Liquid crystal films 11' and 11" are compositions of 90% cholesteryl chloride by weight in cholesteryl oleyl carbonate and 40% cholestery chloride by weight in cholesteryl nonanoate respectively and have respective λ_Θ values of about 5000 A and about 8000 A. The characteristics of the emergent beam 20' obtained from optical filter 10' are illustrated in Fig. 4.

Figure 5 graphically illustrates the λ_o values which can be obtained from the mixture of two cholesteric liquid crystal substances. It has been found that the pitch of two component mixtures of certain cholesteric liquid crystal is a strong function of chemical composition. Over a wide range of materials, the pitch of a mixture can be accurately represented by a weighted average of the ingredients. Further if components with opposite intrinsic screw sense, i.e., right-handed and left-handed, are mixed there will exist one composition corresponding to no net rotation or infinite pitch. In this exemplary case the compositions are made from cholesteryl nonanoate, a left-handed fatty ester and cholesteryl chloride, a right-handed liquid crystal substance. The percentage composition of the two components of the mixture is plotted against λ_o where $\lambda_o=2np$. It can be seen that at a particular composition, the left and right-handed components essentially compensate and the result is an infinite pitch. Moving away from this point the sense of any individual composition is determined by the dominant component.

Example I

A composition containing 85% of anisylidene-p-n-butylaniline (ABUTA) in cholesteryl oleyl carbonate (COC) is prepared. This composition has a λ_o value of about 2.0 microns. A thin layer of the liquid crystalline mixture is placed on a Polaroid sheet and subsequently another Polaroid sheet is placed over the liquid crystal layer so that the axis of polarization of the two Polaroid sheets have an angular relationship of about 90° to each other. This optical filter is then placed in the path of a light beam emitted from a broad band source of infrared radiation. The filter is positioned in a manner such that the incident radiation is normal to the filter. The optical filter substantially completely reflects all of the incident radiation with the exception of a wavelength band contered about a wavelength of about 2.0 microns which is transmitted. Twenty-two further examples are given in the patent.

The optical filter described may be tilted with respect to the incident light beam so that the incident light is not normal to the filter. In this manner the λ_o value of any liquid crystal films in the filter will be shifted. A detailed discussion relating to this procedure is given in U.S. Pat. No. 3,697,152 titled "Tuning Method for Plural Layer Liquid Crystal Filters."

"""""""""""

3,674,338 (U.S.)   Patented July 4, 1972

REAR PROJECTION SCREEN EMPLOYING LIQUID CRYSTALS.

James V. Cartmell and Donald Churchill, assignors to The National Cash Register Company. Application Aug. 5, 1970.

A rear projection screen is disclosed wherein organic nematic mesomorphic compounds, i.e., liquid crystals, are employed. The liquid crystals have a threshold electric field which when exceeded causes the liquid crystals to scatter light. A thin film of the liquid crystal in the light scattering mode can be used for a screen in a rear projection system.

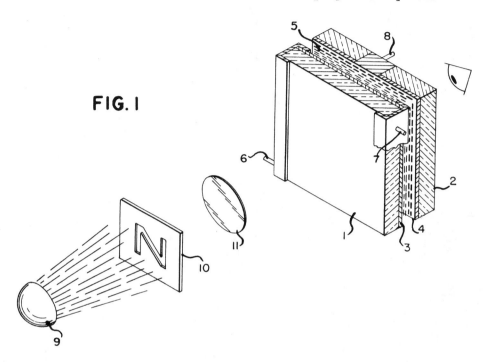

FIG. 1

Substrates 1 and 2 are transparent glass plates measured 4" x 4". The substrates 1 and 2 are aligned so that their adjacent faces 3 and 4 are parallel and spaced about 50 microns apart. The faces 3 and 4 were coated with a thin layer of tin oxide. The substrates 1 and 2 can be positioned apart by means of shims or the spacing can be maintained by means of clamps or a frame-like holder. The space 5 between the two substrates is filled with a liquid crystal. In this example, a 50/50 blend of butyl p-(p-ethoxypenoxycarbonyl)-phenyl carbonate and butyl p-ethylphenyl terephthalatate was employed. Note, the screen is sealed at both ends (not shown) in order to retain the liquid crystal in space 5. Electrical lead wires 6 and 7 are attached to substrate 1 as is electrical lead wire 8 attached to substrate 2. This can be done by conventional techniques such as by silver paint. Electric leads 6 and 7 run full length of substrate 1 and act as a buss bar heater. By applying voltage (not shown) at leads 6 and 8, the liquid crystal can be heated to its liquid crystal state if needed. In this example, the mixture was heated to 40°C by applying 20 volts AC to leads 6 and 7. Leads 6 and 8 are connected to a voltage source (not shown) ranging from 10 to 300 volts. By adjusting the voltage, e.g., 10, 50 or 100 volt increments, the ratio of transmitted light to scattered light can be controlled to suit the viewer. The light rays from the light source 9 pass through the transparency 10, which then are focused by lens 11. The projected image is focused on the liquid crystal material in space 5.

Observance of the screen of this invention demonstrates the high degree of resolution obtained by employing liquid crystals. Fine image detail is delivered by the rear projection screen of this invention as the resolution is 500 lines per inch. Observance of the screen demonstrate uniform brightness thereby indicating desirable light distribution or light spreading characteristics. Viewing the screen at wide angles demonstrates very little screen fall-off.

Not only is this rear projection screen essentially free of sparkle, but it also provides the additional advantages of adjustable gain. The ratio of transmitted light to scattered light can be varied by adjustment of the voltage to the conductive films. In this instance, the variable ratio of transmitted light to scattered light helps to adjust brightness at various angles.

"""""""""""

3,674,341 (U.S.) Patented July 4, 1972

LIQUID CRYSTAL DISPLAY DEVICE HAVING IMPROVED OPTICAL CONTRAST.

<u>Clarence L. Hedman, Jr., and Karl-Dieter S. Myrenne</u>, assignors to SCM Corporation. Application Dec. 8, 1970.

It is a principal object of the present invention to provide a liquid crystal display having improved optical contrast.

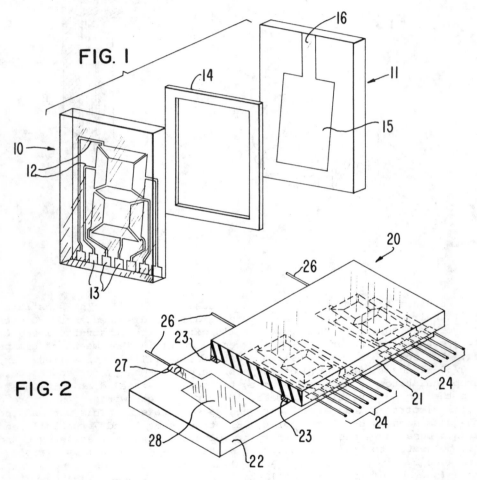

Figure 1 is an exploded view of a liquid crystal cell which may be used in practicing the present invention. Figure 2 is a perspective view of a multi-cell liquid crystal display panel with a portion of one panel member removed.

Figures 4 and 5 illustrate a prepeferred embodiment of the invention with the electrical connections omitted for the sake of clarity. The display device is seen to comprise a parallel-epiped housing, which preferably is opaque, having a front housing member 41 to which is mounted a liquid crystal display panel 42 by means of mounting brackets 43, screws 44 and rubber washers 45. Display panel 42 is mounted in alignment with housing aperture 46 whereby it may be viewed from without the housing through the aperture. To rear housing member 47 is affixed a black, lightabsorbent screen 48 in alignment with aperture 46 and display panel 42. To each side housing member 49 are mounted two light baffles 50 between each set of which is mounted a light source 51. With this arrangement each light source is capable of projecting a beam of light 52 obliquely onto display panel 42. Although two lamps are shown in this embodiment it should, of course, be understood that the use of only one would suffice. The presence of two light sources however provides more uniform illumination of the display panel. Of more importance though is the fact that color change may be effected through the use of two different colored lights. For example, one of the light sources may provide a blue beam of light falling obliquely in display panel 42 and the other light source an oblique beam of red light.

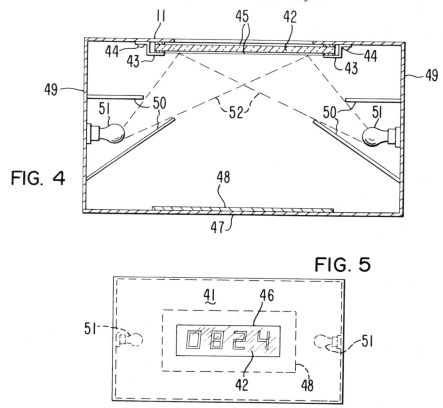

With the display device illustrated in Figs. 4 and 5 display panel 42 is illuminated by light beams 52 which impinge thereon at oblique angles. Ambient light entering the device through aperture 46 and display panel 42 strikes screen 48 and is largely absorbed thereby. This also holds for light emanating from light sources 51 themselves which may be reflected by the display panel or other structural elements within the device onto screen 48. As a result the display panel will appear very dark when viewed from without the housing except for those portions in which an electric field above threshold is impressed through the organic nematic mesomorphic compound. Thus, digits 0824 appear bright in the display panel shown in Fig. 5 while the remaining portions of the display panel appear quite dark. Again, this is due to the fact that light striking the energized portions is forward scattered by the compound while light striking the unenergized portions passes through the transparent panel to each side of the viewer.

" " " " " " " " " " " " " "

3,674,342 (U.S.)   Patented July 4, 1972

LIQUID CRYSTAL DISPLAY DEVICE INCLUDING SIDE-BY-SIDE ELECTRODES ON A COMMON SUBSTRATE.

<u>Joseph A. Castellano</u> and <u>Ronald N. Friel</u>, assignors to RCA Corporation. Application Dec. 29, 1970.

A pair of oppositely disposed substrates sandwich a film of liquid crystal material therebetween. Side by side electrodes, between which a voltage is applied to switch the light transmitting characteristics of the liquid crystal material, are disposed on a single one of the substrates.

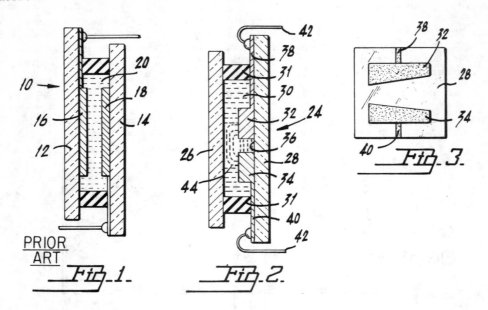

Figure 1 is a side view, in section, of a prior art liquid crystal device. As shown in Fig. 2, a liquid crystal device 24 in accordance with the instant invention comprises a pair of oppositely disposed flat substrates 26 and 28 of, for example, glass, sandwiching a film 30 of a known liquid crystal material therebetween. The substrates 26 and 28 are maintained in spaced apart and sealed together relation by means of a shim 31 of, e.g., a plastic material.

Two electrodes 32 and 34 are provided by means of which a voltage can be applied to the device to alter or "switch" the light transmitting characteristics of the liquid crystal material. That is, depending upon the liquid crystal material used, the liquid crystal material is either normally transparent and is switchable to a light scattering state by the application of a voltage across, or the liquid crystal material is normally light scattering and is switchable to a light transparent state by the application of the voltage.

As shown in Figs. 2 and 3, the two electrodes 32 and 34 are disposed in side by side relation on the inside surface 36 of the substrate 28. The electrodes 32 and 34 comprise thin films of a transparent electrically conductive material, e.g., a 1/25 mil thick film of indium oxide or tin oxide. Also, two connectors 38 and 40, also of a transparent conductive material are disposed on the surface 36 and extend from each of the electrodes 32 and 34, respectively, to exposed peripheral surfaces of the substrate 28, where terminals 42 are soldered to the connectors.

When a voltage is applied between the two electrodes 32 and 34 via the two terminals 42, an electric field, as indicated by the use of dashed lines 44 representative of some of the electric field lines, is provided through the liquid crystal film 30 between the electrodes 32 and 34. As shown, the electric field lines 44 fringe outwardly and away from the two electrodes 32 and 34 and, dependeing upon the dimensions of the device, the liquid crystal material used, and the voltage applied, a portion of the liquid crystal film 30 adjacent to and between the two electrodes 32 and 34 is switched to its alternate light-transmitting or light-scattering state. Owing to the different optical characteristics of this "switched" portion of the film. in comparison with the unaltered optical characteristics of unswitched adjacent portions of the film 30, the two portions are optically distinguishable from each other by a viewer. By suitable shaping of the two electrodes 32 and 34, an image can be displayed.

As shown in Fig. 3, the two side by side electrodes 32 and 34 have a varying spacing therebetween. Since the electrical field between the two electrodes is inversely related to the distance between the electrodes, the response of the liquid crystal material to a voltage applied between the two electrodes 32 and 34 can be made to vary along the length of the electrodes.

Thus, for example, if the liquid crystalline material is one, such as p-methoxybenzylidene-p'-aminobutylbenzene, in which the amount of light scattering is proportional, up to a maximum value, to the electric field through the liquid crystal material, using an applied voltage insufficient to cause maximum light scattering along the entire length of the electrodes results in a light scattering which varies along the electrode length.

The use of side by side electrodes in place of the overlapped electrodes of the prior art provides several advantages. With the side by side electrodes of the instant invention, the spacing between the two substrates is not critical since the spacing between the substrates does not affect the spacing between the electrodes. Devices having short relaxation times can be obtained using relatively closely spaced electrodes and relatively widely spaced apart substrates. Widely spaced apart substrates, as previously noted, provide increased lifetime. One result of the use of side by side electrodes is that much shorter relaxation times than would be expected are obtained. For example, in one series of tests comparing devices having overlapped electrodes with a spacing of 1/2 mil therebetween with devices having side by side electrodes spaced apart one mil, all other parameters being substantially the same, the overlapped electrode devices had a relaxation time of in the order of 1 second. The side by side electrode devices had a relaxation time in the order of 50 milliseconds.

A further advantage of the hereindescribed devices is that, depending upon the particular device being made, only one of the substrates need be provided with electrodes. A still further advantage of the use of side by side electrodes is that the amount of liquid crystal material between adjacent electrodes is generally much reduced in comparison with the amount of liquid crystal material between overalpped electrodes on opposite substrates. This follows because the thickness of the electrodes, in the order of 1/25 mil, is much less than the distance between overlapped electrodes, in the order of 1/2 mil. The smaller amount of liquid crystal material between the electrodes results in the side by side devices being operable with smaller voltages and currents than are required with overlapped electrode devices.

" " " " " " " " " " "

3,675,987 (U.S.)
Patented July 11, 1972

LIQUID CRYSTAL COMPOSITIONS AND DEVICES.

Mary J. Rafuse, assignor to Sperry Rand Corporation.
Application March 29, 1971.

Room temperature nematic liquid crystal compositions suitable for use as scatterers of light in optical devices are presented that include as a basic new component p-methoxyformyloxybenzylidene p-n-butylaniline. Such compositions may also include p-ethoxybenzylidene p-n-butylaniline or p-butoxybenzylidene p-n-butylaniline, along with certain additives including p-toluylidene p-n-butylaniline. The compositions are employed in the form of thin layers within thin cells having transparent electrodes with means for applying electric fields across the layer. The display is formed by electrically controlling lightscattering properties of the liquid crystal layer.

The novel electro-optically active nematic liquid crystal compositions described herein may be employed in the electrically controllable, flat panel display device of Figs. 1 and 2 for the purpose of generating displays in which the size, shape, and location of the two-dimensional display pattern may be changed continuously, as well as in discrete steps. In operation, the apparatus of Figs. 1 and 2 makes significant use of the spatial voltage gradient or variation set up across the transparent high resistance electrode means 14. While electrode means 13 may instead be used as the high resistance electrode, or both electrodes may be of high resistance material, only the electrode 14 will be considered to be high resistivity electrode at this time for the sake of simplifying the discussion. With a potential gradient set up across electrode 14, the potential difference between electrodes 13 and 14 (which is the potential drop seen across the liquid crystal layer 12) varies from one spatial location across layer 12 to a next location. This potential variation gives rise to controllable regions of transparency and translucence within layer 12, providing that the values of V_{13} and V_{14} have been appropriately selected. The dimension of the transition region between transparent and translucent region is relatively sharp when employing the novel liquid crystal materials of the present invention.

Referring to Fig. 3, there is seen a typical display 22 produced according to the present invention within the novel liquid crystal material. The display comprises a rectangular bright area 23 and a rectangular dark area 24 with a common transition boundary 25. Boundary 25 is readily moved to the left or to the right by relative variation of voltages V_{13} and V_{14}.

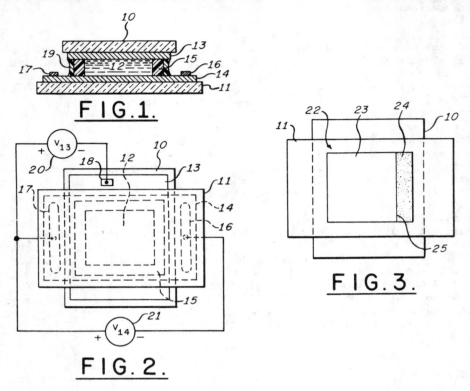

In Figures 1 to 3, the rectangular bright area or bar 22 is changed in width by changing the relative magnitudes of voltages V_{13} and V_{14} according to a desired pattern. The value of voltage V_{13} may be held fixed, while the value of voltage V_{14} may be changed, or vice versa. For example, consider the result when voltage V_{13} is set to zero and voltage V_{14} is increased from zero. This action causes the bright bar or area 23 to increase in width from zero as boundary 25 moves to the right in the drawing, the size of the dark region 24 changing correspondingly.

The multiple component electro-optically active nematic liquid crystal compositions described herein include as a common principal component a novel active electro-optic medium consisting of p-methoxyformyloxybenzylidene p-n-butylaniline. This principal component forms useful room temperature operable nematic liquid crystal compositions when mixed with certain p-alkoxybenzylidene p-n-butylanilines. Ionizable additives may also be added to such mixtures for purposes which will become apparent.

The above principal or primary component is prepared by refluxing the appropriate aldehyde and amine in dry alcohol. Commercially available benzaldehydes and p-n-butylaniline are used after purification. Specifically, p-methoxy formyl benzylidene p-n-butylaniline is prepared by first preparing methyl p-formylphenyl carbonate. To a solution of commercially available p-hydroxybenzaldehyde (2.4 g; 0.02 mole) in 15 ml of dry ether containing 4 ml of pyridine is added methyl chloroformate (2.5 ml), while cooling and stirring the solution. The reaction mixture is stirred substantially at room temperature for two hours and is then filtered. The solvent is removed by evaporation under reduced pressure. The residue is then induced to crystallize by chilling it. The crystal are next redissolved in hexane and again recrystallized. A typical yield using the above proportions is 2.50 g (a 70% yield) of white methyl p-formylphenyl carbonate crystals having a melting point lying between 34 and 36°C. The infrared spectrum of this material in CCl_4 included wave numbers 2820, 2730, and 1710 cm^{-1} stemming from the aldehyde portion of the generated molecule and 1760 cm^{-1} from the carbonate portion. The spectrum confirms the chemical constitution of the resultant material.

Methyl p-formylphenyl carbonate crystals (1.8 g; 0.01 mol) are then refluxed with p-n-butylaniline (1.5 g; 0.01 mol) in 10 ml of dry ethanol for two hours with stirring. The cool reaction mixture is filtered and the precipitate is recrystallized from hexane. With the above proportions, there is a yield of 2.48 g (an 80% yield) of pale yellow crystals of the desired p-methoxyformyloxybenzylidene p-n-butylaniline. The observed infrared spectrum in CCl_4 confirms the chemical constitution of the product, indicating the loss of nitrogen-hydrogen bonds in the new material and the absence of the aldehyde portions.

Tests of the new material yield an operational temperature range between 55 and 61°C displaying nematic characteristics. When cooled, the material forms a monotropic nematic liquid crystal state which converts to a crystalline solid state at 32°C. In the optical scattering condition, the material has a white appearance under white light. It is believed that the benzylidene portion of the molecule is active in conferring liquid crystal characteristics upon the nem material, the general geometry of the molecule being such as otherwise to permit the liquid crystal phase to exist. It is believed that the presence of the methylformyloxy portion of the molecule contributes to the high opacity of this primary material when in its optically scattering state and to its high resistivity and consequent longer lifetime of the state.

The relatively high electro-optically active temperature range of the principal material p-methoxyformyloxybenzylidene p-n-butylaniline is reduced, according to the invention, by the addition of a known alkoxy benzylidene p-n-butylaniline. The added or secondary material may be a p-ethoxy material or a p-n-butoxy material. The phrase "secondary material" is used herein to indicate a second material whose presence plays a role substantially equally as significant as that of the primary material in yielding the desired results exhibited by a binary or ternary composition.

The p-ethoxybenzilidene p-n-butylaniline material is previously known to those skilled in the art as a relatively weak dynamic scattering nematic liquid crystal operable between 35 and 75°C. In the example, equal weights of the primary p-methoxyformyloxybenzylidene p-n-butylaniline material and of the temperature lowering secondary p-ethoxybenzylidene p-n-butylaniline crystals are weighed out and mixed throughly while warming until mutually dissolved and all of the mixture is in its liquid isotropic state. The material is cooled into the nematic liquid crystal state with continued agitation to ensure continued adequate mixing. The dual component composition is utilized by placing it in a cell between closely spaced glass surfaces with a transparent electrode on at least one glass surface, as in the prior art. Application of an electric field across the thin film is accomplished for controlling the transparent and opaque states of the film, as in the prior art. The equal component mixture is found to demonstrate the desired liquid crystal display effects over the wide temperature range of -12° to +60°C. This range is particularly advantageous, extending considerably more above and below ambient temperature, and making the composition useful in many environments for which prior art liquid crystal materials are entirely unsuited. This beneficial result is achieved while still maintaining a contrast ratio (between the transparent and turbulent states) of substantially five to one.

Other than equal proportions by weight of the primary and alkoxybenzylidene p-n-butylaniline materials may be employed. For example, Fig. 4 illustrates the behaviour of the dual component mixture employing the primary material p-methoxyformyloxybenzilidene p-n-butylaniline and the secondary p-ethoxybenzylidene p-n-butylaniline in various ratios.
P-butoxybenzylidene p-n-butylaniline may be mixed with the primary material p-methoxyformyloxybenzilidene p-n-butylaniline to form a second novel electro-optically active material of desirable low temperature nature. (see Fig. 5).

It has furthermore been found that the operating temperature range of the binary material containing p-butoxybenzylidene p-n-butylaniline may be desirably reduced by the addition of a relatively small quantity of a temperature depressant such as the novel material p-toluylidene p-n-butylaniline so that a ternary composition is generated.

One preferred example of a novel ternary liquid crystal composition according the the invention employs 45% by weight of the primary p-methoxyformyloxybenzylidene p-n-butylaniline material, 45% by weight of the secondary p-butoxybenzylidene p-n-butylaniline material and 10% of the p-toluylidene p-n-butylaniline material. This ternary composition displays excellent nematic liquid crystal properties over a range of temperatures from less than 0° to +47°C.

Relatively small additive proportions of easily ionizable materials may be added to any of the foregoing, including ionizable materials such as p-n-butoxybenzoic acid, p-n-butoxyphenol,

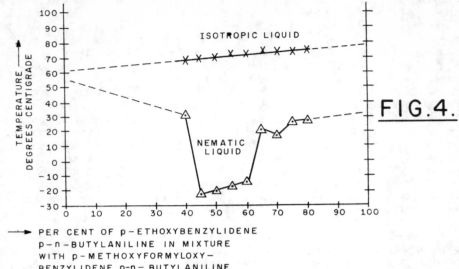

FIG. 4.

PER CENT OF p-ETHOXYBENZYLIDENE p-n-BUTYLANILINE IN MIXTURE WITH p-METHOXYFORMYLOXY-BENZYLIDENE p-n-BUTYLANILINE

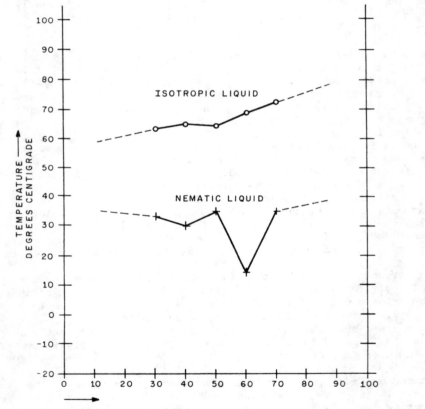

PER CENT OF p-BUTOXYBENZYLIDENE p-n-BUTYLANILINE IN BINARY MIXTURE WITH p-METHOXY-FORMYLOXYBENZYLIDENE p-n-BUTYLANILINE

FIG. 5.

p-methoxyacetophenone, and p-n-butoxybenzaldehyde. The effect of the ionizable additive is to act as ready carriers for electrons injected by the cell electrode, without excessive breakdown of the material, consequently enhancing opacity by increasing turbulence. For example, a compound comprising 48.6 percent by weight of the primary methoxy formyloxybenzylidene p-n-butylaniline and 48.6 percent of p-ethoxybenzylidene p-n-butylaniline with 4.8 percent p-n-butoxybenzoic acid demonstrates generally similar properties to those of the equal portion binary material first described above, but demonstrating liquid crystal characteristics over the modified temperature range including -12° and +55°C, as opposed to -12° and +60°C. Contrast ratios of six to one are demonstrated in such a ternary material. Substitution of p-n-butoxyphenol of p-n-butoxybenzoic acid yields measured contrast ratios as high as nine to one.

3,675,988 (U.S.)   Patented July 11, 1972

LIQUID CRYSTAL ELECTRO-OPTICAL MEASUREMENT AND DISPLAY DEVICES.

Richard A. Soref, assignor to Sperry Rand Corporation.
Application Nov. 25, 1969.

The invention relates to electrically controllable, flat-panel display devices employing liquid crystalline materials as electrically active media and more particularly relates to such display devices in which the size, shape, and location of two-dimensional display patterns can be changed continuously as well as in discrete steps.

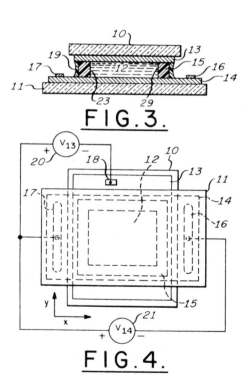

FIG.3.

FIG.4.

Referring to Figs. 3 and 4, a typical construction for the invention is shown utilizing a pair of parallel sided flat glass plates 10 and 11 preferably arranged parallel to each other and separated by a thin layer 12 of electric field sensitive or nematic material of any of the types available from Liquid Crystal Industries, Inc. Plates 10 and plate 11 are coated on their inner surfaces with thin conductive electrode means 13 and 14. A cell containing the nematic material is further defined by a continuous quadrilateral dielectric wall 15. Extended lineal or elongate terminals 16 and 17 are applied in conductive relation to electrode 14 on glass plate 11 at opposite ends of that electrode. By virtue of their relatively low resistance, terminals 16 and 17 have equipotential surfaces. A relatively small terminal 18 may be used in conductive relation with electrode 18 on glass plate 13.

Glass plates 10 and 11 may be made of any suitable glass or other transparent insulating material compatible with the optical requirements of the cell system. The transparent conducting electrodes 13 and 14 may be made of tin oxide, aluminum oxide, or similar materials put down on glass plates 10 and 11 by chemical or evaporative deposition, by sputtering, or by other suitable known methods.

The choice of materials is such that conducting electrode 13 has a low resistivity of the order of 100 ohms per square, for example, so that the whole of electrode 18 may readily reach the same potential level as applied to terminal 13. On the other hand, the material of electrode 14 has a relatively high resistivity of about 500.000 ohms per square, foe example. Other resistivity values may be employed, but a relatively high resistivity is beneficial because ohmic loss within electrode 14 is then minimized, thereby preventing appreciable temperature rise in the liquid crystal layer 12. Also, the current drawn from external power sources is desirably minimized. The resistivity characteristics of the material of electrode 14, which is put down on glass plate 11 (the plate that is normally considered to be the viweing plate of the cell) is of major importance to the operation of the invention, as will be described hereinafter.

So that the liquid crystal layer 12 may be contained in its pure form, protected from contamination, and be of uniform thickness, dielectric wall 15 is formed as a continuous wall; it is readily constructed of a tape available in the market made of a polymerized fluorocarbon resin material sold under the name "Teflon." The tape is available in thicknesses of the order of 1.0 mil, a thickness suitable for use in the invention. The cell may be held together at least in part by a miniscus-shaped film 19 of epoxy material applied to the external free surface of wall 15 so that it bonds to that surface and to the adjacent exterior surfaces of electrodes 13 and 14.

The two elongated terminals 16 and 17 on plate 11 and the small terminal 18 on plate 10 may be constructed in the conventional manner from a silver electrically conducting epoxy material available on the market or by deposition of a strip of low conductivity tin oxide by one of the aforementioned processes. A voltage source 20 for supplying a voltage V_{13} is connected across terminals 18 and 17, while a second voltage source 21 is connected between the terminals 16 and 17 common to electrode 14 for supllying a voltage V_{14} thereacross.

It should be understood in considering the structure of the apparatus of Fig. 3 that the state of the liquid crystal layer 12 may, for instance, be viewed by the observer from above glass plate 11 through transparent electrode 14. It should also be understood that the drawing of Fig. 4 has been made for convenience as if one viswing the drawing is similarly looking through the plate 11 and electrode 14. Below the plane of electrode means 14, the viewer sees the dielectric tape wall 15 and the liquid crystal layer 12. Below the plane in which the latter two items lie, the observer sees the second electrode means 13 and the second glass plate 10.

In operation, the apparatus of Figs. 3 and 4 makes significant use of the spatial voltage gradient or variation set up across the transparent high resistance electrode means 14. While electrode means 13 may instead be used as the high resistance electrode, or both electrodes may be of the high resistance material, only the electrode 14 will be considered to be a high resistivity electrode at this time for the sake of simplifying the discussion. With a potential gradient set up across electrode 14, the potential difference between electrodes 13 and 14 (which is the potential drop seen across the liquid crystal layer 12) varies from one spatial location across layer 12 to a next location. This potential variation gives rise to controllable regions of transparency and translucency within layer 12, rpoviding that the values V_{13} and V_{14} have been appropriately selected. The dimension of the transition region between transparent and translucent regions is relatively sharp if the selected liquid crystal medium has a highly non-linear scattering curve.

In the devices of Fig.s 3 and 4, the potential V_{12} across the liquid crystal or nematic layer 12 may be represented by the graph of Fig. 5, where the parameter x is the right-left position coordinate in layer 12 (x is zero at the left edge of layer 12). The potential distribution on electrode 14 is a function of x only and is independent of y because of the particular orientations of extended terminal means 16 and 17. In Figs. 3 to 6, the rectangular bright area or bar 24 is changed in length (x_o is changed) by changing the relative magnitudes of V_{13} and V_{14} according to a predetermined time pattern. The value of V_{13} may be held fixed, while the value of V_{14} may be changed, or vice versa. For example, consider the result when V_{13} is set to zero and V_{14} is increased from zero to a value above the V_e value shown in Fig. 5. This causes the bright bar to increase in length from x equals zero (to expand from border 23).

The liquid crystal cell arrangement of Fig. 10 illustrates a system in which one can, in essence, create two of the bright moving bars like the one used in Fig. 6 and can cause them to move in cooperative relation so as to expose a movable window or dark area, for example, of constant width. The liquid crystal cell itself may be similar to that employed in Figs. 3 and 4, but the controlling voltages are applied in a new manner which illustrates the versatility of the invention. Accordingly, corresponding reference numerals are used for parts that correspond in Figs. 4 and 10 and the structure of Fig. 10 need not, therefore, be discussed in detail.

However, the Fig. 10 structure is seen to include a pair of parallel-sided plates 10 and 11 separated in parallel relation by a layer of nematic material. Plates 10 and 11 are respectively coated on their inner surfaces with electrodes 13 and 14. The volume of electric field sensitive material is further bounded by a continuous rectangular wall 15 of thin dielectric tape. Lineal or elongate terminals 16 and 17 are applied to electrode 14 at its opposite ends, while a single small terminal 18 is applied to electrode 13. As in Fig. 4, electrode 13 has a relatively low resistance, while electrode 14 may have a relatively high resistivity.

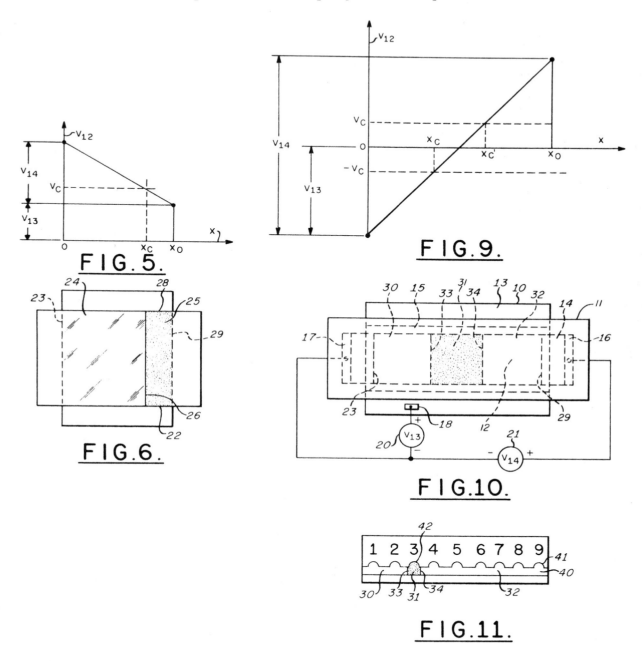

In the case of Fig. 10, the potential V_{12} is arranged to go through a zero value, as shown in Fig. 9, starting at a negative value at terminal 17 and ending at a positive value at terminal 16. As is seen from Fig. 9, the voltage V_{14} is taken to be about twice that of V_{13}, so that the symmetric distribution shown in Fig. 9 is attained. The voltage sources 20 and 21 are connected in a subtractive sense, rather than additively as in Fig. 4. If V_{14} is held constant while V_{13} is varied, a moving window 31 of fixed width is generated. On the other hand, if V_{13} is held constant and V_{14} varied, the width of the window 31 is varied. Combination of the two modes are also possible in which V_{13} and V_{14} are both varied according to a predetermined desired pattern.

Thus, the display of Fig. 10 consists of three portions. First, there may be a bright rectangular area 30 extending from wall surface 23 to the transition point x_c at 33, followed by a dark area 31 extending from boundary 33 to the transition point x_c' at 34. The second bright area 32 extends from boundary 34 to the wall surface 29.

Figure 11 represents the appearance of a window type of voltmeter indicator with an index scale running from 1 to 9 in cooperative relation therewith. Certain elements are similar to those generated in the window presentation shown in Fig. 9 and therefore bear corresponding reference numerals. For example, the presentation 40 includes a bright bar 30 ending at transition boundary 33 and a dark area 31 serving as a pointer and ending at boundary 34 where the second bright area 32 begins and continues to the end of the scale. The usefulness of the display may be enhanced by shaping the electrode 14 (Fig.4) with regularly arranged half-circular extensions. When in the illuminated state, the bright window 31 thus includes a bright pointer-like tip 42, for example, directed toward the index scale number 3.

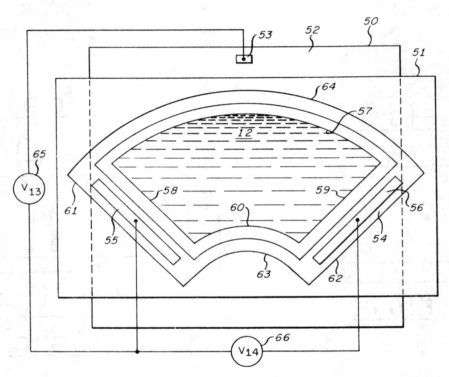

FIG. 12.

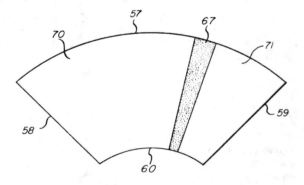

FIG. 13.

One type of indicator that produces a pointer or meter-like radial bar moving about a pivot point is seen in Fig. 12 and 13. Like the apparatus of Figs. 4 and 9, the inventive embodiment uses a first flat glass plate 50 with a rectangular electrode means 52 and a second flat glass plate 51. Plate 51 is coated with a specially shaped transparent electrode means 54; it is seen that electrode 54 has a pair of lineal edges 61 and 62 extending in a radial sense. One set of corresponding ends of edges 61 and 62 is coupled by an arcuate edge 63; the other set of corresponding ends of edges 61 and 62 is coupled by a second arcuate edge 64 of greater radius than edge 63, the two having a substantially common center of curvature. Adjacent edges 61 and 62 are located lineal low resistance terminals 55 and 56, respectively. The field sensitive liquid crystal material 12 is further confined by a thin dielectric wall consisting of sides providing an enclosure substantially similar in shape to the sector shaped electrode 54; namely, a hermetic wall consisting of radially extending wall elements 58 and 59 and cooperating arcuate elements 57 and 60. Electrode means 54 is not only of special shape, but its resistivity characteristics are specially distributed. To achieve the sector window presentation of Fig. 13, it is necessary to make the short radius portion adjacent edge 63 of the transparent conducting sector electrode 54 have higher resistivity than the outer-radius portion adjacent edge 64. Thus, the resistivity is made to taper in value, preferably in a lineal fashion, between edges 63 and 64. The shape and thickness of the electrode 54 can readily be realized using conventional chemical etching or vacuum deposition techniques.

A voltage source 66 is used to supply voltage V_{14} to the terminals 55 and 56. Voltage V_{13} is placed between one of the extended lineal terminals (55) of electrode 54 and the small terminal 53 of the second electrode 53. With the magnitudes V_{13} and V_{14} adjusted as in the instance of the arrangement of Fig. 9 when a movable dark window is produced, the radial bar 67 is presented, as seen in Fig. 13. The indication includes a bright sector 70 bounded by radial wall 58 and partly by arcuate walls 57 and 60, followed by a thin dark sector 67, followed in turn by a bright sector 71 bounded by radial wall 59 and portions of arcuate walls 57 and 60.

Forms of the invention which, like that of Fig. 12, can be said to represent conformal transformations of the devices of Fig. 4 or 9, are illustrated in the two related embodiments of Figs. 14 and 17. The objective is to form a dark ring-like presentation which may be expanded or shrunk about a center point.

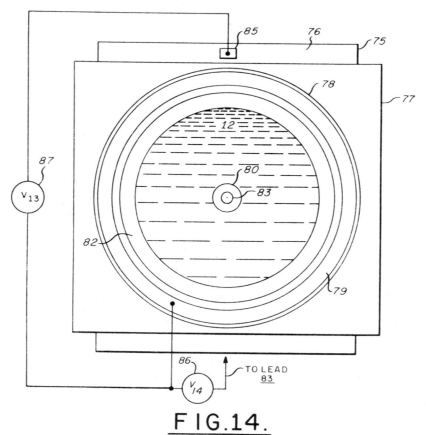

FIG. 14.

The structure of the device of Fig. 14 includes the familiar first flat glass plate 75 and a second flat glass plate 77. Plate 77 is coated on its inner surface with a circular transparent electrode means 78. A circular strip low-resistance terminal 79 is placed at the periphery of circular electrode 78. A second terminal 80 in the form of a small solid circle is placed at the center of circular electrode 78.

The field sensitive liquid crystal material 12 is further confined by a thin dielectric tape in the form of a circular wall 82 of slightly lesser outer diameter than the inside diameter of circular terminal 79. Wall 82 may be fastened to plates 75 and 77 with hermetic seals made by an epoxy or other cement.

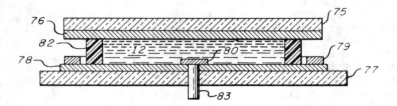

FIG.15.

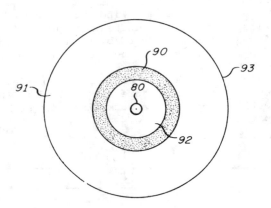

FIG.16.

As seen in Fig. 15, one way of supplying necessary potentials to transparent electrode 78 is by way of the central electrical terminal lead 83 held by a glass-to-metal seal or other known sealing device in the center of plate 77 and electrode 78, through both of which lead 83 projects so as to make electrical contact with low resistance terminal 80.

As is seen in Fig. 14, a low voltage source 86 is used to supply voltage V_{14} between low resistance circular terminal 79 and the terminal lead 83 protruding from the viewing face of plate 77. Likewise, a voltage V_{13} is again supplied to a small low resistance terminal 85 on electrode 76 from voltage source 87 connected to terminal 85 and circular terminal 79. With the magnitudes of V_{13} and V_{14} adjusted according to the previously described method when a movable annular dark window is to be produced, the ring presentation 90 of Fig. 16 is produced. The dark ring 90 is surrounded by concentric annular region 91 which is bright in appearance and ring 90 surrounds a second bright ring which is solid circle about terminal 80.

In the modification shown in the fragmentary view of Fig. 17, there is produced only a major arc sector 90' of the dark circle 90 produced in the arrangement of Fig. 14. The device of Fig. 17 will be understood upon observing that it is similar to that of Fig. 14, with the major exception that a sector has been cut out of the transparent electrode 78 of Fig. 14; this is for the purpose of providing an alternate way of supplying potential to the dot terminal 80 avoiding use of terminal lead 83 projecting through viewing plate 77.

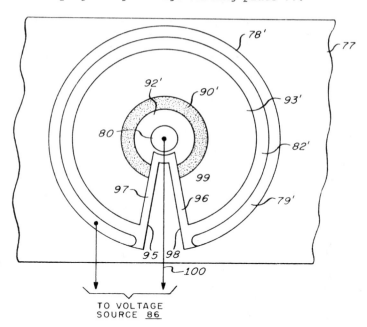

FIG.17.

In Figure 17, transparent electrode 78' has a sectorial area eliminated, being bounded in that area by radial edges 95 and 98. The peripheral low resistance terminal 79' extends in an arc ending at sector edges 95 and 98. The dielectric wall 82' encompassed by termianl 79' has a similar angular extent, but it is now necessary to complete the liquid-crystal-retaining envelope by radial wall elements 96 and 97 and the short wall element 99. These wall elements, together with an arcuate wall 82', combine with suitable sealant means to function as a part of the hermetically sealed envelope enclosing the nematic or field sensitive material. It is to be observed that lead wire 100 attached to terminal 80 is now simply brought out of the cell through dielectric wall element 99.

A view of the arcuate presentation generated is also shown in Fig. 17. Voltages V_{13} and V_{14}, generated and coupled into the cell as in Fig. 14, are adjusted as previously described to cause formation of a movable arcuate dark window 90'. The dark arc 90' is surrounded by a concentric arcuate region 93' which appears bright to the viewer and arc 90' surrounds a second bright region 92' mssing a sector aligned with the sector missing from bright region 93'.

A representative form of the invention in which electrodes with similar resistivity patterns are employed is illustrated in Figs. 18 and 19. In general, two similar high-resistivity electrode means of similar high-resistivity material may be used, each electrode similar, for instance to the high resistivity electrode 14 on plate 11 of the cell of Fig. 4. The electrodes in Fig. 19 are crossed, however, in such a sense that the influence of the voltage gradient in one electrode is at right angle to the influence of the voltage gradient in the other electrode, though angles other than 90 could be employed. It can be seen by considering the mechanism of operation of the apparatus of Fig. 4 that the voltage V_{12} can now be controlled in two spatial dimensions, as well as a function of time.

Referring now to Figs. 18 and 19, it is seen that a novel form of the invention comprises first and second flat glass plates 100 and 101. Each has its inner surface coated with a relatively high resistivity electrode means in the form of electrodes 102 and 103, respectively. A quandrilateral wall 108 of dielectric material further cooperates to form a hermetically sealed cell volume, as before, for protecting and containing the field sensitive liquid crystal 12.

Each electrode means 102 and 103 is supplied with extended lineal low resistance terminals, the pair of lineal terminals associated with one electrode surface being at right angles to the pair of terminals associated with the second electrode surface. For example, electrode 102 has

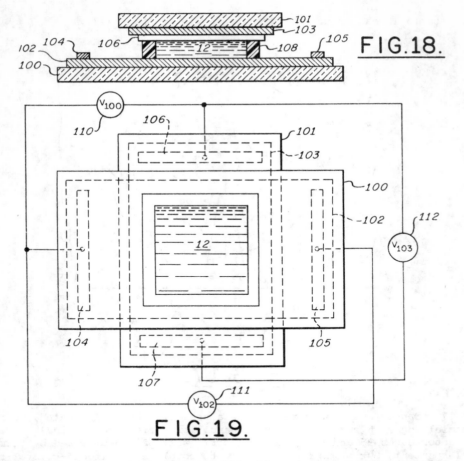

FIG.18.

FIG.19.

has extended low resistance terminals 104 and 105 adjacent and parallel to its opposite edges, which electrode 103 has similar terminals 106, 107 adjacent and parallel to its respective opposite edges. Terminal pairs 104, 105 are at right angles to terminal pairs 106, 107.

It can be analytically predicted and has been experimentally demonstrated that manipulation of the relative values of V100, V102 and V103 produces display patterns that are opaque bars of triangles, or transparent bars oriented in vertical, horizontal, or oblique directions. Desired pattern movements are accomplished by varying any or all of the above voltages according to a predetermined time program. Presentations of many types result by appropriate adjustment of the values of voltages V100, V102, and V103 in the circuit of Fig. 19. For example, not only can window displays similar to those produced in the similar embodiments of the invention be produced (Figs. 20a and 20b), but displays such as those of Figs. 20c and 20d, as well. It will be appreciated that the display shapes of these figures may be continuously interchanged by continuous variation of the relations of the input voltages V100, V102 and V103 and that they may be moved vertically, horizontally, or obliquely.

The versatility of the device of Figs. 18 and 19 is further illustrated by the variety of presentations that can be produced by it. For example, if there is no x electric field gradient (no gradient on electrode 102), a presentation like that of Fig. 20e can be readily produced. With no y field gradient, the Fig. 6 presentation results. Similarly, a dark vertical bar may be moved horizontally across the display; or a dark horizontal bar can be moved up or down.

With values of V102, V103, and V100 appropriately selected, a representation like that of Fig. 20d results. The values of angles A and B in Fig. 20d may widely adjusted by manipulation of the above three voltages. For example, the bar-like presentation of Fig. 20d can be achieved with angles A and B being substantially equal, both being readily variable.

A further aspect of the invention is illustrative of the wide variety of applications in which it may be employed and takes the form of display arrangements with two or more closely spaced but separated layers of field sensitive liquid crystal materials. These multiple-layer

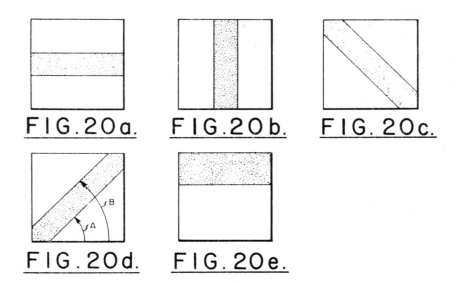

FIG.20a. FIG.20b. FIG.20c.
FIG.20d. FIG.20e.

stacked arrangements offer several features, since their individual visual effects are additive, one being a wide choice of display patterns in which dissimilar display systems may be combined to yield as a cooperative result further new kinds of displays. Secondly, stacked multiple-layer displays may be provided in which liquid crystal layers affording identical scan patterns are overlapped, with the consequent result of correspondingly enhanced boundary dfeinition of the combined pattern. These displays achieve more precisely defined boundaries between the opaque and clear images than do single layer displays. Multiple color displays are also possible in such multi-layer panel devices.

""""""""""

3,675,989 (U.S.)   Patented July 11, 1972

LIQUID CRYSTAL OPTICAL CELL WITH SELECTED ENERGY SCATTERING.

<u>Roy Pietsch and Bernard L. Lewis</u>, assignors to Radiation Incorporated.
Application Nov. 26, 1969.

The present invention is in the field of wireless communications systems, and is directed more particularly to signalling systems wherein information is impressed on a carrier wave as it is reflected back to a remote source of the wave.

A popular type of retrodirective reflector for such purposes has been the corner reflector, which generally comprises a plurality of reflecting surfaces so intersecting one another that a beam or wavefront incident on the corner is successively reflected from two or more of those surfaces in a direction back toward the source of the beam or wavefront.

Control of the reflecting coefficient at one or more of the reflecting planes or boundaries of these corner reflectors is effective to vary the intensity of the returned beam. The present invention provides a technique for exercising control over the reflection coefficient of one or more faces of a corner reflector, and is effective with corners of all types, including the metal type, the partially or fully dielectric-filled type, and the completely dielectric type.

In the specific embodiment of Fig. 3, for use in reflecting visible coherent light beams, the transparent conducting layer includes a thin glass plate 35 on which a transparent conducting film 36, such as tin oxide, has been deposited. The reflective conducting layer may include any rigid substrate 37 to which a thin reflecting, electrically conducting film 38, such as silver, will adhere in smooth, sontinuous, flat contour, Liquid crystals maintained in the mesomorphic state constitute the layer 40 between and contacting conductive films 36 and 38. A spacer gasket 41 (Fig. 4) composed of electrically insulating, liquid-impervious material is cemented in position at the exterior edges of the cell to retain the liquid crystals therein.

Typically, spacer 41 has a thickness in the range from 0.5 mil to 2 mils, this constituting the thickness of the liquid crystal layer 40, for example.

With a one mil spacer 41 between conducting surfaces 36 and 38, a potential difference in the range, for example, from 10 to 130 volts, across the cell is sufficient to produce dynamic scattering mode in the crystalline layer 40, the specific voltages depending on the specific composition of layer 40. The operating potential may be applied to the cell by a modulating signal source 43 connected to the conducting films via leads 44 and 45 (Fig. 3).

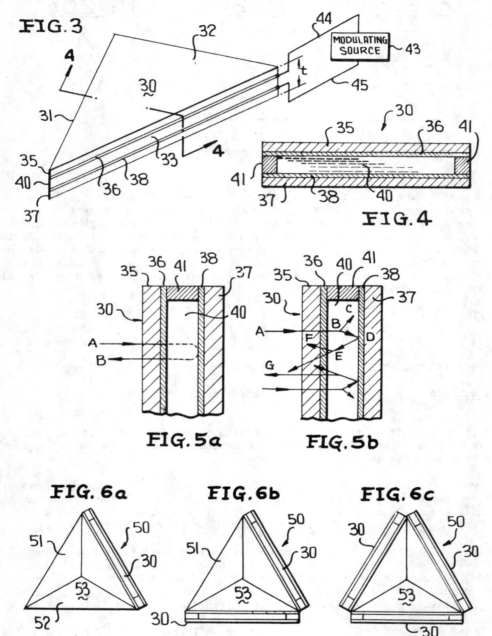

Referring now to Fig. 5a, when the liquid crystals 40 are not in the scattering mode, a light beam represented by ray A is incident on the cell 30, passes through glass plate 35 and transparent conducting film 36, then through liquid crystals 40 and is reflected by surface 38. The reflected ray passes back through the liquid crystalline layer, trabsparent conducting surface, and glass plate, and out of the cell in a direction B (the direction of arrival and direction of reflection neing other than merely opposite, since generally two or three reflecting planes will be encountered between incidence of the beam on the corner reflector and reflection therefrom). Thus, the beam is eventually returned toward its source with virtually no loss in intensity.

With reference to Fig. 5b, the existence of a sufficient potential difference between films 36 and 38 to excite the dynamic scattering mode in the liquid crystals has the effect that the incident beam is diffused as it passes and returns through the liquid crystalline layer. For example, ray A (Fig. 5b) passes through cell layers 35 and 36 and partially through the crystals before diffusion at point B along separate paths BC and BD. At point D on surface 38 a portion of the beam is reflected to point E with additional diffusion, and diffused portions leave the cell along paths EF and EG. Thus, the intensity of the original beam is substantially reduced during passage through the cell when the crystals 40 are in the scattering mode.

As shown in Figs. 6a, b, and c, respectively, a retrodirective corner reflector 50 of tetrahedral configuration may have reflecting planes 51 and 52, and a cell 30, as described above, constituting the plane intersecting planes 51 and 52 to form the corner; or merely one conventional reflecting face 51, with two cells forming the other intersecting reflecting planes; or may have cells 30 constituting all three reflecting planes. Moreover, the open region between the reflecting planes may be partially filled (or completely filled) with dielectric material 53 of different dielectric constant from the surrounding medium. Alternatively, the corner reflector may be basically formed as a solid dielectric, with one or more of its reflecting planes (i.e., boundaries at which total internal reflection can occur) covered with a cell 30.

""""""""""

3,679,290 (U.S.) Patented July 25, 1972

LIQUID CRYSTAL OPTICAL FILTER SYSTEM.

*James E. Adams, Lewis B. Leder, Werner E.L. Haas, assignors to Xerox Corporation.
Application Jan. 6, 1971.*

An optical filter system capable of transmitting light at substantially all wavelengths of incident radiation while simultaneously reflecting radiation at a single wavelength band or plurality of wavelength bands within the incident radiation is described. The system employs liquid crystal films having optically negative properties but which are opposite in intrinsic screw sense.

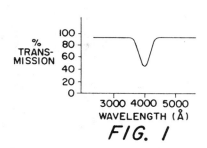

FIG. 1

A right-handed cholesteric liquid crystal substance transmits LHCPL (left-hand circularly polarized light) essentially completely at λ_o whereas the same substance reflects almost completely RHCPL (right-hand circularly polarized light). Conversely a left-handed film is almost transparent to RHCPL at λ_o and reflects LHCPL. Since plane polarized or unpolarized light contain equal amounts of RHCPL and LHCPL, a cholesteric liquid crystal film is approximately 50 percent transmitting at λ_o for these sources when the liquid crystal is in its Grandjean texture. Fig. 1 illustrates the intensity of the output for a liquid crystal substance having a λ_o of 4000A when white light is directed upon it.

A further unique optical property of optically negative liquid crystal films is that contrary to the normal situation when light is reflected, such as by a mirror, where the sense of the circular polarization of the reflected light is reversed, this same phenomenon does not occur with light reflected by these liquid crystal films.

According to the invention the two optically negative liquid crystal films having the same λ_o value but wich are opposite in intrinsic rotatory sense are arranged in combination and incident light is directed upon the device, preferably in a manner such that the direction of light propagation is along the helical axes of the liquid crystal substances. Such combination of liquid crystal films having the same λ_o values but being opposite in intrinsic rotatory sense effectively transmits all of the incident light with the exception of the wavelength band centered around λ_o. This result is illustrated in Fig. 2.

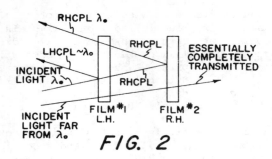

FIG. 2

Referring now to Fig. 3 there is shown an optical filter, generally designated 10, having cholesteric liquid crystal films 11 and 11' with protective members 12 and 12' respectively and separated by optional spacing member 14. The cholesteric liquid crystal films are selected to have the same λ_o value but to be opposite in intrinsic rotatory sense. In this embodiment, for purposes of illustration, films 11 and 11' each have λ_o value of about 5,000A with one being right-handed and the other left-handed. It should be recognized that the position of the two films with respect to each other may be varied without affecting the operation of the device, i.e., either may be positioned in front of the other.

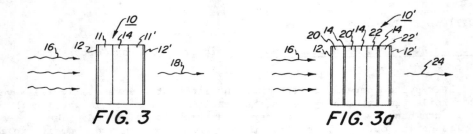

FIG. 3   FIG. 3a

Typically, the optical filter system of the invention will be used at or near room temperature. Generally speaking, the liquid crystal substance will preferably be in the liquid crystal state at the desired operational temperature. For optimum results the liquid crystal films employed in the optical filters of the invention will preferably have a thickness of from about 0.5 to about 20 microns. For liquid crystal films which have a λ_o in the visible region, film thickness of from 3 to about 10 microns will give optimum results. The liquid crystal films 11 and 11' are typically tacky, soft, viscous, glassy or liquid and thus are preferably covered with protective outer layers 12 and 12' respectively to protect the films from foreign matter such as dust, insects or the like. Spacer member 14 is an optional component. It may be any suitable material which is optically isotropic and transparent to the incident light radiation.

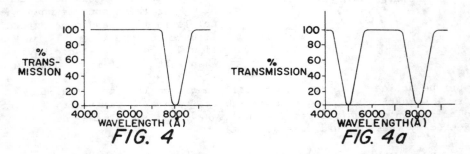

FIG. 4   FIG. 4a

For optimum results the optical filter 10 is preferably arranged in the path of a light beam from a light source in a manner such that the incident radiation, in this case visible light, represented by arrows 16 reaches the filter at normal incidence with liquid crystal films 11 and

11' being preferably disposed so that the helical axes of the liquid crystal films are in the direction of the light propagation. The characteristics of the emergent beam 18 are graphically illustrated in Fig. 4. It is seen that the optical filter 10 essentially completely transmits all wavelengths of incident radiation except for those in the band centered around λ_o (in this case λ_o=5,000A) for the liquid crystal films 11 and 11', whereas the wavelengths centered about λ_o are substantially completely reflected and removed from the emergent beam.

Figure 3a illustrates another preferred embodiment of the invention. There is seen an optical filter having two sets of complementary matched liquid crystal films stacked in series. Thus, liquid crystal films 20 and 20' have λ_o values of about 5,000A and are right-handed and left-handed respectively; and liquid crystal films 22 and 22' have λ_o values of about 8,000A and are left-handed and right-handed respectively. The characteristics of the emergent beam 24 are shown in Fig. 4a. A detailed description of the compositions employed for the liquid crystal films and the preparation of the optical filter is set forth in Example II.

Example I

A composition of 65 percent anisylidene-p-n-butylaniline (ABUTA) in cholesteryl oleyl carbonate (COC) is prepared. This composition is left-handed and has a λ_o value of about 8,000A. A composition of 86 percent cholesteryl chloride (CC) in cholesteryl nonanoate (CN) is prepared. This composition is right-handed and has a λ_o value of about 8,000A. A thin film of the ABUTA in COC composition is applied to a thin glass plate and thin layer of Mylar Sheet is placed over the free surface of the liquid crystal film. A thin film of the CC in CN composition is then applied to the free surface of the thin glass plate and subsequently the free surface of the latter liquid crystal layer is covered with another thin glass plate. Incident light from a broad band incandescent source of visible light is directed upon the optical filter at normal incidence. Transmission spectrum obtained is shown in Fig. 4. It can be seen that the filter transmits substantially completely all of the incident radiation with the exception of the wavelength band centered about 8,000A.

Example II

Two complementary matched pairs of liquid crystal films are incorporated into an optical filter in the following manner. A composition of 57 percent cholesteryl formate (CF) in cholesteryl nonanoate (CN) is prepared. This composition is left-handed and has a λ_o value about 5,000A. A composition of 90 percent cholesteryl chloride (CC) in cholesteryl oleyl carbonate (COC) is prepared. This composition is right-handed and has a λ_o value of about 5,000A. A thin film of the CF in CN composition is applied to a thin glass plate and a second thin glass plate placed over the free surface of the liquid crystal film. A thin film of the CC in COC composition is then applied to the free surface of the second glass plate and subsequently a thin layer of Mylar sheet is placed on the free surface of the latter liquid crystal layer.

Then to the free surface of the first glass plate is applied a thin layer of the 65 percent ABUTA in COC composition (described in Example I). Over the liquid crystal layer is placed a third glass plate and to the free surface of the third glass plate there is applied a thin layer of the 86 percent CC in CN composition (described in Example I). This latter liquid crystal layer in then covered with a thin layer of Mylar sheet. Incident light from a broad band incandescent source of visible light is directed upon the optical filter at normal incidence. The transmission spectrum obtained is shown in Fig. 4a. It can be seen that the filter transmits substantially completely all of the incident radiation with the exception of two wavelength bands centered about 5,000A and 8,000A respectively, these bands being substantially completely reflected. A total of five examples are described.

" " " " " " " " " " " "

3,680,950 (U.S.)   Patented Aug. 1, 1972

GRANDJEAN STATE LIQUID CRYSTALLINE IMAGING SYSTEM.

<u>Werner E.L. Haas, James E. Adams, John B. Flannery</u>, Jr., assignors to Xerox Corporation. Application March 15, 1971.

A system which transforms a liquid crystalline composition having cholesteric optical characteristics from the focal-conic or "undisturbed" texture state into the Grandjean or "disturbed" texture state by an applied AC electrical field, and an imaging system wherein such a liquid crystalline member is imaged in a desired image configuration by the AC electrical field-induced texture transition system. Such transformed compositions exhibit colors and memory characteristics.

In Figure 1 a typical electro-optic liquid crystalline cell 10, sometimes referred to as an electroded sandwich cell, is shown embodying the present invention, wherein a pair of transparent plates 11 having substantially transparent conductive coating 12 upon the contact surface thereof, comprise a parallel pair of substantially transparent electrodes. Cells wherein both electrodes are substantially transparent is preferred where the imaging member is to be viewed using transmitted light; however, a liquid crystalline cell may also be viewed using reflected light thereby requiring only a single transparent electrode while the other may be opaque.

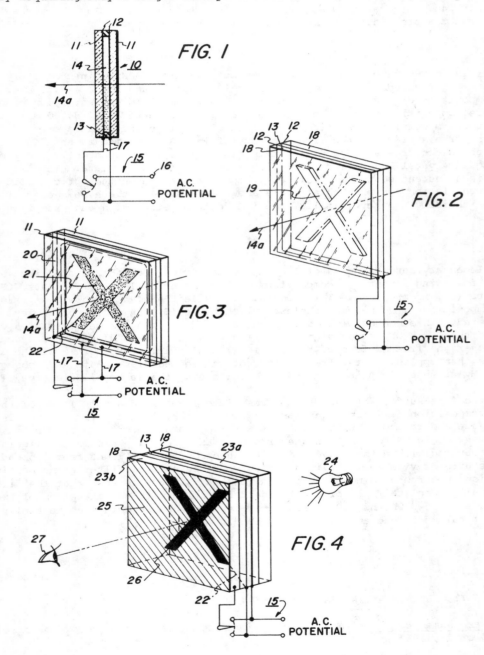

In the advantageous system of the present invention it has been discovered that when the aforementioned liquid crystalline composition is provided in a layered configuration, and the material is provided in its focal-conic, or "undisturbed" texture state, that the application of A.C. electrical fields to the layer of liquid crystalline material causes an A.C. electrical field-induced texture transition to occur wherein the material initially in its focal-conic or "undisturbed" texture state is transformed into the Grandjean or "disturbed" texture state.

The focal-conic texture state is predominantly characterized by its highly diffuse light scattering appearance caused by a distribution of small, birefringent domains. This texture has no single optic axis. The Grandjean texture state is typically characterized by the selective reflection of incident light around a wavelength λ_o where λ_o equals 2np where n equals the index of refraction of the layer of liquid crystalline composition, and p equals the pitch of the liquid crystalline composition. The pitch is the distance between molecular layers having equivalent orientation in compositions having cholesteric liquid crystalline characteristics. The Grandjean state is additionally characterized by optical activity for wavelengths of incident light away from λ_o. Where λ_o is in the visible spectrum, as in the present invention, the liquid crystalline composition layer appears to have the color corresponding to λ_o. The Grandjean texture state is sometimes referred to as the "disturbed" texture state. The focal-conic texture is sometimes referred to as the "undisturbed" texture state. The layer of cholesteric crystalline composition may be provided in its focal-conic or undisturbed texture state by the application of D.C. electrical fields or by low frequency A.C. electrical fields, or, such liquid crystalline compositions may be provided in the focal-conic texture state by heating the compsition to at least about the liquid isotropic transition temperature of the material.

In Figure 2 an embodiment of the electro-optical cell described in Fig. 1 is shown with the desired image defined by the shape of the void areas in the spacer gasket 13. As before, transparent electrodes 18 are separated by the spacer 13; but only the desired image area 19 comprises the liquid crystal film or layer. In operation there is an electrical field across the entire area of the spacer 13, however the image caused by the A.C. electrical field-induced texture transition in the liquid crystalline film causes imaging to occur only in the area 19 where the liquid crystalline film is present.

In Figure 3 another preferred embodiment of the electro-optical cell described in Fig. 1 is shown wherein the desired image is defined by the shape of an electrode, and therefore by the shape of the electrical field created by that electrode.

In Figure 4 an electro-optic imaging cell comprising a pair of substantially transparent electrodes 18 sandwiching the spacer 13 containing a liquid crystalline film is shown being observed between polarizers 23. As illustrated in Fig. 4, light from source 24 is plane polarized by polarizer 23a. In traversing the liquid crystalline film it remains plane polarized in the transformed (Grandjean) areas 26, and ceases to be plane polarized in the non-transformed (focal-conic) areas 25. In passing through polarizer 23b, which is adjusted to a suitable angle, the Grandjean areas 26 appear dark if monochromatic light is used, or appear colored if white light is used. The focal-conic areas 25 appear bright in either case.

Example I

An electro-optical liquid crystalline cell is prepared by providing a tin oxide coated glass slide, providing on the conductive side of said slide an about 1/2 mil thick spacer gasket of Tedlar, a polyfluoride resin film available from Dupont, having a small, square opening cut therein, and a second tin-oxide coated glass slide over the spacer gasket and enclosing a liquid crystalline composition having cholesteric optical characteristics between the conductively coated slides and within the opening of the spacer gasket. The liquid crystalline composition is prepared by mixing about 50 percent cholesteryl oleyl carbonate and about 50 percent anisylidene p-n-butylaniline (ABUTA), and the composition is placed in the cell during cell fabrication. This is a room temperature liquid crystalline composition. The electrically conductive tin oxide coatings of the two slide electrodes are electrically connected to circuitry including a Hewlett-Packard 150 signal generator and a Bogen amplifier. This circuit is suitable for providing A.C. or D.C. electrical fields across the electrodes. The liquid crystalline composition having cholesteric optical characteristics is provided in the focal-conic texture state by a D.C. electrical field. An A.C. electrical field of field strength of about $2.5 \times 10^5 V_{rms}/cm$, and of frequency of about 100Hz, thereby transforming the composition into the Grandjean texture state

The advantageous effects of the transformation are observed by placing the electro-optic cell between polarizers and observing it in transmitted collimated or convergent light in a Leitz polarizing microscope. In collimated white light the field of view becomes uniformly colored at the onset of optical activity characteristic of the transformed Grandjean state. In convergent light a uniaxial negative interference figure is observed. The arms of the uniaxial cross do not reach the central area which retains the same hue observed in the collimated light.

The transformed composition is observed in reflected light by the naked eye, and it exhibits the selective reflection color which is characteristic of the same composition which is placed in the Grandjean texture state by another means, here mechanical shearing. The 50-50 percent mixture of cholesteryl oleyl carbonate and ABUTA exhibits a green reflection color in the transformed Grandjean state. The transformed composition exhibits memory characteristics by maintaining the transformed Grandjean state. The patent describes six additional examples.

" " " " " " " " " " " "

3,687,515 (U.S.)   Patented Aug. 29, 1972

ELECTRO-OPTIC LIQUID CRYSTAL SYSTEM WITH POLYAMIDE RESIN ADDITIVE.

<u>Werner E.L. Haas, James E. Adams and John B. Flannery, Jr.</u>, assignors to Xerox Corporation. Application Jan. 6, 1971.

An electro-optic system wherein a layer of nematic liquid crystalline composition which is optically uniaxial with the optic axis normal to the plane of the layer has an electric field applied perpendicular to the optic axis of the composition layer thereby inducing optical biaxiality in the composition layer. Electro-optic cells and imaging systems are disclosed using the optic retardation accompanying a field induced change from the uniaxial to the biaxial state.

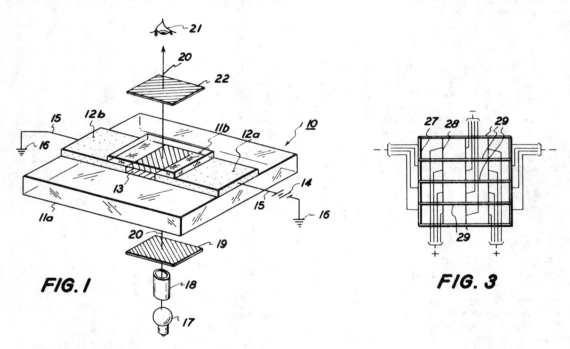

Individual electro-optic cells or valves are shown in conjunction with Figs. 1 and 2, and grids or bid-matrices of coplanar electro-optic cells embodying the present invention are shown in Figs. 3 and 4.

Example I.

A nematic liquid crystalline electro-optic cell is provided by placing a pair of aluminum strips having square cut ends on a standard microscope slide and adjusting the square ends of the strips so that they are about 4 millimeters apart. The aluminum strips are of thickness of

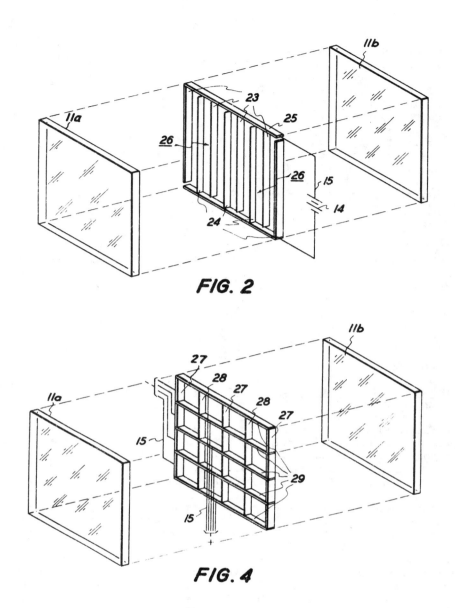

FIG. 2

FIG. 4

about 15 microns. In the space between the aluminum electrode strips on the substrate slide, a layer of (ABUTA) p-methoxy-benzilidene-p'-n-butylaniline is provided, and is leveled to about the thickness of the aluminum electrodes by placing a standard microscope cover slide over the layer of nematic liquid crystalline electro-optic composition. The nematic composition assumes the nematic homeotropic texture wherein the optic axis of the composition layer is aligned normal to the plane of the layer. This imaging cell is then placed under a Leitz Dialux microscope between crossed polarizers. In parallel or collimated transmitted light, the cell appears dark in all stage positions, and in convergent light the conoscopic figure is a uniaxial cross. The optic sign of the composition is positive, i.e., light travels fastest in the direction of the optic axis.

A D.C. field of about 200 volts/cm is applied to the sample, and during the application of the field the transmitted light is extinguished only if the direction of the field, which is normal to the line of transmission and are parallel to optic axis of the layer of electro-optic composition, is parallel to the direction of vibration of either the polarizers or the analyzer. It is observed that during rotation of the microscope stage through 360°, the extinction occurs 4 times. Maximum transition is observed to occur when the field direction is at an angle of about 45° from the direction of vibration of the polarizer or analyzer, which are crossed at

175

about 90°. The sample is observed to transmit light uniformly and transmission increased with increasing electrical fields. Initially, the color of the transmitted light is white, but as the field strength is increased uniform interference hues become visible. Field reversal does not effect the result. When the electric field strength exceeds about 1,500volts/cm of sample width, the sample uniformity decreases. At higher fields of about 3,500 volts/cm of sample width, dynamic scattering is observed to occur.

In convergent light the consocopic figure is clearly uniaxial before the application of the field. The interference figure is exceptionally sharp. The application of the D.C. field results in the breaking of the uniaxial figure, and when the film is observed in the 45° position (electrical field direction of about 45° to the direction of vibration of the polarizer or analyzer) two hyperbolic isogyres are seen, the separation of which increases with field strength.

Example II

A nematic liquid crystalline electro-optic cell is provided as described in Example I, however in this cell the nematic composition is doped with about 0.5 percent of Versamid 100, a polyamide resin available from General Mills, Inc. The dopant enhances the spontaneous alignment of the composition to the nematic homeotropic texture. Patent describes three more examples.

" " " " " " " " " " " "

3,689,131 (U.S.)   Patented Sept. 5, 1972

LIQUID CRYSTAL DISPLAY DEVICE.

Richard I. Klein, Sandor Caplan and Ralph T. Hansen, assignors to RCA Corporation. Application June 29, 1970.

With reference to Figure 1, a liquid crystal animated display device 10 is shown. The device 10 comprises a liquid crystal cell 12, a frame 14 within which the cell 12 is mounted, a printed circuit board connector assembly 16 for making electrical connections to the various electrodes on the cell 12, and a programmer assembly 18 for applying voltages to the electrodes of the liquid crystal cell 12 via the connector assembly 16.

In the instant embodiment, the cell 12 comprises a pair of oppositely disposed transparent front and rear glass substrates 20 and 22, respectively, sandwiching a film 24 (Fig. 2) of a liquid crystal material therebetween. The inside surface 26 of the front substrate 20 is provided with three vertically aligned electrodes 28, 30 and 32, each electrode having a circular shape (Fig.1) and each electrode including a connector 34, 36, and 38, respectively, extending to an edge portion 40 of the front substrate 20. As shown in Fig. 1, the edge portion 40 of the front substrate 20 extends beyond the edge of the rear substrate 22, whereby the ends of the connectors 34, 36 and 38 on edge portion 40 are exposed.

The inside surface 42 (Fig. 2) of the rear substrate 22 is provided with a single continuous electrode 44. To provide means whereby an electrical connection can be made to the electrode 44, a conductive pin 50 is provided extending between the spaced apart substrates 20 and 22 electrically connecting the electrode 44 on the substrate 22 to a connector 52 on the substrate 20. The connector 52 extends to an edge portion 53 of the substrate 20 extending beyond the substrate 2, whereby the end of the connector 52 is exposed.

The various electrodes and the connectors therefor are of a transparent conductive material, e.g., tin oxide.

When a voltage is applied between the electrode 44 and one or more of the electrodes 28, 30, and 32, as described hereinafter, the portions of the liquid crystal film 24 between the energized electrodes are switched to the light scattering mode. All other portions of the film 24 remain transparent. Thus, using a light source 54 above and behind the device, with the light rays from the source directed downwardly through the device and away from the viewer 56 in front of the device, the light passing through the portions of the film 24 in the light scattering mode is scattered in directions including directions toward the viewer. Thus, luminous images, corresponding in shape to the shape of the various ones of the energized electrodes 28, 30, and 32, become visible to the viewer. The frame 14 comprises a pair of flat members 60 and 62 which cooperate with one another to sandwich therebetween the liquid crystal cell 12 and portions of the connector assembly 16.

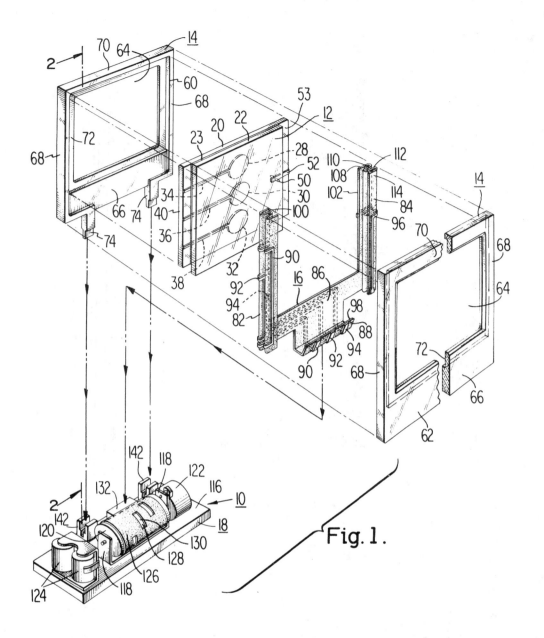

Fig. 1.

The printed connector assembly 16 is formed of a sheet of a stiff but flexible and resilient insulating material, e.g., a known plastic polymer, such as "Mylar" or "Kapton", the sheet having a thickness in the order of 3 mils. The connector assembly 16 has a generally U-shape, including a pair of elongated return bent arms 82 and 84, and an arm connector 86 including a dependent flap 88. Disposed on one surface of the sheet are a plurality of spaced conductive strips 90 and 92, and 94, and 96, e.g., of copper, each strip extending from between a different preselected position along edge 98 of the flap 88 to a different preselected position along an edge 100 or 102 of the two arms 82 and 84, respectively. Along the arm edges 100 and 102, the ends of the strips 90, 92, 94, and 96 are so disposed as to register with individual ones of the exposed ends of the connectors 34, 36, 38, and 52 respectively, of the liquid crystal cell 12 when the device 10 is assembled. An electrically conductive cement, such as a silver epoxy, is used to bond the ends of the various strips 90, 92, 94, and 96 to the corresponding registered exposed ends of the connectors 34, 36, 38, and 52.

The programmer assembly 18 comprises a base plate 116, serving as the support plate for the entire device 10, a pair of vertical brackets 118 mounted on the plate 116, and a cylinder 120 mounted for rotation about the axis thereof between the brackets 118. A motor 122 is mounted on the plate 116 to rotate the cylinder 120, and batteries 124 are provided to drive the motor.

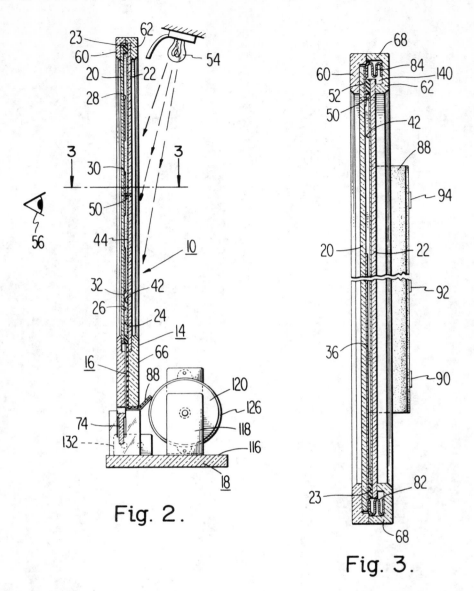

Fig. 2.

Fig. 3.

The wall of the cylinder 120 is made of an electrical conductive material, e.g., copper, covered by a layer 126 of insulating material, e.g., "Mylar". Openings 128 are provided through the insulating layer 126 exposing three narrow strips 130 of the conductive material of the cylinder wall, each strip 130 having parallel side edges, and each strip being parallel to a plane perpendicular to the cylinder axis. A battery 132, of a voltage suitable to activate the liquid crystal film 24, is mounted on the plate 116, one terminal of the battery 132 being connected to the conductive material of the cylinder wall via a commutator, not shown, and the other terminal of the battery being connected to the strip 96 on the connector assembly arm 84 by a means of a wire, not shown. Each of the ends of the various strips 90, 92, and 94 (Fig.1) on the connector assembly flap 88 is aligned with a different one of the cylinder 120 conductive strips 130, whereby, as the cylinder rotates, electrical contact is established between each of the connector strip ends and corresponding ones of the cylinder strips, i.e., the flap 88 serves as a commutator. By selection of the location of the cylinder strips 130 relative to one another, the battery 132 voltage can be applied to the various electrodes on the liquid crystal cell in any desired sequence.

In assembly of the device 10, the liquid crystal cell 12 is first provided by known means. The connector assembly 16 is then aligned with the liquid crystal cell 12 and the ends of the connector strips 90, 92, 94, and 96 are bonded to the exposed ends of the connectors 34, 36, 38, and 52, respectively, on the substrate 20 of the liquid crystal cell. Then the liquid crystal cell 12 and the arms 82 and 84 of the connector assembly 16 attached thereto are disposed within the seat formed by the two frame members 60 and 62, and the frame members are locked together. As shown in Fig. 3, slots 140 are provided in the inside surface of the side portions 68 of the frame member 62 to receive the folded arms 82 and 84. The frame sub-assembly is then mounted on the programmer assembly 18 by force fitting the two dependent tabs 74 on the frame member 60 into brackets 142 on the base plate 116 of the programmer assembly. This automatically disposes the flap 88 of the connector assembly 16 in proper tensioned registry against the cylinder 120 of the programmer assembly.

" " " " " " " " " "

3,690,745 (U.S.)   Patented Sept. 12, 1972

ELECTRO-OPTICAL DEVICES USING LYOTROPIC NEMATIC LIQUID CRYSTALS.

Derick Jones and Sun Lu, assignors to Texas Instruments Incorporated.
Application March 3, 1970.

A display device in which a thin layer of a lyotropic nematic mesomorphic composition is utlized to diffuse light from a source toward an observer by applying a suitable voltage, for example, 20 volts for a 1 mil layer. One form of the device transmits light (Fig. 2) through the layer. A second form of the device (Fig. 3) uses light diffused and reflected to the observer.

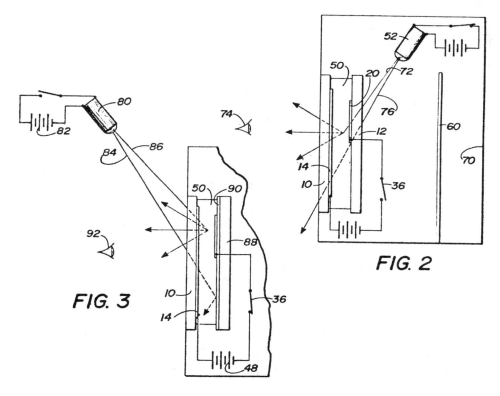

Lyotropic nematic mesomorphic compositions which can be utilized with the present invention will include a solvent and a solute. The types of molecules which will form the required lyotropic mesophase are usually of an elongated, relatively straight and in some cases flattened structure. This shape favors the parallel arrangement of molecules characteristic of the nematic mesophase. In addition these compounds preferably are nematogenic at room temperature, for example in the range of from 15° to 30°C. Examplary solutes are as follows:

a)   butyl-p-(p-ethoxyphenoxycarbonyl)phenyl carbonate,

b)   p-(p-ethoxyphenylazo)phenyl heptanoate,
c)   p-[N-(p-methoxybenzylidene)amino]phenyl acetate,
d)   p-(p-ethoxyphenylazo)phenyl undecylenate,
e)   4,4'-bis(heptyloxy)azoxybenzene,
f)   p-(p-ethoxyphenylazo)phenyl hexanoate,
g)   4,4'-bis(pentyloxy)azoxybenzene,
h)   4,4'-bis(hexyloxy)azoxybenzene,
i)   4,4'-dibutoxyazoxybenzene,
j)   4,4'-dipropoxyazoxybenzene,
k)   p-[N-(p-octyloxybenzylidene)amino] n-butyl benzene, etc. etc.

Preferred solute molecules are compounds a) and c) above, and more preferably a mixture of about 80 percent by weight (a) and about 20 percent by weight of (c).

A preferred solvent is:
m)   p-[N-(p-methylbenzylidene)amino]n-butyl benzene.

Most preferably about 60 mole percent of the solvent (m) is mixed with 40 mole percent of the aforementioned preferred mixture of (a) and (c).

" " " " " " " " " "

3,691,755 (U.S.)   Patented Sept. 19, 1972

CLOCK WITH DIGITAL DISPLAY.

<u>Pierre Girard</u>, assignor to Manufacture des Montres Rolex S.A.
Application Oct. 16, 1970; prior application in Switzerland Oct. 21, 1969.

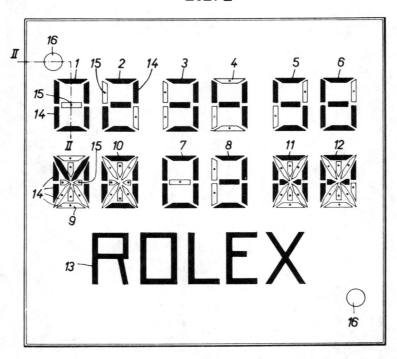

FIG. 1

The present invention concerns a clock with digital display composed of a visible panel provided with display cells, each one of which is formed of a group of fixed display elements which are capable of chnaging their optical properties when excited by an electrical pulse, an electronic logic circuit, the outputs of which are connected to the display elements in order to transmit these pulses to them, a time base controlling the electronic logic circuit, and a power source, all being arranged in such a manner that at each moment the indication of the time formed by the cells occurs as a result of the excitation of various sub-assemblies of their elements.

The clock according to the invention is characterized by the fact that at least a part of the electronic logic is connected to the panel so that the latter comprises a monolithic unit equipped with the display elements and a minimum number of input connections connecting the output of the time base to the electronic logic circuit built into the panel, the latter being attached in a detachable fashion in the timepiece.

Figure 1 shows essentially the arrangement and shape of the display cells 1 to 13 imbedded in a panel which forms the dial of a clock according to the invention. These cells are intended for representing the letters and numbers of a complete time indication including not only the hours, the minutes and the seconds, but also the day of the week and the date, as well as an additional permanent indication. These cells are of three different kinds: The cells 1 to 8 represent letters indicating the day of the week and the month and the cells 13 can be grouped into a single one giving a permanent indication, for example, an indication of the brand.

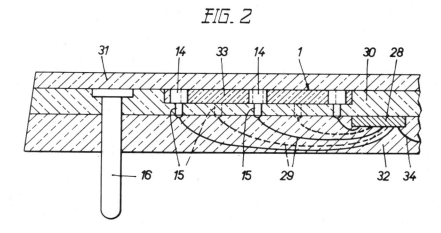

Figure 2 shows schematically a view following the longitudinal axis of the cell 1 through the panel of Fig. 1.

The electronic logic circuit provided for controlling the cells 1 to 12 can include nine circuits such as the element 28 attached to the other side of the panel 30. In effect, it will be necessary to have one decoding circuit for each one of the cells 1 to 6, but one circuit can be used for each one of the pairs of cells 9-10, 7-8 and 11-12. These various integrated circuits are connected to one another by connections such as the connection 34 and each integrated circuit can include a frequency divider so that it will only be neccessary to equip the panel described with an input plug similar to the plug 16 which is intended for receiving the signal coming from the time base and with a third plug for the feed. The panel will thus constitute a monolithic unit provided with a minum number of connections and can be attached removably to the inside of the wall of the timepiece described. It will be the timepiece's dial provided with all of the elements necessary for displaying the time.

""""""""""""

3,693,084 (U.S.)   Patented Sept. 19, 1972

METHOD AND APPARATUS FOR DETECTING MICROWAVE FIELDS.

<u>Carl F. Augustine</u>, assignor to Bendix Corporation.
Application June 17, 1969.

The present invention provides a substantially real time microwave field detector having a film of liquid crystals or the like in close proximity with a thin, continuous resistive layer. A microwave field incident on the composite structure generates alternating currents in the resistive layer which are in accord with the intensity of the microwave energy (Figs. 1 and 2). The alternating electrical currents generate a heat pattern through dissipation in the resistive layer which corresponds to the intensity distribution of the incident microwave field. The heat pattern is imposed on the adjacent liquid crystals through thermal transfer. Since the liquid crystals assume particular colors in accordance with the temperature of the crystals, they will respond chromatically to incident microwave energy levels. More particularly, a visible color pattern is produced representative of a map of the intensity distribution of the microwave field.

The invention also provides a microwave field detecting system which is particularly suited for use as a device for plotting the radiation pattern of an antenna (Fig.3).

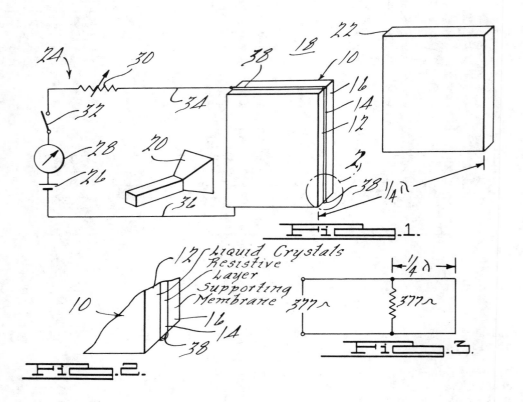

This invention further provides an apparatus for nondestructive internal examination of an object which is permeable to microwave radiation having a first source of radiation directed at the object for modulation thereby, a second source of radiation which interferes with the modulated radiation from the first source, and a microwave field detector positioned to display the resulting interference pattern (Figs. 4,6). The display on the microwave detector reveals imperfections in the object under examination.

This invention still further provides a method and apparatus for determining standing wave ratios and matching the impedance of a load to the impedance of a source for optimum power transfer (Fig.8). Particularly, a transmission line is used between the source and the load which radiates a portion of the transmitted energy. A microwave detector of this invention is positioned to receive the radiated portion. If the impedances of the source and load are not equal, standing waves result which are displayed by the detector. The impedances can be matched by minimizing the standing waves.

As a yet another embodiment, this invention provides a power density meter which is particularly suited for detecting and measuring leakage of microwave energy around the doors and the like of microwave ovens. In this regard, it has been discovered that commercial microwave ovens may leak potentially dangerous levels of microwave energy. The power density meter includes a first microwave detector according to this invention which records both microwave and infrared radiation and a similar second detector recording only infrared radiation (Fig.9).

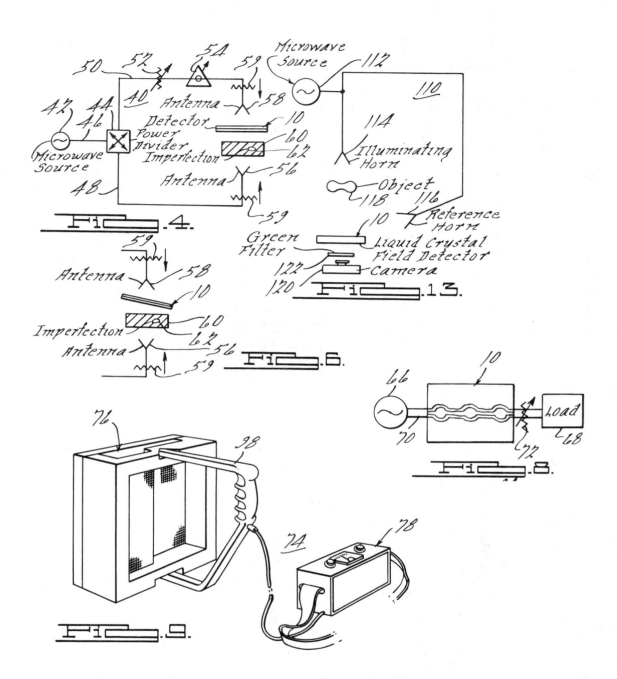

Through measuring only the difference in color levels of the detectors, the infrared energy contribution is effectively "cancelled" since it is the same for both. A first detector is positioned one-quarter wavelength from the first detector to cause all of the incident microwave energy to be dissipated in the resistive film. A second reflector is positioned one-half wavelength from the second detector such that all of the incident microwave energy is reflected and none is absorbed by the resistive film. A circuit is provided for balancing the color levels of both detectors to provide a quantitative measurement of the incident microwave energy.

As a still another embodiment, this invention provides an apparatus for forming holograms using microwave sources. A first source of coherent microwave energy is modulated by an object. A second source of coherent microwave energy is directed to interfere with the first source. A microwave detector according to this invention is positioned in the region of interference of the two sources for displaying the interference pattern. The interference pattern represents a hologram of the object (Fig. 13).

" " " " " " " " " " " "

3,694,053 (U.S.)

NEMATIC LIQUID CRYSTAL DEVICE.

Frederic J. Kahn, assignor to Bell Telephone Laboratories, Inc.
Application June 22, 1971.

A device characterized by an electronically-tunable optical birefringence over the range of 0.0 to 0.2 for applied voltages below 25 volts rms includes a thin film of a nematic liquid crystal. The tunable birefringence occurs below the threshold for dynamic scattering and is the result of electric-field-induced spatially uniform molecular reorientation in a ell-aligned nematic liquid crystal. Applications for such a cell include display elements, tunable retarders, color modulators and variable density filters.

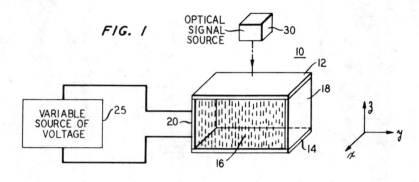

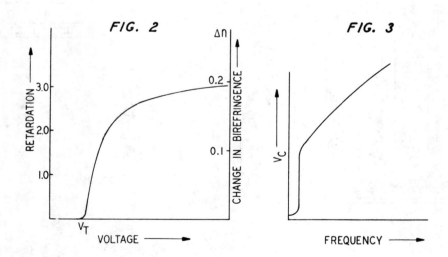

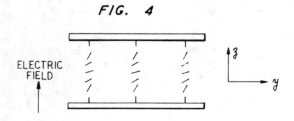

The specific illustrative liquid crystal device shown in Fig. 1 includes two parallel glass plates 12 and 14 having a thin film 16 of nematic liquid crystal sandwiched between them. Thin conductive coatings on the facing surfaces of the plates 12 and 14 provide electrodes by means of which a uniform electric field can be established across the depicted film. Polymeric spacers 18 and 20 made, for example, of Mylar maintain the thickness of the liquid crystal material at any desired value in a range of approximately 1 to 200 microns measured in the z direction. Thus, the liquid crystal is in effect a thin film lying in the xy plane.

In accordance with the principles of the present invention, the active material 16 included in the device 10 of Fig. 1 comprises a nematic liquid crystal of a specified type (MBBA).

The rodlike molecules of the nematic liquid crystal 16 shown in Fig. 1 are designated by short lines. These lines represent the long axes of the molecules. In accordance with this invention, the long axes of the molecules of the thin film are initially oriented parallel to the z axis. Such an orientation or homeotropic ordering is achieved either by mixing a surfactant additive with the nematic liquid crystal material or by directly coating the facing electrode surfaces of the planar members 12 and 14 with an appropriate surfactant.

Surfactants which have been found to induce good homeotropic order in nematic liquid crystals include hexadecyltrimethyl ammonium bromide (Hereinafter identified as HMAB), stearic acid, lecithin and polyamide resins such as, for example, Versamide 100.

Homeotropic ordering of MBBA may be achieved by adding thereto a small amount of HMAB. The resulting mixture is flowed between the planar members (by capillary action, for example). It is believed that the HMAB molecules then plate out on the planar members. This plating action can be accelerated by momentarily heating the liquid crystal material (to about 50°C., for example) and then allowing it to cool to room temperature. For 12-micron-thick samples the optimal concentration of HMAB in MBBA is about 0.005 percent by weight. If all the HMAB is assumed to plate out, this would correspond to a uniform coating on the order of a monolayer on each electrode surface.

Alternatively, homeotropic ordering of a material such as MBBA may be achieved by directly depositing a thin layer of, for example, HMAB on the aforementioned electrode surfaces. This direct deposition method has been found to provide more uniform homeotropic ordering than the aforementioned additive method. A uniform surfactant layer may be obtained by dissolving the surfactant in an appropriate organic solvent, spraying the solution on the electrode surfaces and allowing the solvent to evaporate.

That the above-described procedures are indeeed effective to establish homeotropic ordering of the molecules of the thin film 16 can be experimentally verified. By viewing the film 16 between crossed polarizers in convergent light directed along the z axis, one sees the characteristic conoscopic figure corresponding to that of an optically uniaxial crystal which has its optic axis perpendicular to the electrode surfaces coated on the members 12 and 14.

To create a preferred anisotropic direction in the nematic liquid crystal thin film, which in turn will give rise to a predetermined molecular orientation in the presence of an electric field applied thereto, a further preparation step is required in the course of fabricating the device represented in Fig. 1. This step involves selectively treating the electrode layers that are respectively deposited on the planar members 12 and 14. Illustratively, the treatment comprises rubbing the electrodes, foe example with lens paper, in a direction parallel to the direction in which the molecular axes of the thin film are to lie in the presence of an applied electric field. Rubbing each of the electrode surfaces in the y direction about six times with moderate pressure has resulted in imposing the desired anisotropic direction in the thin film 16. In carrying out this step neither the rubbing material, the number of strokes, nor the amount of pressure appears to be critical. The rubbing as described above is carried out prior to deposition of the surfactant.

One other illustrative way in which to establish preferred direction for molecular reorientation is to utilize a permanent magnetic film for one of the electrodes of the depicted device. In that case the magnetic field established thereby, directed for example along the y axis, will determine the preferred reorientation direction.

Figure 3 illustrates for a particular sample the manner in which V_c, the voltage threshold for hydrodynamic instability, increases as the frequency of the voltage applied to a nematic liquid crystal is increased. Thus, by applying ac electric fields to the thin film 16 shown in Fig. 1, it is possible to make devices which are tunable over virtually the full available birefringence characteristic of the nematic liquid crystal material, but which operate at all times below the threshold for hydrodynamic instability.

A source 25 is shown in Fig. 1 for applying variable voltages between the electrode surfaces respectively disposed on the members 12 and 14. By that means, variable electric fields can be established in the nematic liquid crystal material 16. In the presence of an applied field in the $\underline{z}$ direction, some of the homeotropically-oriented molecules of the material are rotated in a spatially-uniform manner (spatially uniform in the $\underline{xy}$ plane) toward the $\underline{y}$ axis. This reorientation, which is represented schematically in Fig. 4, gives rise in turn to a change in the birefringence characteristic of the device 10. Thus, optical signals directed at the device 10 by a source 30 (Fig. 1) "see" a birefringence characteristic whose value is a function of the voltage applied across the thin film 16.

A typical curve of birefringence versus applied voltage for a specific 16-micron thick film of nematic liquid crystal excited at 1kHz with an electric field in the $\underline{z}$ direction is shown in Fig. 2. The voltage threshold at which a change in birefringence is first observed is well defined. Fig. 2 also indicates the variation of optical retardation with applied voltage for the above-specified example. The retardation is zero at zero applied voltage and approaches 3 microns at an optical wavelength of 546 nanometers when the applied voltage reaches about 24 volts rms. Just above the voltage threshold the differential change in retardation is about 1.25 microns per volt rms. The advantageous low-voltage nature of the described device is apparent. For example, retardations greater than 1 micron can be obtained with applied voltages less than 5 volts rms.

"""""""""""

3,697,150 (U.S.) Patented Oct. 10, 1972

ELECTRO-OPTIC SYSTEMS IN WHICH AN ELECTROPHORETIC-LIKE OR DIPOLAR MATERIAL IS DISPERSED THROUGHOUT A LIQUID CRYSTAL TO REDUCE THE TURN-OFF TIME.

Joseph J. Wysocki, assignor to Xerox Corporation.
Application Jan. 6, 1971.

A system for transforming an optically negative liquid crystalline mesophase composition to an optically positive liquid crystalline mesophase composition by an applied electrical field, and imaging compositions which facilitate the relaxation of the transformed optically positive composition into the initial optically negative state. The invention also encompasses imaging systems wherein this electric-field-induced transition images a liquid crystalline imaging composition.

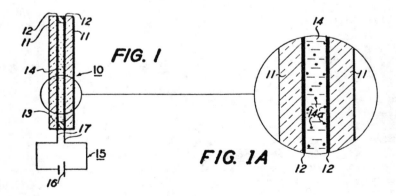

Figure 1 illustrates a liquid crystalline, electro-optic cell 10 in partially schematic, cross-sectional view. Fig. 1A is a partially schematic magnification of a portion of the cross-sectional view of Fig. 1 showing the support members 11 with the conductive or substantially transparent conductive coating 12 thereon with the liquid crystalline film or layer 14 enclosed between the conductive electrodes. In addition, in the present invention the liquid crystalline composition also comprises or includes the advantageous additive 14A which is dispersed throughout the liquid crystalline electro-optic composition matrix.

The electro-optic effect of the liquid crystalline imaging members of the present invention, which may also be used as a liquid crystalline electro-optic imaging system, is in part described in U.S. Pat. No. 3,652,148, the entire disclosure of which is incoprorated by reference in the present specification. Any of the address means suggested in U.S. Pat. No. 3,652,148 may be used, including electron beam address systems, and electron beam address systems using other included electrical field generating systems such as pin tubes, or layers of secondary emission materials; X-Y grid address systems; electrostatic latent images, for example, electrostatic latent images on any sort of insulating supports which are brought into close proximity with layers of the electro-optic compositions of the present invention, such as an electrostatic latent image on a photoconductor layer; combined electrical and thermal address systems; and a variety of multiple cell, and coplanar multiple cell address systems, as well as any other suitable means for providing the appropriate electrical fields across the electro-optic composition layer. It is again noted that in various embodiments the electrical fields may be provided by Ac or DC potential sources or any suitable combinations thereof.

EXAMPLE I.

A cholesteric liquid crystalline composition comprising about 35 percent cholesteryl chloride and about 65 percent cholesteryl nonanoate is prepared, and an about 50-50 percent mixture of $PbCrO_3$ and a copolymer of styrene and n-butyl methacrylate pigmented with carbon black, is added in an amount somprising about 5 percent of the total resultant composition. This composition is mixed to suspend the additive in the liquid crystalline composition, and a layer of the composition is placed into a cell comprising a pair of substantially transparent electrodes, the electrode surface being a transparent chromium coating on a glass substrate, and the electrodes enclose the composition in an about 1 mil thick Mylar spacer gasket.

A voltage of about 900 V. DC is provided across the thickness of the composition layer, thereby causing the cholesteric-nematic phase transition. Microscopic observation reveals that the included particulate additives bound back and forth between the electrode plates. Recovery or off-times are measured for the above cell and for a control cell which does not contain the additive. The present cell exhibits off-times which are about one-half the duration of off-times in the control cell.

EXAMPLE II.

A cholesteric liquid crystalline composition comprising about 59 percent cholesteryl chloride, about 39.4 percent cholesteryl nonanoate, and about 1.6 percent oleyl cholesteryl carbonate, is prepared, and polysulphone is added and dispersed therein in an amount comprising about 15 percent of the total resultant composition. A layer of this composition is provided in a cell as in Example I, except that the spacer gasket is of a thickness of about one-half micron.

A voltage of about 1,500 V. DC is provided across the thickness of the composition layer, thereby causing the cholesteric-nematic phase transition to occur. Under a microscope, the dispersed additive is observed to move vigorously throughout the composition. Recovery or off-times are measured for the above cell and for a control cell which does not contain the additive. The present cell exhibits off-times which are about one-third the duration or off-times in the control cell.

EXAMPLE III.

A cholesteric liquid crystalline composition comprising about 60 percent cholesteryl chloride and about 40 percent cholesteryl nonanoate, is prepared, and cottonseed oil is added and dispersed therein in an amount comprising about 10 percent of the total resultant composition. A layer of this composition is provided in a cell as in Example II.

A voltage of about 100 V. DC is provided across the thickness of the composition layer, thereby causing the cholesteric-nematic phase transition to occur. Recovery or off-times are measured for the above cell and for a control cell which does not contain the additive. The present cell exhibits off-times which are about one-fourth the duration of off-times in the control cell.

Five more examples are given in the patent...

" " " " " " " " " " "

3,697,152 (U.S.) Patented Oct. 10, 1972

TUNING METHOD FOR PLURAL LAYER LIQUID CRYSTAL FILTERS.

<u>James E. Adams, Lewis B. Leder</u>, assignors to Xerox Corporation.
Application Jan. 6, 1971.

A method for tuning optical devices constructed with one or more liquid crystal films possessing optically negative characteristics which comprises rotating the liquid crystal films with respect to the beam of incident light.

When a substance or a film is said to be right-handed it is meant that it reflects right hand circularly polarized light (RHCPL) and when it is said to be left handed it is meant that it reflects left hand circularly polarized light (LHCPL).

A right-handed cholesteric liquid crystal substance transmits LHCPL essentially completely at λ_o whereas the same substance reflects almost completely RHCPL at λ_o. (λ_o equals $2np$ with n representing the index of refraction of the liquid crystal substance and p the pitch or repetition distance of the helical structure). Conversely, a left-handed film is almost transparent to RHCPL at λ_o and reflects LHCPL. Since plain polarized light or unpolarized light contain equal amounts of RHCPL and LHCPL, a cholesteric liquid crystal film is approximately 50% transmitting at λ_o for these sources when the liquid crystal is in its Grandjean or "disturbed" texture. These properties make such cholesteric liquid crystal substances extremely valuable when utilized in optical devices. The geometry for this dispersive reflection phenomenon is shown in Fig. 1.

Rotating a liquid crystal film with respect to an incident light beam constitutes a tuning technique. In the case of a band pass filter which has one liquid crystal film and transmits only one band of wavelengths, the transmitted band can be shifted to another region of the light spectrum. For multiband pass filters which have more than one liquid crystal film and transmit a plurality of bands of wavelengths corresponding to the number of films present, the individual transmitted bands may be shifted independently of each other by rotating one film while maintaining the others in their original position or they may be shifted in tandem by rotating the filter itself.

With respect to optical notch filters which utilize complementary matched pairs of liquid crystal films a variety of results may be achieved by resort to the method of the invention. By a complementary matched pair of liquid crystal films is meant two individual films arranged in a manner such that they both have the same λ_o value with one being a left-handed film and one being a right-handed film. Of course, both films may have the same intrinsic λ_o value and thus could both be arranged in the path of the incident light beam in a manner such that the light is directed on the films at normal incidence. Consider, however, a situation where the two films do not have the same intrinsic λ_o value. In such a case, the reflection spectra of the two individual films may be brought into coincidence by rotating one film while the other is held fixed. At this point the two films would then constitute a complementary matched pair. Of course, with respect to any complementary matched pair of films, whether the individual films which make up the pair are instrinsically matched or matched in the manner described above, the notch itself, i.e., the band of wavelengths which is substantially completely reflected by the filter, can be shifted by rotating the matched pair of films in tandem. Moreover, where this type of filter has a plurality of complementary matched pairs of liquid crystal films so that the filter substantially completely reflects a plurality of wavelength bands the respective reflected bands may be shifted independently or in tandem.

Referring now to Fig. 2 there is seen an exemplary optical band pass filter 10 consisting of a liquid crystal film 12 having optical negative properties, and positioned between a linear polarizer 14 and a linear analyzer 16, the latter two members having a predetermined angular relationship between their axes of polarization such that the cooperative action of these two members is effective to prevent the transmission of incident unpolarized light through the linear analyzer.

Preferably, to obtain optimum results, the optical filter 10 is arranged in the path of the incident light beam, represented by arrows 18, in a manner such that the incident light beam is normal to the liquid crystal film 12. Thus, the emergent beam 20 contains only a wavelength band centered about some wavelength λ_o, the remainder of the wavelengths of the wavelengths of radiation within the incident light beam having been extinguished by the cooperative effect of the linear polarizer and the linear analyzer. Of course, λ_o in any instance is determined by the particular liquid crystal substance, mixture of liquid crystal substances or composition compris-

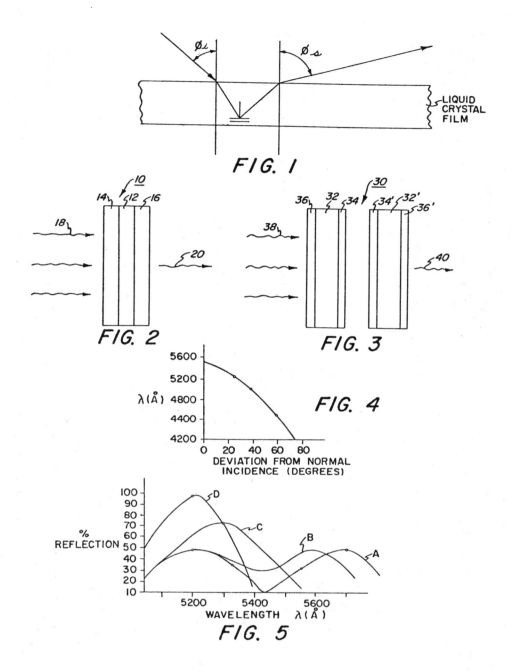

FIG. 1

FIG. 2

FIG. 3

FIG. 4

FIG. 5

ing a liquid crystal substance employed for liquid crystal film 12. In this exemplary instance liquid crystal film 12 is comprised of a composition of 23 percent cholesteryl chloride by weight in cholesteryl nonanoate.

Consider now what occurs when the liquid crystal film 12 is rotated with respect to the incident light beam such that the angular incidence of the light beam on the film deviates from normal incidence. The results obtained are illustrated by Fig. 4. It is seen that at normal incidence λ_0 for the liquid crystal film is about 5500 A. When the film 12 is rotated to an extent such that the angular incidence of the incident light beam upon the film deviates from normal incidence by 20° the λ_0 value for the film is shifted to about 5350A. When rotating the liquid crystal film the λ_0 value thereof can be shifted to various other wavelengths as is clearly shown.

Referring now to Fig. 3 there is seen an exemplary optical notch filter 30. The filter 30 is shown for purposes of illustration as having two components. One component has a thin liquid crystal film 32 comprising a composition of 24.5 percent by weight of cholesteryl chloride (CC) in cholesteryl decanoate (CD) positioned between a thin glass plate 34 and thin Mylar sheet 36. This liquid crystal composition is left-handed and has a λ_o of about 5,200A. The other component has a thin liquid crystal film 32' comprising a composition of 20 percent by weight CC in cholesteryl bromide (CB) positioned between a thin glass plate 34' and a thin Mylar sheet 36'. The liquid crystal composition of the latter component is right-handed and has a λ_o of about 5,700A. While it is clearly apparent that the two liquid crystal films could be selected to have the same intrinsic λ_o values, i.e., for incident light directed upon them at normal incidence they would reflect the same wavelength band, for purposes of illustrating the invention in this exemplary instance they have been chosen to have different instrinsic λ_o values.

The reflection spectra of the optical filter of Fig. 3 for various conditions are shown in Fig. 5 where the percent reflection of the filter is plotted against wavelength (A). Curve A shows the condition of the reflected beam when both films 32 and 32' are positioned in a manner such that the incident light beam, represented by arrows 38 strikes both films at normal incidence. There are seen two discreet wavelength bands, which are partially reflected by the filter, one centered around a wavelength of about 5,200A and corresponding to liquid crystal film 32 with the other centered around a wavelength of about 5,700A and corresponding to liquid crystal film 32'. When liquid crystal film 32' is rotated by 15° and film 32 is held fixed the reflection spectrum of the filter changes. The spectrum for this situation is illustrated by Curve B. Rotating film 32' by 30° while keeping film 32 normal to the light beam produces the reflection spectrum shown in Curve C. Curve D illustrates the result which occurs when film 32' is rotated by 42°. It can be seen that the filter now substantially completely reflects light in the wavelength band centered around 5,200A since the individual reflection spectra of films 32 and 32' have now been brought into coincidence, i.e., both films now reflect the incident light most strongly in the region around 5,200A. Thus, the filter now constitutes a typical notch filter since it transmits substantially all of the incident light while simultaneously rejecting substantially completely the wavelengths of light in the narrow band centered around 5,200A. At this condition the emergent beam 40 for optical filter 30 contains all the wavelengths of radiation of the incident light except for the reflected band centered around 5,200A.

" " " " " " " " " " " "

3,697,297 (U.S.)   Patented Oct. 10, 1972

GELATIN-GUM ARABIC CAPSULES CONTAINING CHOLESTERIC LIQUID CRYSTAL MATERIAL
AND DISPERSIONS OF THE CAPSULES.

Donald Churchill, James V. Cartmell, and Robert E. Miller, assignors to The National Cash Register Company.   Application Oct. 22, 1970.

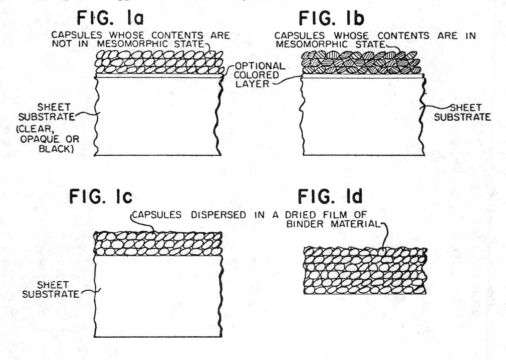

A temperature-responsive display device or composition, comprising mesomorphic materials contained in minute capsules as the temperature-responsive component, is provided. The minute capsules, when placed in heat-conductive relation with the heat source together with an associated thermal gradient pattern to be displayed, exhibit an iridescence indicative of a certain characteristic temperature range, which temperature range varies according to the types and mixtures of mesomorphic materials affords physical and chemical protection from various kinds of degradation and allows their use in practical and efficient thermal-gradient display systems.

Up to this time, the utility of cholesteric mesomorphs in temperature-sensitive systems has been severely limited due to a number of reasons, the following points being some of the assigned faults:

1) Such systems containing a mixture of one or more intermingled mesomorphic compounds as a film are subject to crystallization of large areas at the desired working temperatures.

2) Mesomorphic cholesterol derivatives are oily liquids at and above their melting temperatures. When they exist as a film on any surface, the film, being wet, is subject to injury of many types from aging and contact with the environment.

3) The subject wet-film mesomorphic systems are prone to anomalous color change behavior upon the event of only slight contamination of the system by various solutes.

Isolation and protection of these thermotropic liquids as droplets as the cores of minute transparent-walled capsules, and the distribution of those capsules in a film-layer onto a desired eligible substrate for use as a temperature sensing device, represent the predominant substance of this invention.

Fig. 1a is a diagrammatic edge view of a substrate sheet having a coating of capsules, on a surface, with the cholesterol derivative contents not in the mesomorphic state but either in an isotropic liquid state or in a crystalline solid state. The capsules depicted have been applied from a slurry wherein the slurrying liquid is a liquid in which the capsules were manufactured or another, dryable, liquid without dissolved solid materials. The capsules in these showings are adhered together, to other capsules and to the substrate sheet by residual capsule wall material which remained in solution in the manufacturing vehicle during encapsulation, or by the tacky capsule wall material itself.

Fig. 1b shows the sheet of Fig. 1a with the capsule contents shown to be in the mesomorphic state. Fig. 1c is a diagrammatic edge view of a substrate sheet having a coating wherein capsules are embedded in a polymeric binder material. Fig. 1d is a diagrammatic edge view of a self-supporting temperature-responsive display film comprising a matrix of polymeric binder material throughout which are embedded minute capsules containing mesomorphic materials.

Following are specific examples of an encapsulating procedure and the mesomorphic materials used for preparing capsules of the novel temperature-responsive sensing system.

Example I

1.25 grams of acid-extracted pigskin gelatin, having a Bloom strength of 285 to 305 grams and an isoelectric point of pH 8-9, and 1.25 grams of gum arabic were stirred with 125 grams of distilled water at 55 degrees centigrade in a Waring Blendor to yield a solution which was about pH 4.5. When the solution was formed, the pH was adjusted to 6.0 by the drop-by-drop addition of 20%, by weight, aqueous sodium hydroxide solution. To the above system was added a solution of 2 grams of cholesteryl propionate in 8 grams of cholesteryl oleate. The cholesterol derivative materials were emulsified in the Waring Blendor vessel to an average droplet size of 10 to 50 microns, and the pH of the system was slowly reduced by the dropwise addition of 14%, by weight, aqueous acetic acid solution. Addition of acid was continued until the single, liquid-walled, capsules clustered to form aggregates having diameters of about 25 to 100 microns. At this final state, the system pH was about 5.

The entire liquid system was then chilled with an ice bath to below 10 degrees centigrade, while the agitation was continued. At 10 degrees centigrade, 0.6 milliliter of a 25%, by weight, aqueous solution of pentanediol, a chemical hardening agent for the gelatin, was added to the blender vessel, and the system was allowed to stir for about 12 hours while slowly returning to room temperature. At the end of that time, the capsule walls were firm and hardened, and the capsular system was poured through a wire mesh sieve having opening of 74 microns. That which passed the screen was suitable for coating the intended substrate. Capsules, along with the equilibrium liquid which passed through the sieve, were coated, using a drawdown applicator, to a wet thickness of more than 2 but less than 10 mils onto a blackened 5-mil-thick, polyethylene terephthalate film (sold as Mylar) and then were dried in air at about 25 degrees centigrade. If temperature was raised at a point on the coated layer, concentric rings representing a temperature gradient pattern appeared outwardly from a central ring of blue, thence spectrally in order through red, as the temperature gradient decreased. As the temperature of the point was increased or decreased, the effect was evidenced by expansion or contraction of the rings. The system of Example I first exhibited the mesophase coloration transition at about 25 degrees centigrade and continued to about 25 to 30 degrees centigrade. Systems of adjacent expanding and contracting rings intersected, and the resultant color was a function of the temperature of a given area.

Eight additional specific examples are described in the patent.

"""""""""""

3,700,306 (U.S.)   Patented Oct. 24, 1972

ELECTRO-OPTIC SHUTTER HAVING A THIN GLASS OR SILICON OXIDE LAYER BETWEEN THE ELECTRODES AND THE LIQUID CRYSTAL.

James V. Cartmell, Donald Churchill, and Donald E. Koopman, assignors to The National Cash Register Company.   Application Sept. 22, 1971.

A electro-optic shutter employing nematic liquid crystal material is disclosed. The electrodes of the shutter of this invention are protected with a thin overcoat of an insulating material such as glass. This thin overcoat essentially prevents rapid electrode failure due to conduction of current through the liquid crystal material.

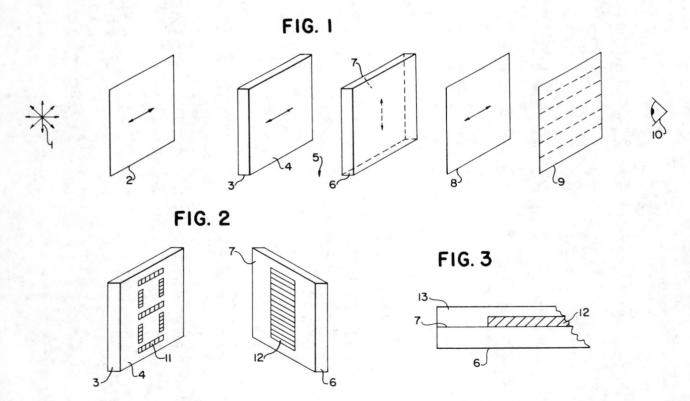

A major problem with electro-optical devices employing nematic liquid crystals is rapid electrode failure due to electrolysis of the electrode material as a result of current conduction through the cell. One solution to this problem has been the employment of high purity liquid crystal material.

It now has been found that the electrodes can be protected with a thin overcoat of an insulating material such as sintered or fused glass, silicon oxide or the like. This thin overcoat removes the criticality of the purity of the liquid crystal material as current conduction is prevented even if the liquid crystal material is not pure.

In Fig. 1, light rays (not shown) pass from light source 1 through polarizer 2. Substrates 3 and 6 are aligned so that their adjacent faces 4 and 7 are parallel and spaced about 0.5 to 1.0 mil apart. Shims (not shown) can be used to maintain this spacing. Substrates 3 and 6 then are sealed (not shown). Prior to sealing, space 5 is filled with the desired liquid crystal materials. Light transmitted through substrates 3 and 6 then is transmitted through polarizer 8 and is seen on light scattering screen 9 by viewer 10.

In Fig. 2, electrodes 11 and 12 are shown. Note, electrode 11 is a 7-bar matrix and electrode 12 is a coating. Electric leads (not shown) are attached to the electrodes. It is these electrodes which are coated with the protective overlay of this invention.

In Fig. 3, a side view of substrate 6 is shown with electrode 12 and face 7 coated with thin overcoat 13, i.e., the protective overlay of this invention.

A feature of this invention is the cell construction. After the plates have been etched and rubbed, a thin ribbon of Etylene Acrylic Acid copolymer is laid around the edge of the plates on the rubbed side. The liquid crystal is placed between the plates and they are brought together with pressure and heat (to 100°C.) The copolymer wets the cell surfaces near the edges and effectively seals the cell.

Typically, light passing through the first polarizer is polarized in the horizontal plane (arrow). As it passes through the cell it is rotated 90° so that it emerges polarized in a vertical direction. If polarizer 2 is aligned to transmit light only in the horizontal plane, the light is extinguished and the observer sees no light. When the field is turned on, the cell does not affect the plane of polarization and the light emerges from the cell polarized in the horizontal direction, is transmitted by the second polarizer and the viewer sees light on the screen. A light image of the character in a dark background is observed. Alternately, the second polarizer may be rotated 90° and the field off mode will transmit light so that when the field is applied a dark image on a lighted background is observed. Contrast values of 30:1 are typically obtained.

The scattering screen is used to improve the quality of the image. Also, since the effect is that of a shutter, the angle of view is very restricted and the screen acts as a diffuser. Without the screen, and with proper lenses, the system could be used as a dynamic projection image.

The preferred mode of operation is with a pulsed, unipolar DC bipolar DC or AC potential. The exact potential required is a function of the thickness and dielectrics constant of the insulating layer. Typical threshold voltages are 4 volts for 250 pulses per second unipolar DC and 2 volts for 60 Hz AC. Since this is a field orientation the current is very low. The AC frequency can be varied from 20 Hz to several thousand Hz. The switching time is about 10 milliseconds on and 40 milliseconds off.

Additional features of the insulating overlay are the optical uniformity of the cell when the overlay is present. Without the overlay, the liquid crystal electrode interface has an optical reflectivity different from that of the interface of the liquid crystal and the surrounding glass areas. Thus the electrodes are visible. With the overlay, the electrode areas are essentially indistinguishable from the surrrounding surface. Also the overlay protects the electrode areas duriing rubbing and a more vigorous rubbing procedure may be applied without electrode damage. The advantages of this invention are further illustrated by the following examples.

Example I

Two identical electro-optic shutters were operated according to this invention on a 10 volt pulsed 250 cycles per second, unipolar DC potential. A shutter such as described in Figs. 1 and 2 was employed Substrates 3 and 6 were 4x4 inches glass plates. Electrode 11 was etched as a 7-bar matrix and electrode 12 was etched panel. The electrodes were indium oxide. In one shutter, a thin, approximately 1 micron, overlay of vacuum deposited SiO was placed over the electrodes, while no overlay was employed in the second shutter. Employing a polishing cloth, the interior surface of substrate 3 was rubbed in a horizontal direction and the interior surface of substrate 6 was rubbed in a vertical direction. Space 5 of both shutters was filled with a mixture containing by weight 40 percent methoxybenzilidene-n-butylaniline (MBBA), 40 percent ethoxybenzilidene-n-butylaniline (EBBA) and 20 percent p-n-butoxybenzilidene-p'-aminobenzonitrile and then sealed. Cross-polarizers 2 and 8 were oriented in a horizontal direction.

In the shutter with no protective overlay, the electrodes failed after 20 hours of continuous operation. In the shutter with the SiO protective overlay, the electrodes showed no sign of failure after 2 months of continuous operation. Note, the only difference between the two shutters was the SiO overlay.

Example II

A second cell was constructed using transparent tin oxide electrodes. A layer of Vita 1001 glass frit was deposited over the electrode areas by a silk-screen technique. The plates were then heated to 615°C. to fuse the fritt to a clear layer of glass 1 mil thick. The surfaces were then rubbed and the nematic mixture of Example I was sealed in the cell. The lifetime and operating characteristics were similar to the cell of Example I having the protected electrodes.

"""""""""

3,700,805 (U.S.)   Patented Oct. 24, 1972

BLACK-AND-WHITE IMAGE CONTROL BY ULTRASONIC MODULATION OF NEMATIC LIQUID CRYSTALS.

Thomas F. Hanlon.
Application Aug. 26, 1971.

A modulator in which ultrasonic waves are generated by a video carrier potential applied to a piezoelectric transducer mounted in a glass-sided cell containing a nematic liquid crystal, is positioned in the path of light from a source consisting of a stroboscopically pulsed (at the rate of the speed of sound) visible light laser or non-coherent light source, to produce a complete black-and-white television image.

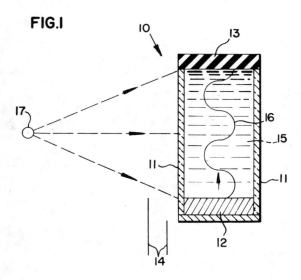

The modulator 10 has a number of applications in imaging systems, including television; it consists of a container or cell having opposed transparent, or glass sides 11, a piezoelectric transducer 12 shown positioned in the bottom of the container, an attenuator 13 made, for ex., of neoprene, at the opposite end of the container, and a nematic liquid crystal 15 filling the space between the transducer and the attenuator. Leads 14 serve to apply potentials to electrodes, not shown, on the opposed sides of the transducer. The numeral 16 designates a train of ultrasonic waves traveling through the liquid crystal in the direction of the arrow.

The cell 10 in an arrangement similar to that shown in Fig. 1 will provide black-and-white density control in a black-and-white receiver. Application of high frequency video voltage to the leads 14 causes the ultrasonic waves through the liquid to compress, rarify and shear the ordered clear molecular structure of the liquid crystals. The light emerging from the cell, the source of which is shown at 17, varies in density accordingly as the high frequency carrier voltage is modulated by the video signal, whereby light scattering or density centers are formed; the density changes are dependent upon changes in the high frequency voltage.

If, instead of attempting to produce a spot of light of element size, we shall have at the transducer end of the cell an amount of light proportional to the instantaneous value of the video signal received, and stretched out across the image of the cell we have at any and every instant a complete record, in light, of the signals sent out by the television transmitter during the preceding (for example) 50 microseconds. The details of light and shade move along the length of this image at a speed corresponding to the velocity of sound in the nematic liquid crystal and therefore are invisible to the eye. If, however, the light source 18, Fig. 2, is pulsed visible light laser or a pulsed non-coherent light, pulsed at the rate of speed of sound, the whole image is made to move across the screen (right-hand side of cell 10) at the same speed in the opposite direction, and those details will be immobilized and will become apparent to the eye. Each picture detail will then be illuminated from the moment the ultrasonic waves corresponding to it leave the transducer 12 until these same waves pass out of the illuminated portion and are terminated at the attenuator 13. The output of the modulator, or cell, 10, which has been stroboscopically stopped by the pulsed laser or pulsed non-coherent light source 18 is focused on to a slowly rotating mirror or galvanometer 19 that produces the vertical scan when synchronized with the vertical scan signal. Thus a complete television image is produced and this can be projected on to a screen 20 for viewing, or to photographic film for recording.

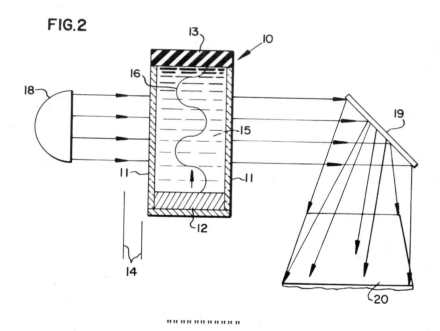

3,702,723 (U.S.)

Patented Nov. 14, 1972

SEGMENTED MASTER CHARACTER FOR ELECTRONIC DISPLAY APPARATUS.

<u>Howard C. Borden, Jr</u>., assignor to American Micro-Systems, Inc.
Application Apr. 23, 1971.

An improved highly readable character comprised of segments which can be energized in various combinations to define each of the arabic numerals in an electronic display apparatus. In one embodiment several characters are formed as thin film conductive members on the front plate of a liquid crystal display apparatus.

In Figure 2 is shown a transparent front plate 10 of a typical visual display or readout device having a series of identical, spaced-apart, segmented master characters 12 embodying the principles of the present invention. Each of these characters comprises nine segments formed by

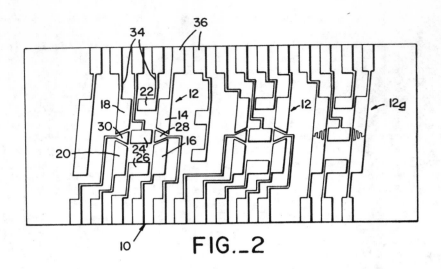

FIG._2

areas of a deposition layer of a transparent conductive material on the plate 10, which are separated by narrow areas of no deposition. These include a first aligned pair of segments 14 and 16, and a second similar pair of aligned segments 18 and 20 that are parallel but spaced apart from the first pair. Three spaced apart horizontal segments 22, 24, and 26 extend between the pairs of vertical segments at various levels. At their junction the vertical segments 14 and 16 on one side are each beveled at an angle (e.g. 45°) to form a triangular gap within which is situated a slightly smaller triangular segment 28. The vertical segments 18 and 20 are similarly beveled at their adjacent ends to form a similar triangular gap within which is a triangular segment 30. Thus, when all of the nine segments are viewed together the master character 12 has the appearance of a complete numeral 8 in a highly readable form.

Each master character with its separate and distinct segments may be formed of any suitable transparent conductive material which can be applied as by a deposition process to the inner surface 32 of the plate 10. Attached to and extending along the surface from each segment is a lead member 34 and these members terminate at bonding pads 36 spaced along an edge of the display plate. The lead members are rendered invisible by elimination of the electric field in these selective areas. To the pads 36 may be connected the various leads from a signal supplying control circuit of whatever apparatus the displayer is used on.

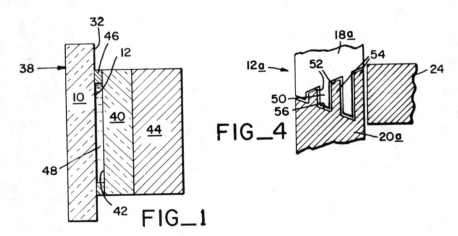

FIG_1   FIG_4

In Fig. 1 is shown schematically a typical display device 38 which can utilize the front plate 10 with its series of spaced apart master characters 12 to provide a digital clock readout. In this arrangement for a so-called liquid crystal type display unit the front plate 10 is installed adjacent and parallel to another rear plate 40. This rear plate is also made of glass or a clear plastic and has a uniform coating 42 of a transparent conductive material on its outer

196

surface. Preferably, this rear plate is fixed to an opaque mounting fixture 44 such as an epoxy light sink. The two plates 10 and 40 are held apart a small distance by a peripheral sealing member 46 such as a plastic or structural glue which retains a quantity of a transparent organic liquid crystal commonly referred to as "liquid crystal" 48.

Light shining through the front glass plate 10 passes through the plate 40 and is absorbed by the rear opaque fixture 44. Whenever any segment of a master character 12 is supplied with an electrical signal so as to produce a voltage across the liquid at that location, the segment will be energized to reflect or diffuse light and thus become visible to the human eye.

It will be seen that, for each master character energizing all segments in the aforesaid manner will produce a perfect figure eight. For any numeral from 0 to 9 the triangular segment 28 will always be energized, but the other triangular segment 30 is used only for the numerals 0, 4, 5, 6, 8 and 9.

A modified version of the master character, designated 12a is shown on the right end of the group of characters in Fig.2 and in greater detail in Fig.4. Here, vertical portions 18a and 20a are configured at their adjacent end portions so as to eliminate the need for the triangular portion 30. This simplifies the character because it also eliminates the associated input leads for the triangular portions 28 and 30 and the attached pads on the plate 10, as well as the logic circuitry necessary to activate each triangular portion.

The finger-like extensions 52 and 54 of the two character portions 18a and 20a interdigitate, so that when both of these portions are energized, a substantially solid vertical bar appears.

" " " " " " " " " " "

3,703,329 (U.S.)   Patented Nov. 21, 1972

LIQUID CRYSTAL COLOR DISPLAY.

<u>Joseph A. Castellano</u>, assignor to RCA Corporation.
Application Dec. 29, 1969.

A color display system capable of producing essentially all the colors of the visible spectrum comprises three cells each including a solution consisting of a pleochroic dye in a nematic liquid crystal composition. Each of the solutions can change its transmission of polarized white light in response to an electric field so as to change the color appearance of the solution. One solution of the system can change in appearance from colorless to magenta, another from colorless to cyan, and a third from colorless to yellow. The system includes means for applying an electric field separately to each of the solutions and means for passing polarized white light successively through each solution.

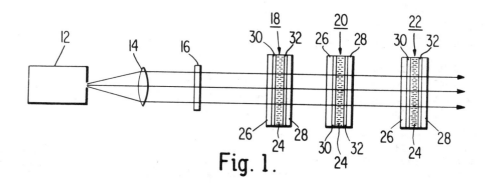

Fig. 1.

Referring to Fig. 1, there is shown a source of white light 12 which is preferably collimated by means of a collimating lens 14. The collimated white light then passes through a polarizer 16 and thence in tandem through three liquid crystal electro-optical cells 18, 20 and 22 respectively. Each liquid crystal cell comprises a solution 24 of a pleochroic dye dissolved in a nematic liquid crystal composition which aligns in an electric field. The solution 24 is contained between transparent support plates 26 and 28 having transparent conductive coatings 30 and 32 respectively on the inner surfaces thereof. In operation, each of the conductive coatings is connected to a voltage source (not shown) so as to separately control and vary the electric field applied across the liquid crystal solutions of each liquid crystal cell. In this way the degree of alignment of each of the pleochroic dyes can be separately controlled. The specific pleochroic dye composition in each of the three solutions is different from that of the others.

In operation, by varying the electric field of each cell from between zero to that which is necessary to produce maximum alignment of the dye molecules, one can obtain substantially every color in the visible spectrum, including black and white, when starting with polarized white light which passes in tandem through all three cells.

The system may be made more compact and the number of glass surfaces from which reflective losses may occur may be reduced by providing a structure as shown in Fig. 2 wherein two inner support plates 42 and 44 each help support two solutions and electrodes therefor at the same time. In this instance, both sides of each of the inner support plates 42 and 44 are provided with transparent conductive coatings, 46 and 48 on plate 42 and 50 and 52 on plate 44.

Typically, the solutions consist of nematic liquid crystal host material of the type which aligns in response to an electric field, into which is dissolved from about 0.5 to 5 weight percent of a pleochroic dye. A preferred dye concentration is from 1-2 percent. It should be pointed out that the preferred pleochroic dyes are non-ionic in the nematic liquid crystal solution.

The specific solutions employed in the preferred system consists of:

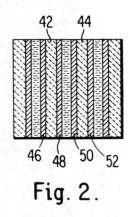

Fig. 2.

a) cyan to colorless;
0.65 wt. percent indophenol blue and 0.35 wt. percent 4-butoxybenzylidene-4'-amino-4"-
-nitroazobenzene dissolved in an equimolar nematic solution of p-hexoxybenzylidene-p-
-aminobenzonitrile, p-hexoylbenzylidene-p'-aminobenzonitrile, and p-heptoylbenzylidene-
-p'-aminobenznitrile.

b) magenta to colorless:
1 wt. percent of 2-amino-8-hydroxy-1-azonaph-thyl-4' benzonitrile dissolved in the same equimolar nematic solution as given in (a).

c) yellow to colorless;
1.5 wt. percent of 4-butoxybenzylidene-4'-
-amino-4"-nitroazobenzene dissolved in the same equimolar nematic solution as in (a)

" " " " " " " " " " "

3,703,331 (U.S.)   Patented Nov. 21, 1972

LIQUID CRYSTAL DISPLAY ELEMENT HAVING STORAGE.

<u>Joel E. Goldmacher</u> and <u>George H. Heilmeier</u>, assignors to RCA Corporation.
Application Nov. 26, 1971.

A liquid crystal element exhibiting storage of its light scattering state after removal of the electric current initiating the state comprises a mixture of cholesterol, a cholesterol derivative or a cholesteric liquid crystal compound with a nematic liquid crystal of the type that exhibits non-destructive turbulent motion when an electrical current of sufficient magnitude is passed therethrough.

Example I

Fig. 3 is an example of a preferred novel liquid crystal element in the form of a crossed grid optical display device 30. The novel device consists of back and front transparent support plates 31 and 32 respectively. The plates 31 and 32 are parallel and are separated by a distance

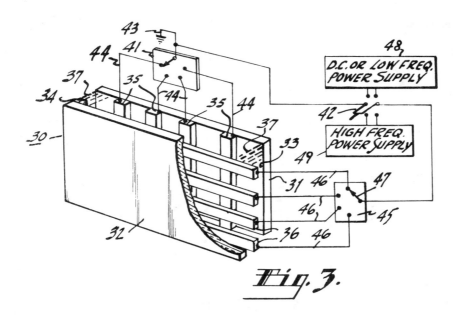

Fig. 3.

of about one-fourth mil. On the inner face 33 of the back plate 31 is an array of parallel spaced transparent conductive back electrode strips 35. On the inner surface 34 of the front plate 32 is an array of parallel, spaced transparent conductive strips 36. The front and back conductive strips 35 and 36 are mutually perpendicular.

The space between the front and back plates 31 and 32 is filled with a liquid mixture 37 comprising 80 wt. percent of a nematic liquid crystal composition and 20 wt. percent of cholesterol derivatives. The nematic composition consists essentially of an equal weight ratio of anisylidene-p-aminophenylacetate, p-n-butoxybenzylidene-p-aminophenylacetate and anisylidene-p--aminophenylbutyrate. The cholesterol derivatives consist essentially of 23 weight percent of cholesteryl chloride and 77 weight percent of cholesteryl oleate. The mixture may be sealed in the device 30 by using epoxy cement around the edges of the device 30.

Also shown in Fig. 3 is a schematic representation of a selection circuit which may be used for operating the novel device. In operation, the device 30 is normally substantially transparent to light incident upon it. By applying a D.C. voltage, of for example 50-100 volts, or a low-frequency AC voltage, of for example 50-100 volts at 30-120 Hertz, across the intersection of any of the conductive strips 35 and 36, the region of the liquid mixture 37 in this intersection will scatter light incident upon it. Upon removal of the voltage, light will still be scattered in this region until the mixture thermally relaxes and returns to its unexcited state or until a high frequency AC voltage of for example 100 volts at 1,200 to 2,000 Hertz is applied to the intersection whereupon the mixture 37 returns to its transparent unexcited state within the order of about 10 milliseconds. Complete thermal relaxation to the transparent state may take several weeks or longer at room temperature. This relaxation time is substantially decreased at increased operating temperatures. After three days at room temperature the contrast ratio of this cell was reduced by about 25 percent.

Example II

In this example, the device 30 or a similar device is comprised of a mixture of the same compounds as described in Example I except that the nematic liquid crystal compositions comprises 90 weight percent of the total mixture and the cholesterol derivatives comprise 10 weight percent of the mixture. With this mixture, scattering can be produced with only 35 volts DC or 60-110 volts AC at 30 Hertz. Erasure of the light scattering can be accomplished with from 60-100 volts AC at about 600 Hertz. The contrast ratios between the light scattering region and an adjacent transmitting region is in the order of about 6:1. This value depends on the particular mixture being used and the voltage and temperatures to which it is subjected.

Figure 4 is a graph showing the thermal decay of scattering as a function of the time after the removal of the voltage which initiated the scattering effect. Brightness due to scattering is plotted as the ordinate in arbitrary units and time in minutes is plotted on the abscissa.

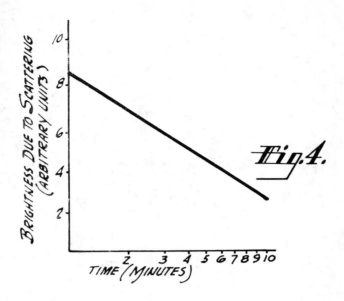

Fig. 4.

A test cell was comprised of parallel glass plates separated by a 1/4-mil thick layer of a mixture of 10 weight percent cholesteryl nitrate, 30 weight percent anisylidene-p-aminophenyl-acetate, 30 weight percent p-anisylidene-p-aminophenylbutyrate and 30 weight percent p-n-butoxy-benzylidene-p-aminophenylacetate. The inner face of one glass plate was coated with aluminum and the inner face of the other glass plate was coated with conductive transparent tin oxide. These coatings served as conductive contacts. The cell, maintained at 28°C. was excited to its light scattering state by a DC voltage. The voltage was removed shortly thereafter and the brightness of scattered light from the cell was measured as a function of time. The brightness of this particular mixture at 28°C is shown to be reduced by about 75 percent of its initial brightness in about 10 minutes. If one compares this to the relaxation rate of the mixture of Example I, it can be seen that the relaxation rate and hence the storage time is greatly dependent upon the particular composition of the mixture.

" " " " " " " " " " "

3,704,056 (U.S.)   Patented Nov. 28, 1972

IMAGING SYSTEM.

Joseph J. Wysocki, James E. Adams and Robert W. Madrid, assignors to Xerox Corporation.
Application Aug. 31, 1971.

This patent is a division of U.S. Patent No. 3,642,348 and discloses a liquid crystal imaging system similar to the one described in U.S. Pat. No. 3,652,148. However, the advantageous process of the present invention produces a different effect and the voltages and field stregths used for imaging the liquid crystal imaging members in the present invention have lower values than those used in the previous system.

In Figure 4 a liquid crystal imaging member comprising a pair of substantially transparent electrodes 18 sandwiching a spacer 13 containing a liquid crystal film is shown being observed between polarizers 23. Cholesteric liquid crystals are typically selectively dispersive and when placed in the electrical field characterstic of the advantageous system of the present invention, the Grandjean to focal-conic texture change occurs thereby rendering the transformed portion of the liquid crystalline film more diffusely light scattering in transmitted or reflected light. At the same time, those areas remaining in the Grandjean texture state continue to be selectively dispersive. When such a liquid crystalline imaging sandwich is observed between polarizers, light from source 24 is plane polarized while passing through polarizer 23a, selectively dispersed by the liquid crystals in the Grandjean texture in non-image areas 25, and is more fully transmitted by the more diffusely scattering effect of the focal-conic textured image areas 26. A viewer 27

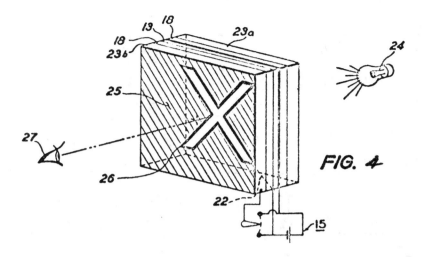

FIG. 4

then sees the planar polarized light which passes through polarizer 23b which originated from source 24 and was scattered and passed through the image portion of spacer 13. Although the light was polarized by polarizer 23a in a plane crossed with the plane of polarizer 23b, the scattering effect of the cholesteric liquid crystal between the substantially transparent electrodes scattered sufficient amounts of the originally planar-polarized light to allow some of it to pass through polarizer 23b. However, in the non-image areas 26 where the liquid crystals remain in their Grandjean texture state, the effect of polarizers 23, when said polarizers have their respective planes of polarization crossed, is to cut off some or all of the light transmitted through polarizer 23a and transformed non-image areas 26 so that those areas 26 appears colored or dark to the observer.

" " " " " " " " " "

3,704,625 (U.S.)   Patented Dec. 5, 1972

THERMOMETER USING LIQUID CRYSTAL COMPOSITIONS.

<u>Hiroshi Seto, Mituka Ueda and Humio Segawa</u>, assignors to Sankyo Keiryoki Kabishiki Kaisha. Application Dec. 28, 1970.

A thermometer comprising a plurality of temperature responsive sections formed of liquid crystal compositions and which change their color in response to a specific range of prime temperature.

Referring to Figs. 1 and 2, the reference humeral 1 indicates a support comprising a base sheet 2 formed of metal, such as aluminum, or synthetic resin, and a front cover 3 formed of a substantially transparent synthetic resinous material and having dimensions substantially equal to those of the base sheet 2. Between the base sheet 2 and the front cover 3 are arranged a plurality of temperature responsive sections 4A, 4B, 4C... formed of liquid crystal compositions. Seven such sections are illustrated in Fig. 1 in alignment with each other with spacings, and accompanied by numerical indications 6A, 6B... indicating threshold temperatures at which the temperature responsive sections 4A, 4B, 4C... react.

The temperature responsive sections may be formed by suitably depositing or spraying on the bottom surface of the front cover 3 or on the upper surface of the base sheet 2 seven different liquid crystal compositions each having a color changing property at predetermined temperatures. Namely, the temperature responsive section 4A is responsible to a specific range of temperature or a temperature between 35° and 36°C., the section 4B between 36° and 37°C., the section 4C between 37 and 38°C and so forth, so that the thermometer as illustrated may be used to measure a temperature between 35°C and 42°C. For better visual observation, it is preferred that a black or dark backing 5 is deposited to directly underlie the temperature responsive sections, so that the sections will appear black in the non-responsive state.

When the section 4A whose prime temperature range is between 35° to 36°C is heated first at 35°C and the temperature is gradually elevated to 36°C, it will first be colored red and the

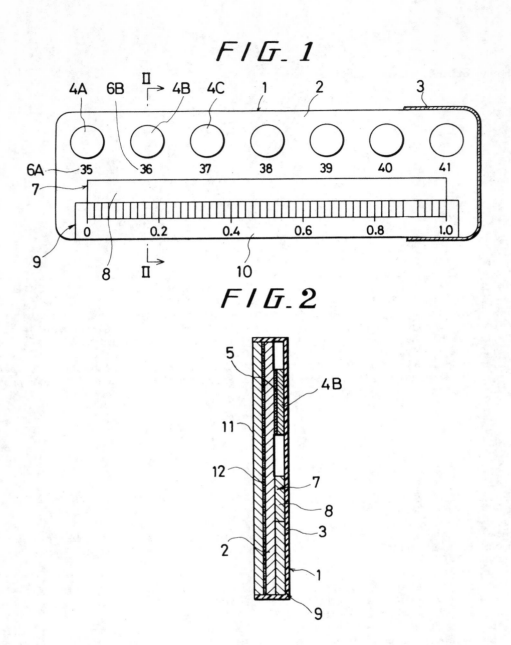

FIG_1

FIG_2

color will then change successively from red to orange, yellow, green, light blue, dark blue and purple in the order mentioned.

The thermometer of this invention further comprises a first strip 8 carrying a chromatic pattern 7 thereon and a second strip 10 which bears a graduation 9 with numerical indications. The first and second strip 8 and 10 may be bonded to the surface of the base sheet 2. Alternatively the chromatic pattern 7 and the graduation 9 may be reproduced directly on the base sheet 2 or either side of the front cover 3, for example, by printing.

The chromatic pattern 7 represents a rainbow, like pattern including red, orange, yellow, green, sky blue, dark blue, and purple. These colors correspond to the colors that can be produced by any temperature responsive section when the temperature to which said section is subjected falls in its prime temperature range. The entire length of the chromatic pattern 7 which covers 1° Centigrade is graduated as at 9 so as to indicate, for example, a fiftieth of a degree. In the embodiment shown, the graduation 9 is numbered in steps of 0.2°C., which will practically assist reading of substantially a tenth of a degree Centigrade.

In operation, when the device is exposed to an object whose temperature is to be measured and the temperature responsive section, say 4C, produces a color other than black or different from the other sections, it can be read that the temperature is 37°C or above, but less than 38°C. The specific color produced should then be identified on the chromatic pattern. If the color is orange and the visually equivalent color in the chromatic pattern is read to be indicated by 0.2 of the graduation, the measuring temperature should be 37.2°C. When the temperature is say exactly 37°C, two adjacent sections, such as sections 4B and 4C will operate to be responsive. In such a case, either one of them may be optionally selected as an operating section, and the color thereof compared with the chromatic pattern.

On the rear surface of the base sheet 2 may be deposited an adhesive film 12 which in turn may be covered with a thin protection sheet 11 made of a material capable of being easily peeled. The adhesive film 12 permits the device to be so positioned as to be properly exposed to an object or a patient whose temperature is to be measured.

If a reversible-type liquid crystal composition is employed instead of a memory type composition, repeatedly usable thermometers may be provided.

" " " " " " " " " " "

3,705,310 (U.S.)   Patented Dec. 5, 1972

LIQUID CRYSTAL VOLTAGE DISPLAY DEVICE HAVING PHOTOCONDUCTIVE MEANS TO ENHANCE THE CONTRAST AT THE INDICATING REGION.

<u>Peter Wild</u>, *assignor to Brown, Boveri & Company Limited.*
Application May 24, 1971; prior application in Switzerland May 27, 1970.

The present invention relates to means for displaying measured voltages, and particularly to arrangements for displaying measured voltages with an electro-optic indicating medium extending two-dimensionally between two energized electrodes.

Figures 2, 3a, and 5 illustrate two electrodes 1 and 2 sandwiching and indicating medium 3 that extends two-dimensionally between them. Fig. 1 illustrates the electrode 1 separately.

As shown in Fig. 3b a voltage gradient is produced by voltages applied to left and right end terminals on the electrode 2 to produce a voltage gradient changing from left to right. A voltage substantially equal to the sum of the threshold voltage necessary for changing the characteristics of the medium 3 and the voltage to be displayed is applied to the electrode 1 which is composed of three parts 1', 1", and 1'". These voltages produce a voltage that varies approx. linearly along the medium 3. When no signal to be displayed is applied to the electrode 1 the voltage across the medium 3 at no point exceeds the threshold value. Thus, at no time does an optically distinguished zone appear. When the signal to be displayed is added to the voltage on electrode 1 the voltage across the medium exceeds the threshold value over a zone depending upon the value of the signal to be measured. The boundary of this zone shifts with the variation in the voltage to be measured. The zone then has an optical characteristic different from the optical characteristic of the remaining zone in the medium 3.

The electrode 2 is formed from a transparent layer of constant surface resistance such as a layer of SnO_2. Potentials of V_2 and zero are applied to the terminals 6 and 7 respectively to form a voltage gradient $V_2(x)$ in the x direction. This gradient is indicated in the right hand part of Fig. 3b.

As shown in Figs. 1 and 3, electrode 1 comprises bands 28a...28z extending substantially perpendicularly to the direction x of motion of the boundary X_M between zones. The bands are electrically connected by means of two bus bars 30 and 31. The bands have a first generally opaque section 1' of resistance R_1, adjoined by a second, transparent, electrically highly conductive section 1". This latter section is followed by a third section 1'". The latter is composed of a resistive layer whose resistance R_3 depends upon the incident light arising from the indicating medium 3'. The dependence is such that the original high resistance becomes low when the section is illuminated. The resistance R_2 of the conductive section 1" is much lower than R_1 of the section 1'. The resistance R_1 is substantially smaller than the total resistance of the indicating medium 3.

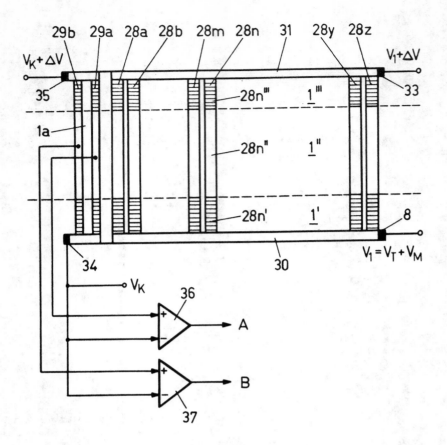

Fig.1

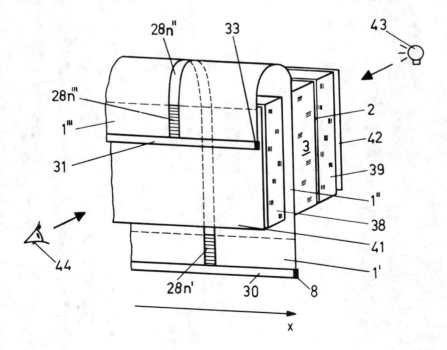

Fig.2

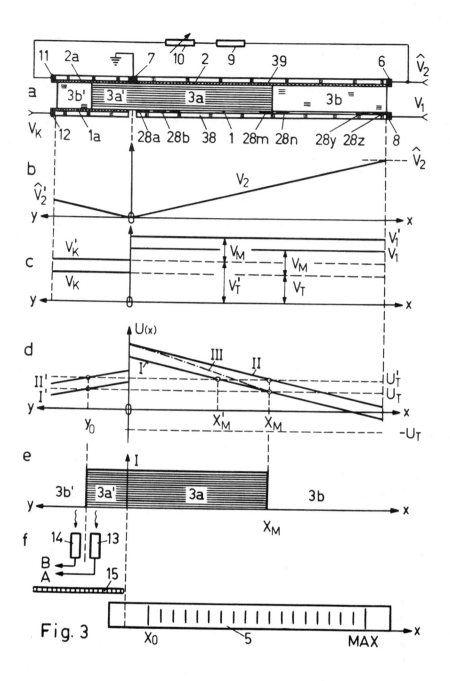

Fig. 3

A terminal 8 connects the bus bar 30 to the voltage $V_1 = V_T + V_M$. The voltage V_T is slightly higher than the threshold voltage U_T of the indicating medium 3. The voltage V_M is the measured voltage. That is to say, it is the voltage which is to be displayed by the display according to the invention. A terminal 33 connects the bus bar 31 to a voltage. $V_1 + \Delta V$ which represents the voltage V_1 increased by the voltage ΔV.

The potential of the central band section 1" is then $\quad V_1'' \approx V_1 + \Delta V \cdot R_1/R_1 + R_3$

The band sections 1", as shown in Fig. 2, are in contact with the indicating medium 3 and serve as electrodes. Thus the indicating medium 3 is initially at the voltage

$U(x) = V_1 - V_2(x)$ which decreases linearly in the direction x. This voltage passes through the threshold level U_T at X_M. $U(x)$ is plotted in the form of curve 1 in Fig. 3d. The zone marked by 3a in Fig. 3e is optically distinguished. That is to say, it is the zone established by the voltage across the medium greater than the threshold level.

The increased light output from the indicating medium in the zone 3a reduces the resistance R_3 of the third section 1''' of any band located in the zone 3a. Consequently, the potential V_1'' of the central section 1'' rises so that the local voltage $U(x)$ present in the indicating layer is increased and the light intensity intensified further. Thus, a positive feedback effect is produced. The local voltage distribution $U(x)$ then has substantially the form of the curve III in Fig. 3d. The optical effect follows a steep non-linear characteristics.

The indicating medium 3 should be subjected to the potential differences $U(x)$ brought about by the second section 1''. At the same time the third section 1''' should be able to receive the light from the indicating medium 3. To accomplish these ends electrode 1 can take the form of a band-like, flexible, transparent synthetic material foil, coated with tin dioxide (SnO_2). In accordance with Fig. 2, a foil of this kind is secured to the indicating medium by a glass plate 38 and curved at the top so that the third section 1''' partially overlaps the second indicating zone 3. According to one embodiment of the invention the electrode 2 is applied as a tin dioxide layer to the glass plate 39.

The third section 1''' is shielded from environmental light. Inadvertent overlapping onto neighboring bands is avoided by optical decoupling, for example, by the provision of optical partition walls.

If liquid crystals are used for the indicating medium 3, then the entire indicating area, including the coupling section covered by the foil area 1''' must be illuminated. According to one embodiment of the invention this is achieved by a light source 43 disposed at the side of the indicating unit opposite the observer 44. (see Fig.2). Two crossed polarizers are used. One polarizer 42 is inserted between the light source 43 and glass plate 39 for passing light only in the x direction. The other polarizer 41 is located between glass plate 38 and the section 1''' of the foil electrode 1. This passes light only in the direction perpendicular to the direction x. If no optical scattering takes place in the indicating medium 3 then only a minimum of light can pass the crossed polarizer arrangement. Thus the display is substantially dark as far as the observer is concerned. However, the higher the scattering power of the medium 3 the greater will the amount of light which is transmitted due to depolarization to the side at which the observer is stationed. The light coupling section is also masked off vis-a-vis the observer.

In Fig. 3, the band electrode 1 and the electrode 2 with the uniform resistive layer, each adjoin a secondary electrode 1a of transparent electrically conductive material and 2a with a homogeneous resistive layer, respectively. Main and secondary electrodes 1, 2 and 1a, 2a are preferably made of the same material.

The voltage drop $V_2(x)$ alon- electrode 2 is produced by application of the voltage V_2 to the terminal 6 and grounding the terminal 7 as shown in Figs. 3a and 3b. The potential distribution in the section 1'' as shown in Fig. 1, of the electrode 1, produced by application of the voltage V_1 to the terminal 8 is illustrated at the right of Fig. 3c. For simplicity's sake the band structure of Figs. 1 amd 2 has been omitted. That is to say the illustration represents the limiting case of an infibite number of bands. The local voltage distribution $U(x)$ in the indicating medium is reproduced by Fig. 3d at the right, curve I. The indicating medium 3 assumes an optical characteristic in the zone 3a different from the zone 3b as shown in Figs. 3a extends from x=0 to x=x_M for $U(x)$ greater than U_T. The particular value of the measured voltage V_M is obtained from a comparison of the optically distinguished zone 3a with the measurement scale 5. That is to say the value of the voltage V_M to be measured and displayed is indicated by comparing the optically distinguished zone 3a with the measurement scale 5.

When the indicating medium 3 is in the form of a liquid crystal layer the voltage V_2 should not exceed twice the value of the threshold voltage U_T of the indicating medium. This avoids localized potential differences such as $U(x)$=MAX exceeding U_T when the voltage V_M=0.

The specific resistance of the electrode 2 is selected so that the current flow between the the electrodes 1 and 2 through the indicating medium 3, modifies the linearity of the voltage $V_2(x)$ to an insignificant extents. According to one embodiment of the invention a current

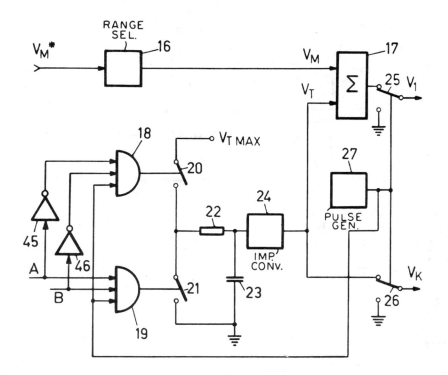

Fig. 4

density of $2\mu A/cm^2$ may be regarded as typical in nematic liquid crystals in the saturating contrast change range.

A secondary electrode 2a on one side of the medium 3 adjacent the electrode 2 is grounded at terminal 7. A terminal 11 connects the secondary electrode to the voltage V_2 through resistors 9 and 10. A terminal 12 connects a secondary electrode 1a to a chopped voltage V_k whose amplitude corresponds to the voltage V_T on the electrode 1.

These voltages result in the voltage characteristics of Figs. 3b and 3d at the left. In Fig. 3c secondary electrode 1a is at the potential V_k. The chopping is not indicated. The secondary electrode 2a exhibits the linearly rising voltage characteristics shown in Fig. 3b and extending from zero to V'_2. This develops a local voltage characteristic $U(x)$ in the indicating medium 3. This characteristic is of the kind indicated by the curve 1' at the left in Fig. 3d.

If chemical processes within the indicating medium change the threshold value U_T to U_T' as shown in Fig. 3b then the limit x_M of the optically distinguished zone 3a shown in Fig. 3e will shift to the value of x_M'. This would result in a false indication of the measured value. To prevent such false indication two photo-detectors 13 and 14 are arranged close together in the area of secondary electrodes 1a and 2a. The photo-detectors 13 and 14 are connected to a circuit of the kind shown in Fig. 4. The photo-detectors produce respective logic signals A and B. If the two photo-detectors are illuminated then A=1 and B=1. If they are not illuminated then A=0 and B=0.

The mask 15 in Fig. 3f serves to mask off the control photo-detector 13 and 14 and the accompanying control section of the display apparatus from the observer. Varying the setting of the potentiometer 10 alters the magnitude of the voltage gradient along the secondary electrode 2a.

In Fig. 5 a heating resistor 40 heats the indicating medium across the electrode 2/39. The indicating medium 3 includes a cholesteric crystal material in the zone 3c. Such cholesteric liquid crystals alter their spectral reflectivity as a function of a temperature in constant to some nematic liquid crystals which alter their scattering properties as a function of an applied voltage. In this embodiment of the invention, as shown in Fig. 5 the display operates in reflection mode in contrast to Fig. 2 where it operates in transmission mode. Instead of effecting heating by means of an electrical resistor the same effect can be produced by directing light onto the system from a source whose intensity is regulated.

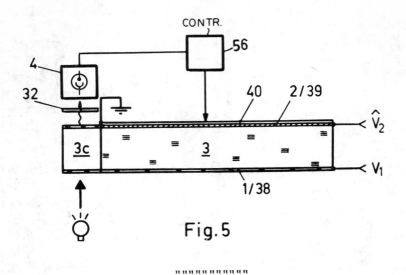

Fig.5

3,707,322 (U.S.) Patented Dec. 26, 1972

ELECTROSTATIC LATENT IMAGING SYSTEM USING A CHOLESTERIC TO NEMATIC PHASE TRANSITION.

Joseph J. Wysocki, James E. Adams, James H. Becker, Robert W. Madrid and Werner E.L. Haas, assignors to Xerox Corporation. Application Aug. 5, 1971.

This is a division of application Ser. No. 821,565, filed May 5, 1969 now U.S. Patent Number 3,652,148.

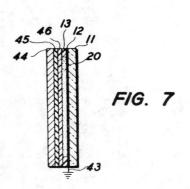

FIG. 7

In Fig. 7 the advantageous liquid crystalline imaging member of the present invention is shown in a system where said member is imaged by an electrostatic latent image on an electrostatic latent image support surface such as a photoconductive layer. In Fig. 7 typically transparent, support substrate 11 is shown with typically substantially transparent conductive coating 12 which is grounded at 43. Adjacent substantially transparent conductive coating 12 spacer-gasket 13 contains liquid crystalline imaging layer 20 between the substantially transparent grounded electrode and an electrostatic latent image support surface. The electrostatic latent image support surface may be any substantially insulating surface which is capable of supporting an electrostatic latent image.

The electrostatic latent image supported by said surface, when taken in conjunction with grounded substantially transparent conductive layer 12 creates an imageswise electrical field across liquid crystalline imaging layer 20, and said field causes the advantageous optical negative-positive, cholesteric-nematic phase transition of the advantageous system of the present invention. A typical electrostatic latent image support surface is a photoconductive plate, for example as typically used in xerography. As illustrated, conductive supporting layer 44 supports a layer of photoconductive material 45, typically comprising selenium, or any other photoconductive insulating layer. Between the photoconductive insulating layer 45 and liquid crystalline imaging layer 20 there may be an insulating protective layer 46 which may retain the liquid crystalline substance in some embodiments, and at the same time layer 46 protects both the liquid crystalline layer and the photoconductive insulating layer 45.

3,707,323 (U.S.)   *Patented Dec. 26, 1972*

LIQUID CRYSTAL DEVICES AND SYSTEMS FOR ULTRASONIC IMAGING.

<u>Lawrence W. Kessler and Samuel P. Sawyer</u>, *assignors to Zenith Radio Coproration.*
Application Nov. 6, 1970.

A device is disclosed which includes a layer comprising a nematic liquid crystal toward which ultrasonic wavefronts are directed. A marked change over any particular area of the layer occurs from a clear state to a light-scattering state when the intensity of the ultrasound applied to that area exceeds a threshold value. The degree of light scattering obtainable is found to be functionally related to the intensity of the applied ultrasound above the threshold. A matrixed array of such devices is arranged in a system with suitable ultrasonic addressing means and scanning means to provide a real-time image display system having a gray-scale. A system for visualizing ultrasonic image information either conventionally or holographically is also provided by causing a sound beam carrying image information as well as a reference beam to impinge on a common area of such a device. The resulting pattern of ultrasonic intensity variations over the area causes light to be variably scattered from different points in the area in accordance with the ultrasonic intensity pattern, thereby rendering visible the analogue optical hologram of the image.

The device 10 of Fig. 1 includes two spaced opposed plates 11 and 12 which are made from, for example, glass, one of which has an optically reflective coating 13 which may be on either side of that plate. Spacers 14 and 15 maintain the plates parallel to each other and at a separation of the order of 1 mil. A sound transducer 16 of a piezoelectric material, such as barium titanate or quartz, is mounted on the back plate 11 which bears the reflective coating and is itself connected to a signal source 17.

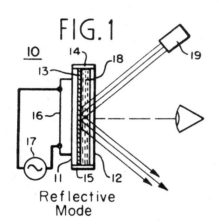

The space between the opposed plates 11 and 12 is filled with a material which comprises liquid crystal of the nematic type, such as p-methoxy-benzylidene, p-n-butyl-aniline 4-methoxy 4' n-butyl-benzylidene-aniline; in this case, the pure crystal was used. Such nematic crystal must be kept within their nematic temperature range, which for this particular crystal is 10°C to 47°C. Although the layer of liquid crystal 18 thus formed is preferably of the order of 1 mil in thickness, this is an examplary magnitude and is not crucial to the operation of the device; however, when the thickness is a multiple half-wavelength of sound, the opposed plates form a resonant cavity about the layer which will enhance the effect of the sound.

The faceplate 12 is transparent to light but preferably reflective to sound; glass is a suitable material. It receives either ambient light or for greater contrast, a quasi-collimated beam of light from light source 19 which illuminates the device 10. In either case, however, the light rays should predominantly impinge at an angle such that the reflective coating 13 of the back plate 11 reflects away most of the light from the eye of a viewer positioned in the normal viewing position; i.e., in front of the plate 12 on a line normal to that plate, as illustrated.

The liquid crystal layer 18 is at this point quiescent and essentially transparent, and the observer perceives the device as comparatively dark. With the source 17 generating an AC signal of appropriate frequency and of variable intensity, the liquid crystal layer 18 is stimulated by an ultrasonic field generated by transducer 16, which vibrates in response to the signal. Although the transducer is preferably driven in the compressional mode, other excitation modes such as shear, transverse or surface will also serve. The frequency of the driving signal from source 17 must be of the order of 10 MHz or higher, in the present embodiment, a 10 MHz signal was used. An observer viewing with the unaided eye sees no change in the optical properties of the device 10 until a certain threshold value of ultrasonic field intensity in the frequency range indicated above is reached. This value will vary with the type of liquid crystal compound and the thickness of the liquid crystal layer 18, in the present embodiment it was found to average that ultrasonic

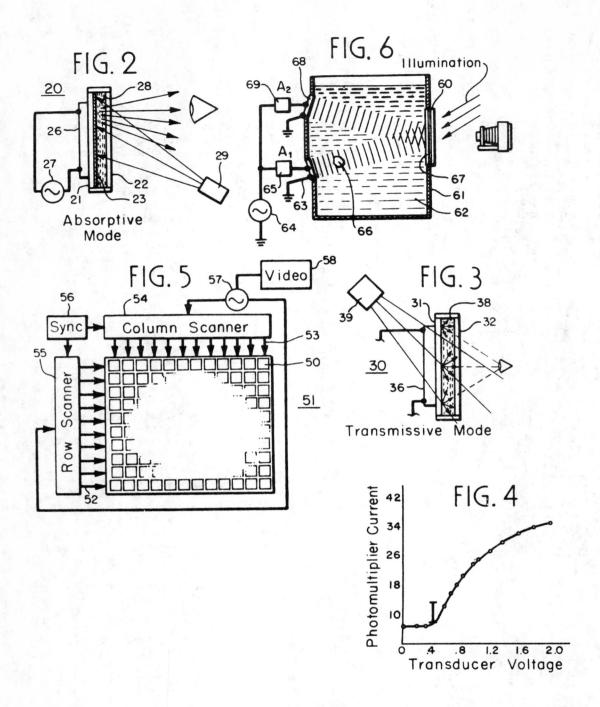

power which will be produced by a piezoelectric transducer driven at an electrical power of 15 milliwatts/cm^2.

Once the ultrasonic field intensity applied to the crystal layer 18 reaches threshold, a sudden and marked increase in the light-scattering capability of the layer 18 takes place, and the observer now sees as bright the entire ultrasonically-stimulated area which hitherto was dark. This is, of course, because much less of the light initially directed to the device is now reflected away; rather, a substantial fraction of this previously-lost light is now scattered in many directions by the stimulated layer 18 so that much more light is now received by the eye of the observer directly as well as by virtue of being reflected from reflective coating 13. Thus the foregoing has described an embodiment wherein the reflectivity to incident light is controlled, but also of interest are embodiments in which the control of light absorption, as well as the control of light transmission is effected.

Fig. 2 is a side view of an elemental device according to the invention, adapted for the absorptive mode;

Fig. 3 is a side view of an elemental light modulating device according to the invention, adapted for the transmissive mode of operation;

Fig. 4 is a graph showing the relationship between the intensity of ultrasonic stimulation applied to the device and the degree to which it scatters light;

Fig. 5 is a schematic illustration of a basic image display system utilizing a matrix of any of the above devices according to the invention; and

Fig. 6 is a schematic illustration of a basic system for visualizing the ultrasonic field bearing image information in the form of a hologram.

" " " " " " " " " "

1,302,482 (British)   Patented Jan. 10, 1973

VISUAL DISPLAY DEVICE.

The National Cash Register Company (Inventor's name not given).
Application April 19, 1971; prior U.S. application February 11, 1970.

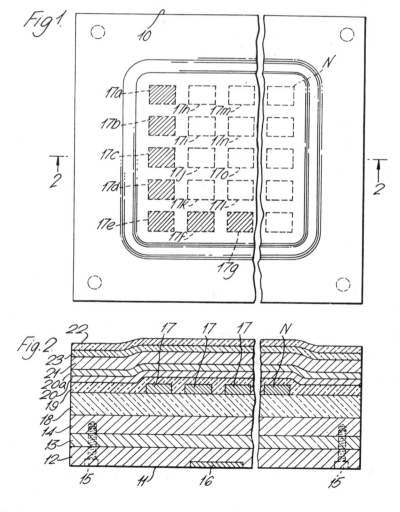

A display device comprises a layer of encapsulated liquid crystals, means for applying an electrostatic field to render the layer translucent, heating means for maintaining the layer at a constant temperature below the clear point and means for heating selected areas of the layer above the clear point, this being the temperature at which the isotropic state occurs, the crystals becoming opaque and remaining so on cooling. A device comprises an array of Ta heating elements 17 deposited on a glass or alumina substrate 18 covered by a glass insulating layer 19; a Ta layer 20 forming one field electrode, oxidized on one side 20a to form a dark background; a layer of encapsulated liquid crystals 21; and a flexible glass plate 22 having a conductive Au or tin oxide layer 23 forming a second field electrode. The temperature is maintained constant by a heating element 13 between Al layers 12, 14 operated by a thermistor 16. Each element 17 is connected at an intersection of a crossed array of conductors, current to each conductor being controlled via transistors and Si rectifiers by a digital computer or encoding device. A diode in series with each element 17 prevents current flow in unselected elements. The liquid crystal layer is initially rendered translucent by a potential pulse applied between electrodes 20, 23 then selected elements 17 are actuated to raise the local temperature above the clear points e.g., 75°C which become and remain opaque on subsequent cooling to the constant temperature, thereby providing a display. The liquid crystal layer 21 is formed by spraying a dispersion including a polymeric binder and capsules containing 70% cholesterol nonanoate, 25% cholesterol chloride and 5% cholesterol cinnamate.

""""""""""

3,711,181 (U.S.) Patented Jan. 16, 1973

OPTICAL NOTCH FILTER.

James E. Adams and John L. Dailey, assignors to Xerox Corporation.
Application March 8, 1971.

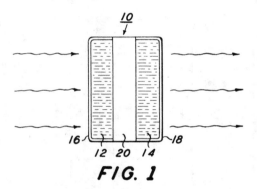

FIG. 1

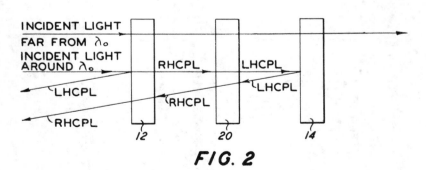

FIG. 2

According to the invention there is provided an optical filter system employing at least one pair of optically negative liquid crystalline films in conjuction with means for converting circularly polarized light of one sense to circularly polarized light of the opposite sense. Each pair of liquid crystal films comprises two individual films which have the same instrinsic rotatory sense and which are arranged in the path of an incident light beam in a manner such that they reflect substantially the same wavelength band within the incident radiation. By positioning each film of any film pair employed in a particular apparatus on either side of the means for converting the circularly polarized light from that of one sense to that of the opposite sense a highly effective and efficient optical filter is constructed.

Referring now to Fig. 1, there is shown a typical optical filter, generally designated 10, of the invention comprising in this illustrative instance, optically negative liquid crystalline films 12 and 14, with optional protective elements 16 and 18 respectively, positioned on either side of element 20 for converting circularly polarized light of one sense to that of the opposite sense.

Referring now to Fig. 2, it is seen that within the incident light which strikes the optical filter, the wavelengths of incident light which are far from λ_o (where $\lambda_o=2np$ with n representing the index of refraction of the liquid crystal substance and p the pitch or repetition distance of the helical structure) pass through the filter substantially completely unattenuated. Consider however what occurs with respect to those wavelengths of incident light which are around, and including λ_o. Within this wavelength region the LHCPL component of the incident light is substantially completely reflected by lefthanded liquid crystal film 12 and the RHCPL component of the incident light is substantially completely transmitted by film 12. When the RHCPL component contacts element 20 it passes through substantially completely unattenuated but now emerges from element 20 as LHCPL. Subsequently, the now LHCPL strikes left-handed liquid crystalline film 14 and is substantially completely reflected with the reflected light remaining LHCPL. The light beam then is directed back on element 20 which again reverses the sense of the polarization of the light while transmitting it therethrough substantially completely unattenuated, the light now becoming RHCPL again. The RHCPL is then substantially completely transmitted by liquid crystal film 12 thus resulting in the substantially complete reflection by the filter of the wavelength band centered around λ_o.

Example I

A first right-handed optically negative liquid crystalline composition containing about 50% cholesteryl iodide and about 50 % cholesteryl chloride and having a λ_o value of about 5900A is prepared. A second right-handed optically negative liquid crystalline composition containing about 10% cholesteryl chloride and about 90% cholesteryl bromide and having a λ_o value of about 5900 A is also prepared. A thin film of the first liquid crystal composition is applied to one surface of a half wave plate available from Polaroid Corporation and a glass plate is placed over the free surface of the liquid crystal film. A thin film of the second liquid crystal composition is then applied to the other surface of the half wave plate and a glass plate is placed against the free surface of the second liquid crystal film. This optical filter is placed in the path of a light beam emitted from a broad band incandescent source of visible radiation so that the radiation strikes the filter at normal incidence. The optical filter substantially completely transmits all of the incident radiation with the exception of a wavelength band centered about a wavelength of about 5900A which is reflected.

The procedure described in Example I is repeated with the exception that the liquid crystalline composition listed for each example is used for both liquid crystal films of the optical filter. Twenty additional examples are described in the patent.

" " " " " " " " " " "

3,711,713 (U.S.)

Patented Jan. 16, 1973

ELECTRICALLY CONTROLLED THERMAL IMAGING SYSTEM USING A CHOLESTERIC TO NEMATIC PHASE TRANSITION.

Joseph J. Wysocki, James E. Adams, James H. Becker, Robert W. Madrid and Wener E.L. Haas, assignors to Xerox Corporation.   Application Aug. 5, 1971.

This is a division of application Ser. No. 821,565, filed May 5, 1969 now U.S. Pat. No. 3,652,148.

In Fig. 8 a liquid crystalline imaging member is shown imaged by a thermal image projection address system. Here the liquid crystalline imaging member comprises substantially transparent electrodes 18 separated by spacing gasket 13 which encloses a cholesteric liquid crystalline composition within space 47 comprising substantially all of the area within spacer gasket 13. In this embodiment a source of a thermal image 48, here a heat source in the desired image configuration, is shown in position with conventional means 49 for focusing and projecting a thermal or optical image, and the thermal image 50 appears in the liquid crystalline substance is heated into the temperature range required for the optical negative-positive, cholesteric-nematic phase transition, while at the same time, the liquid crystalline imaging member is biased by external circuit 15 so that the field across the cholesteric liquid crystalline film is sufficient to cause the phase transition when the imaged area of the film reaches the transition threshold temperature. An observer 51 sees the transformed image area 50 become transparent to transmitted light from source 52, or go blank in reflected light from source 53, as the phase change occurs in response to the thermal image. If the imaging member is observed while between polarizers, the imaged area of the liquid crystal imaging member would appear dark to the observer 51 while the non-imaged areas 47 would remain substantially unchanged as a result of light scattering.

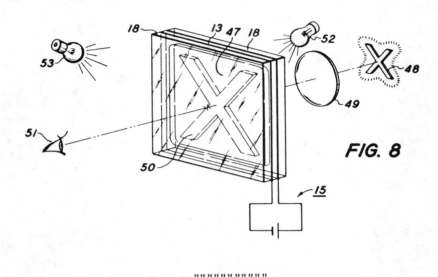

FIG. 8

3,712,047 (U.S.)   Patented Jan. 23, 1973

TIME DISPLAY DEVICE FOR TIMEPIECES.

Pierre Girard, assignor to Manufacture des Montkes Rolex S.A. Bienne.
Application May 11, 1971; prior application in Switzerland May 11, 1970.

With the time display device according to the invention the plates sandwiching the liquid crystal layer and the electrodes carried by those plates are transparent at least in part and the device, moreover, comprises a light reflecting or emitting means, which is arranged behind the liquid crystal cell. In a preferred embodiment of the invention the solid rear plate of the cell is made translucent and a source of light is arranged behind that plate.

Figure 1 shows a portion of a cell provided for displaying a complete time indication. The electrode segments 3a, 3b, etc. of the cell portion shown in Fig. 1 are arranged so as to enable displaying any one of the digits from 0 to 9 upon applying a predetermined potential through conductors 6a, 6b, etc. to some of the segments represented. The segments 3 are coated on the lower rear plate 4 of the cell. Plate 4 carries the device and constitutes the bottom of the cell. The peripheral area of this plate is bent upwards. Plate 4 is located opposite an intermediate plate 2, the front and rear face of which each is provided with a conducting coating constituting a second electrode. This second electrode may be constituted either by a thin coating covering the whole surface of the plate or by a net of threads being thin enough in order to remain invisible. If these threads are close to each other, they produce the same effect as a homogeneous coating covering the whole plate. The double cell represented in Fig. 1 further comprises a front upper plate 1 similar to plate 4. However, the rear inner face of plate 1 is covered by a transparent metallic coating 5, which extends over the whole surface of plate 1 with the exception of areas corresponding to segments 3.

To obtain the desired structure coating 5 is firstly provided on the entire rear face plate 1 and then removed from all the portions of this plate, which are opposite segments 3. Two

FIG. 1

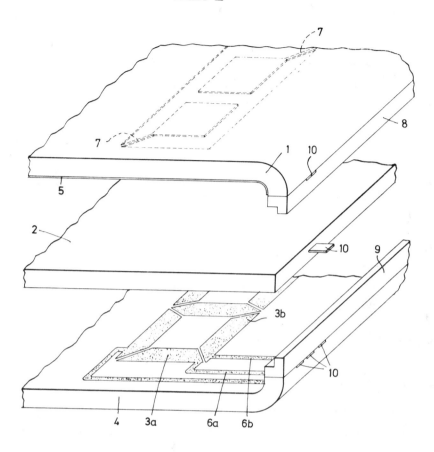

narrow tongues 7 are, however, felt. They electrically interconnect the areas of coating 5 enclosed by the loops of the eight and the remaining areas of coating 5. It should be understood that coating 5 is provided with as many openings like that represented in Fig. 1 as the time indication to be displayed comprises digits.

The cell described is tightly closed by means os spacers 8 and 9 as well as by means of end walls (not shown). The terminals 10 are provided in such a manner that they can easily be connected to a control unit, which may consist of an integrated circuit emitting the impulsions required for displaying the time indication. This electronic unit is diagrammatically represented at 11 in Fig. 2, which is a longitudinal sectional view of a cell capable of displaying three digits.

The device disclosed hereabove operates as follows:

The middle electrode constituted by the intermediate plate 2 may be considered as an electrode of reference. A constant potential is applied thereto. Moreover, a constant potential, which differs from that of electrode 2, can similarly be applied to electrode 5, so that the whole area of the dial, with the exception of the portions in which the digits are displayed, appears opaque and gives the dial its ground appearance. As regards segments 3, some of them, i.e., those which do not participate in the formation of the digit to be displayed, are set by the electronic control unit 11 to a potential which differs from that of the intermediate plate 2. Accordingly, the liquid portions comprised between the electrodes 2 and 3 becomes opaque opposite the segments in question of electrode 3, while the remaining liquid portions comprised between these electrodes are transparent.

It appears from the foregoing description that the segments 3, which form the digits to be

215

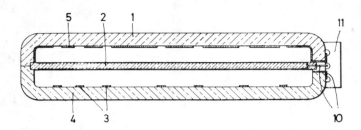

FIG. 2

displayed, are the only areas of the front surface of rear plate 4 which are not covered by an opaque liquid portion either of the layer sandwiched between plates 1 and 2 or of the layer sandwiched between plates 2 and 4. The digits themselves are thus visible because the light illuminating the cell is reflected by the segments 3 forming the digits to be displayed.

Tests have shown that the structure disclosed displays the digits in a sharper and accordingly more easily readable manner than a cell comprising a single layer of liquid crystal.

" " " " " " " " " "

3,713,156 (U.S.) Patented Jan. 23, 1973

SURFACE AND SUBSURFACE DETECTION DEVICE.

Robert G. Pothier.
Application Oct. 12, 1970.

This invention relates to detection apparatus and in particular to apparatus for detecting substances which are concealed from view. By way of example, detection apparatus of this type can be employed to detect and identify weapons which are concealed or hidden by a person's clothing or by suitcases or other containers for security inspection at governmental facilities, military facilities, transportation terminals, industrial plants, commercial facilities, and the like.

In Fig. 1, the target is shown as a passenger 10 who is illuminated by or subjected to incident microwave energy beamed from a microwave antenna element 11. A wide angle focussing element 12 serves to focus the reflected microwave rediation pattern to or on a focal or image plane 13. Located in the image plane 13 is an energy conversion element 14 which converts the variable density thermal pattern. A display means 15 which is preferably a layer of liquid crystals converts the variable density thermal pattern to a visual image pattern of the passenger 10 and of any concealed objects such as the piston 16. A viewing channel 18 permits the visual image to be viewed as indicated by the eye 19 of an observer. A light illuminator (not shown) may be required to view the image in low ambient light conditions.

For a more detailed showing of energy converter 14 and liquid crystal display 15, reference is made to the enlarged cross-sectional view of Fig. 3 which is taken along the lines 3-3 of Fig. 1. The energy converter element 14 is a layer, sheet or film of resistive material, which according to one design may have a resistivity of 377 ohms per square. The liquid crystal display includes a transparent plastic material 15a in which a layer of liquid crystals 15b is embedded. A plane metallic backing sheet 17 is spaced one fourth of wavelength ($\lambda/4$) behind the resistive sheet 14. The backing sheet helps to assure the conversion of substantially all the incident energy to heat. The intervening space is filled with a low-loss dielectric filter material 52, such as polyfoam. The sandwich structure of the transparent plastic 15a, resistive sheet 14, filler material 52 and backing plate 17 can be affixed to one another as by lamination or bonding.

The reflected microwave radiation pattern which is focussed on the image plane 13 passes through the plastic layer 15a and embedded liquid crystal layer 15b and sets up currents in sheet 14. The current density at each point of sheet 14 is proportional to the intensity of the radiation incident thereto such that the heat dissipated results in a thermal pattern having a spatially distributed intensity. Because the color of light scattered from the liquid crystal layer varies with temperature, distinct color lines surround those areas of the liquid crystal layer through which microwave energy of substantially uniform intensity passes so as to form images of the target and any concealed objects.

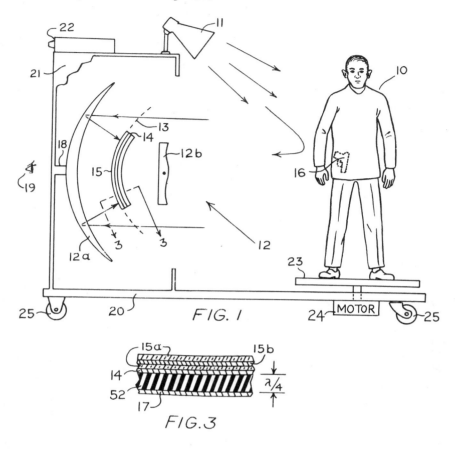

The target or passenger 10 is positioned on a revolvable plate 23 driven by a motor 24. The motor 24 is mounted on the underside of platform 20. By rotating plate 23, the passanger 10 can be illuminated from more than one direction. The platform 20 can be made mobile as by means of wheels 25.

" " " " " " " " " " " "

1,303,947 (British)

Patented Jan. 24, 1973

OPTOELECTRICAL DISPLAY ELEMENT.

<u>Compagnie des Montres Longines, Francillon S.A</u>. (Inventor's name not given).
Application May 4, 1971; prior application in Switzerland May 25, 1970.

A time indicating device comprises two plates 1, 5, at least one of which is transparent on either side of a layer os insulating material 7 having apertures 8 containing liquid crystal material, the plates being conducting in areas 2 covering the apertures. The plates comprise a glass sheet or metal coated transparent sheet 5, in the latter case a light source is placed behind the device. The apertures comprise radial strips on a circular panel or alternating squares and strips on an elongated scale the device is wholly operated by electronic components including a crystal oscillator. The plate 5 may be of a semi-conductor material and form a substrate for an integrated circuit. The layer 17 is of transparent or colored plastics material.

(see illustration the following page...)

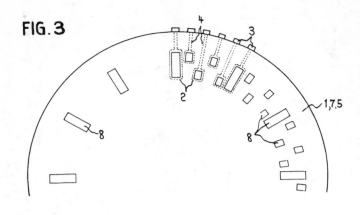

FIG. 3

1,304,268 (U.S.)   Patented Jan. 24, 1973

IMAGE DISPLAY DEVICE.

<u>Derek, H. Mash</u>, assignor to Standard Telephones and Cables Limited.
Application Dec. 11, 1970.

In a display device electrodes 4 disposed either side of a liquid crystal layer 1 apply a field to any selected one A of a set of parallel (row) strips of the layer to render it opalescent, and light is directed along a selected one B of a set of parallel (column) paths within the layer which intersect the strip regions, thereby providing a bright spot only at the selected row and column intersection 13. Such an arrangement is said to avoid cross talk which occurs in devices using X and Y addressing electrodes, where other parts of the selected row and column may also tend to modulate light which is simultaneously directed over the whole display.

A row of semiconductor injection lasers 7 (Al Ga As or Ga As P for instance providing visible light) are selectively operated to provide a collimated beam in the required column. If the light is not in the visible region, a phosphor payer is disposed over the device, e.g. an anti-stoke phosphor for infra-red. The number of lasers may be reduced by providing each one with a light deflector, e.g. a piezo-electrically actuated mirror, an electro-optic deflector, or, as described with reference to Fig. 2, an arrangement comprising magneto-optic polarization switches and composite birefringent prisms for providing four parallel beams per laser 7. At least one electrode is light transmissive i.e. transparent or in the form of fine wire. The electrode may consist of two parallel sets of strips, or one set of strips and a continuous sheet, when at least one electrode is transparent. If both electrodes are transparent, a plurality of devices may be stacked to provide a three-dimensional image, and the beam deflectors may then serve to select a particular device. The lasers may be scanned sequentially, or groups may be simultaneously activated, and the voltage applied to the lasers and/or the electrodes 5 is modulated, e.g. for half-tone effects.

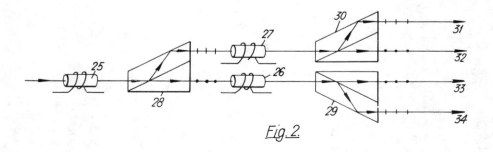

Fig. 2.

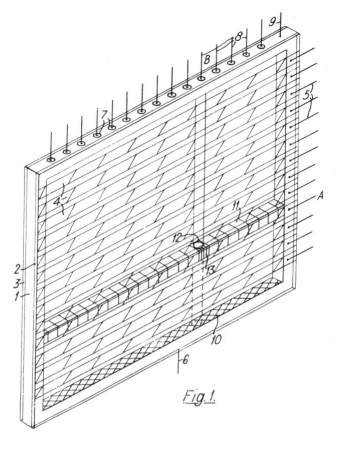

Fig. 1.

1,304,554 (British)   Patented Jan. 24, 1973

LIQUID CRYSTAL DISPLAY DEVICE.

<u>RCA Corporation</u> (Inventor's name not given).
Application Nov. 6, 1970; prior U.S. application Nov. 6, 1969.

A liquid crystal display device has glass substrates 12, 14, conducting layers 26, 44, a dielectric layer 36 and film 20 of liquid crystal material. The conducting layer 26 is divided into two sections (30, 32), Fig. 3, by opening 28, and the layer 44 is divided into two sections (48, 46), Fig. 5, by opening 50. The dielectric layer has a gap 38 in the form of a vertical bar.

When light, e.g. from a fluorescent bulb 78, is passed through the device, three different images, which appear dark against an illuminated background, are formed as follows: an image having the shape of section 30 is formed by application of an AC voltage between section 30 and sections 46, 48, this voltage switching the liquid crystal material adjacent section 30 from the transparent to the light scattering state. Similarly an image having the shape of gap 38 is formed by application of a DC voltage between sections (30, 32) and sections 46 and 48 and an image having the shape of section 48 is formed by application of an AC voltage between sections 30, 32 and the section 48.

An illuminated image against a dark background can be obtained by having a specularly reflective surface on the substrate 14 and having the light source on the same side of the device as the viewer 76 in a position such that no light is reflected in the direction of viewing when the film 20 is in its transparent state.

In an alternative embodiment, Fig. 8 more than one image is defined in the layer 26.

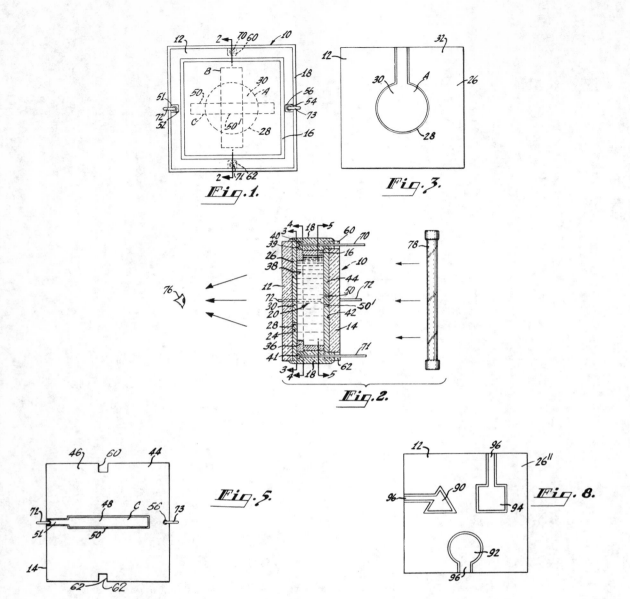

3,716,289 (U.S.)   Patented Feb. 13, 1973

ELECTRO-OPTICAL DISPLAY DEVICES USING SMECTIC-NEMATIC LIQUID CRYSTAL MIXTURES.

<u>Linda T. Creagh, Derick Jones and Sun Lu</u>, assignors to Texas Instruments, Inc.
Application Aug. 31, 1970.

   A display device includes a layer of a smectic-nematic liquid crystal composition that is normally transparent to light. The composition is effective to scatter light in response to a voltage applied thereacross. The smectic-nematic compositions exhibit the mesomorphic state through a broad range of temperatures that preferably includes room temperature.

   Fig. 1 is an exploded schematic view of a display device utilizing the composition of the present invention.

220

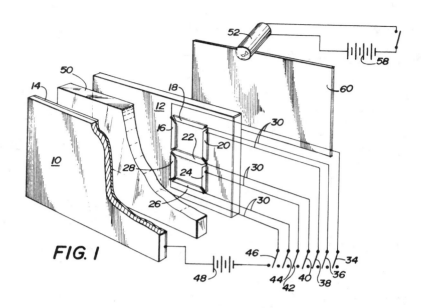

FIG. 1

The electrical energy or impressed voltage across the liquid crystal layer must be sufficiently large to reach or exceed a threshold voltage at which the nematic-smectic mesomorphic composition will scatter light. It has been found that for layers having a thickness of about 1 mil, the threshold voltage for most compositions occurs at around 7 volts while preferably a voltage on the order of 20 volts is utilized. For best results it has been found that the layers should be relatively thin, preferably less than 20 mils.

Examples of specific nematic-mesomorphic compositions which may be utlized in the present invention are set forth in Table I below.

TABLE I

Ex. No.	*Composition	Weight Per-Cent	Crystal to Mesomorphic Temperature	Mesomorphic to Isotropic Temperature
1	20% C	80%	72°C	111°C
2	40% C	60% A	71°C	108°C
3	60% C	40% A	58°C	85°C
4	80% C	20% A	28°C	62°C
5	20% C	80% A	58°C	112°C
6	40% B	60% A	55°C	106°C
7	60% B	40% A	59°C	95°C
8	80% B	20% A	35°C	89°C
9	90% B	10% A	30°C	78°C
10	60% B	40% D	67°C	96°C
11	80% B	20% D	43°C	90°C
12	90% B	10% D	22°C	77°C
13	10% D	90% C	19°C	49°C
14	20% D	80% C	22°C	55°C

*Examples of Materials Utilized:
A — 4,4'-bis(heptyloxy)azoxybenzene
B — p-ethoxybenzylidene-p-n-butylaniline
C — p-methoxybenzylidene-p-n-butylaniline
D — ethyl p-hexyloxybenzal-p-aminocinnamate

These compositions exhibit a wide transition temperature range and several of the compositions exist in the mesomorphic state at room temperature. Specific compositions were prepared by placing the appropriate weight of each component in a sealed vial. The vial was warmed until the materials were isotropic and was then placed in an ultrasonic bath at that temperature for 2 to 3 hours. The vial was allowed to cool slowly to room temperature in the ultrasonic bath.

As noted in Table I, a composition including by weight 90 percent p-ethoxybenzylidene-p-n-butylaniline and 10 percent ethyl p-hexyloxybenzal-p-aminocinnamate (example 12) exists in the mesomorphic state at room temperature. This composition was tested in a 1-mil electro-optical display cell at 25°C and exhibited dynamic scattering when a potential of about 22 volts was applied across the cell.

When the above compositions are doped with about 1-10 percent by weight of cholesteric liquid crystal such as cholesterol nonanoate, the compositions exhibit memory, and may be erased with an alternating current signal of about 7 kHz.

" " " " " " " " " " "

3,716,290 (U.S.)

LIQUID-CRYSTAL DISPLAY DEVICE.

Joseph Borel and Jacques Robert, assignors to Commissariat a l'Energie Atomique. Application Oct. 18, 1971.

An electrooptical display device comprising a film of liquid crystals between two systems of electrodes. One system comprises a plurality of integrated circuits, forming a supporting layer on a substrate and a plurality of flat metallic electrodes deposited on the supporting layer, each electrode being connected electrically to one circuit. The other system comprises a transparent insulating plate provided on one face with a uniform coating of electrically conductive and semi-transparent material.

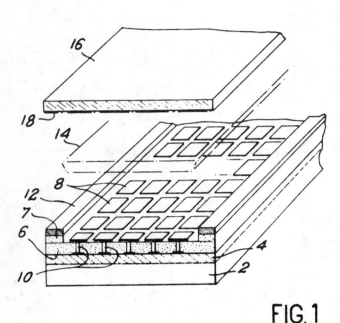

FIG. 1

In Fig. 1, provision is made on the top face of a substrate 2 for a layer of integrated circuits 4 which are not shown in detail; provision is made on the top face of an insulating layer 6 for a plurality of metal plates or "pads" 8 forming electrodes which are joined to said integrated circuits by means of electrical connections 10; a second insulating layer 7 supports a metallic wall 12; a thin film of liquid crystals 14 is interposed between the plurality of electrodes 8 and a transparent insulating plate 16 which is provided on one face with an electrically conductive and semi-transparent coating 18; the plate 16 rests on the metallic wall 12 by means of its face 18.

In one advantageous embodiment of the invention, the device has a rectangular shape such as that shown in Fig. 1. The different pads are arranged in n-columns consisting of p-lines so as to form a chequerboard pattern which, in the case of Fig. 1, comprises 5 columns made up of 7 lines, namely a total of 35 pads. This rectangular shape is convenient in the construction of units for the display of alphanumeric characters. The pads can have dimensions of the order of 300 microns by 300 microns and can be formed of aluminum.

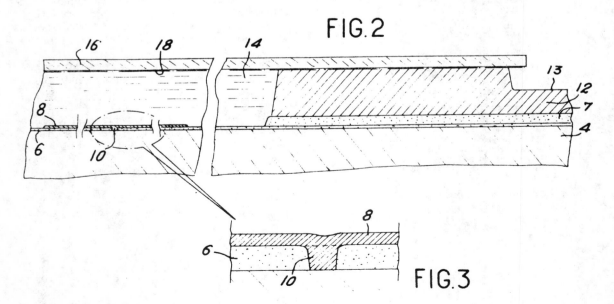

FIG. 2

FIG. 3

The sectional view of Fig. 2 shows certain constructional details of the device, these details being drawn to scale.

Fig. 3 is an enlarged view of the zone which is surrounded by a circle in Fig. 2 and corresponds to the central portion of a pad 8. There are again shown in cross-section in this figure the insulating layer 6, a portion of a pad 8 and the electrical connection 10.

In Fig. 4, there is shown a diagram of connection of three display units which are designed in accordance with the invention.

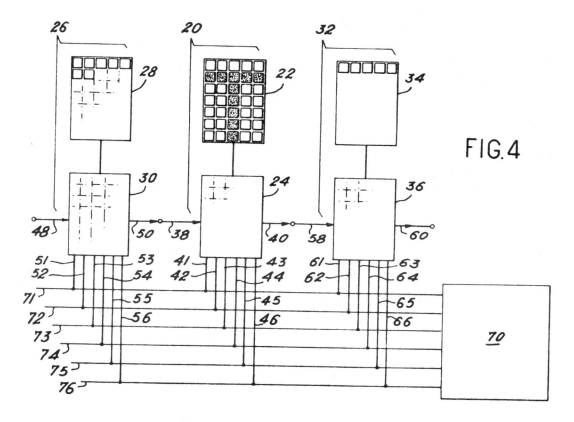

FIG.4

Each circuit assembly comprises eight connections. The principle of interconnection of said circuit assemblies is as follows:

The supply unit 70 which is not shown in detail permits the excitation of the shift registers which are formed by the integrated circuits corresponding to each display unit. Said shift registers are supplied in known manner by means of six interconnections leading from the supply unit 70. Said connections comprise one ground connection 71, two voltage connections 72 and 73, one alternating-current voltage connection 74 and two connections 75 and 76 for producing shift pulses which are derived from electronic clocks. Depending on the mode of operation which is chosen, each circuit assembly 24, 30 and 36, therefore receives either a direct-current voltage or an alternating-current voltage in addition to the different signals derived from the clocks contained in the supply unit 70. By means of the fabrication technique and especially by virtue of the presence of an electrical connection 13 which is readily accessible at the level of the wall surrounding each display unit, the units 20 and 32 can be readily connected in series by means of the connections 40 and 58. Similarly, the display units 20 and 26 can be connected in series by means of the connections 38 and 50. Thus, irrespective of the number of display units (three in Fig. 4), the number of connections which are necessary for the supply of electric current to said display units is always eight as shown in the arrangement employed in Fig. 4. It is therefore apparent that, in contrast to the devices of the prior art, the display device in accordance with the invention makes it possible to employ an electronic circuit of very simple design.

" " " " " " " " " " " "

3,716,658 (U.S.)   Patented Feb. 13, 1973

LIQUID-CRYSTAL TELEVISION SYSTEM.

<u>Michael I. Rackman</u>.
Application Apr. 9, 1970. Continuation in part of U.S. Pat. No. 3,513,258.

    A television system having a horizontal and vertical arrays of parallel conductors with a liquid crystal film interposed between them. The vertical conductors are connected at spaced intervals to a delay line having a delay sufficient for the delay line to represent an entire line of the picture signal. Successive lines of the video signal are fed into the right end of the delay line. The signal travels down the delay line to the left end. Just at the instant when a complete line of the picture signal is within the delay line between the leftmost and the rightmost vertical conductors, one of the horizontal conductors is momentarily energized. This causes a complete line of the picture signal to be displayed. The horizontal conductors are sequentially energized to form the complete display. Each line of the display is formed as a unit at the end of the reception of a complete line of the video signal.

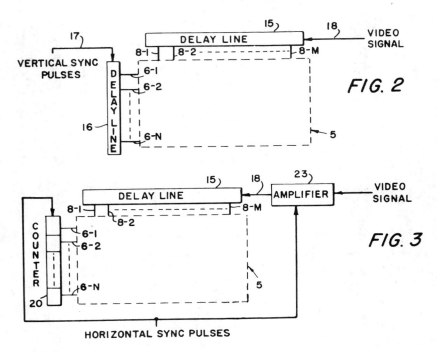

    Figures 2 and 3 show two illustrative embodiments of the invention.

            " " " " " " " " " " " " "

1,306,912 (British)   Patented Feb. 14, 1973

LIQUID CRYSTAL CELLS.

<u>Turrell Uleman</u>, assignor to PPG Industries, Inc.
Application Jan. 15, 1971; prior U.S. application Jan. 19, 1970.

    In a liquid crystal electro-optic cell (see illustration) at least one of the electrodes comprises a transparent conductive cathodic sputtered metal oxide film deposited on a transparent substrate. Cathodic sputtering occurs when a glow discharge is maintained between an anode and cathode in an atmosphere of chemically reactive gas (such as oxygen) so that cathode metal is deposited as an (oxide) layer on the anode or another adjacent member. In the present case the metal may be In, Sn, Sb, Cd, or the mixtures thereof. A dopant of higher valency state included in the metal gives an oxide layer of higher conductivity, and addition of metal of a higher atomic weight enhances the sputtering rate, as does the presence of an inert gas such as a noble gas in the atmosphere. As particularly described, the liquid crystal exhibits dynamic scattering,

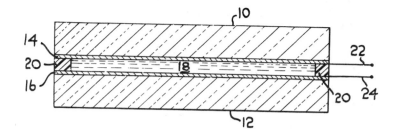

and the other electrode may be identical to the first, or opaque, or reflective. Parts of the sputtered metal oxide film may be removed, e.g. by etching, grinding or removal of previously applied resist material, to give a patterned electrode, for example, a predetermined shape or a grid, lattice or mosaic, with elements which are independently charged by a scanning signal.

""""""""""

1,307,809 (British) Patented Feb. 21, 1973

CAPRONYLHYDROXY-BENZOIC ACID ESTERS, THEIR USE AS NEMATOGENIC COMPOUNDS AND NEMATOGENIC MIXTURES CONTAINING THEM.

Farbwerke Hoechst Aktiengesellschaft (Inventor's name not given).
Application Apr. 19, 1971; prior application in Germany Feb. 28, 1970.

A mixture exhibiting the dynamic scattering effect comprises (a) 4-capronylhydroxy-benzoic acid-4'-ethoxy or 4'-butoxy phenyl ester, (b) 4-methoxy or 4-ethoxy-benzylidene-4'-n-butyl aniline, and (c) 4-methoxy-benzylidene-4'0-n-butyryl-amino phenol.

Example I

A solution of 21 grams of 4-hydroxy-phenetole in 300 millilitres of benzene were added to a solution containing 49.3 grams of 4-capronylhydroxy-benzoyl chloride in 250 millilitres of benzene, 15 millilitres of pyridine in 100 millilitres of benzene were added dropwise to this solution. The reaction mixture was then boiled under reflux for 3 hours. After cooling, the pyridine hydrochloride precipitated was filtered off. The benzenic solution was then washed with dilute sodium hydroxide solution and subsequently with water, dried over sodium sulphate and finally the benzene was distilled off. The distillation residue was recrystallized from hexane. The 4-capronyl-hydroxy-benzoic acid-4'-ethoxy-phenyl ester melts at from 64°C to 65°C, to yield a nematic melt having a n/i-transition point from 100°C to 101°C.

Example II

4-capronylhydroxy-benzoic acid-4'-butoxy-phenyl ester is prepared in analogous manner using 4-capronylhydroxy-benzoyl chloride an 4-hydroxy-phenyl-butyl ether. It has a melting point of from 59°C to 60°C and an n/i transition point of from 89°C to 90°C (from hexane).

Example III

The mixture having the lowest melting point and the widest range of stability of the nematic liquid has the following composition:

```
    i) 2 moles of compound I)   melting point          : -3°C
       2 moles of compound III) n/i-transition point   : +80°C
       1 mole of compound IV)
or ii) 2 moles of compound I)   melting point          : -2°C
       2 moles of compound III) n/i-transition point   : +75°C
       1 mole of compound V)
```

Example IV

A mixture consisting of...

```
3 moles of compound I)   melting point            :   -3°C.
2 moles of compound III) n/i-transition point     :   +62°C.
1 mole of compound IV)
```

has likewise a melting point below 0°C, but a less wide range of stability. Of course, mixtures containing compound II have higher melting points.

Compound I)   4-methoxy-benzylidene-4'-n-butyl aniline;
Compound II)   4-thoxy-benzylidene-4'-n-butylaniline;
Compound III) N-(4-methoxy-benzylidene)-4'-O-n-butyryl aminophenol;
Compound IV)   4-capronyl-hydroxy-benzoic acid-4'-ethoxy-phenyl ester;
Compound V)   4-capronylhydroxy-benzoic acid-4'-butoxy-phenyl ester.

" " "" "" "" "" "

3,718,380 (U.S.) Patented Feb. 27, 1973

IMAGING SYSTEM IN WHICH EITHER A LIQUID CRYSTALLINE MATERIAL OR AN ELECTRODE IS SHAPED IN AN IMAGE CONFIGURATION.

<u>Joseph J. Wysocki, James E. Adams, James H. Becker, Robert W. Madrid and Werner E. Haas</u>, assignors to Xerox Corporation.   Application Aug. 5, 1971.

This is a division of application Ser. No. 821,565, filed May 5, 1969 now U.S. Pat. No. 3,652,148.

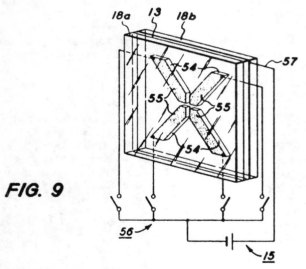

FIG. 9

Figure 9 illustrates a uni-planar, multiple cell, liquid crystalline imaging member suitable for use with the electric field-induced phase transition imaging system of the present invention. In the imaging member of Fig. 9 the substantially transparent electrodes 18 are separated by the typical spacer gasket 13 which contains voids 54 which contain the liquid crystal film. Corresponding areas 55 are the shaped, substantially transparent, conductive coating on the inner surface of electrode 18a. Each one of these cells 55 is capable of being selectively imaged either individually, or jointly, as desired, through the use of switching system 56 in external cicuit 15 which is also connected to the substantially transparent conductive layer on the inner surface of substantially transparent electrode 18b, by electrical lead 57.

It will be appreciated that uni-planar, multiple cell, imaging members such as the one in Fig. 9 may be designed so that various combinations of the desired image cells may be imaged to create any desired figure or character in any language or number system.

" " "" "" "" "" "

3,718,381 (U.S.) Patented Feb. 27, 1973

LIQUID CRYSTAL ELECTRO-OPTICAL MODULATORS.

<u>Georges J. Assouline, Michel Hareng, Eugene Leiba</u>, assignors to Thomson-CSF. Application March 12, 1971; prior application in France March 17, 1970.

The present invention relates to electro-optical modulators wherein dynamic scattering of a layer liquid crystal is controlled for transmitting a variable amount of radiant energy.

The liquid crystal modulator according to the invention comprises a cell associated with an AC generator. Under the action of this generator the delay time of the cell is reduced without introducing any misalignment of the liquid crystal molecules in the absence of an electrical control signal.

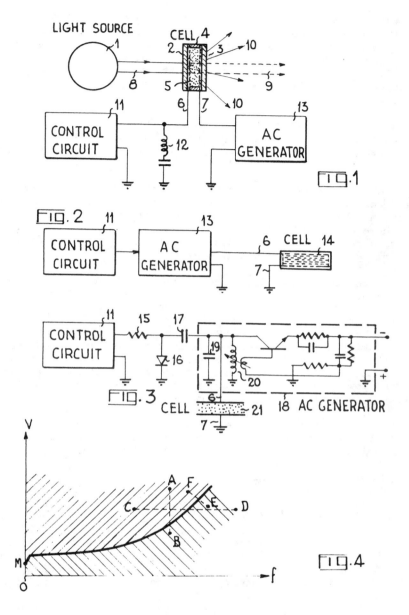

In Fig. 1, there can be seen the diagram of a dynamic scattering electro-optical modulator in accordance with the invention. A light source 1 illuminates a diffuser cell constituted by two transparent plates 2 and 3 attached to a support 4. The internal faces of the plates 2 and 3 are thinly metallized in order to exhibit good transparency while being conductive at the same time; they form the electrodes 6 and 7 which are in contact with the liquid crystal 5 filling the interior of the cell. The crystal 5 is in the nematic phase. The electrode 6 connected to an electrical source 11 designed to provide the control signal for the modulator; the electrode 7 is connected to an AC voltage generator 13; a bypass element 12 can be provided in order to decouple the input terminals of the modulator from the generator 13.

In the absence of any voltage across terminals 6 and 7, the molecules of the liquid crystal 5 can align themselves in relation to the internal faces of the plates 2 and 3 in such a way as not to diffuse the light which they receive from the source 1. If a direct or alternating voltage is applied between said electrodes 6 and 7, then it will be observed that beyond a certain voltage

227

threshold, the molecular alignment undergoes a substantial disturbance which has the effect of scattering the light received by the cell.

Experience shows that the voltage threshold beyond which electrical excitation can give rise to the phenomenon of dynamic scattering varies depending upon the nature of the voltage applied. The lowest threshold corresponds to the application of a direct voltage. If the voltage is an alternating one, the threshold rises and indeed to a progressively greater extent the higher the frequency of said voltage. The law of variation of the threshold voltage V as a function of the frequency f is sketched by the curve shown in Fig. 4. This curve splits the quadrant VOf into two zones.

The region in which the cell scatters the radiant energy, corresponds to the upper zone since each point in this zone defines a voltage amplitude which exceeds the threshold fixed by the curve; the zone located between the curve and the axis Of, corresponds to the transparent state of the cell.

In accordance with the invention, the generator 13 applies to the liquid crystal cell an AC voltage the amplitude of which is higher than the DC voltage threshold OM of Fig. 4; however, the frequency of the generator 13 is adjusted so that no misalignment of the liquid crystal molecules takes place. When the electrical source 11 produces a control signal, dynamic scattering takes place but as soon as said signal disappears, rapid realignment is brought about under the action of the alternating voltage produced by the generator 13.

In the absence of the generator 13, the molecular realignment would take much too long a time. By way of example, a liquid crystal cell in the nematic phase, capable of spontaneous realignment in 250 ms, has this time reduced to 10 ms when it is associated with a generator 13 producing an AC rms voltage of 140 V at 5 KHz/s.

In the circuit of Fig. 1, the control signal is simply superimposed upon the alternating voltage produced by the generator 13; it is possible to provide at the terminals of the source 11 a rejection filter 12 tuned to the frequency of the alternating voltage in order to prevent any parasitic AC from flowing through the control circuit. It goes without saying that the generator 13 can supply several liquid crystal cells.

In the foregoing, the amplitude and frequency of the AC voltage produced by the generator 13 have been selected in order that the point having the corresponding coordinates is plotted below the curve of Fig. 4 defining the molecular misalignment threshold.

In Fig. 2, the diagram of a variant embodiment can be seen. The control signal source 11 is connected to an amplitude modulated generator 13 which produces an alternating voltage of fixed frequency at the terminals 6 and 7 of the liquid crystal cell 14. When the source 11 is not producing any signal the generator 13 is adjusted so that the point marked B whose coordinates correspond with the frequency and amplitude of the supplied AC voltage, is located beneath the threshold curve of Fig. 4. The control signal acts upon the generator 13 to cause the AC amplitude to increase so that the working point is located at A in Fig. 4; thus, molecular misalignment is brought about and produces dynamic scattering of the light. As soon as the control signal disappears, the AC amplitude reverts to its original value and molecular realignment takes place rapidly.

In Fig. 3, a second variant embodiment of the electrooptical modulator can be seen. The liquid crystal cell 21 has its electrodes 6 and 7 connected to the output terminals of the conventional oscillator 18. The tank circuit of the oscillator 18 comprises an inductor 20 and a capacitor 19 chosen in such a way that the alternating voltage applied to the cell 21 has a frequency of such value as not to bring about molecular misalignment of the liquid crystal; the amplitude of the alternating voltage produced by the oscillator 18 is constant, enabling rapid realignment of the molecules of the crystals. The control signal source 11 is connected through the medium of the resistor 15 to a diode 16; in the absence of any signal, the diode 16 is blocked and becomes conductive as soon as the control voltage supplied by the source 11. The capacitor 17 is placed in parallel with the capacitor 19 when the diode 16 is conductive, which has the effect of reducing the frequency of oscillation to below the threshold frequency at which the amplitude of the alternating voltage enables a transition from one to the other of the cross-hatched zones of Fig. 4, to be made.

In the absence of a control signal, the oscillator 18 supplies the cell 21 with an alternating voltage whose amplitude and frequency are determined by the coordinates of the point D in Fig. 4. When a control signal is present, the oscillator 18 retains the same amplitude of oscillation but the frequency of oscillation falls; in Fig. 4, the point defining the new al-ernating voltage, is located at C, that is to say in the zone in which the liquid crystal diffuses the light.

""""""""""

3,718,382 (U.S.)   Patented Feb. 27, 1973

LIQUID CRYSTAL IMAGING SYSTEM IN WHICH AN ELECTRICAL FIELD IS CREATED BY AN X-Y ADDRESS SYSTEM.

<u>Joseph J. Wysocki</u>, <u>James E. Adams</u>, <u>James H. Becker</u>, <u>Robert W. Madrid</u>, <u>Werner E.L. Haas</u>, assignors to Xerox Corporation. Application Aug. 5, 1971.

This is a division of U.S. Pat. No. 3,652,148 (application May 5, 1969).

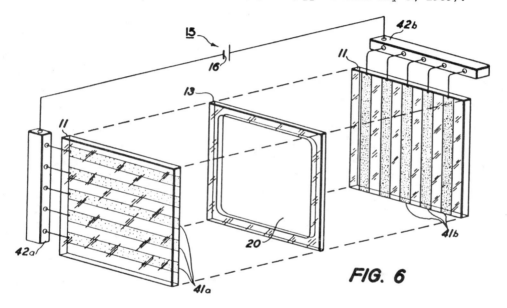

FIG. 6

In Fig. 6 an X-Y address system suitable for imaging a liquid crystalline imaging member is illustrated in exploded isometric view. The liquid crystalline imaging film is placed in void area 20 within the transparent and substantially insulating spacer-gasket 13. The liquid crystalline film and spacer 13 are sandwiched between a pair of substantially transparent electrodes comprising transparent support plates 11 upon which strips of substantially transparent, conductive material 41 is coated. The substantially transparent electrodes are oriented so that conductive strips 41b and conductive strips 41a on the respective electrodes cross each other in an X-Y matrix or grid. Each conductive strip in each set of parallel strips 41a and 41b, is electrically connected to a circuit system 42 which is suitable for selective or sequential operation. Through selection systems 42 and external circuit 15 including source of potential 16, an electric field suitable for creating the optical negative-positive, cholesteric-nematic phase transition of the advantageous system of the present invention can be created across selected points or a selected sequence of points in the illustrated imaging system. It will be understood that substantially transparent conductive strips 41 may vary in width from a very fine wire-like structure to any desired strip width. In addition, one support plate 11 may be opaque where the imaging system is to be observed from one side, using only reflected light.

""""""""""

3,718,842 (U.S.)   Patented Feb. 27, 1973

LIQUID CRYSTAL DISPLAY MOUNTING STRUCTURE.

<u>Charles P. Abbott III</u>, and <u>John M. Reilly</u>, assignors to Texas Instruments Incorporated. Application Apr. 21, 1972.

A liquid crystal display mounting structure wherein the electrical contacts on the back side of a liquid crystal display panel are held a pressure contact with a flat flexible cable which provides electrical connection to a matrix of diodes within a pair of diode boats or holders; interconnections between the diodes and other circuits are provided by etched circuit boards.

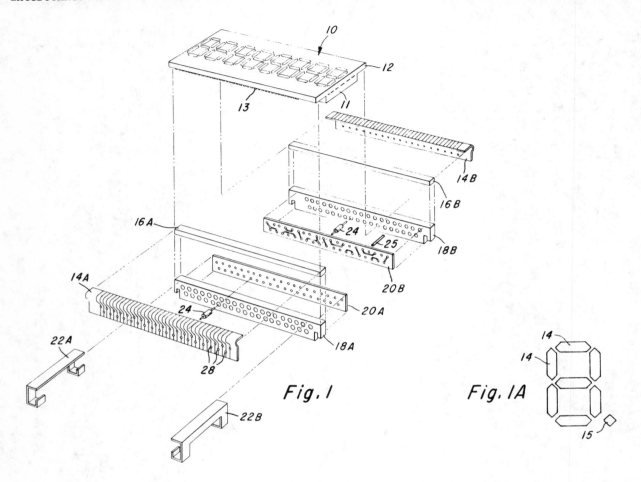

Illustrated in Fig. 1 is a liquid crystal display panel 10. The panel comprises a back plate 11 and a front plate 12 which is slightly wider than the back plate. The plates are separated from each other a short distance by a spacer around the periphery of the back plate. Between the plates is a thin cavity filled with a crystalline liquid of the type now well-known. On the inside surface of the front and back plates are conductive regions or electrodes which are arranged in patterns and groups so that numbers or letters are displayed by the panel upon excitation of the crystalline liquid by application of electrical current to selected ones of the electrodes. Electrical contacts to the various electrodes is by means of a conductive lead pattern on the back side of the front plate shown in phantom at 13. Each lead of the pattern extends from one of the electrodes within the panel through the seal between the plates to one of the overhanging ledges of the front plate. Two electrical contacts are required for excitation of each segment of each figure of the panel. For example, any of the numerals 0 to 9 may be displayed using selective excitation of the electrodes 14 arranged as shown in Fig. 1A; one additional electrode 15 may be used for a decimal point. Two electrodes are required for each segment of the figure but one of the electrodes may be a common electrode for all of the segments of a single figure, usually the back electrode. Thus, it may be seen that nine contacts are required for each digit of a numerical-decimal display. An eight-digit display panel then requires a seventytwo individual connections to the liquid crystal panel.

It can be seen that the area to which electrical contact can be made to each of the conductive leads is quite small and that the conductive leads must be very closely spaced on a panel which has overall dimensions of only 2-3/4" x 1-1/10". Further, when the panel is to be excited by a strobing or multiplexing technique, that is, each digit excited in turn sequentially repeatedly for short periods of time, it is often desirable or necessary to connect a diode in series with each of the front electrode segments.

Shown in the figures is the liquid crystal display mounting structure of the present invention. As shown, the mounting structure comprises the liquid crystal display panel 10 having a conductive pattern of leads 13 on the underside of the top plate, a pair of flat flexible cables 14a and b, a pair of rubber spacers 16a and b, a pair of diode holders or boats 18a and b, a pair of etched circuit boards 20a and b, and a pair of spring clips 22a and b.

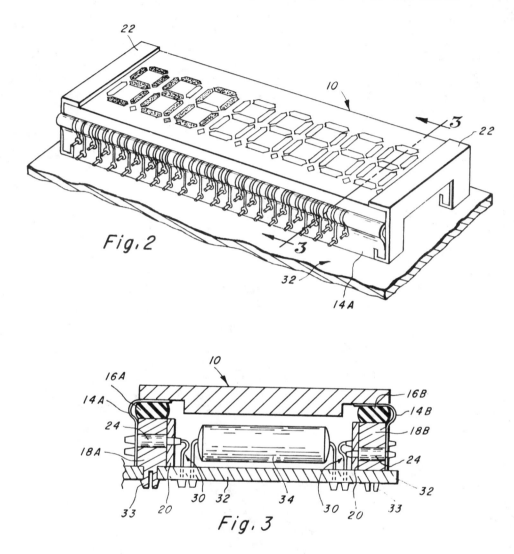

When assembled, the liquid crystal panel 10 rests on top of the flat flexible cables 14 with the metallization pattern of the cable in matching contact with the conductive lead pattern 13 of the liquid crystal panel. Such flexible cables are commercially available or can be custom-made from a metallized sheet of plastic such as Mylar by etched circuit techniques. The spacers 16 which may be of rubber or other suitable resilient material are positioned under the flexible cables and rest on the diode boats or holders 18. The diode boats are made of a suitable insulating material such as phenolic or insulating plastic. Small glass diodes 24 are positioned in the holes 26 of the boat holders. Plain wire segments 25 may be used for through connections when diodes are not a circuit requirement. The other ends of the flexible cables have small holes 28 in the leads of the metallization pattern arranged to fit over and make contact with the diode leads or wires which protrude past the outer surfaces of the diode holders as shown in Fig. 2. Good electrical and structural contact between the diode leads and the flat flexible cable leads can be achieved either by soldering or by a spot of conductive epoxy glue at each contact. Etched or connecting circuit boards 20 having an appropriate metallization pattern and hole pattern are placed against the inner surfaces of the diode boats with the diode leads extending through the holes in the etched circuit board as best shown at 30 in Fig. 3. The diode leads may be conductively connected to the circuit board metal pattern by solder or conductive epoxy glue in the same manner used to connect the diodes to the flexible cables. Spring clips 22 hook under each end of each of the diode boats and over the liquid crystal panel to hold the entire assembly together.

The liquid crystal display assembly can then be mounted by appropriate means, such as screws threaded into the bottom surfaces of the diode boats. As shown in Fig. 2, the display assembly is mounted onto a main circuit board 32 by detent fastener 33 leaving room beneath the diplay for the integrated circuits such as shown at 34 which provide the electronic functions of the system. System interconnections are provided by a metallization pattern on the main circuit board. Electrical connections between the display assembly and the main circuit board may be provided through a connector socket (not shown) in the main circuit board or by direct connection of the diode leads extending through the display assembly circuit boards to the main circuit board as illustrated.

" " " " " " " " " "

1,308,208 (British)   Patented Feb. 28, 1973

DISPLAY SYSTEM.

RCA Corporation (Inventor's name not given)
Application July 28, 1970; prior U.S. application Aug. 7, 1969.

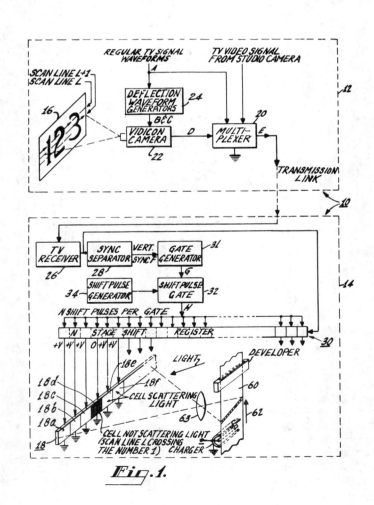

Fig. 1.

In a system wherein individual line scans of a message image are multiplexed in sequential vertical blanking periods of a television signal, each line scan is separated at a receiver and converted to parallel form in a shift register of N stages to control the light transmission or reflection of corresponding ones of a row of N liquid crystal cells until reception of the following line scan. Successive lines may be recorded on an element moving in a direction normal to the row of liquid crystal cells, e.g. photographically or electrophotographically, to synthesize the complete image.

" " " " " " " " " " "

1,308,237 (British)   Patented Feb. 28, 1973

IMINES AND THEIR USE IN ELECTRO-OPTIC COMPOSITIONS AND DEVICES.

<u>RCA Corporation</u> *(Inventor's name not given).*
Application May 26, 1970; prior U.S. application June 2, 1969.

The novel compounds of the present invention are represented by the general formula

$$RO-\bigcirc-\underset{C}{\overset{H}{C}}=N-\bigcirc-OM$$

where OR is an alkoxy radical of from 1 to 8 carbon atoms and OM is a branched chain acyloxy radical having from 4 to 10 carbon atoms. The radical OM may be a branched chain acyloxy radical having from 5 to 10 carbon atoms branching occurring at the beta carbon atom.

It has been discovered that by providing branched chain acyloxy radicals, the lifetime of the device can be increased by as much as an order of magnitude.

The compounds useful in the novel device can be prepared by reacting the p-alkoxy-benzylidene-p'-aminophenol with the branched acyloxy acid anhydride in pyridine.

Example I

A solution of 54.5 grams of p-aminophenol, 68.0 grams of anisaldehyde and 0.1 grams of benzenesulfonic acid in 200 milliliters of benzene is refluxed for 4 hours at which time about 9.0 milliliters of water is collected in a Dean Stark trap. The product, p-methoxybenzylidene--p'-aminophenol is collected and recrystallized from a 50/50 ethanol: benzene solution to give colorless crystals. A mixture of 2.1 grams of these crystals, 50 milliliters pyridine and 1.9 grams of 3-methylvaleric anhydride is stirred for about 1 hour. The mixture is poured onto crushed ice stirred and filtered with suction and the resulting filter cake is recrystallized from isopropanol. Colorless crystals of p-methoxybenzylidene-p'-aminophenyl 3-amthylvalerate are formed.

Example II

p-methoxybenzylidene-p'-aminophenyl-3-methylhexanoate is formed by a procedure similar to that given in Example I. The only change is that in place of the 3-methylvaleric anhydride is the anhydride of 3-methylhexanoic acid.

Example III

p-ethoxybenzylidene-p'-aminophenyl-5-methylhexanoate is prepared by reacting p-ethoxybenzylidene-p'-aminophenol with 5-methylhexanoic anhydride.

Example IV

p-propoxybenzylidene-p'-aminophenyl-3-methylvalerate is prepared by reacting p-propoxybenzylidene-p'-aminophenol with 3-methylvaleric anhydride.

Additional examples and many compounds are listed in the patent.

" " " " " " " " " " "

3,720,456 (U.S.)   Patented March 13, 1973

METHOD FOR NARROWING THE BANDWIDTH OF AN OPTICAL SIGNAL.

James E. Adams and Werner E.L. Haas, assignors to Xerox Corporation.
Application April 29, 1971.

A method for narrowing the bandwidth of an optical signal comprising directing a linearly polarized optical signal through an optically negative liquid crystal film and a linear analyzer arranged in tandem is disclosed. The optically negative liquid crystal film is chosen to have characteristics such that some of the light is extinguished whereas the remainder escapes extinction and is transmitted.

In the field of optics, there are many instances where optical signals having one or more discrete wavelength bands are required. Band pass filters such as the well known interference filters can be used for this purpose. U.S. Pat. No. 3,669,525 describes a method and devices which involve the use of optically negative liquid crystal films. However, in some instances, the bandwidth of the discrete wavelength bands may be too broad for the particular requirements needed. According to the method of the invention, the bandwidth of the optical signal can be narrowed to substantially any desired width.

According to one embodiment of the invention, the optically negative liquid crystal film is chosen to have a pitch and thickness such that incident light of wavelength λ_o is rotated by $(2m\pm1)\pi/2$ radians (where m is any integer) and escapes extinction while incident light of wavelengths $\lambda_o \pm \Delta\lambda s/2$ is rotated by $m\pi$ radians and is substantially completely extinguished. According to this embodiment of the invention the halfwidth, $\Delta\lambda s$, of the original optical signal is narrowed by approximately a factor of two and two small side bands appear. It is not necessary to narrow the bandwidth by a factor of two. By choosing the film pitch and thickness properly, it is possible to narrow the bandwidth of the signal to substantially any desired extent. There is, however, a trade off between the extent of narrowing and side band amplitude. The respective axes of polarization of the linear polarizer and linear polarizer elements are preferably positioned at right angles to each other. Alternatively, by rotating one element with respect to the other it is possible to conveniently shift the position of the maxima and minima thus providing a method for "tuning" the device.

According to another embodiment of the invention, further narrowing of the optical signal can be accomplished by employing additional optical devices such as has been previously described. Arranging a plurality of stages in series does not cause excessive intensity loss since the center wavelength always passes through the linear analyzer element unattenuated. The principal losses in intensity caused by the optical stages are reflection losses.

The invention is described in detail by means of specific examples.

Example I

A cholesteric liquid crystalline composition comprising about 10% of cholesteryl chloride and about 90% of cholesteryl oleyl carbonate is prepared. This composition has a λ_o value of about 5,300 A. An approximately 12 micron thick film of the liquid crystalline composition is applied to one surface of a Polaroid sheet and subsequently another Polaroid sheet is placed over the liquid crystal film so that the axes of polarization of the respective Polaroid sheets have an angular relationship of about 90° to each other.

An optical signal is provided by directing the output from a Cary 14 Spectrometer through a Gelatin Filter Roscolene No. 874 Medium Green. The optical signal is then directed upon the device described above at normal incidence. The optical signal transmitted by the device has a halfwidth approximately one-half that of the original signal and also has two small side bands. Twenty additional examples are described in the patent.

" " " " " " " " " " " "

1,309,558 (British) Patented March 14, 1973

METHOD OF DETECTING ELECTROMAGNETIC RADIATION AND HOMOGENEOUS COMPOSITIONS FOR USE IN PERFORMING THE METHOD.

<u>Westinghouse Electric Corporation</u> *(Inventor's name not given).*
Application Feb. 27, 1970; prior U.S. application Apr. 30, 1969.

Ultraviolet radiation is detected by a liquid crystal composition which changes color when exposed thereto. The composition is in the form of film of 0.25-50 microns thickness disposed on a substrate, and comprises a mixture of cholesteryl nonanoate, cholesteryl oleyl carbonate, and either beta-carotene or cholesteryl p-phenylazophenyl carbonate.

Image converters for ultraviolet, visible and infra-red radiation of the type disclosed in U.S.A. Specification 3,114,836 comprise, as the film of thermally sensitive material, a homogeneous composition consisting of (A) a material of the cholesteric-phase liquid crystal type and (B) a sensitizing agent which responds to said radiation by absorption thereof or by a change of color. Component (A), which preferably comprises not less than 40 wt. percent of the composition is suitable a mixture of oleyl cholesteryl carbonate and cholesteryl nonanoate, with cholesteryl benzoate as an optional third ingredient. Component (B), the choice of which depends on the frequency of the radiation to be converted may be: (i) for I.R. radiation an oil soluble dye of the azo, diphenylmethane, triphenylmethane, phthalocyanine, anthraquinone, ecridine or indigo series; (ii) for visible radiation a phototropic substance which may be a Schiff's base, hydrazone, osazone, semicarbazone, stilbene derivative or fulgide; (iii) for U.V. radiation beta-carotene or an azobenzene compound such as cholesteryl p-phenylazophenyl carbonate.

Example I

To a piece of polyethylene terephthalate film about 0.00025 inch thick, there was applied on one side a coating of black spray enamel. Then, on the other side of the film there was applied a liquid made by mixing into 500 milliliters of petroleum ether the following materials: 5 grams of brilliant green dye, 10 grams of oleyl cholesteryl carbonate, 10 grams of cholesteryl benzoate, and 25 grams of cholesteryl nonanoate. This was permitted to dry, to leave a stratum about 20-30 microns thick, on the one side of the film, and at a temperature of about 50°C this film appeared green. Maintained at about such temperature, with electromanetic radiation of infrared frequency of sufficient amplitude incident thereon, the film turned bluish-green.

Six additional examples are given in the patent.

" " " " " " " " " " "

3,722,998 (U.S.) Patented March 27, 1973

LIQUID CRYSTAL APPARATUS FOR REDUCING CONTRAST.

<u>John E. Morse</u>, *assignor to Eastman Kodak Company.*
Application Oct. 19, 1970.

An apparatus is provided for masking a projected image from a slide transparency which apparatus includes interposing in the projection path a photoconductor-liquid crystal sandwich comprising a layered structure having in order, a first transparent electrode, a transparent layer, a liquid crystal layer and a second transparent electrode, and including means for applying a potential between the transparent electrodes. Upon projection of the transparency image, the lighter areas will cause the photoconductive material to become more conductive than will the darker areas so that the corresponding areas of the liquid crystal layer will become diffuse thereby diffusing some of the light away from the optical system in the brighter areas to provide an image on the photosensitive surface which has less contrast than the priginal transparency image.

Fig. 1 is a perspective view of a projection device for projecting a slide transparency image onto a photosensitive material utilizing a photoconductor-liquid crystal sandwich masking device constructed in accordance with this invention; and

Fig. 2 is a horizontal section through the photoconductor-liquid crystal sandwich, taken along

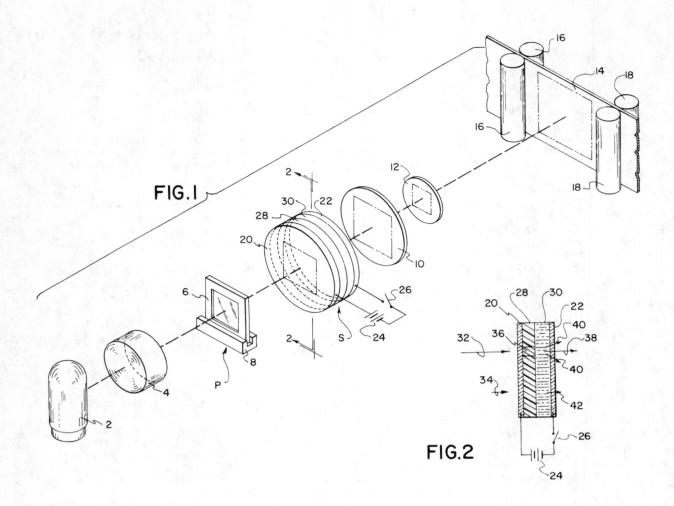

FIG.1

FIG.2

line 2-2 of Fig. 1, and showing the effect of the masking device on the projected image.

As seen in Fig. 2, the photoconductor-liquid crystal sandwich comprises a pair of spaced transparent electrodes 20 and 22 such as NESA glass, connected to a source of potential 24 which can be applied by means of a switch 26. A photoconductive layer 28 is provided adjacent electrode 20 in contiguous relationship therewith and a liquid crystal layer 30 is provided between the electrode 22 and the photoconductive layer 28, as shown. It will be understood that the photoconductive material 28 is also transparent so that the image on slide transparency 6 may be projected through the sandwich 8 onto photosensitive surface 14. Since a slide transparency has a relatively long exposure scale, the contrast between the lighter and darker areas thereof is greater than that permissible for making a high quality print, as on photographic paper. In Fig. 2, the projection of light through one of the lighter areas is indicated symbolically by long arrow 32 whereas projection of light through a darker area is indicated by shorter arrow 34. When little or no illumination is projected onto photoconductive member 28 it is insulating and therefore most of the potential drop between electrodes 20 and 22 is across the photoconductive layer 28 and very little potential drop is across liquid crystal layer 30. In this condition, the liquid crystal layer is transparent so that light may be projected therethrough. However, upon exposure of the photoconductive layer it becomes more conductive so that a greater potential drop exists across liquid crystal 30 which causes the nematic crystal material to diffuse some of the light. Photoconductive layer 28 can be made selectively conductive by projecting the image from slide transparency 6 through it. In areas of high illumination such as that represented by arrow 32, the photoconductive material becomes conductive in area 36 but remain insulative in areas where little or no light such as the light represented by arrow 34 strikes. As a result, the liquid crystal layer 30 becomes light diffusing in areas corresponding to the exposed areas of the photoconductive layer but remain transparent in other areas. Thus, the light as represented by arrow 32 is diffused to some extent so that the amount of light which is transmitted through the liquid crystal layer to be projected onto the photosensitive layer 14 is of lesser magnitude and is represented by a somewhat shorter arrow 38. Some of the light is diffused as indicated by arrows 40 and does not strike the objective lens 12 and therefore does not expose the photosensitive material. The lesser light as represented by arrow 34 travels through the liquid crystal as arrow 42 which is substantially undiminished in magnitude from arrow 34. It will be noted that arrow 33 has a greater magnitude than arrow 42 but that this differential is less than between arrows 32 and 34 so that the light striking the photosensitive layer 14 will produce a resulting image having less contrast than the image of the original slide transparency 6.

A suitable thickness for the liquid crystal layer is 12 microns and is 10 microns for the photoconductive layer. The two layers conveniently are separated by a layer of cellulose nitrate of one micron or less in thickness which layer is to inhibit adverse chemical reactions between the liquid crystal and the photoconductive material.

A potential of 250-400 volts across electrodes 20 and 22 is satisfactory. A light intensity of 1000-foot candles is suitable, although this may vary depending on the photographic characteristics of surface 14.

When it is desired to project another slide transparency, the liquid crystal layer 30 can be made transparent throughout again by opening switch 26 momentarily to remove the potential across the liquid crystal layer. For a subsequent projection of another image slide transparency 6 may be replaced with a second slide transparency and this image projected through the sandwich while switch 26 is closed. This masking device is provided in the form of a photoconductive-liquid crystal sandwich which requires no registration with the slide transparency nor is any developing solution required to form the mask image.

" " " " " " " " " "

3,723,346 (U.S.)   Patented March 27, 1973

TEMPERATURE INDICATOR USING THE SMECTIC C PHASE OF A LIQUID CRYSTAL.

Ted R. Taylor and James L. Fergason, assignors to International Liquid Xtal Company.
Application May 24, 1971.

This invention relates to novel organic chemical compositions that exhibit a variable-tilt Smectic C phase and to a method of sensing or mapping temperatures of an object with the use of such compositions.

It has been unknown, prior to the present invention, that there are compounds of the Smectic C phase that exhibit a temperature-dependent tilt angle. These compounds of the Smectic C phase may be used for the measurement of temperature, at least within the temperature domain wherein the compound involved exhibits the Smectic C phase. A prerequisite to the practice of the invention is that the molecules that are within the Smectic C phase are oriented in such a way that the layers are all parallel to the surfaces that contain the liquid crystal material. This parallelism may be accomplished in various ways, such by rubbing the containing surfaces with a cotton cloth, or with paper or other material that has an orienting material is in contact. Liquid crystal material is then placed between the two rubbed surfaces in such a way that the rubbed directions of the surfaces are parallel. In some instances, a desired parallel orientation of the layers may be achieved by containing the liquid crystal material between two surfaces, only one of which has been so unidirectionally rubbed. There is thus obtained a liquid crystal unit, and when circularly polarized light is caused to become incident upon the stratum of liquid crystal material, the liquid crystal material gives a uniform retardation thereacross, provided that the temperature is constant across the sample. If the temperature throughout the domain of the stratum of temperature-dependent variable-tilt Smectic C liquid crystal material is not uniform, the liquid crystal material then does not have a tilt angle that is uniform, but rather has a tilt angle that varies in accordance with the temperature in various portions of said domain, and there is thus obtained an optical retardation that varies from one location to another within the sample, with the value of said optical retardation depending upon the tilt angle at any particular location within the sample. Thus, if the sample is observed in transmission of a circular polarizer having a sense opposite to that of the incident light, there is observed an interference color that is characteristic of the temperature of the liquid crystal material at a particular point within the above-mentioned domain.

One example of a material that exhibits the properties that are discussed above is terephthal-bis-(4-n-butyl) aniline, which has a variable-tilt Smectic C phase in the temperature range of 172.5°C. to 144.1°C. In that temperature range, the tilt angle changes from 0 degree to 26 degrees. The rate of change of the tilt angle is not linear over the temperature range. In the temperature range of 172.5°C to 166°C, the tilt angle changes rather rapidly at a rate exceeding 3 degrees per degree centigrade, and from 166°C to 144.1°C the rate of change is on the order of 0.5 degree per degree of centigrade. This change of tilt angle is reversible with temperature, so long as the liquid-crystal material is not allowed to crystallize into a different phase.

Terephthal-bis-(4-n-butyl)aninline may be prepared as follows: One mol of terephthaldehyde in 8 mols of ethanol is mixed with 2 mols butyl aniline. The mixture is refluxed for 16 hours. Terephthal-bis(4-n-butyl)aniline is obtained in 85 percent yield, after recrystallization from ethanol.

We have found moreover, that the above-mentioned tilt angle is repeatable or reproducible with temperature variations within a range of + or -0.2 degree centigrade.

There are other liquid crystal materials or compounds that have exhibited properties that indicate that they possess a variable-tilt Smectic C phase. Such compounds are terephthal-bis-(4-n-butylaminobenzoate); terephthal-bis-(4-n-hexyloxyaniline); and terephthal-bis-(4-n-octyloxy-aniline). Their Smectic C temperature ranges are, respectively; 91°C to 134°C; 175°C to 230°C; and 173.7°C to 231.5°C.

It may be stated that the compounds that exhibit the variable-tilt Smectic C phase appear for the most part to be products of the reaction of one mole of terephthaldehyde with 2 moles of an organic amine or amide of the class having the structure:

$$R-\bigcirc-A$$

where A is an amine ($-NH_2$), or an amide ($-COHN_2$) group and R is a saturated and preferably straight tail aliphatic alkyl or alkoxy group containing 1 to 12 carbon atoms. As will be apparent, the group R is located para to the reactive group A by means of which it is joined to the terephthal-dehyde to form the compound in question. There is thus obtained a compound of the structural formula:

$$R-\bigcirc-X-CH-\bigcirc-CH-X-\bigcirc-R$$

where X is a radical selected from the group consisting of $=N-$ and $=N-\overset{O}{\overset{\|}{C}}-$

In accordance with the further embodiment of the present invention, a Smectic C variable-tilt liquid-crystal material is caused to be oriented upon a metallic reflecting surface with the liquid crystal material then being observed in reflected and circularly polarized light that is transmitted through the liquid crystal material undergoea a phase change upon reflection so that it is converted into circularly polarized light os a sense opposite to its original sense. Thus, when the reflected light is observed through the original circular polarizer, there is obtained a result that is equivalent to the viewing of a sample in transmission that is positioned between two polarizers of opposite sense.

If the temperature is constant throughout the domain of the stratum of liquid crystal material the interference color of the sample viewed in accordance with this embodiment is uniformly the same, but, if there is a variation in temperature from one location to another throughout the stratum comprising the domain, there is a corresponding variation of the interference colors observed. The kind of configuration discussed above is especially applicable to thermal imaging.

Devices made with the use of the variable-tilt Smectic C liquid-crystal material of the present invention has the advantage of responding substantially more rapidly than the cholesteric-phase liquid-crystal material described by Fergason in U.S. Pat. No. 3,114,836 since no molecular rearrangement is involved.

Yet another embodiment of the instant invention comprises the observation of the liquid-crystal material by means of transmitted plane-polarized light. When linear polarizers are used, the interference colors that are observed depend upon the orientation of the direction of the long axes of the molecules with respect to the privileged directions of the polarizers.

" " " " " " " " " " "

3,723,651 (U.S.)   Patented Mar. 27, 1973

OPTICALLY-SCANNED LIQUID-CRYSTAL PROJECTION DISPLAY.

<u>Istvan Gorog</u>, assignor to RCA Corporation.
Application Dec. 27, 1971.

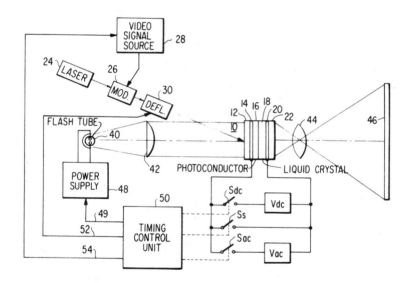

A display system is disclosed which includes a multilayer light control panel having a first transparent electrode, a photoconductor, a normally-transparent liquid crystal, and a second transparent electrode. A laser light beam is modulated with video information and raster scanned to the photoconductor. A direct-current potential is applied across the electrodes during the scanning of a frame, so that spatial variations are created in the light transmissivity of the liquid crystal. During a first portion of a vertical retrace period, a short circuit is placed across the electrodes, and a flash-lamp is projected through the light control panel to a display screen. During a second portion of the vertical retrace period, an alternating current potential is applied to the electrodes to restore the liquid crystal to its transparent condiction.

" " " " " " " " " " "

3,725,899 (U.S.)   *Patented Apr. 3, 1973*

DATA EXHIBITING SCREEN DEVICE WITH A LIQUID-CRYSTAL LAYER, AND METHOD OF MANUFACTURE.

<u>Waldemar Greubel</u>, assignor to Siemens Aktiengesellschaft.
<u>Application Oct. 21, 1970</u>; prior application in Germany July 29, 1970.

    A data exhibiting screen device comprises a layer of liquid crystals whose light scattering or transparency is controllable by applying an electric field. A ferroelectric layer of ceramic material extends face-to-face in parallel proximity to the liquid crystal layer and constitutes a capacitance controllable by the magnitude of the applied electric field. Electric field means in the form of a cross-bar arrangement supply the field excitation.

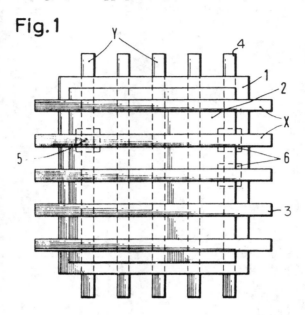

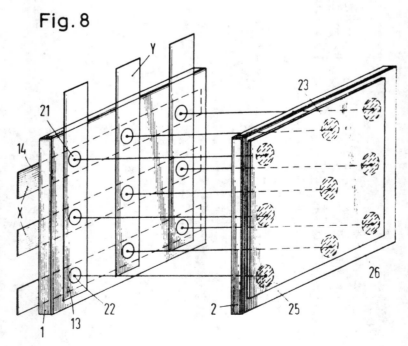

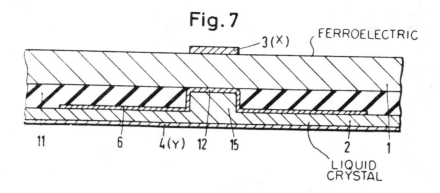

Fig. 7

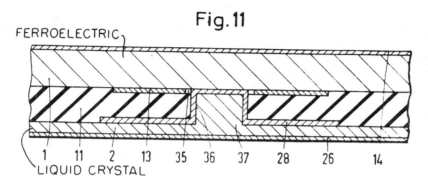

Fig. 11

The indicating screen illustrated in Fig. 1 for slow writing of data comprises a layer 1 of ferroelectric ceramic material, and a layer 2 of liquid crystals extending parallel and face-to-face to the layer 1 (see also Figs. 7, 8 and 11). In Fig. 1 the ceramic layer 1 is shown larger than the layer 2 for illustrative reasons. In reality, the two layers preferably have the same length and width. The double layers 1, 2 carry on the respective outer faces the X-electrodes 3 and Y-electrodes 4 (see also Fig. 8). Each intersection of an X-electrode with a Y-electrode constitutes a raster point 5 of the screen. An intermediate electrode 6 (see also Fig. 7) is interposed at each raster point 5 between the ceramic layer 1 and the liquid crystal layer 2. It will be understood that the liquid crystal layer 2 in Fig. 1 (and Fig. 7) must be kept confined in the layer space by a transparent front pane of glass or plastic such as by a transparent front electrode member as shown at 26 in Fig. 11. Any liquid crystal substance is applicable which retains its liquid crystal qualities in a relatively large temperature range and does not necessarily require auxiliary heating means. Among the known ferroelectric materials, the use for the purpose of the invention of lead-circonite-titanate ceramic has been found preferable. Up to a thickness of about 60 microns this material is transparent if the surface is polished, and remains translucent up to a thickness of a few hundred microns.

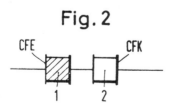

Fig. 2

Fig. 2 is an electrical substitute diagram for an individual raster point 5 in a screen device as shown in Fig. 1. The substitute diagram comprises a series connection of the capacitance CFE of a ferroelectric element 1 and the capacitance CFK of a liquid crystal element 2. When an AC voltage is impressed upon this series connection, the voltage at the capacitance CFK of the liquid crystal element 2 is proportional to this charge QFE imposed upon the capacitance CFE of the ferroelectric element 1 since the charges of both capacitors CFE, CEF are always equal. The capacitance CFK of a liquid crystal element 2 is by a multiple larger than the capacitance CFE of a ferroelectric element 1 so that an applied voltage U is virtually placed entirely upon the capacitance CFE of the ferroelectric element. 1.

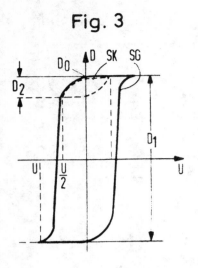

Fig. 3

Fig. 3 schematically illustrates the hysteresis characteristic of a ferroelectric ceramic element 1 that exhibits a nearly rectangular configuration. When applying an AC voltage of the amplitude U to the above-described series connection of the capacitances CFE and CFK, the large loop S_g is traversed, whereas when AC voltage of the amplitude U/2 is applied, the small loop S_k shown in Fig. 3 by broken lines is traversed. Assume that the initial state of polarization of the ceramic elements 1 is at point DO in both cases. The maximal voltages UFK 1, UFK 2 at the liquid-crystal element 2 during both cycles represented respectively by the full-line and the broken-line hysteresis curves in Fig. 3 are released to each other like the two changes in displacement density D1 and D2 of the ceramic elements relative to each other.

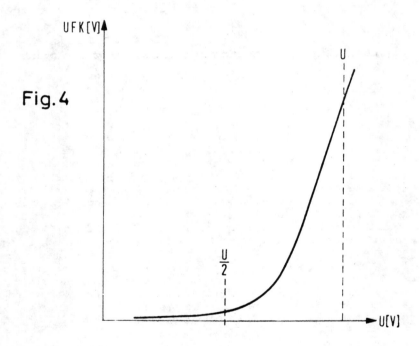

Fig. 4

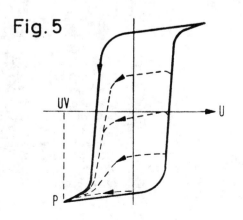

Fig. 5

The diagram in Fig. 4 represents the relation between the voltage U applied to the above-described series connection (Fig. 2) on the one hand, and the voltage amplitude UFK at the liquid crystal elements 2 on the other hand. This curve exhibits a diode characteristic which becomes the more pronounced the more the hysteresis characteristic approaches a rectangular configuration.

Since the electric resistivity of the applicable ferroelectric ceramic materials 1 is several orders of magnitude higher than the resistivity of the liquid crystals 2, the above-described arrangement can also be operated with a bias voltage UV. This can be done by applying a direct voltage UV to the series connection of the capacitances CFE, CFK, which direct voltage is then virtually impressed entirely on one ceramic element 1. The bias voltage UV is so chosen that the ceramic elements 1 are saturated in the initial state. When operating with such a bias voltage, the writing pulses are direct-voltage pulses whose polarity is reversed relative to the bias voltage UV.

Fig. 5 exemplifies hysteresis curves of the ceramic elements 1 when applying a bias UV and direct-voltage pulses of respectively different amplitudes. In the initial state, the respective ceramic elements 1 are saturated according to point P in Fig. 5. When operating with a bias voltage UV, then, if it should become necessary for other reasons, a ceramic material of the type having a less or not propnounced rectangular hysteresis curve can be used without foregoing an improved diode characteristic as compared with alternating-voltage operation. On the other hand, when using a ferroelectric ceramic material with a sufficiently rectangular hysteresis loop, a graduated scale of brightness values of the indicating elements can be obtained by varying the pulse amplitudes.

The circuit diagram of Fig. 6 relates to the control of the raster points 5 of an indicating screen device according to the invention which is operated with a bias voltage UV and which is suitable for the slow writing of the data onto the screen. The Y and X-electrode paths denoted by X and Y in Fig. 6 are connected to a direct voltage source UV through respective decoupling diodes of very slight capacitance. The resistors R are very high-ohmic and serve, inter alia, to prevent a too rapid discharging of the capacitances. The schematically represented electronic switch GS which when operating with bias voltage UV needs to switch pulses of only one polarity, must possess a very high blocking (inverse) resistance. A switching circuit JSK is shown in Fig. 6 by a broken line, this circuit connecting the electronic pulse switch JS with one of the respective Y-electrode conductors.

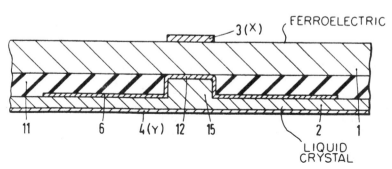

Fig. 7 (above) shows the cross section of one of the raster elements 5 of an indicating screen for slow writing operation. Since the effective dielectric constant of the ceramic layer 1 is always greater than the dielectric constant of the liquid crystals 2, and since the thickness of the ceramic layer 1 should be kept as small as possible in order to operate with small control pulses UE, the electrode areas pertaining to a raster element 5 are smaller on the ceramic layer 1 than on the liquid crystal layer 2. This feature readily permits meeting the capacitance conditions. For this purpose an insulating layer 11 is disposed between the ceramic layer 1 and the liquid crystal layer 2. The insulating layer 11 has a circular hole 12 at each raster point 5, the diameter of the hole 12 being equal to the width of the column electrode 3 on the outer side of the ceramic layer 1. Intermediate electrodes 6, for example of circular shape, are vapor-deposited in concentric relation to the openings 12. Cylindrical bulges 15 of the liquid crystal layer 2 enter into, or pass through, the respective holes 12 into the insulating layer 11. At these localities the liquid crystal layer 2 cannot be excited on account of the layer thickness being too large; but this does not cause any disturbance or detriment because of the very small hole diamater. The line (Y) electrode 4 on the outer side of the liquid crystal layer 2 has a thickness approximately equal to the diameter of the intermediate electrode 6. For satisfying the capacitance conditions, the width of the column (X) electrodes 3 is made smaller than the width of the line (Y) electrode 4.

The principle of a screen device that affords rapid writing will be explained with reference to Fig. 8. The X-electrode paths are arranged on one side of the ceramic layer 1, the Y-conductor paths being located on the opposite side of the same layer. A very small electrode spot 22 electrically insulated from the Y-paths is located at each intersection point 21 of the X and Y-electrode paths, the spot 22 being situated on the Y-conductor side of the ceramic layer 1. A corresponding electrode spot 23 is located on the rear side of the liquid crystal layer 2. The two spots 22 and 23 at each intersection are electrically connected with each other. A common, uniform and transparent front electrode 26 is provided on the front side 25 of the liquid crystal layer 2.

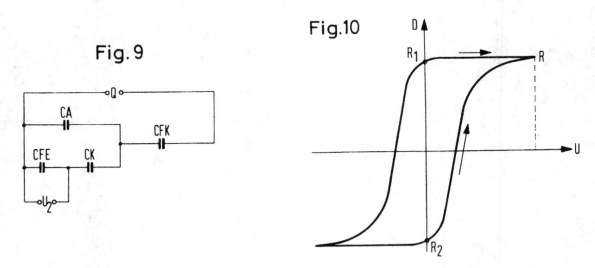

The substitute circuit diagram in Fig. 9 represents one of the raster elements in an indicating screen which affords a rapid writing operation and thus corresponds to one of the raster elements in a device according to Fig. 8. In the diagram of Fig. 9 the capacitance of a ceramic element 1 between the X and Y-conductor path electrodes 13, 14 is denoted by CFE. The capacitance of a ceramic element 1 with an X-path electrode 14 and an electrode spot 22 is denoted by CA, the capacitance of a liquid crystal element 2 between an electrode spot 23 and the front electrode by CFK, and the capacitance between an electrode spot 22 and the appertaining Y-path electrode 13 by CK. At the beginning of the writing operation, the ceramic elements of all raster points 5 are in the same condition of polarization in accordance with point R1 in Fig. 10. A negative DC voltage pulse U1 is applied between the line (X) electrodes 14 and the column (Y) electrodes 13 of the raster points that are to be excited. This negative pulse controls the ceramic elements to flip to a remanence condition R2 (Fig. 10). In this manner, the information is rapidly written into the ceramic layer 1 since the ceramic material 1 affords being readily and very rapidly polarized, the order of magnitude of the polarizing time being 1 microsecond.

Accordingly, the pulse sequence at rapid writing is about 1 microsecond. That is, whenever a character element is written, the time needed is in the order of 1 microsecond. Likewise, an entire line can be written within this interval of 1 microsecond. In contrast, the slow writing the completion of a character element or line requires about 1 millisecond.

A current source Q issuing periodic voltage pulses is connected between all line electrodes X and the common front electrode 26 upon which the liquid crystal layer 2 is located. By proper dimesioning of the capacitances CFE, CFK, CK, CA it can be made certain that these excitation pulses, at those raster elements whose ceramic elements 1 are in the state R1 and hence not to be excited, are virtually entirely applied to the ceramic layer 1, whereas at the other raster elements that are excited and hence are in the other state R2, these excitation pulses virtually are entirely applied to the liquid crystal layer 2. A voltage pulse from source Q therefore has the effect that the ceramic elements are tripped to flip from the state R2 to the opposite remanence state R1. This cabcels the information previously stored in the ceramic layer 1; but since simultaneously the corresponding liquid crystal elements 2 have become excited, the information is

Fig. 6

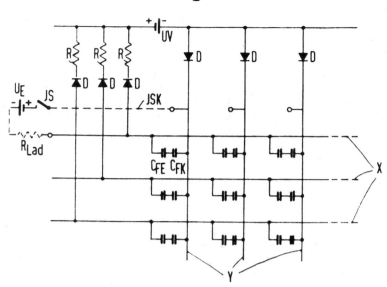

rapidly written into the ceramic layer 1 where it is temporarily memorized; and a voltage pulse from source Q applied between the common front electrode 26 and all of the line (X) electrodes causes the entire picture contents or set of data to be made visible by the changing color, transparency or reflectivity within the liquid crystal layer. These phenomena are rapidly repeated as long as the information is to be indicated on the screen panel.

According to Fig. 11 the cross section of a raster element in a panel structure as shown in Fig. 8 comprises an insulating layer 11 between the ceramic layer 1 and the liquid crystal layer 2. The line electrodes 14 are located on the outer side of the ceramic layer 1. The column electrodes 13 are disposed between a ceramic layer 1 and the insulation layer 11. The column electrodes 13 have a hole 35, for example of circular shape, at each raster point. Accordingly, the insulating layer 11 has at each raster point a hole 26 of corresponding circular shape, and a cylindrical projection or bulge 37 of the liquid crystal layer 2 extends into the hole 26. The above-described electrode spots 22 and 23 represented in Fig. 8 are preferably combined in a circular read-out electrode illustrated in Fig. 11. The read-out electrode 28 is vapor-deposited upon the insulating layer 11 on the side pacing the liquid crystal layer 2, the read-out electrode being concentric to the cylindrical bulge 37 of the liquid crystal layer 2. A large common front electrode plate 26 of transparent material is situated on the outer side of the liquid crystal layer 2. The plate 26 consists of glass or plastic upon which the transparent front electrode is deposited. Examples of suitable mechanical dimensions of the components in devices according to the invention are...

about 100 microns thickness of the ceramic layer,
about 20 microns thickness of the insulating layer,
about 5 to 30 microns of the liquid crystal layer.
The mutual spacing of the X- and Y-electrodes, as well as the overall area of the screen can be chosen at will. The resistivity of the liquid crystal is in the order of 10^{10} dm cm, the resistivity of the ceramic layer in the order of at least 10^{14} sm cm. The operating voltages are preferably in the range of about 100 to 200 volts

In devices according to the invention the ceramic layer 1 need not necessarily consist of a single coherent body but may be composed of individual component pieces, the gaps between these pieces being filled with insulating material.

The above-mentioned devices according to the invention for the indication of data can be operated in transmission or in reflection.

" " " " " " " " " " " " "

3,726,584 (U.S.) Patented Apr. 10, 1973

LIGHT MODULATION SYSTEM.

James E. Adams and Werner E.L. Haas, assignors to Xerox Corporation.
Application May 28, 1971.

A system for modulating light is disclosed comprising directing a beam of light (linearly polarized and monochromatic) on an optically negative liquid crystal film and subsequently varying the pitch of the liquid crystal film in response to some stimulus to which the pitch of the film is sensitive. Specific applications in which the system can be utilized are also described.

Optically negative liquid crystalline substances possess anomalously high optical activity which changes sign at some critical wavelength λ_0 where $\lambda_0 = 2np$ with n representing the index of refraction of the liquid crystal substance and p the pitch or repetition distance of the helical structure of the liquid crystal material. Thin films of optically negative liquid crystal substances, i.e., from about 0.5 to about 50 microns in thickness rotate the plane of polarization of linearly polarized light by substantial amounts depending upon both the wavelength of the incident light and the pitch of the liquid crystalline substance. The pitch of these optically negative liquid crystal materials can be varied by the presence of some stimulus to which their pitch is sensitive, e.g., electric fields, magnetic fields, foreign chemicals, shear, temperature, pressure, etc. Thus varying the film pitch in response to some stimulus results in a change in the amount of rotation imparted to the incident light by the liquid crystal film.

The invention takes advantage of this phenomenon associated with optically negative liquid crystalline substances to provide an advantageous system for modulating a monochromatic light beam.

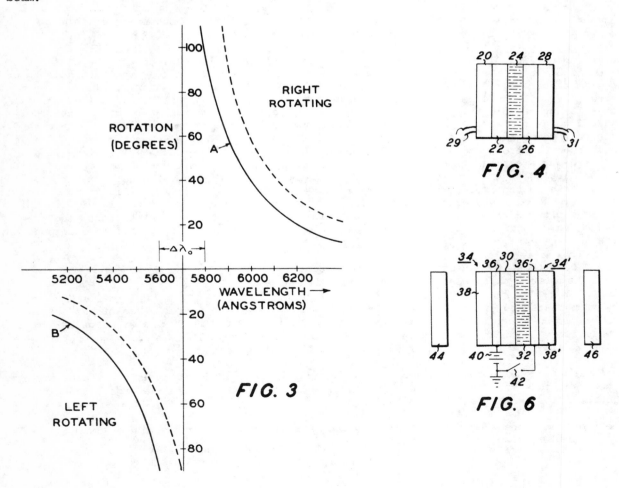

Referring to Fig. 3 there is illustrated the optical activity of a cholesteric liquid crystalline film comprised of about 30% by weight cholestryl chloride and 70% by weight cholesteryl nonanoate and having a λ_o value of about 5700A. The plot shows the optical rotation in degrees (referred to hereinafter as θ) imparted to incident linearly polarized monochromatic light of various wavelengths by the liquid crystal film.

Solid lines A and B represent the rotation imparted to the incident light by the liquid crystal film when the liquid crystalline composition has its intrinsic λ_o value, in this instance about 5700A. It will be clearly seen that the optical activity of the liquid crystalline composition changes sign at the wavelength region around λ_o, designated herein as $\Delta\lambda_o$. Curve A represents wavelengths of linearly polarized monochromatic light which are rotated to the right by the film whereas Curve B shows those which are rotated to the left.

It can be seen that for incident linearly polarized light having a wavelength of 6200A., θ is about 20° rotation to the right. Consider now what occurs when the λ_o value of the liquid crystalline composition is changed in response to some external stimulus applied thereto. For purposes of illustration it will be assumed that there occurs a 100A shift toward the red region of the visible spectrum in the λ_o value of the composition, the new value λ_F, being about 5800A. It is theorized that when such a shift in λ_o takes place the optical activity of the liquid crystal film, represented initially by solid lines A and B also shifts by the same magnitude in a corresponding direction. Thus the optical activity of the liquid crystal film when $\lambda_F=2np_2$ (where p_2 is the pitch of the composition after the external stimulus has been applied thereto) is shown by broken lines C and D. It can be clearly seen that the θ value obtained for an incident wavelength of 6200A is now about 28°. Therefore it will be appreciated that when a linear analyzer element is arranged behind the liquid crystal film, and its axis of polarization is adjusted so that the incident light is extinguished when λ_o for the liquid crystalline composition is equal to 2np (where p is the pitch of the composition before any external stimulus is applied thereto) some light will then be seen when an external stimulus acts upon the composition and causes a change in its pitch. Details relating to the preparation of Curves A and B are given in Example I.

Referring now to Fig. 4 there is shown a temperature detection apparatus comprising a light source 20, optional linear polarizer element 22, optically negative liquid crystalline film 24, linear analyzer element 26 and optional light energy analyzing element 28. Light source 20 is connected to a potential source (not shown) by means of wires 29. The device is connected to a voltage measuring device (not shown) such as a voltmeter or potentiometer by means of wires 31.

The light modulation system of the invention can also be adapted to function in an imaging mode. Referring now to Fig. 6, there is seen a photoconductive insulating layer 30, which may be any suitable photoconductive insulating material which is transparent to oncident electromagnetic radiation, and an optically negative liquid crystal film 32 arranged between transparent electrodes 34 and 34' which in this illustrative instance comprise a thin optically transparent conducting layer of tin oxide 36 and 36' respectively coated on a layer of optically transparent glass 38 and 38' respectively. The conducting surface 36 of the electrode 34 is connected to one side of the potential source 40. The conducting surface 36' of electrode 34' is connected to the other side of potential source 40 through switch 42 so that when switch 42 is closed an electric field is applied across the photoconductive insulating layer 30 and liquid crystal film 32 through electrodes 34 and 34'. The outer elements of the device are optional linear polarizer element 44 and the linear analyzer element 46.

Switch 42 is closed thus causing a uniform electric field to be applied across photoconductive insulating layer 30 and liquid crystal film 32. The device then is exposed to flood illumination with essentially monochromatic light of wavelengths to which the photoconductive insulating material is not sensitive and the linear analyzer element is positioned in a manner such that the incident light is extinguished. Subsequently activating electromagnetic radiation, i.e., that to which the photoconductive insulating material is sensitive, is directed upon the device in imagewise fashion thereby forming an electrostatic latent image on the photoconductive insulating layer. The formation of the electrostatic latent image on the photoconductive insulating layer 30 causes the electric field across liquid crystal layer 32 to be varied in image configuration with a resultant change in the pitch of the liquid crystal film in the areas thereof corresponding to the areas of the photoconductive insulating layer struck by activating radiation. Thus the plane of polarization of the incident monochromatic light is rotated through a different angle by the image areas of the liquid crystal film and the image is read out of the device.

Example I

A mixture of about 30 percent cholesteryl chloride and about 70 percent cholesteryl nonanoate is prepared having a λ_o value of about 5700A at 23°C. An approximately 5-10 micron thick film of the mixture is applied to the surface of a glass slide and another glass slide is then placed over the free surface of the liquid crystal film.

The liquid crystal film is placed in a polarizing microscope and the optical rotation measured for various wavelengths of linearly polarized monochromatic light.

Three additional examples are given in the patent.

""""""""""

3,727,527 (U.S.)   Patented Apr. 17, 1973

PHOTOGRAPHIC APPARATUS WITH LIQUID CRYSTAL VOLTAGE INDICATOR.

<u>Kurt Borowski and Alfred Kubitzek</u>, assignors to Agfa-Gevaert Aktiengesellschaft.
Application June 8, 1972; prior application in Germany June 11, 1971.

Figure 1 illustrates an electrical exposure control circuit in a still camera or motion picture camera.

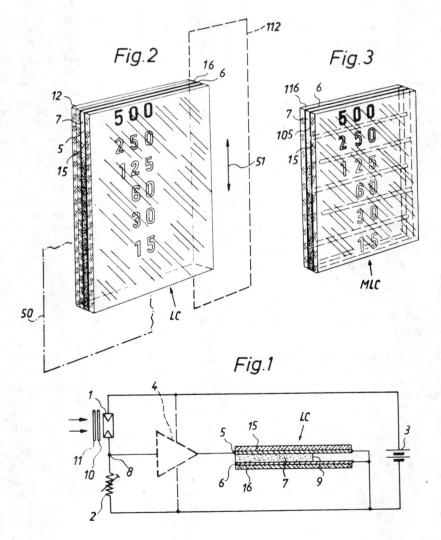

The circuit comprises a voltage divider including a photoelectric resistor 1 and a variable resistor 2. The voltage at the top 8 of the voltage divider is amplified by an output amplifier 4 and is applied to the leftmost point of a transparent electrode layer 5 forming part of a liquid crystal cell LC. The resistors 1 and 2 are connected in series with a battery 3 or an analogous energy source whose negative pole is connected with the right-hand end of the electrode 5 as well as with the right-hand end of a second transparent electrode 6 which registers with and is separated from the electrode by a liquid crystal layer 7. The electrodes 5 and 6 are respectively applied to the inner sides of plate-like carriers 15, 16 of transparent glass.

The resistor 1 is exposed to scene light so that its ohmic resistance varies as a function of scene brightness. Thus, the voltage at the output 8 also varies as a function os scene brightness to thereby change the position of the boundary 9 between the light-transmitting and opaque or differently colored portions of the liquid crystal layer 7. The position of the boundary 9 between the left-hand and right-hand ends of the electrodes 5,6 can be further influenced by means of light-weakening elements 10, 11 which can be placed into the path of incoming scene light in front of the photoelectric resistor 1.

The position of the boundary 9 between the light-transmitting and opaque portions of the liquid crystal layer 7 can be read with reference to a scale 12 shown in Fig. 2 to allow for appropriate adjustment of the shutter (not shown) so that the selected exposure time will be best suited for the making of an exposure with the preselected f/stop and with a film whose speed is indicated by the position of the filter 11.

The entire cell LC can be mounted in the view finder 50 so that, by looking through the eyepiece of the viewfinder, the user of the camera can immediately detect the position of the boundary 9. The graduations located below the boundary 9 are obscured because the liquid crystal layer portion below the boundary is assumed to be opaque whereas the liquid crystal layer portion above the boundary 9 transmits light.

When the exposure control of Fig. 1 is in use, the voltage between the electrodes 5 and 6 decreases in a direction from the left to the right due to resistance of the electrode 5 to the flow of electric current. The dynamic scattering in the liquid crystal layer 7 takes place at a certain voltage so that the position of the boundary 9 will invariably reflect the voltage at the output 8 and hence the intensity of light which reaches the resistor 1. Thus, the position of the boundary 9 relative to the scale 12 will indicate the prevailing scene brightness modified by the selected f/stop and the film speed. If the layer 7 does not contain any coloring agents, the right-hand position of the layer 7 will remain transparent whereas the other portion of the layer 7 (to the left of the boundary 9) will either obscure or completely conceal the graduations therebehind. If the layer 7 does contain a coloring agent, the liquid crystal layer portion to the left of the boundary 9 will be colored whereas the remaining portion of the layer 7 will remain transparent or will exhibit a different color.

Figure 3 illustrates a modified liquid crystal cell MLC wherein the scale 12 or 112 is omitted and the gradations representing various exposure times are applied directly to (for ex., etched into) the glass plate carrier 116. Furthermore, the electrode 105 at the inner side of the glass plate carrier 15 is of sinusoidal or meandering shape in order to insure a desirable mathematical relationship between the applied voltage and the movement of boundary between the light-transmitting and opaque portions of the liquid crystal layer 7. As the voltage at the point 8 increases, the opacity of the liquid crystal layer 7 increases along the meandering electrode 105. Therefore, the transition of boundary between the opaque and light-transmitting portions of the layer 7 in the direction of graduations on the plate 116 is more readily discernible than in the longitudinal direction of the electrodes 105 and 6.

"""""""""""

3,728,007 (U.S.)

Patented Apr. 17, 1973

REFLECTIVE TYPE LIQUID CRYSTAL DISPLAY DEVICE HAVING IMPROVED OPTICAL CONTRAST.

<u>Karl-Dieter S. Myrenne and Clarence L. Hedman, Jr.</u>, assignors to SCM Corporation. Application May 27, 1971.

It is a principal object of the present invention to provide a reflective type liquid crystal display device having improved optical contrast.

As may be seen by reference to Figs 5 and 5A, panel 20 is seen to be mounted at an incline to light baffling panels 32 and 34. Planar, transparent front electrode 40, which is patterned in the form of alphanumeric segments in display area 41, and planar, reflective rear electrode 42 are thus oriented facing light baffling panel 34, liquid crystal layer 43 being sandwiched between electrodes 40, 42.

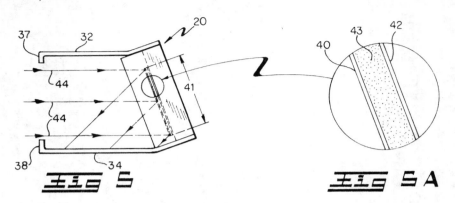

With this orientation ambient light rays 44 entering the device through the spacing between front panels 37 and 38 strike and are reflected by the inclined, rear electrode 42 onto the upper surface of light baffling panel 34 which surface is black. To an observer the display panel thus appears quite dark. When one or more segments forming front electrode 40 are energized the layer of liquid crystal therebehind goes into its light scattering mode causing some of the ambient light incident thereon to be reflected back out of the device between front panels 37 and 38 for observation while light passing adjacent the energized segments continues to be reflected toward and absorbed by panel 34. Scattered light which is not reflected to the exterior is also absorbed by panel 32 and 34 or a side panel, not shown. In this manner the optical contrast of the reflective type liquid crystal display panel is greatly enhanced.

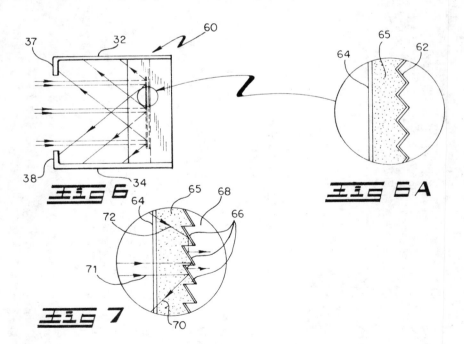

Figures 6 and 6A illustrates another embodiment of the invention having the same assembly of light absorbing panels as that just described. Here, however, the liquid crystal display panel 60 is mounted normally to top and bottom panels 32 and 34. Furthermore, rear, reflective electrode 62 is seen to be serrated. Front, transparent electrode 64 however remains planar. With this configuration and orientation light entering the device in the light transmissive air medium between front panels 37 and 38 is reflected both to top and bottom light absorbing panels 32 and

250

34 as shown by the illustrated light rays. This embodiment offers space saving advantages over that shown in Figs. 5 and 5A at the price of somewhat more expensive liquid crystal panel.

Figure 7 illustrates yet another embodiment of the invention. Here, the liquid crystal display panel is mounted to the light absorbing panels in the same manner as shown in Fig. 6. The assembly itself of light absorbing panels is again the same as that of Figs. 5 and 5A and 6 although the presence of panels 32 and 37 is optional. The principal distinction over the device shown in Figs. 6 and 6A is that of the rear liquid crystal electrode which again is located adjacent front, transparent electrode 64 with liquid crystal 65 disposed therebetween.

The electrode (rear electrode) here is seen to comprise a plurality of juxtapositioned parallel, planar, reflective segments 66 formed by coating every other surface of a jigsaw-shaped edge of glass support 68. The glass surfaces between each of the coated segments are not coated. The parallel reflective segments face bottom light baffling panel 34. With this configuration and orientation light rays which pass through unenergized portions of liquid crystal 65 are either reflected to panel 34 as ray 70 does or pass through glass support 68 behind the reflective electrode. In addition, some stray light rays such as ray 72 are likewise reflected by segments 66 on behind the reflective electrode. All this, of course, serves to enhance the optical contrast of the display device.

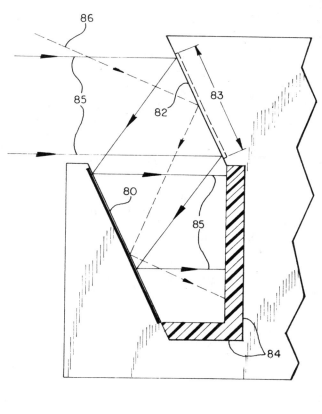

Fig 8

The display device shown in Fig. 8 utilizes a black-backed light absorbing mirror 80, i.e., a specularly reflective good light absorber mounted parallel to a reflective type liquid crystal display panel 82 having a display area 83. The light absorbing mirror 80 faces a body of black, light absorbing plastic foam 84, i.e., a diffusely reflective, good light absorber. With this embodiment ambient light rays 85 are reflected from the display panel onto the black-backed mirror which absorbs a substantial portion of the incident light. That light which is not absorbed is reflected onto the body of black foam which absorbs still more light. Ambient light rays from above the device, such as the ray depicted by dashed line 86 are likewise reflected the display panel and absorbed by mirror 80 and foam 84. This configuration has been found to provide remarkably high contrast.

" " " " " " " " " " "

3,728,008 (U.S.)   Patented Apr. 17, 1973

LIQUID CRYSTAL DISPLAY.

Frank V. Allan and Paul Y. Hsieh, assignors to Ing. C. Olivetti and Co., S.p.A. Application Dec. 1, 1971.

A silane derivative is chemisorbed on the interior surfaces of the substrates of a liquid crystal display to eliminate cloudiness of the display and to increase contrast.

A problem which has been experienced with liquid crystal devices is that, even when not energized, they appear somewhat cloudy and not highly transparent due to the random orientation of liquid crystal molecules parallel to the surface. This cloudiness is apparent even if the surfaces of the substrate and electrodes have been carefully cleaned.

According to the invention, the substrate and the electrodes deposited on the surface thereof are rendered hydrophobic by chemisorbing a very thin layer of a cationic silane derivative on the surface.

The chemisorbed coating of the silane derivative may be deposited by immersing the electrode coated substrate in a solution of the silane derivative in water or an organic solvent or in a silane derivative vapor.

A very important effect of having the exposed surfaces of the substrate and electrodes hydrophobic is a great improvement in contrast. This occurs not only because the unenergized (background) portion of the liquid crystal is much clearer, but also because liquid crystal on the hydrophobic surface forms a close-knit honeycomb domain structure at a voltage slightly above the threhold voltage, instead of the usual sausage-like structure.

Example I

Two pieces of NESA glass are cleaned in a chromic acid solution or a caustic soda solution at pH 8-9 and rinsed thoroughly, first with clean tap water and then distilled water. The cleaned glass is immersed in a 1 percent aqueous solution of Siliclad (Clay Adams Division of Becton, Dickinson and Co.) for about 5 seconds. After complete immersion, they are removed and rinsed thoroughly with distilled water to remove any excess Siliclad from the surface, and the coated surfaces are air dried. The length of immersion time is not critical. A longer immersion may give a thicker layer of the coating; otherwise, there is no detrimental effect on the surface properties.

Air drying is adequate to get fully cured treatment, but heat is often employed to speed up the curing process. The treated surfaces may be heated to a temperature of 100°C for 10 minutes. Any temperature up to 260°C may be employed.

The treated surfaces have a smooth hard film which is hydrophobic and gives a high contact angle with the liquid crystal.

The NESA glass is used to prepare a liquid crystal display panel containing liquid crystal film of 1 mil thickness which is perfectly clear and which has high contrast when energized at 30V DC or AC. The panel shows no spotty appearance after a series of heating and cooling cycles.

Five additional examples are given in the patent.

" " " " " " " " " " "

3,730,607 (U.S.)   Patented May 1, 1973

INDICATOR SCREEN WITH CONTROLLED VOLTAGE TO MATRIX CROSSPOINTS THEREOF.

Josef Grabmaier, Hans Krueger and Ulrich Wolff, assignors to Siemens Aktiengesellschaft. Application July 12, 1971; prior application in Germany July 23, 1970.

    This invention provides an indicator screen comprised of two vertically superimposed systems of parallel transparent conductor paths, a nematic liquid crystalline layer positioned between the two conductor systems, and an interrupted barrier-free layer of a nonlinear resistor material arranged between one of the conductor systems and the liquid crystalline layer.

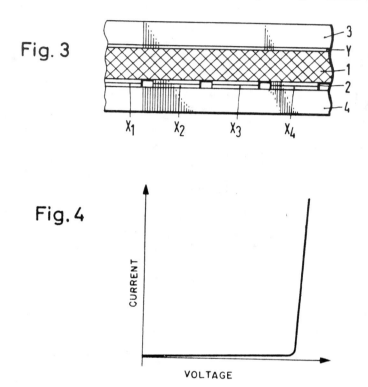

    Figure 3 shows a partial section of the matrix of an indicator screen construction according to the present invention. In this screen construction, an interrupted barrier-free layer 2 of a non-linear resistor material is arranged between one of the conductor systems, i.e., the system having transparent conductor webs $X_1...X_4$, and the nematic liquid crystalline layer 1. As can be seen in Fig. 3, the interraupted layer 2 is placed directly over the individual conductive transparent webs X_1, X_2, X_3 and X_4 of the conductive system arranged in the glass plate 4. Such an arrangement provides a greater resistance of the voltage that may be applied to the indicator screen matrix.

    In Figure 4, a graph illustrates the current voltage characteristic of the non-linear resistant material of layer 2 shown in Fig. 3. According to the present invention, when voltages are applied to the conductor paths of the indicator screen matrix construction shown in Fig. 3 (and illustrated by the circuit diagram in Fig. 5) in such a manner that the energizing voltage not only occurs at the intended and selected cross point but also in amounts of one-third of this energizing voltage at the remaining cross points, the energizing of the cross points other than the selected cross point can be avoided by the layer of non-linear resistor material 2 arranged between the conductor system and liquid crystal layer 1. This is because of the non-linear characteristics of the resistor material of layer 2 which elements are illustrated in the circuit diagram of Fig. 5 as resistor elements R_{11}, $R_{21}...R_{43}$.

    Referring to the circuit diagram shown in Fig. 5, voltage may be applied to the conductor path X_2 while no voltage is applied to the conductor path Y_2 and only one-third of the energizing voltage applied to X_2 is applied to the non-selected conductor paths X_1, X_3 and X_4 as well as the non-selected conductor paths Y_1, and Y_3. With this arrangement, by measuring the energizing voltage, the current-voltage characteristics of the non-linear resistor material, and the threshold voltage of the liquid crystalline layer 1 in a suitable manner, a light-straying effect or glowing effect can be obtained by application of the full energizing voltage at a selected cross point but not with the application of one-third of the energizing voltage at the other non-selected cross points.

Various non-linear resistor materials may be utilized according to the present invention, however, aluminum oxide and tantalum oxide have proven to be effective non-linear resistor materials in the present invention. Also, cadmium selenide layer may be utilized as the resistor material 2.

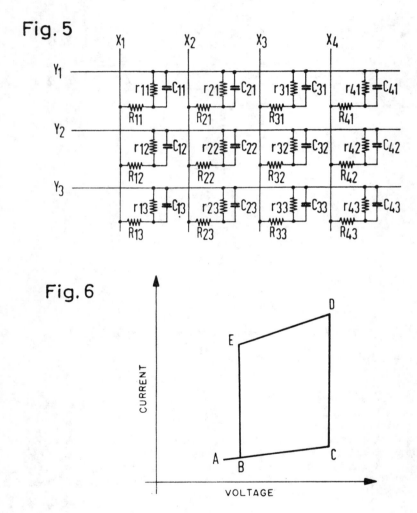

Fig. 5

Fig. 6

As illustrated in Figure 6 by the current-voltage characteristic of a bistable layer in the present invention has proven to be quite advantageous. Such a current-voltage characteristic is shown, for example, by a cadmium selenide layer. As indicated in the graph of Fig. 6, if a liquid crystalline layer is in contact or connected in series with a bistable layer, a low current will flow through the matrix only until a voltage is applied to the bistable layer that exceeds a certain threshold voltage. As shown in Fig. 6, where a low voltage is applied as at the level shown by point "A", the conductivity of the bistable layer remains low and constant. However, if the voltage applied to the bistable element exceeds the certain threshold voltage corresponding to point "C" in Fig. 6, the conductivity of the layer increases suddenly to a level indicated by point "D", i.e., greater than 1,000 times, and the liquid crystalline layer is energized. The liquid crystal layer 2 remains energized until the applied voltage drops to a value represented by the point "E" at which value, and point, the conductivity of the layer rapidly decreases to a low value as indicated by point "B".

" " " " " " " " " "

3,731,986 (U.S.)   Patented May 8, 1973

DISPLAY DEVICES UTILIZING LIQUID CRYSTAL LIGHT MODULATION.

<u>James L. Fergason</u>, *assignor to International Liquid Xtal Company. Application Apr. 22, 1971.*

Optical display devices are disclosed for converting electrical intelligence into optical images. The invention has particular utility in computer and calculator read-outs, for example, since the display can be energized at a voltage level compatible with that used to drive the integrated circuitry used in such devices without the necessity for relatively high voltage driving circuitry.

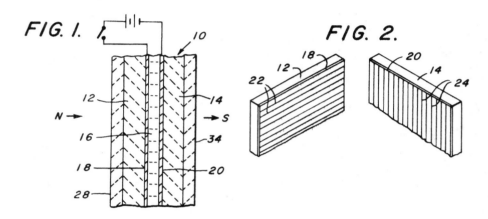

With reference to the drawings, and particularly to Fig. 1, there is shown a liquid crystal unit 10 comprising first transparent plate 12, preferably of glass, and a second transparent plate 14, also of glass, and extending parallel to the plate 12. The plates 12 and 14 are spaced apart by suitable spacers, not shown. This space is filled with a nematic-phase liquid crystal material with a positive dielectric anisotropy such as 20 percent to 80 percent each of bis-(4'--n-octyloxybenzal-2-chlorophenylendiamine and p-methylbenzal-p'-n-butylaniline, these making up about 60 percent of 97 percent of the total composition and p-cyanobenzal-p'-n-butylaniline comprising the remaining 3--40 percent.

Disposed on the interior surfaces of the transparent plates 12 and 14 and in contact with the liquid crystal layer 16 are coatings 18 and 20 of thin transparent electroconductive material, such as tin oxide or indium oxide coatings.

In Figure 2, there is shown a view of the plates 12 and 14 which may comprise flat glass on the order of about one-eight inch thick having layers 18 and 20 of transparent conducting material deposited on the facing surfaces thereof. In the preparation of a liquid crystal unit, the layers of transparent conducting material that are in contact with the nematic-phase liquid crystal material must be prepared by being stroked or rubbed unidirectionally with, for example, a cotton cloth. The direction of rubbing on the respective plates 12 and 14 is indicated by the lines 22 and 24 in Fig. 2; and it will be appreciated that the directions of rubbing on the respective plates are at right angles to each other. The effect of this rubbing is to produce a twisted nematic structure. A property of the nematic-phase liquid crystal materials is that the molecules in the vicinity of a rubbed surface tend to align themselves with it. Thus, the molecules nearest the surface of the plate 12, for example, are inclined to orient themselves parallel with the lines 22, and those nearest the surface of plate 14 are inclined to orient themselves parallel to the lines 24. The structure is fluid and active; and under condicitions of no applied voltage, the molecules in the various layers that are parallel to the surfaces of plates 12 and 14 arrange themselves in what may be considered a number of layers of suitable intermediate "mainstream" diections, ranging from one close to parallel to the lines 22 (a short distance from the surface of plate 12) through one at about a 45° angle with respect to both the lines 22 and 24 (at about the midpoint of the distance between the plates 12 and 14), and on to one close to parallel with the lines 24 (a short distance from the surface of plate 14).

The effect of the liquid crystal unit on polarized light directed through the plates 12 and 14 and polarized parallel to the lines 22, foe example, is that the unit effects a rotation of the plane of polarization of the light as it passes through the unit, so that the light emanating from the surface of the plate 14 is plane polarized parallel to the lines 24. The extent of rotation does not need to be 90°. Any desired extent of rotation may be obtained, merely by properly orienting the unidirectionally rubbed surfaces on the plates 12 and 14. However, when the directions of rubbing are at right angles to each other, the extent of rotation is 90°.

The effect of the crystal unit 10 on polarized light is schematically illustrated in Fig. 3. Thus, a source of unpolarized or natural light at 26 impinges upon a conventional polarizer 28 which polarizes the light indicated by the broken lines 30. This polarized light, as it passes through a liquid crystal unit such as unit 10 shown in Fig. 1, will be rotated through 90° so that the polarized light is then polarized in a plane indicated by the broken lines 32. This polarized light will then pass through a second polarizer 34 adapted to pass polarized light in a plane which is rotated at 90° with respect to the plane of polarization of polarizer 28, as indicated by the broken lines 36. Hence, under the conditions described, the polarized light passing through polarizer 28 will be rotated through 90° in unit 10 and will then pass through the polarizer 34. On the other hand, if the plane polarizer 34 should be rotated such that the plane of polarization indicated by broken lines 36 is parallel to the plane of polarization of polarizer 28, then no light will pass through polarizer 34.

Now if an electric potential, on the order of 5 volts or greater, is applied between the conducting films 18 and 20, the liquid crystal unit 10 will no longer rotate the plane of polarization through 90°. In the arrangement shown in Fig. 3, for example, application of a suitable potential to the conducting films 18 and 20 will cause the polarizer 34 to block the transmission of light. It can thus be seen that the device acts as an optical shutter. On the other hand, if the polarizer 34 is oriented 90° with respect to that shown in Fig. 3, no light will be transmitted in the absence of a potential applied between the films 18 and 20; whereas light will be transmitted when a potential is applied thereacross.

In Fig. 1, the means for applying an electric field between the conducting films 18 and 20 is shown as a conventional battery 38 adapted to be connected into the conducting films 18 and 20 through switch 40. Alternatively, however, the same effect can be achieved with the use of a magnetic field in which the lines of flux extend perpendicularly to the surfaces of the plates 12 and 14 as indicated by the north and south pole indications of Fig. 1.

With reference to Fig. 4, one type of optical display which can be provided with the liquid crystal device 10 of Fig. 1 is shown. In this case the conducting films 46 and 48 are in the form of the numeral 4. Assuming that the plates 42 and 44 are assembled with polarizers in the arrangement of Fig. 3 and that switch 52 is closed to apply a potential from battery 50 across the films 46 and 48, the area covered by the films will be opaque while the area around the conducting films 46 and 48 will transmit light. assuming that a white background is behind the assembled plates 42 and 44 with liquid crystal material and that the plate is viewed from the side opposite the white background, the effect will be to produce the numeral 4 in black-on-white. Of course, when the switch 52 is again opened, the device will be totally light transmitting and no numeral or other optical image will appear to the eye of the observer.

A system for producing any desired numeral, letter or other image within the same area is shown in Figs. 5-7. The system again includes two plates 54 and 56 (Fig. 5) having facing surfaces which are rubbed at right angles with respect to each other, the space between the two surfaces being filled with a layer of nematic liquid crystal material of positive dielectric anisotropy. The mating surfaces of the plates 54 and 56 are again coated with a conducting film; but in this case, the plate 54, for example, is etched, utilizing conventional photoresist masking techniques, to provide five vertical columns 58 each having seven enlarged areas 60 spaced along its length. In a somewhat similar manner, the plate 56 is coated and then etched to provide seven horizontal rows 62 each provided with five enlarged area sections 64 of conducting film material between its ends. The plates 54 and 56, when facing each other with a layer of liquid crystal material therebetween, are positioned such that the enlarged area portions 60 on

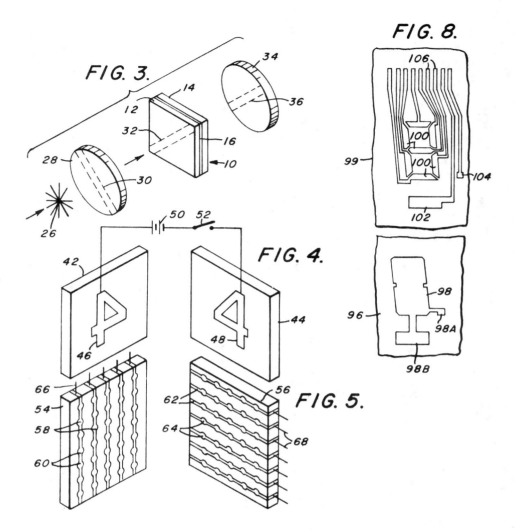

the plate 54 are aligned with or overlie the enlarged area portions 64 on the plate 56. The ends of the stripes or columns 58 on the plate 54 are connected to five electrical leads 66. Similarly, the ends of the strips or horizontal rows 62 on plate 56 are connected to a second set of seven electrical leads 68.

The manner in which an assembly formed of the plates of Fig. 5 can be used to produce various images is shown in Figs. 6 and 7. Clock pulses for the display are supplied from an oscillator 72 typically having a frequency of about 960 hertz. These pulses are applied to a flip-flop circuit 73, the output of the flip-flop circuit being fed to a conventional three-bit ring counter 74 which produces pulses on leads 76, 78, and 80, those on lead 76 being divided by two, those on lead 78 being divided by four and those on lead 80 being divided by eight. The pulses on leads 76-80 are applied to a decoding matrix 82 in accordance with well known techniques to produce pulses on output leads 84 which are displaced in phase with respect to each other. These are applied through inverters 86 and leads 68 to the respective horizontal rows 62 which are identified by the letters A-G. The outputs of the inverters 86 appearing on leads 68 are identified as waveforms A through G in Fig. 7. Note that the pulses in eaveforms A-G are of negative polarity. These pulses are applied to the rows in succession continuously regardless of the optical image, such as a numeral or letter, which it is desired to produce.

The pulses on leads 76-80 are also applied to a read-only memory unit 88 connected, for example, to computer circuitry 90, or the like. The pulses on leads 76-80 activate the read-only memory unit 88 to apply to leads 92 a succession of pulses representative of a particular numeral, letter or other image to be displayed. These are applied through inverters 94 and capacitors 96 to the vertical rows 58 which are identified by the letters H-M. In order to produce the numeral "2" for example, only those areas colored black in Fig. 6 between the strips 58 and 62 should

257

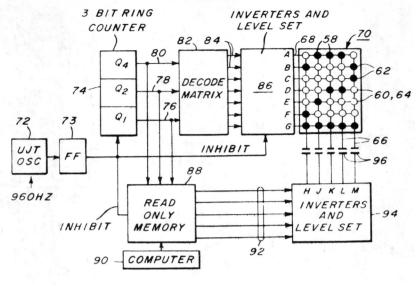

FIG. 6.

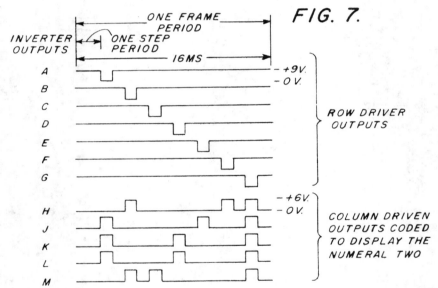

FIG. 7.

have electrical potentials applied therebetween, whereby these areas will be opaque and appear black when viewed by an observer. In order to accomplish this effect, the waveform H-M of Fig. 7 are applied to the leads 66. Note that in order to produce the numeral 2, the second, sixth and seventh areas 60, 64 between strips 58 and 62 in column H must have potentials applied there-across. The various areas forming the numeral 2 of Fig. 7 will not be continually opaque; however the sweeping action will occur sufficiently rapidly so that a continual image will appear to the naked eye. Any flicker effect appearing to the observer can be reduced by shortening the frame period and increasing the scanning frequency.

In Figure 8, still another embodiment of the invention is shown wherein one of two transparent plates 96 is provided with a continuous layer of transparent conducting material 98; while the other transparent plate 99 is provided with a series of mutually insulated strips of transparent conducting material 100. The total configuration, when opaque, represents the numeral 8. Beneath the configuration 100 is a line or bar 102 and to the right of the configuration is a dot 104 which forms a decimal point when a plurality of the arrays of Fig. 8 are placed side-by-side. The dot 104 is aligned with area 98A of layer 98; while area 98B is aligned with the bar 102. The various mutually insulated conductive strips forming the configuration 100, in turn, are connected through a plurality of mutually insulated strips of transparent conducting material 106 to external leads, not shown. Assuming, for example, that it is desired to form the numeral 3, the plate 98 on one side of the layer of liquid crystal material will be connected to a source of positive potential while the transparent strips on the other side forming a 3 will be connected through leads 108 to a source of negative potential. If it is desired to place a decimal point beside the numeral, the lead connected to the spot 104 will be connected to the same source of negative potential; and if it is desired to provide a line beneath the numeral, the lead connected to strip 102 will be connected to the source of negative potential. In any case, only those portions on the plate 99 will appear opaque on a white background (or vice versa) which are connected to a source of potential of polarity opposite to that applied to the plate 98. As will be appreciated, a series of the displays shown in Fig. 8 can be assembled in side-by-side relationship to provide any desired number of digits.

" " " " " " " " " "

3,732,119 (U.S.)   *Patented May 8, 1973*

TEMPERATURE SENSITIVE VISUAL DISPLAY DEVICE.

<u>Donald Churchill, James V. Cartmell and Robert E. Miller</u>, *assignors to The National Cash Register Company.   Application Nov. 5, 1968.*

A temperature-responsive display device or composition, comprising liquid crystal materials contained in minute capsules as the temperature sensitive component, is provided. Encapsulation of the liquid crystal materials affords physical and chemical protection from various kinds of degradation and allows their use in practical and efficient thermal-gradient display systems.

The patent offers nine specific examples of an encapsulating procedure and the mesomorphic materials used for preparing capsules of the novel temperature-responsive sensing system.

See U.S. Pat. No. 3,697,297 of Oct. 10, 1972 by D. Churchill *et al.*

" " " " " " " " " " "

3,732,429 (U.S.)   *Patented May 8, 1973*

LIQUID CRYSTAL DEVICE.

<u>Morris Braunstein and William P. Bleha, Jr.</u>, *assignors to Hughes Aircraft Company. Application Nov. 12, 1971.*

The liquid crystal device comprises a liquid crystal layer, a thin photoconductor layer, and a thin insulator layer sandwiched between the transparent supporting covers having a transparent electrically-conductive layer thereon. The photoconductor layer is cadmium sulphide (CdS) and the insulator layer is zinc sulphide (ZnS).

Figure 1 illustrates a first embodiment 10 of a basic cell arrangement. The cell is a laminar structure. Glass substance 12 is ordinary clear glass. Transparent, electrically conductive coating 14 is applied to one side of the glass. The electrically conductive coating 14 is of such thickness as to give resistance in the range from 1 to 10^3 ohms per square.

Vacuum deposited onto the coating 14 is a thin layer as the insulator zinc sulphide, illustrated of the layer 16. Thin insulator films or materials such as silicon dioxide or silicone monoxide could also be used. The insulator 16 is in the range from 10 to 10^2 nm in thickness, and the presently considered optimum thickness is 50 nanometers. Over the zinc sulphide layer 16 is vacuum deposited layer 18 of the photoconductor cadmium sulphide. Photoconductor films of cadmium selenide could also be used. The photoconductor layer 18 is vacuum deposited to a thickness of from 2.0 to 12.5 microns.

On top of the photoconductor layer 18 is deposited a liquid crystal layer 20. The liquid crystal layer 20 is a nematic-cholesteric liquid crystal. The thickness of the liquid crystal layer is typically 12.5 microns, and as in conventional liquid crystal practice can conveniently range from 4 to 25 microns. The thickness is contained by spacers positioned around the edges

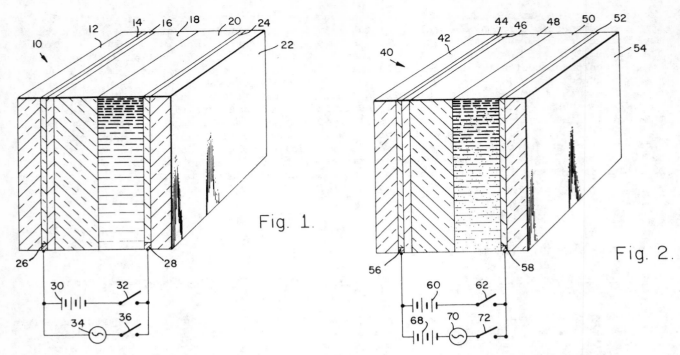

Fig. 1.

Fig. 2.

of the cell 10, to maintain the volume in which the liquid crystal 20 is positioned.

The cell is completed by glass 22 on which is coated transparent conductive layer 24. This structurally correspond to the glass 12 in transparent conductive layer 14, described above. The two glass layers 12 and 22 thus define the outer surfaces of cell 10.

Electric contacts 26 and 28 are respectively connected to the electrically conductive substantially transparent layers 14 and 24. Source 30 of direct current, together with its series control switch 32, are connected across the contacts 26 and 28. The closure of switch 32 applies a DC field across the layers between the electrically conductive layers 14 and 24. AC source 34 and its serially connected switch 36 are also connected across contacts 26 and 28, in parallel to the DC source and switch. The AC source provides AC current from 1,000 to 100,000Hz.

Figure 2 illustrates a cell 40 which is a second embodiment of the liquid crystal device in accordance with this invention. Cell 40 is built off of a plurality of layers, comprising glass layer 42, transparent conductive layer 44, zinc sulphide insulating layer 46, cadmium sulphide photoconductive layer 48, liquid crystal layer 50, transparent conductive layer 52 and glass cover 54. The layers 42, 44, 46, 48, 52 and 54 are respectively of an equal thickness to layers 12, 14, 16, 18, 24 and 22 of the cell 10 shown in Fig. 1 and described with respect thereto. Furthermore, the method of formation of these various layers is the same in both cells. The principle difference between the cells is that the layer 50 is a nematic liquid crystal.

Electric contacts 56 and 58 are respectively connected to substantially transparent conductive layers 44 and 52. Electrically connected to these contacts are the series combination of DC voltage source 60 and its switch 62. This is identical to the DC source 30. In addition, in the embodiment of Fig. 2, a combined DC and AC source is connected across the contacts 56 and 58, in parallel to the DC sources 60. Thus, DC source 68 is connected to one of the contacts, and through AC source 70 and switch 72 to the other of the contacts. By closure of switch 72, an AC dield with a superimposed DC field is applied between the substantially transparent layers 44 and 52. Thus, the cells 10 and 40 are the same, except for the difference between the liquid crystal layers, and the fact that the nematic liquid crystal layer 50 also has the capability of having a combined field applied thereacross. (The method of manufacture of the insulating layer 16 and the photoconductive layer 18, and their corresponding layers 46 and 48 is given in detail).

In the use, a light having an intensity pattern in accordance with the desired image is projected onto the cell. After imaging is completed, the liquid crystal stored image can be viewed. This viewing can be accomplished either by a projected display light, or by reflective observation of the image by frontal illumination. In either case, a display light is also utilized.

Considering the structure of Fig. 1, both switches 32 and 36 are open. Thereupon, the imaging light is turned on. The imaging light projects the desired image into the cell, both through the nematic-cholesteric misture liquid crystal and the photoconductive layers. With the imaging light on, the switch 32 is closed for a short time, for example, from 0.1 to 1 second. When the voltage, typically from 20 to 150 volts, is applied, the direct current potential field is divided across the liquid crystal layer 20, and the insulator layer 16 and photoconductive layer 18. The imaging light is of such wavelength where the photoconductor is sensitive. For cadmium sulphide, the imaging light is thus between 400 and 520 nanometers. Where there is no imaging light, the photoconductor and insulator layer totally comprise a higher impedance than the liquid crystal. Where the imaging light is projected onto the cell, the resistance of the photoconductor decreases, so that it is less than the liquid crystal. Thus, in the exposed area of the photoconductor, the current is caused to flow to induce change in the scattering properties of the liquid crystal. In the regions where there was no imaging light, the voltage pulse caused by closure of switch 32, the voltage division of such that the principle voltage drop was across the resistor and the photoconductor and the liquid crystal adjacent thereto is uneffective. Thus, this area remains clear. After switch 32 is open, and the imaging light removed, the image is retained in the liquid crystal, because of the scattering storage properties of the nematic-cholesteric mixture.

At a later time, when it is desired to remove the image stored in the liquid crystal, the liquid crystal layer can be brought to its non-scattering state in all regions by closing the switch 36 and applying the AC field across the liquid crystal, from source 34. A frequency greater than about 1,000 cycles per second is required to effect erasure and return the liquid crystal to the clear state. Closure of the switch 36 for 0.5 to 2 seconds is adequate to obtain this erasure.

In the structure of Fig. 2, the liquid crystal 50 is a nematic material. There is no optical memory, in the sense of scattering condition of such a material, in the absence of a continuously applied field. The employment of the DC field provided by the Dc source 60 and closure of its switch 62 is alternative to the application of the DC field superimposed upon the AC field, by closure of switch 72. The DC field provided by source 60 is from 10 to 100 volts. On the other hand, the combined field comprises a DC potential from source 68 of from 10 to 100 volts, tohether with an AC potential from source 70 of from 10 to 100 volts rms at a frequency in the range of from 1,000 to 20,000 cycles per second. With either of the switches 62 or 72 closed, the imaging light is turned on. The imaging light is of such a wavelength that it affects the photoconductor. In the case of the cadmium sulphide, the imaging light has a wavelength from 400 to 520 nanometers. The illumination light, either for frontal or projective viewing of the images of the liquid crystal, is of such a wavelength that the photoconductor is not photosensitive thereto. In tha case of cadmium sulphide, the projection light has a wavelength above 520 nanometers. In this mode, the information supplied by the imaging light is displayed in real time by the viewing light. This is accomplished because the regions where the photoconductor is of lower impedance due to the incidence of light, the voltage drop is substantially taken across the liquid crystal, and this causes dynamic scattering in that region. In the regions where the imaging light does not occur, the principle part of the voltage drop is across the photoconductor, so that the voltage drop across the nematic crystal is not sufficient to cause scattering. Therefore, in those areas the liquid crystal remains transparent.

In order to achieve real time imaging, the use of the combined DC and AC field is preferred to the ordinary DC field provided by the DC source 60. The normal decay time of the nematic liquid crystal under the influence of the DC field, and without illumination, may not be sufficient to provide the real time display. Therefore, the superimposed AC wave quickly erases the image in areas where the imaging light has been removed.

Referring to Figure 3, the cell 78 could be either of the cells 10 or 40, equipped with the electric application means described with respective Figs. 1 and 2. The cell 78 is positioned between a source of visible light 80 which serves as a projection light and screen 82.

A source of visible imaging light 84 provides visible radiation for photoactivating the photoconductor layer 18 inside the cell 78, the photoconductor layer in this instance being on the left side of the cell facing the sources 80 and 84. By this arrangement, images formed inside the liquid crystal layer of the cell may be portrayed on the screen 82. As is typical, the imaging light from the source 84 is of a shorter wavelength than the display light from the source 80. If the nematic liquid crystal layer is transparent to imaging light, the cell 78 can be turned around so that the photoconductor layer is on the right side of the cell 78 facing the screen 82 with the source 84 to the left of the cell thus being arranged to expose the photoconductor material layer by transmission of light from source 84 through the liquid crystal layer to the photoconductor material layer.

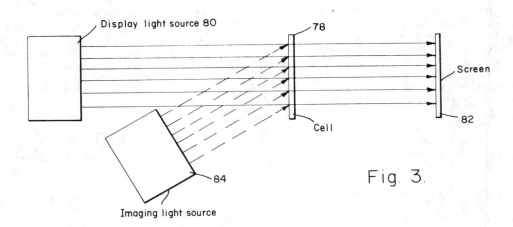

Fig. 3.

The benefits of the invention to be accrued are improved device performance consisting of improved contrast ratio in the projected image, and improved sensitivity to the input (write) light signal.

" " " " " " " " " " "

1,316,213 (British)   Patented May 9, 1973

REAR PROJECTION SYSTEM.

<u>The National Cash Register Company</u> (Inventor's name not given).
Application July 20, 1971; prior U.S. application Aug. 5, 1970.

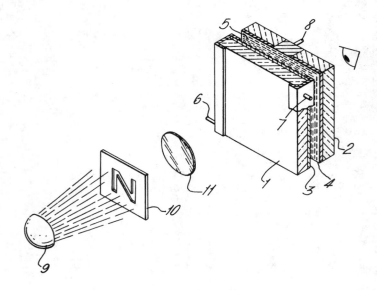

A rear projection system includes a light source 9, a projector and a screen, the projector being arranged to project on to the screen a focused image of an object 10 disposed between the light source 9 and the projector lens 11 the screen having two transparent parallel plates 1, 2 spaced less than 500 microns apart, the adjacent surfaces of each plate being coated with a transparent electrically conductive film 3, 4 an organic nematic mesomorphic material 5, which scatters transmitted light in response to an electric field above a certain threshold value, filling the space between the plates, and means for applying a voltage across the films 3, 4 so that in use an electric field exceeding the threshold value is established. The nematic material is, inter alia, a 50/50 blend of butyl p(p-ethoxy phenoxy carbonyl) phenyl carbonate and butyl p-ethoxy phenyl terephthalate and includes up to 1 weight percent of an ion producing material e.g. octoic acid. The voltage applied across the films, which are preferably indium oxide or tin oxide, is between 10 and 300 volts.

"""""""""""

1,316,497 (British)   Patented May 9, 1973

METHODS AND APPARATUS FOR VISUAL REPRESENTATION, IN A MODEL, OF A SURVEYED MEDIUM.

<u>Gerard Grau and Andre Fontanel</u>, assignors to the Institut Francais du Petrole des Carburants et Lubrifiants.   Application May 18, 1970; prior application in France Dec. 11, 1969.

Recordings of waves received at several spaced reception points from one or more transmission points. The waves may be acoustic or radio waves and the invention is applicable to, e.g., seismic surveying or biological inspection.

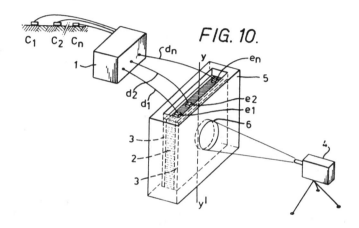

The optical model of Fig. 10 comprises a liquid crystal layer 2 sandwiched between transparent (glass or plastics) plates 3 in a regulated-temperature jacket 5. Signals from the wave receivers $C_1....C_n$ have their timescale compressed and reversed in unit 1 (e.g. by transferring a digital recording of them to magnetic disc storage) and then drive respective piezo-electric transducers $e_1....e_n$ to produce ultrasonic waves of propagation velocity V' in the model. By making the model dimensions represent a vertical section through the surveyed zone with a scale reduction factor α, and by making the time-scale compression factor $V'\alpha/V_m$, the ultrasonic waves will produce energy concentrations, and hence color changes, at points in the liquid crystal representing the diffraction points and reflection images within the surveyed zone. These changes are recorded by camera 4.

If the liquid crystal and transparent plate thicknesses are less than the ultrasonic wavelength, V' is a function of the relative thicknesses, which thicknesses can be tailored to compensate for changes of V_m with depth in the surveyed zone.

"""""""""""

3,733,485 (U.S.)   Patented May 15, 1973

EXPOSURE METER FOR THERMAL IMAGING DEVICES.

<u>Joseph Gaynor and Terry G. Anderson</u>, assignors to Bell & Howell Company. Application March 18, 1971.

An exposure meter to permit the determination of optimum thermal exposure of a document to

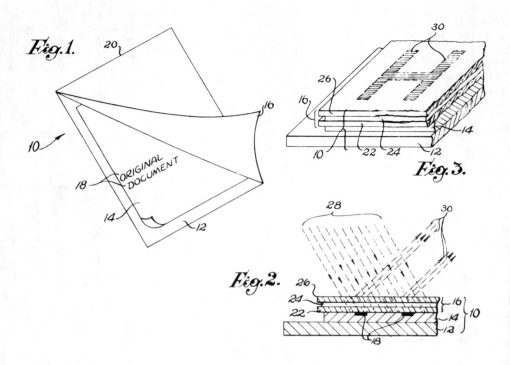

be thermographically reproduced. A sheet of thermotropic material is assembled with the document and exposed to thermal radiation until a change in form is discerned corresponding to the document information. Exemplary thermographic materials include cholesteric liquid crystalline material, thermochromic compounds, heat-fusible materials, and materials having electrical resistance which is sharply temperature dependent.

Referring to Fig. 1, an assembly 10 is illustrated which includes a base sheet 12 of relatively thick paper supporting an original document 14 and a sheet 16 of thermotropic material. In this illustration, the sheet 16 includes a layer of cholesteric liquid crystalline material. The original document 14 contains infrared-absorptive indicia 18 on its face and is placed face-up on the base sheet 12 beneath the thermotropic sheet 16. The top edge of the thermotropic sheet 16 is secured by adhesive or the like to the top edge 20 of the base sheet 12 so as to provide an easily manipulated assembly.

Referring to Fig. 2, the assembly 10 is processed to determine the optimum level of heat exposure to which to subject the original document 14 during thermographic duplication. The original document 14 is placed on the supporting base sheet 12 with its infrared-absorptive indicia 18 facing upwardly. The indicia 18 are constituted of carbon, heavy metal or any material which upon exposure to infrared will absorb more infrared radiation than the surrounding non-image areas so as to convert the absorbed energy to a thermal pattern corresponding to the visible image pattern on the document 14.

The thermotropic sheet 16 is formed with a substrate 22 sufficiently thin, about 1/2-2 mils thick. Mylar film can be used. The substrate 22 is coated with a thin layer 24, also about 1/2-2 mils thick, of cholesteric liquid crystalline material, simulating the thickness of a thermographic layer so that the total thickness results in the same heat transport rate as the thermal imaging material. In this case, the cholesteric liquid crystalline material is a mixture, by weight, of about 5 percent cholesteryl nonanoate 50 percent cholesteryl chloride and 45 percent oleyl cholesteryl carbonate. This material has a cholesteric-isotropic transition temperature of about 33°C and the mixture remains in liquid crystalline form at room temperature, the material undergoes a color play, visible against a black background, from red to violet over a temperature range of 33-35°C.

In this example, another sheet 26 of Mylar, about 2 mils thick, is provided to protect the layer of liquid crystalline material 24 from contamination. The sandwiched assembly of Mylar layers 22 and 26 and liquid crystalline material 24 constitutes the thermotropic sheet 16. The assembly of base sheet 12, document thermotropic and sheet 16 is sandwiched together so that directly opposing surfaces are contiguous with one another. The assembly 10 is then exposed to radiation 28 which is rich in infrared rays and which contains substantial amounts of visible light. The radiation 28 is directed onto the sheet 16 so as to penetrate the sheet 16 and impinge onto the original document 14. The infrared portions of the radiation 28 generate a temperature rise in the indicia 18 portions of the document 14, resulting in a thermal pattern emanating from the document 14 which selectively raises the temperature of the sheet 16 in regions corresponding to the indicia 18.

Exposure radiation 28 can be utilized containing visible light including all of the wavelengths of the visible spectrum and one can simply look for a color play to determine the fact that a temperature range of 33°-35°C has been reached.

As the sheet 16 is progressively exposed to the radiation 28, the temperature of selected portions is raised in accordnace with the heat pattern generated by the indicia 18. Upon reaching the predetermined temperature of 33°C, an image is observed as a result of reflection of wavelengths 30 of red light, thereby signaling a temperature of 33°C.

Referring to Figure 3, there is illustrated a portion of the top surface of the sheet 16 following the thermal exposure step of Fig. 2, wherein the image 30 corresponding to the indicia 18 is seen as a reflection pattern from the surface of the liquid crystalline layer 24.

The time required from initial exposure to discernment of the reflected image is measured and recorded. This time in conjunction with the known and constant intensity of the infrared radiation and spatial relationship between the infrared source and surface of the thermotropic sheet 16 can be correlated with the exposure required in any particular instrument for optimum thermographic reproduction from that document 14.

The cholesteric material can be coated from a chloroform solution onto the Mylar sheet 22. Alternatively, the cholesteric material can be placed in liquid form along one edge of the Mylar substrate 22 and the cover Mylar sheet 26 placed thereon. The sandwich can then be squeezed between rollers to spread the cholesteric material as the layer 24 evenly. In place of the Mylar substrate, one can utilize any infrared translucent material, such as glass, rigid plastic, and the like. It is also known to encapsulate liquid crystalline material in spheres approx. 40-50 microns in diameter. The encapsulated material can be used in the same manner as the pure material.

" " " " " " " " " "

3,734,597 (U.S.)
Patented May 22, 1973

PROCESS FOR PRODUCING A COLOR STATE IN A DISPLAY DEVICE.

Donald Churchill, James V. Cartmell, assignors to The National Cash Register Company.
Application Apr. 9, 1971.

This patent is a division of application Ser. No. 707,706 filed Feb. 23, 1968 and now U.S. Pat. No. 3,600,060, found elsewhere in this volume.

" " " " " " " " " "

3,734,598 (U.S.)
Patented May 22, 1973

LIQUID CRYSTAL DISPLAY DEVICE HAVING AN INCLINED REAR REFLECTOR.

William Ross Aiken, assignor to Display Technology Corporation.
Application Apr. 30, 1971.

A display device settable between transparent and translucent or opaque conditions comprising a layer of a material changeable between light-transmitting and light-scattering conditions in response to electric fields, such as the liquid crystal materials, front and rear covers of transparent material containing said layer, each having a transparent film of a conductive material, means for setting up an electric field between the said films through said layer, a source of light in front of the front cover of said layer, and an inclined mirror arrangement located behind the rear cover of said layer for directing light from said source obliquely through said layer from its rear face to its front face in such a manner, that the source of light is concealed from view by an observer before the front face of said layer when said layer is in transparent light-transmitting condiction. This inclined mirror arrangement reflects the light from the source of light obliquely through said layer in a direction in which it misses the observer's eyes. In this manner light from the source of light is removed from view by the observer when the layer is in transparent light-transmitting condition, yet is available for scattering and thus gives the layer a frosted or milky appearance when the layer of light transmission changing material is in light-scattering condition.

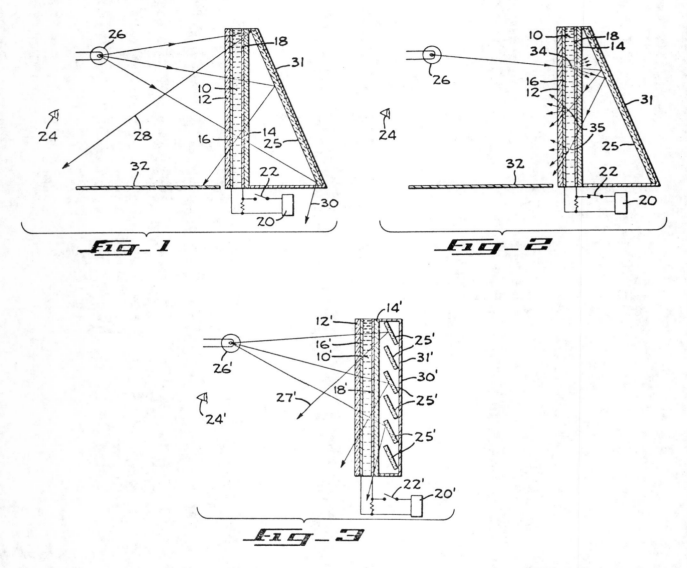

Fig. 1 is a schematic side elevation of a vertical section through a display device embodying the invention in its transparent condition;

Fig. 2 is a schematic side elevation similar to Fig. 1 of the same device in its opaque or translucent condition; and

Fig. 3 is a schematic side elevation of a section through a modified embodiment of the invention.

The embodiment of the invention illustrated in Fig. 3 is essentially the same and operates on the same principles as the embodiment presented by Figs. 1 and 2. It comprises a sandwich composed of a layer 10' of liquid crystal material contained by and between transparent covers 12' and 14' each provided with a transparent film of a conductive material 16' and 18' respectively. Means are provided in the form of a source of electric power 20' to establish an electrostatic field through the layer of crystal material 10' by closure of a switch 22'.

The embodiment of the invention illustrated in Fig. 3 differs from the embodiment represented in Figs. 1 and 2 in respect of the mirror means which is intended to direct light received from a source of illumination 26' in front of the device in forward direction from points behind the rear cover 14' obliquely through the layer of crystal material 10' in such a manner that the light avoids, and is, therefore, not visible to, the observer's eyes. In the embodiment of the invention illustrated in Fig. 3 this mirror means is subdivided into a plurality of vertically superposed parallel, slanted mirrors in the form of relatively spaced, reflective slats 25' of sufficient inclination that even the light reflected by the uppermost slat represented by line 27' will pass through the layer of crystal material 10 in forward direction but will miss the observer's eye 24 when the device is in transparent condition. It will, however, scatter and present a frosted opaque appearance of layer 10' to the observer when the switch 22' is closed and an electrostatic field is set up between the electrically conductive films 16' and 18'. As in the embodiment of the invention illustrated in Figs. 1 and 2, the mirror means 25' are covered by an enclosure 31' whose inner surfaces are dark so that the field presented to the observer when the switch 22' is open, and the layer of crystal material 10' is transparent, will appear to be black or at least dark. Over the embodiment represented by Figs 1 and 2, the embodiment of the invention illustrated in Fig. 3 has the advantage that it is of greater compactness due to the vertically superposed, subdivided mirror arrangement, which can be covered by a much more compact enclosure 31' than the enclosure 31 that is necessary to cover the inclined, continuate mirror 25 of Figs. 1 and 2.

" " " " " " " " " " "

1,318,007 (British)   Patented May 23, 1973

ARRANGEMENT FOR LOCALLY ALTERING THE BRIGHTNESS AND/OR COLOR OF AN INDICATOR MEDIUM.

Peter Wild, assignor to Brown, Boveri and Company Ltd.
Application Apr. 19, 1971; prior application in Switzerland March 26, 1970.

A device for displaying in analogue form the value of an applied voltage V_M consists of glass substrates 1 and 2 supporting film electrodes (e.g. tin oxide) 4 and 4', between which is a layer 3 of a liquid crystal medium having an optical characteristic I which shows abrupt variation with potential difference above a threshold value.

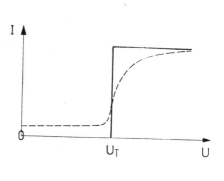

Fig.1

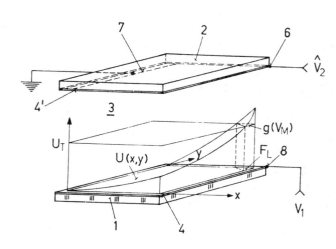

Fig.2

The film 4' has a much greater surface resistivity than the film 4, and opposite ends are connected to earth and to a fixed potential V_2 respectively, so that the potential distribution has the form shown in Fig. 3(a). To the conductive film 4 is applied a fixed potential V_T plus

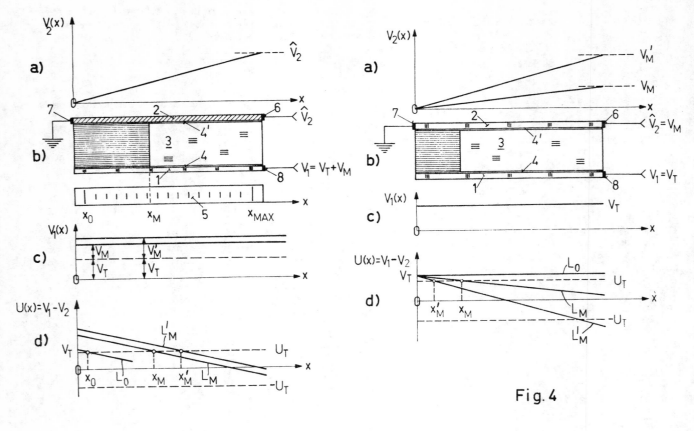

Fig. 3

Fig. 4

the voltage V_M, giving a potential distribution across film 4 as shown for two values of V_M in Fig. 3(c).

The material 3 is such that for a threshold voltage U_T, I abruptly changes; thus all the material with $x<x_M$ (Fig. 3(d)) will, for an applied voltage V_M, have a value of I considerably different from that of the material for $x<x_M$.

Optionally the voltage V_M may be applied to contact 6, (Fig. 4), or the potential on one film may vary two-dimensionally (Fig. 2).

" " " " " " " " " " " "

1,318,011 (British)   Patented May 23, 1973

NEMATIC LIQUID CRYSTAL MATERIAL.

<u>Kabushiki Kaisha Daini Seikosha and Kabushiki Kaisha Hattori Tokeiten</u> (Inventor's name not given. Application Apr. 23, 1971; prior application in Japan May 2, 1970.

A nematic liquid crystal material comprises at least one compound of the formula

$$R_1O\text{-}\bigcirc\text{-}OC\text{-}\bigcirc\text{-}OCOR_2$$

wherin R_1 and R_2 are alkyl groups, in admixture with at least one compound of the formula

$$C_2H_5\text{-}\langle\text{-}\rangle\text{-OC(O)-}\langle\text{-}\rangle\text{-OCR}_3\text{(O)}$$

wherein R_3 is a n-butyl or n-amyl group. These compositions may be prepared by mixing one mole of a compound of each of the above formulae. Examples are given which illustrate the preparation of such compositions. Such liquid crystals may be used in digital display devices, e.g. disposed between conductive oxide coated glasses and subjected to a field of more than 10^4 V/cm; strong light scattering was observed. (Three examples are described in the patent.)

" " " " " " " " " "

1,318,012 (British) Patented May 23, 1973

NEMATIC LIQUID CRYSTAL COMPOUNDS AND MIXTURES.

<u>Kabushiki Kaisha Daini Seikosha and Kabushiki Kaisha Hattori Tokeiten</u>, (Inventor's name not given) Application Apr. 19, 1971; prior application in Japan Apr. 17, 1970.

An optical display device for a digital computer contains a nematic liquid crystal mixture comprises in admixture either (1) at least two compounds each of the formula

$$R_1O\text{-}\langle\text{-}\rangle\text{-OC(O)-}\langle\text{-}\rangle\text{-OCOR}_2 \quad (I)$$

wherein R_1 and R_2 are alkyl groups, or each of the formula

$$C_2H_5\text{-}\langle\text{-}\rangle\text{-OC(O)-}\langle\text{-}\rangle\text{-OCOR}_3 \quad (II)$$

wherein R_3 is an alkyl group; or (2) at least one compound of Formula I and at least one compound of Formula II. The mixture exhibits dynamic scattering when an electric field is applied thereto. (Four examples are described in the patent).

" " " " " " " " " "

3,736,047 (U.S.) Patented May 29, 1973

LIQUID CRYSTAL DISPLAY DEVICE WITH INTERNAL ANTI-REFLECTION CASTING.

<u>Robert M. Gelber and Edward A. Small, Jr.</u>, assignors to Optical Coating Laboratory, Inc. Application Aug. 13, 1971.

Liquid crystal display device having first and second substantially transparent insulative plates having front and rear surfaces with said plates being arranged so that the rear surface of the first plate and the front surface of the second plate face each other. An antireflection coating is disposed on at least one of said front and rear surfaces of said front plate. A conducting coating is carried on said rear surface of said first plate and said front surface of said second plate. Means secures the first plate to the second plate so that there is a space provided between said conducting layers. A layer of liquid crystal material is disposed in said space.

In the method, an antireflection coating is provided on a surface so that the pattern carried by the display device is invisible when the display device is not activated.

The liquid crystal display device consists of substantially transparent insulative front and rear plates 11 and 12. The front plate includes a substrate or body 13 and the rear plate 12 includes a substrate or body 14. Both of the substrates 13 and 14 are formed of a suitable transparent material which has insulating properties as, for example, glass. A typical glass which can be utilized for this purpose is a soda lime glass having an index of refraction of approx. 1.52. By way of example, one-eight inch float glass can be utilized because its surfaces are quite flat.

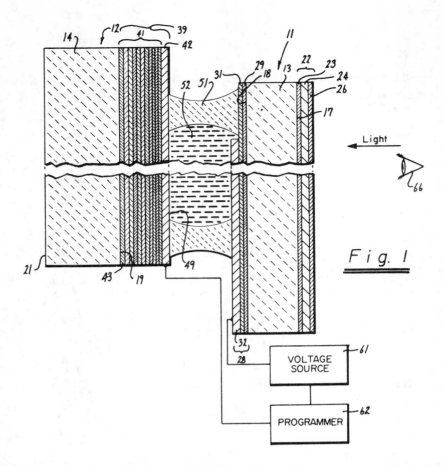

Fig. 1

The substrate of body 13 is provided with front and rear spaced generally planar parallel surfaces 17 and 18, whereas the substrate or body 14 is provided with spaced generally parallel planar surfaces 19 and 21. An antireflection coating 22 is provided on the front surface 17 of the substrate 13. The antireflection coating 22 can be of the type described in Thelen U.S. Pat. No. 3,185,020. Another antireflection coating which would be suitable for use on the surface 17 is described in Rock U.S. Pat. No. 3,434,225.

A combination antireflection and conducting coating 28 is provided on the surface 18 of the substrate 13 and consists of three layers 29, 31 and 32 which can be identified as first, second and third layers counting from the substrate. The first layer 29 is preferably formed of a high index material having an index of refraction ranging from 1.9 to 2.3. Materials having indices of refraction within this range are zirconium oxide, Neodymium oxide, tantalum oxide, indium oxide, titanium oxide, silicone dioxide and certain mixed oxides of the type disclosed in Kraus et al. U.S. Pat. No. 3,034,924. The second layer is formed of a low index material having an index of refraction in the vicinity of 1.38. Materials such as magnesium fluoride and zinc cryolite can be used.

The layers 29 and 31 form a high-low index combination which is formed of non-conducting or insulating materials. The layer 32 also forms a part of the antireflection coating and is formed of a conducting material such as indium oxide or tin oxide. The conducting layer is deposited to a suitable thickness as, for example, to a thickness having a conduction between 500 to 1,000 ohms/square. Suitable photolithographic or silk screen resist methods of a type well known to those skilled in the art are utilized for forming the desired conducting pattern in the conducting layer 32.

One of the principal characteristics desired for the front plate 11 of the liquid crystal display device is that the pattern carried by the front plate 11 be invisible or substantially invisible to the naked eye when the device is not activated. To accomplish this objective, it is necessary that both the conducting and non-conducting parts of the combination anti-reflection and conducting coating 28 be relatively low in reflectance and that the two layers 29 and 31 have substantially the same appearance to the visible eye as the three layers 29, 31, and 32.

The rear plate 12 consists of a substrate 14 having a multilayer dielectric coating or stack 41 deposited on the nromal light reflecting surface 19. A conducting layer 42 is provided on the multilayer stack 41. It is a requirement that all of the coatings carried by the front and rear plates 11 and 12 be able to withstand a temperature of at least 500°C. In addition it is a requirement that the multilayer dielectric coating 41 be continuous, nonconducting, substantially neutral in color throughout the visible spectrum and reflect at least 90% of the visible light striking the same.

It is also a requirement for the coatings carried by the front and rear plates 11 and 12 that they not react with the liquid crystal material which is to be utlized in the device.

The dielectric layers 31 and 29 and the multilayer dielectric coating 41 both must be capable of supporting a conducting layer which is relatively invisible. The dielectric reflector should approach as close as possible the reflectivity of an aluminum layer and its neutrality in color.

It is also a requirement that the dielectric layers which are utilized be able to resist the etch which is utilized for etching the conducting layer deposited on the dielectric layers.

In order to achieve the desired amount of reflectance, it is desirable that the dielectric coating 41 be formed of a plurality of layers of high and low index materials. At least seven of such layers should be utilized although it is preferable that 11 or more layers be utlized. Thus, the multilayer coating 41 is formed of a plurality of separate layers 43 with the odd numbered layers counting from the medium being formed of a high index material and even numbered layers being formed from a low index material. Titanium dioxide and zirconium oxide can be utilized as high index materials, whereas silicon dioxide or Vycor or magnesium fluoride can be utilized as low index materials. It is desired that the multilayer stack be designed such that the high index layer is next to the surface 19.

After the multilayer stack hes been formed, the conducting coating 42 is formed on the stack. This conducting coating can be formed of a suitable conducting material such as indium oxide or tin oxide. The desired pattern is formed in the conducting layer 42 in the same manner as the pattern is formed in the conducting layer 32. For example, photolithographuc techniques in combination with a suitable etch can be provided to remove the undesired portions of the conducting layer 42 to provide the necessary pattern.

One design which has been found to be satisfactory for the multilayer stack 41 and the conducting coating 42 is set forth below:

Layer:	Code:	Index:	Thickness in nm:
1	C	2,000	1100
2	H	2,315	470
3	L	1,450	470
4	H	2,315	470
5	L	1,450	470
6	H	2,315	545
7	L	1,450	620
8	H	2,315	620
9	L	1,450	620
10	H	2,315	620
11	L	1,450	620
12	H	2,315	620

For the above, the index of refraction for the medium was 1,5, whereas the index of the substrate was condidered to be 1,52. The reference or design wavelength was 550 nanometers. The first layer counting from the medium which is a conducting layer and bearing the code C was formed of indium oxide having an index of refraction of 2,0 and a half-wave optical thickness of 1,100 nanometers. The high index material was titanium dioxide and the low index material was silicon dioxide; however, Vycor could be used in place of silicon dioxide.

The combination antireflection and conducting coating 39 functions in a manner similar to the combination antireflection and conducting coating 28 in that it will have a substantially uniform

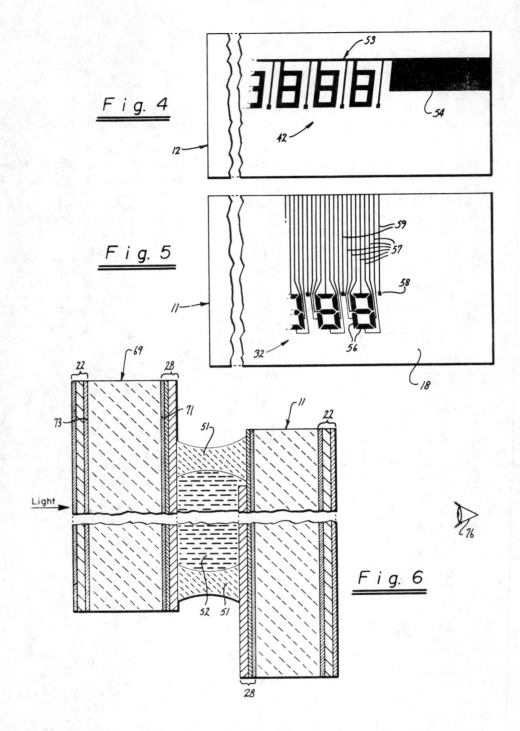

Fig. 4

Fig. 5

Fig. 6

appearance and the pattern will not be visible when the device is not activated.

As soon as the necessary coatings have been provided on the front and rear plates 11 and 12, a conventional glass frit, such as supplied by Corning Glass Works, is placed on the outer margins of the combination coatings 28 and 39 and then the two front plates are placed so that the surfaces 18 and 19 face each other. The patterns carried by the conducting layer 32 and the conducting layer 42 must be in registration with a spacing between the coatings carried by the front and rear plates of 5 to 25 microns and preferably approximately 0.001 of an inch. The glass frit is then fired for an appropriate period of time at a temperature of approximately 500°C to form a glass frit seal 51 which circumscribes a predetermined area on both of the front and rear plates

11 and 12 to provide an enclosed space 49. The glass frit seal 51 provides an excellent seal which bonds the front and rear plates to each other and which also establishes a seal which is impervious to moisture. Since the front and rear plates are also impervious to moisture, moisture cannot contaminate a liquid crystal material 52 which is placed within the hermetically sealed space 49. The liquid crystal material 52 can be placed in the space 49 by drilling one or more small holes in one of the front or rear plates 11 and 12 and then filling the space 49 through these holes and then plugging the holes with some inert material such as a silicon type compound. Since the liquid crystal layer is about 0.001 inch thick, it is held between the front and rear plates by capillary action.

Let it be assumed that the front and rear plates 11 and 12 have patterns on them as shown in Figs. 4 and 5 so that the liquid crystal display device can serve as a display device for displaying numerals. When such is the case as shown in Fig. 4, the pattern 53 is continuous and is connected to a single pad 54. As shown in Fig. 5, the pattern on the front plate 11 is formed of discontinuous segments 56 which are utilized to form the numerals arranged in a pattern well known to those skilled in the art and which are connected to leads 57 which extend downwardly to the bottom edge of the plate 11. In addition, there are provided periods 58 between each numeral which are connected by leads 59 also extending downwardly to the bottom edge. When the front and back plates are fritted together as shown in Fig. 1, the lower extremity of the front plate 11 extends below the rear plate so that contact can be made to the various leads 57 and 59.

If it is desired to have a liquid crystal display device work other than by ambient light and using a reflective type cell or device, it is necessary to provide a different type of cell or device, for example, one which is of the transmission type. A cell of this type is shown in Fig. 6. In such a case, the back plate does not utilize a reflector. It just carries a pattern. The front plate 11 would be identical to the front plate utilized in the previous embodiment. The back plate 69, however, would be significantly different from the back or rear plate 12 in that a reflector has been eliminated from the coating facing the liquid crystal material 52. The back or rear plate 69 is very similar to the front plate 11 and has a combination antireflection and conducting coating 28 identical to that hereinbefore described and is formed on the front surface 71 of the glass substrate 69. The only difference is that the pattern which is provided in the conducting layer 32 of the coating 28 is the pattern carried by the conducting layer 42 of the combination antireflection and conducting layer 39. If desired, an antireflection coating 32 can be provided on the rear surface 73.

The front plate 11 and the rear plate 69 are formed into a cell in the same manner as the cell or device hereinbefore described. In order to make the pattern visible, it is necessary to provide light as, for example, from a lamp, on the rear side of the cell as indicated by the arrow in Fig. 6 so that it would be visible to the viewer 76 on the front side of the liquid crystal display device.

"" "" "" "" "" ""

3,737,567 (U.S.)

Patented June 5, 1973

STEREOSCOPIC APPARATUS HAVING LIQUID CRYSTAL FILTER VIEWER.

Shunsei Kratomi.
Application Feb. 28, 1972; prior application in Japan Oct. 25, Nov. 17, Nov. 24, 1971; Jan. 11, '72.

A stereoscopic apparatus comprises at least one viewing device having a pair of liquid crystal filters (worn by an observer), the transparency of the liquid crystal filters being dependent on the intensity of an electric field applied thereto; at least one viewer controller; and a stereopair reproducing means, such as a motion picture projector and screen, a television receiver, a videotape player, or the like, having a synchronizing signal generator to generate signals to be transmitted to said viewer controller operating said viewer in such a manner that a pair of liquid crystal filters, covering the right and left eyes respectively, alternately exchange trnsparency and translucency in synchronism with each other, the alternation of right-eyed pictures and left-eyed pictures of stereopairs being displayed by the stereopair reproducing means.

Figure 1 is a perspective view of an embodiment of the stereoscopic apparatus incorporating a television receiver or a video player, in accordance with the invention;

Figure 2 is a perspective view of an embodiment of the stereoscopic apparatus incorporating

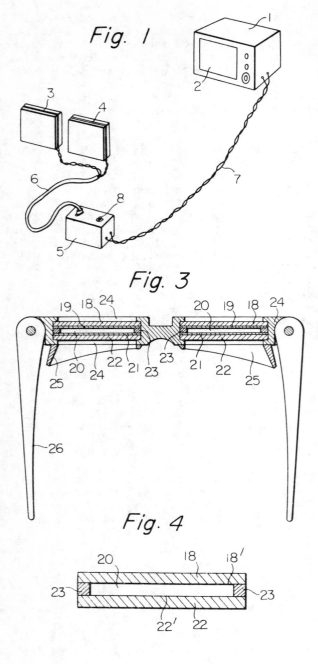
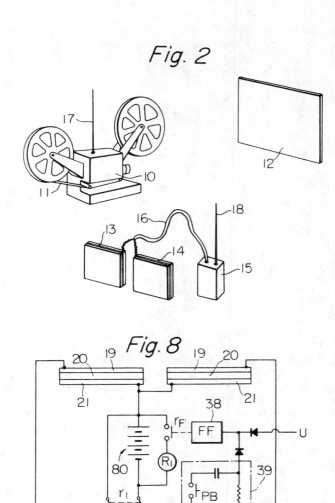
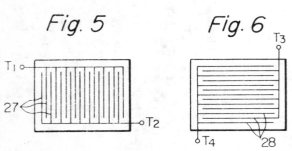

a motion picture projector and screen, in accordance with the invention;

Figure 3 shows a cross section of an embodiment of a viewer device in accordance with the invention;

Figure 4 shows a cross section of an embodiment of a liquid crystal filter for use in a viewer device in accordance with the invention;

Figs. 5 and 6 show the patterns of the parallel line electrodes provided on each of the pair transparent plates of a liquid crystal filter of an embodiment of the viewer device;

Figure 8 shows a schematic wiring diagram of a viewer controller in accordance with the invention.

" " " " " " " " " " " " "

3,740,717 (U.S.) Patented June 19, 1973

LIQUID CRYSTALS DISPLAY.

Robert C. Huener, Stanley J. Niemiec, and David K. Morgan, Assignors to RCA Corporation. Application Dec. 16, 1971

Multiplexed operation of a plurality of liquid crystal numeric indicators is achieved by applying a unipolarity, alternating exciting voltage to the backplate of a selected indicator, applying to all other backplates a unipolarity alternating voltage in a frequency range above that at which light scattering occurs, and applying to each segment of all indicators a unipolarity alternating exciting voltage which may be either in phase with or 180° out-of-phase with the exciting voltage applied to the backplate of the selected indicator.

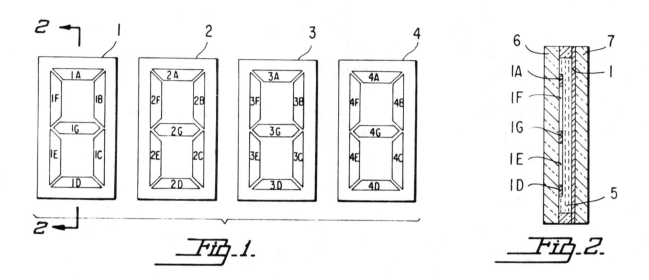

Fig. 1.   Fig. 2.

Figure 1 illustrates a four digit numeric indicator panel useful, for example, for watches, calculators and the like. Each indicator includes seven segments such as 1A, 1B, 1C, 1D, 1E, 1F and 1G and each also includes a backplate, such as 1. Liquid crystal 5, shown in the cross-section of Fig. 2, is located between the seven segments on the one hand the backplate on the other hand. The liquid crystal preferably is of the nematic type which operates on the principle of dynamic scattering.

In an arrangement of the type shown in Figs. 1 and 2, in the interest of being able to simplify the logic and to reduce the number of drive circuits and the number of leads extending from the panel to the circuits for driving the segments, it is desirable to excite only one indicator at a time. If the indicator is placed in the on condition for a sufficient interval of time and this excitation recurs at a sufficiently high rate, the normal relaxation time of the material and the persistence of vision are such that the indicator will appear to remain on (the liquid crystal will appear to remain in its light scattering condition) in between excitation intervals.

Figure 3 shows how the segments may be interconnected to achieve the results above. Each row of the matrix shown consists of one of the indicators. The row conductors correspond to the four backplates 1 through 4 and are similarly legended. The A segments of all four indicators are connected to a common conductor, the A column conductor of Fig. 3. Similarly the B segments are connected to a common conductor - the B column conductor, the C segments to the C conductor and so on. The liquid crystal elements at the intersections of the rows and columns are identified by the row and column numbers. For example, the elements in row 1 are 1A, 1B, 1C and so on and the elements in column 1 are 1A, 2A, 3A and so on, all corresponding to like legended elements of Fig. 1. Each element consists of a volume of liquid between a segment and the backplate and its equivalent circuit would be a relatively high value of resistance, and a relatively low value of capacitance associated with the resistance.

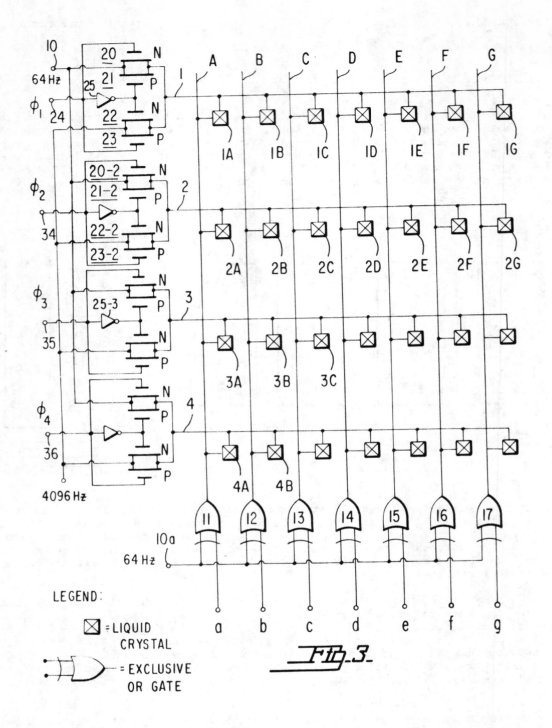

Fig. 3

LEGEND:
⊠ = LIQUID CRYSTAL
⟩ = EXCLUSIVE OR GATE

EXCLUSIVE OR gates 11-17 are connected one to each column. Each such gate receives at one input terminal the 64 Hertz (Hz) unipolarity alternating driving voltage applied to terminal 10a and receives at its other terminal a segment control voltage - the voltages a through g, respectively. The unipolarity driving voltage varies in amplitude between two levels such as zero voltage representing binary 0 and +15 volts representing binary 1. The control voltages a through g are direct voltage levels which have a value of zero volts or +15 volts.

A group of two dual transmission gates connects to each row of the array. As the groups are identical, only the group for row 1 will be described. Each dual transmission gate consists of an n-type metal-oxide-semiconductor (MOS) transistor such as 20 and a p-type MOS transistor such as

21. The gate electrode of the n-type transistor 20 and of the p-type transistor 23 are connected to terminal 24. A character select signal ϕ_1 is applied to this terminal. This character select signal is also applied via logical inverter 25 to the gate electrodes of p-type MOS transistor 21 and n-type MOS transistor 22.

The same 64 Hz exciting voltage as applied to terminal 10a is applied also to terminal 10. This latter terminal connects to one end of the parallel connected transmission paths of n-type transistor 20 and p-type transistor 21. A 4,096 Hz erase voltage is applied to one end of the parallel connected transmission paths of n-type MOS transistor 22 and p-type MOS transistor 23. The other end of these parallel connected transmission paths is connected to row 1. Similarly, the corresponding end of the parallel connected transmissions paths of transistors 20 and 21 is connected to row 1.

As mentioned above, the ϕ_1 character select voltage is applied to terminal 24. This selects the first numeric indicator (the row 1 indicator). The ϕ_2 character select voltage is applied to terminals 34 for selecting the second indicator, that is, the indicator of row 2 and similarly the other character select voltages are applied to terminals 35 and 36 for selecting rows 3 and 4, respectively.

The operation of the circuit of Fig. 3 may be understood by referring both to Fig. 3 and to the waveforms of Fig. 4. During the period t_0 to t_1, the character select voltage ϕ_1 is +15 volts representing binary 1. The character select voltages $\phi_2 \phi_3 \phi_4$ are at zero volts representing binary 0. The ϕ_1 voltage is applied directly to the gate of n-type transistor 20 turning the transistor on and through inverter 25 to the gate of p-type transistor 21 turning that transistor on. Accordingly, the dual transmission gate 20-21 is on, and the unipolarity 64 Hz drive signal is applied through this gate to row 1.

The ϕ_1 signal is also applied to the gate of p-type transistor 23 turning that transistor off and through inverter 25 to the gate of n-type transistor 22 turning that transistor off. Accordingly, the dual transmission gate 22-23 is off and the 4,096 Hz erase signal is prevented from being applied to row 1.

The situation is just the reverse of the above at rows 2,3 and 4. For example, at row 2 the ϕ_2 character select signal is at zero volts. This causes the dual transmission gate 20-2, 21-2 to be off and the dual transmission gate 22-2, 23-2 to be on. Accordingly, the 4,096 Hz erase signal is applied through the dual transmission gate 22-2, 23-2 to row 2 and the off dual transmission gate 20-2, 21-2 prevents the 64 Hz exciting voltage from being applied directly to row 2. In similar fashion, the 4,096 Hz erase voltage appears ar rows 3 and 4.

An EXCLUSIVE OR gate, as is well understood in the art, produces a 1 output (an output signal representing binary 1) when one of the input signals represents a 1 and the other represents a 0 and produces a 0 output of both input signals represent 1 or if both input signals represent 0. Assume, for example, that the segment control voltage a represents a 1 (is +15 volts). In this case when the 64 Hz signal represents a 1, the EXCLUSIVE OR gate 11 produces an output representing a 0 and when the 64 Hz signal represents a zero the EXCLUSIVE OR gate 11 produces an output representing a 1. In other words, when a segment control voltage such as a represents a 1, the EXCLUSIVE OR gate produces an output complementary to the 64 Hz square wave it receives. It can also be shown that when a segment control voltage such as a represents a 0, the EXCLUSIVE OR gate produces an output which is the same as, that is, which is in phase with, the 64 Hz signal applied to the second input terminal to the EXCLUSIVE OR gate.

In operation, the segment control voltages a through g applied to the EXCLUSIVE OR gates are made to have values to turn on those segments of the indicator selected corresponding to the number it is desired to display. For example, if it is desired to display the number "3" when the row 1 indicator is selected, then segments A, B, C, D, and G must be turned on and segments E and F must be turned off. This operation is achieved by causing the segment control voltages, a, b, c, d, and g to have a value representing binary 1 (+15 volts) and the two remaining control voltages e and f to have a value representing binary 0 (0 volts). In the first case, that is, for example, when the control voltage c=1 and when row 1 is selected, the 64 Hz signal appearing at column C will be complementary to the 64 Hz signal appearing at row 1. The situation will be that depicted in Fig. 4, during the first time interval t_0 to t_1, in the last four waveforms. The result, shown in the last waveform, is that for one hlaf cycle t_0-t_{oa}, 15 volts in one sense (+15 volts) appears across the crystal and for the other half cycle t_{oa}-t_1, -15 volts appears across the crystal. These voltage levels are sufficient to cause the liquid crystal to scatter

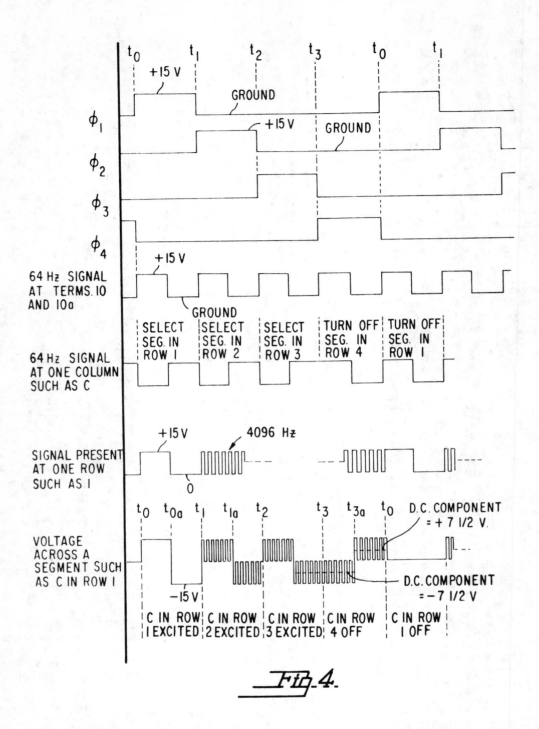

Fig. 4.

light. The application of voltages in this way is "push-pull" excitation of the liquid crystal element and achieves the effect of bipolar alternating voltage (positive and negative half-cycles) excitation and there is zero average current through the liquid crystal.

If it is desired not to select a segment of a selected indicator, the segment control voltage is made to be zero. For example, suppose the control voltage f=0 during the time the 64 Hz signal appears at row 1. In this case, the 64 Hz signal appearing on column F will be in phase with the 64 Hz signal at row 1 and the net voltage appearing across the liquid crystal element 1F will be zero. The situation will be similar to that depicted in the last four waveforms of Fig. 4 during the second t_0-t_1 period. This can be considered as "push-push" operation with the actual voltage across the liquid crystal zero as shown in the last waveform during the second period

278

t_0-t_1.

Summarizing the above, when an indicator is selected, any particular segment of the indicator may be either "on" (light scattering occurs at the segment location). When a segment is on, a 15 volt unipolarity alternating voltage is applied across the crystal in push-pull. In other words, the effect achieved is that of a 30 volt peak-to-peak bipolar alternating voltage with average direct current through the liquid crystal zero. When a segment of a selected indicator is not selected, the 15 volt unipolarity signal is applied across the liquid crystal in push-push so that zero volts (both alternating and direct) appears across the liquid crystal.

When an indicator is not selected, then a 4,096 Hz unipolarity voltage appears across the liquid crystal in the manner illustrated at the bottom of Fig. 4. Assume, for example, that the indicator of row 2 is selected, that the indicator of row 1 is not selected, and that the C segment 2C of row 2 is selected - is on. The situation now will be as depicted in the time period t_1-t_2 of the last waveform of Fig. 4 which illustrates the voltage across liquid crystal element 1C of a non selected indicator.

The voltage present at row 1 is the 4,096 Hz erase signal. The voltage present on column C is the 64 Hz signal which is out-of-phase with the 64 Hz signal at terminal 10a. During The time column C is at zero volts, row 1 will vary in amplitude between zero and +15 volts at the 4,096Hz rate as shown for the first half t_1 to t_{1a} of the time period t_1 to t_2. During the time column C is at +15 volts, the voltage at row 1 will vary between zero and +15 volts at the 4,096 Hz rate to produce across the liquid crystal element 1C the voltage shown during the last half t_{1a}-t_2 of the period from t_1-t_2.

The 4,096 Hz frequency is in the "erase" voltage frequency range, that is, it tends to cause the liquid crystal to assume a non-light scattering condition.

According to one theory, the ions which would tend to cause turbulent motion of the liquid crystal domains, cannot travel any great distance in a rapidly alternating field. Instead, They oscillate around an equilibrium position moving only a very small distance each half cycle without substantially disturbing the liquid crystal domains. However, as the voltage across the liquid crystal during the period, for example, of t_1-t_{1a} is a 0 to 15 volt unipolarity alternating voltage signal, it develops across teh liquid crystal a direct voltage component of +7-1/2 volts amplitude. This direct voltage component would tend to cause light scattering if of sufficient amplitude (say greater than 10 or so volts) but is chosen to be lower than the threshold for light scatteting for the liquid crystal so that light scattering does not accur. As already mentioned, the threshold for light scattering depends upon such parameters as the thickness of the liquid crystal layer and the liquid crystal material, etc. and appropriate voltages can be chosen to make the average direct voltage level lower than this thrshold.

One problem still remains and that is that the average direct voltage level of +7-1/2 volts, if not compensated for, would tend to lessen the life of the liquid crystal. But the present arrangement does provide such compensation. Note that in the following period t_{1a} to t_2 the effect achieved is that of a voltage across the liquid crystal which varies in amplitude from zero volts to -15 volts at a 4,096 Hz rate. Here, the direct current component through the liquid crystal is -7-1/2 volts. This -7-1/2 volts direct voltage across the liquid crystal during the period t_{1a}-t_2 is exactly equal to and opposite in sign to the +7-1/2 volt direct voltage component across the liquid crystal during the equal period t_1 to t_{1a}. Accordingly, for the entire period t_1 to t_2 the average direct voltage level across the liquid crystal is zero.

The description above is for the case in which there is a segment in a non-selected row which is being erased at the time that a corresponding segment (one in the same column) in a selected row is being excited. The time period t_3 to t_0 in the last waveform illustrates the case in which the C element 1C in row 1 is being erased at the same time the C element 4C of the selected row 4 is being maintained in the off condition. Here, the situation is exactly the same as the situation for element 1C during the period t_1 to t_2 except that the phasing is different. During the period t_3 to t_{3a}, the average direct voltage level across liquid crystal element 1C is -7-1/2 volts whereas during the second half of the period, that is, t_{3a} to t_0, the average direct voltage level across the liquid crystal is +7-1/2 volts. The net direct voltage level across the liquid crystal for the entire period t_3 to t_0 is zero volts.

Summarizing the above, any element for a non-selected indicator, that is, for a non-selected row has continuously applied thereto a 4,096 Hz signal. In terms of the indicator of Fig. 1, this signal is applied to the backplate. During one half of each indicator non-select period, an average DC level of +7-1/2 volts appears across the segment on a non-selected row and during the other half period the average DC level across such a segment is -7-1/2 volts. The average DC level therefore for the entire period is zero volts. For proper operation, the +7-1/2 volt component (and the -7-1/2 volt component) should be lower than the threshold voltage for light scattering for the liquid crystal.

Summarizing the overall operation of the system, with an arrangement such as described, the four indicators shown in Fig. 1 may be operated in multiplexed fashion, that is, one indicator is turned on at a time. The turn-on period (the time the liquid crystal is in its light scattering state) should be sufficiently long and the interval between turn on periods sufficiently short and flicker is minimized. This is a matter of engineering design.

The arrangement of the present invention has the advantage that the number of output leads from the indicator panel of Fig. 1 is reduced. Note that there are a total of 28 segments and four backplates and these are arranged in such a way (shown in Fig. 3) that only 11 leads for the entire panel (of seven columns and four rows) are needed.

" " " " " " " " " " "

Patent Information

Regarding U.S. Patents: Copies of U.S. Patents may be obtained from:

The Commissioner of Patents
Washington, D.C. 20231
U.S.A.

50¢ per copy paid in advance by Post Office Money Order or Certified Check (made payable to the Commissioner of Patents). Coupons in pads of ten for $5.00 and 50 for $25.00 are available. Allow 35-90 days for delivery of ordered patent specifications.

Regarding British Patents: To obtain complete patent specifications address request to:

Comptroller-General, The Patent Office, Sale Branch
Block "C", Station Square House, St. Mary Cray
Orpington, Kent
England

25p per copy payable by International Money Order or bank draft. Send extra amount for shipping via Airmail. Overpayment will be promptly refunded. Excellent, reliable service. Airmail orders usually filled within 14 days.

" " " " " " " " " " "

LIST OF PATENTS

3,499,112	(U.S.)	ELECTRO-OPTICAL DEVICE...	Mar.	3, 1970
3,499,702	(U.S.)	NEMATIC LIQUID CRYSTAL MIXTURES FOR USE IN LIGHT VALVE...	Mar.	10, 1970
3,503,672	(U.S.)	REDUCTION OF TURN-ON DELAY IN LIQUID CRYSTAL CELL...	Mar.	31, 1970
3,503,673	(U.S.)	REDUCTION OF TURN-ON DELAY IN LIQUID CRYSTAL CELL...	Mar.	31, 1970
3,505,804	(U.S.)	SOLID STATE CLOCK...	Apr.	14, 1970
1,194,544	(British)	LIQUID CRYSTAL DETECTOR...	Jun.	10, 1970
3,519,330	(U.S.)	TURNOFF METHOD AND CIRCUIT FOR LIQUID CRYSTAL DISPLAY...	Jly.	7, 1970
1,201,230	(British)	CHROMATIC DISPLAY OR STORAGE DEVICE...	Aug.	5, 1970
3,524,726	(U.S.)	SMECTOGRAPHIC DISPLAY...	Aug.	18, 1970
3,527,945	(U.S.)	MOUNTING STRUCTURE FOR A LIQUID CRYSTAL THERMAL IMAGING DEVICE...	Sep.	8, 1970
3,529,156	(U.S.)	HYSTERETIC CHOLESTERIC LIQUID CRYSTALLINE COMPOSITIONS AND RECORDING DEVICES UTILIZING SUCH COMPOSITIONS...	Sep.	15, 1970
3,532,813	(U.S.)	DISPLAY CIRCUIT INCLUDING CHARGING CIRCUIT AND FAST RESET CIRCUIT...	Oct.	6, 1970
3,540,796	(U.S.)	ELECTRO-OPTICAL COMPOSITIONS AND DEVICES...	Nov.	17, 1970
3,551,026	(U.S.)	CONTROL OF OPTICAL PROPERTIES OF MATERIALS WITH LIQUID CRYSTALS...	Dec.	29, 1970
1,219,840	(British)	REDUCTION OF TURN-ON DELAY IN LIQUID CRYSTAL CELL...	Jan.	20, 1971
1,220,169	(British)	TURN OFF METHOD AND CIRCUIT FOR LIQUID CRYSTAL DISPLAY ELEMENT...	Jan.	20, 1971
3,569,614	(U.S.)	LIQUID CRYSTAL COLOR MODULATOR FOR ELECTRONIC IMAGING SYSTEM..	Mar.	9, 1971
3,569,709	(U.S.)	THERMAL IMAGING SYSTEM UTILIZING LIQUID CRYSTAL MATERIAL...	Mar.	9, 1971
3,572,907	(U.S.)	OPTICAL CELL FOR ATTENUATING, SCATTERING AND POLARIZING ELECTROMAGNETIC RADIATION...	Mar.	30, 1971
1,227,616	(British)	LIQUID CRYSTAL ELECTRO-OPTIC DEVICE...	Apr.	7, 1971
1,228,606	(British)	ELECTRO-OPTIC LIGHT VALVE...	Apr.	15, 1971
3,575,491	(U.S.)	DECREASING RESPONSE TIME OF LIQUID CRYSTALS...	Apr.	20, 1971
3,575,492	(U.S.)	TURNOFF METHOD AND CIRCUIT FOR LIQUID CRYSTAL DISPLAY ELEMENT	,Apr.	20, 1971
3,575,493	(U.S.)	FAST SELF-QUENCHING OF DYNAMIC SCATTERING IN LIQUID CRYSTAL DEVICES...	Apr.	20, 1971
3,576,364	(U.S.)	COLOR ADVERTISING DISPLAY EMPLOYING LIQUID CRYSTAL...	Apr.	27, 1971
3,576,761	(U.S.)	THERMOMETRIC COMPOSITIONS COMPRISING ONE MESOMORPHIC SUBSTANCE, ONE CHOLESTERYL HALIDE, AND AN OIL SOLUBLE DYE SELECTED FROM THE GROUP CONSISTING OF DISAZO, INDULENE, AND NIGROSINE DYES...	Apr.	27, 1971
3,578,844	(U.S.)	RADIATION SENSITIVE DISPLAY DEVICE CONTAINING ENCAPSULATED CHOLESTERIC LIQUID CRYSTALS...	May	18, 1971
3,580,864	(U.S.)	CHOLESTERIC-PHASE LIQUID-CRYSTAL COMPOSITIONS STABILIZED AGAINST TRUE-SOLID FORMATION, USING CHOLESTERYL ERUCYL CARBONATE...	May	25, 1971
3,585,381	(U.S.)	ENCAPSULATED CHOLESTERIC LIQUID CRYSTAL DISPLAY DEVICE...	Jun.	15, 1971
3,588,225	(U.S.)	ELECTRO-OPTIC DEVICES FOR PORTRAYING CLOSED IMAGES...	Jun.	28, 1971
3,590,371	(U.S.)	METHOD UTILIZING THE COLOR CHANGE WITH TEMPERATURE OF A MATERIAL FOR DETECTING DISCONTINUITIES IN A CONDUCTOR MEMBER EMBEDDED WITHIN A WINDSHIELD...	Jun.	29, 1971
3,592,526	(U.S.)	MEANS FOR ROTATING THE POLARIZATION PLANE OF LIGHT AND FOR CONVERTING POLARIZED LIGHT TO NONPOLARIZED LIGHT...	Jly.	13, 1971
3,592,527	(U.S.)	IMAGE DISPLAY DEVICE...	Jly.	13, 1971
3,597,043	(U.S.)	NEMATIC LIQUID-CRYSTAL OPTICAL ELEMENTS...	Aug.	3, 1971
3,597,044	(U.S.)	ELECTRO-OPTIC LIGHT MODULATOR...	Aug.	3, 1971
3,600,060	(U.S.)	DISPLAY DEVICE CONTAINING MINUTE DROPLETS OF CHOLESTERIC LIQUID CRYSTALS IN A SUBSTANTIALLY CONTINUOUS POLYMERIC MATRIX...	Aug.	17, 1971

3,600,061	(U.S.)	ELECTRO-OPTIC DEVICE HAVING GROOVES IN THE SUPPORT PLATES TO CONFINE A LIQUID CRYSTAL BY MEANS OF SURFACE TENSION...	Aug. 17, 1971
3,604,930	(U.S.)	METHOD AND APPARATUS FOR DISPLAYING VISUAL IMAGES OF INFRARED BEAMS...	Sep. 14, 1971
1,246,847	(British)	LIQUID CRYSTAL DISPLAY ELEMENT HAVING STORAGE...	Sep. 22, 1971
3,612,654	(U.S.)	LIQUID CRYSTAL DISPLAY DEVICE...	Oct. 12, 1971
3,613,351	(U.S.)	WRISTWATCH WITH LIQUID CRYSTAL DISPLAY...	Oct. 19, 1971
3,614,210	(U.S.)	LIQUID CRYSTAL DAY/NIGHT MIRROR...	Oct. 19, 1971
1,251,790	(British)	IMAGE CONVERTER...	Oct. 27, 1971
3,617,374	(U.S.)	DISPLAY DEVICE...	Nov. 2, 1971
3,619,254	(U.S.)	THERMOMETRIC ARTICLES AND METHODS FOR PREPARING SAME...	Nov. 9, 1971
3,620,889	(U.S.)	LIQUID CRYSTAL SYSTEMS...	Nov. 16, 1971
3,622,224	(U.S.)	LIQUID CRYSTAL ALPHA-NUMERIC ELECTROOPTIC IMAGING DEVICE...	Nov. 23, 1971
3,622,226	(U.S.)	LIQUID CRYSTAL CELLS IN A LINEAR ARRAY...	Nov. 23, 1971
3,623,795	(U.S.)	ELECTRO-OPTICAL SYSTEM...	Nov. 30, 1971
3,625,591	(U.S.)	LIQUID CRYSTAL DISPLAY ELEMENT...	Dec. 7, 1971
3,627,408	(U.S.)	ELECTRIC FIELD DEVICE...	Dec. 14, 1971
3,627,699	(U.S.)	LIQUID CRYSTAL CHOLESTERIC MATERIAL AND SENSITIZING AGENT COMPOSITION AND METHOD FOR DETECTING ELECTROMAGNETIC RADIATION...	Dec. 14, 1971
3,628,268	(U.S.)	PURE FLUID DISPLAY...	Dec. 21, 1971
1,258,739	(British)	OPTICAL INFORMATION STORING SYSTEM...	Dec. 30, 1971
3,637,291	(U.S.)	DISPLAY DEVICE WITH INHERENT MEMORY...	Jan. 25, 1972
1,263,277	(British)	DISPLAY APPARATUS...	Feb. 9, 1972
1,263,278	(British)	LIQUID CRYSTAL DISPLAY APPARATUS...	Feb. 9, 1972
3,642,348	(U.S.)	IMAGING SYSTEM...	Feb. 15, 1972
3,645,604	(U.S.)	LIQUID CRYSTAL DISPLAY...	Feb. 29, 1972
3,647,279	(U.S.)	COLOR DISPLAY DEVICES...	Mar. 7, 1972
3,647,280	(U.S.)	LIQUID CRYSTAL DISPLAY DEVICE...	Mar. 7, 1972
3,648,280	(U.S.)	THERMOCHROMIC LIGHT-FLASHING SYSTEM...	Mar. 7, 1972
3,650,603	(U.S.)	LIQUID CRYSTAL LIGHT VALVE CONTAINING A MIXTURE OF NEMATIC AND CHOLESTERIC MATERIALS IN WHICH THE LIGHT SCATTERING EFFECT IS REDUCED WHEN AN ELECTRIC FIELD IS APPLIED...	Mar. 21, 1972
3,650,608	(U.S.)	METHOD AND APPARATUS FOR DIAPLAYING COHERENT LIGHT IMAGES...	Mar. 21, 1972
3,652,148	(U.S.)	IMAGING SYSTEM...	Mar. 28, 1972
3,653,745	(U.S.)	CIRCUITS FOR DRIVING LOADS SUCH AS LIQUID CRYSTAL DISPLAYS...	Apr. 4, 1972
3,654,606	(U.S.)	ALTERNATING VOLTAGE EXCITATION OF LIQUID CRYSTAL DISPLAY MATRIX...	Apr. 4, 1972
3,655,269	(U.S.)	LIQUID CRYSTAL DISPLAY ASSEMBLY HAVING INDEPENDENT CONTRAST AND SPEED OF RESPONSE CONTROLS...	Apr. 11, 1972
3,655,270	(U.S.)	ELECTRO-OPTICAL DISPLAY DEVICE USING NEMATIC MIXTURES WITH VERY WIDE TEMPERATURE RANGES...	Apr. 11, 1972
3,655,971	(U.S.)	IMAGING SYSTEM...	Apr. 11, 1972
3,656,834	(U.S.)	ADDITIVE FOR LIQUID CRYSTAL MATERIAL...	Apr. 18, 1972
3,661,444	(U.S.)	COMPOUNDED LIQUID CRYSTAL CELLS...	May 9, 1972
1,273,779	(British)	ELECTRO-OPTIC LIGHT MODULATOR...	May 10, 1972
3,663,086	(U.S.)	OPTICAL INFORMATION STORING SYSTEM...	May 16, 1972
3,663,390	(U.S.)	METHOD OF CHANGING COLOR PLAY RANGE OF LIQUID CRYSTAL MATERIALS...	May 16, 1972
3,666,881	(U.S.)	ELECTRO-OPTICAL DISPLAY DEVICE EMPLOYING LIQUID CRYSTALS...	May 30, 1972
3,666,947	(U.S.)	LIQUID CRYSTAL IMAGING SYSTEM HAVING AN UNDISTURBED IMAGE ON A DISTURBED MACKGROUND AND HAVING A RADIATION ABSORPTIVE MATERIAL DISPERSED THROUGHOUT THE LIQUID CRYSTAL...	May 30, 1972
3,666,948	(U.S.)	LIQUID CRYSTAL THERMAL IMAGING SYSTEM HAVING AN UNDISTURBED IMAGE ON A DISTURBED BACKGROUND...	May 30, 1972
3,667,039	(U.S.)	ELECTRICITY MEASUREMENT DEVICES EMPLOYING LIQUID CRYSTAL MATERIALS...	May 30, 1972
1,276,523	(British)	LIQUID CRYSTAL DISPLAY ASSEMBLY HAVING INDEPENDENT CONTRAST AND SPEED OF RESPONSE CONTROLS...	Jun. 1, 1972
3,668,861	(U.S.)	SOLID STATE ELECTRONIC WATCH...	Jun. 13, 1972
3,669,525	(U.S.)	LIQUID CRYSTAL COLOR FILTER...	Jun. 13, 1972
3,674,338	(U.S.)	REAR PROJECTION SCREEN EMPLOYING LIQUID CRYSTALS...	Jly. 4, 1972

Number	Country	Title	Date
3,674,341	(U.S.)	LIQUID CRYSTAL CRYSTAL DISPLAY DEVICE HAVING IMPROVED OPTICAL CONTRAST...	Jly. 4, 1972
3,674,342	(U.S.)	LIQUID CRYSTAL DISPLAY DEVICE INCLUDING SIDE-BY-SIDE ELECTRODES ON A COMMON SUBSTRATE...	Jly. 4, 1972
3,675,987	(U.S.)	LIQUID CRYSTAL COMPOSITIONS AND DEVICES...	Jly. 11, 1972
3,675,988	(U.S.)	LIQUID CRYSTAL ELECTRO-OPTICAL MEASUREMENT AND DISPLAY DEVICES...	Jly. 11, 1972
3,675,989	(U.S.)	LIQUID CRYSTAL OPTICAL CELL WITH SELECTED ENERGY SCATTERING...	Jly. 11, 1972
3,679,290	(U.S.)	LIQUID CRYSTAL OPTICAL FILTER SYSTEM...	Jly. 25, 1972
3,680,950	(U.S.)	GRANDJEAN STATE LIQUID CRYSTALLINE IMAGING SYSTEM...	Aug. 1, 1972
3,687,515	(U.S.)	ELECTRO-OPTIC LIQUID CRYSTAL SYSTEM WITH POLYAMIDE RESIN ADDITIVE...	Aug. 29, 1972
3,689,131	(U.S.)	LIQUID CRYSTAL DISPLAY DEVICE...	Sep. 5, 1972
3,690,745	(U.S.)	ELECTRO-OPTICAL DEVICE USING LYOTROPIC NEMATIC LIQUID CRYSTALS...	Sep. 12, 1972
3,691,755	(U.S.)	CLOCK WITH DIGITAL DISPLAY...	Sep. 19, 1972
3,693,084	(U.S.)	METHOD AND APPARATUS FOR DETECTING MICROWAVE FIELDS...	Sep. 19, 1972
3,694,053	(U.S.)	NEMATIC LIQUID CRYSTAL DEVICE...	Sep. 26, 1972
3,697,150	(U.S.)	ELECTRO-OPTIC SYSTEMS IN WHICH AN ELECTROPHORETIC-LIKE OR DIPOLAR MATERIAL IS DISPERSED THROUGHOUT A LIQUID CRYSTAL TO REDUCE THE TURNOFF TIME...	Oct. 10, 1972
3,697,152	(U.S.)	TUNING METHOD FOR PLURAL LAYER LIQUID CRYSTAL FILTERS...	Oct. 10, 1972
3,697,297	(U.S.)	GELATIN-GUM ARABIC CAPSULES CONTAINING CHOLESTERERIC LIQUID CRYSTAL MATERIAL AND DISPERSIONS OF THE CAPSULES...	Oct. 10, 1972
3,700,306	(U.S.)	ELECTRO-OPTIC SHUTTER HAVING A THIN GLASS OR SILICON OXIDE LAYER BETWEEN THE ELECTRODES AND THE LIQUID CRYSTAL...	Oct. 24, 1972
3,700,805	(U.S.)	BLACK-AND-WHITE IMAGE CONTROL BY ULTRASONIC MODULATION OF NEMATIC LIQUID CRYSTALS...	Oct. 24, 1972
3,702,723	(U.S.)	SEGMENTED MASTER CHARACTER FOR ELECTRONIC DISPLAY APPARATUS...	Nov. 14, 1972
3,703,329	(U.S.)	LIQUID CRYSTAL COLOR DISPLAY...	Nov. 21, 1972
3,703,331	(U.S.)	LIQUID CRYSTAL DISPLAY ELEMENT HAVING STORAGE...	Nov. 21, 1972
3,704,056	(U.S.)	IMAGING SYSTEM...	Nov. 28, 1972
3,704,625	(U.S.)	THERMOMETER USING LIQUID CRYSTAL COMPOSITIONS...	Dec. 5, 1972
3,705,310	(U.S.)	LIQUID CRYSTAL VOLTAGE DISPLAY DEVICE HAVING PHOTOCONDUCTIVE MEANS TO ENHANCE THE CONTRAST AT THE INDICATING REGION...	Dec. 5, 1972
3,707,322	(U.S.)	ELECTROSTATIC LATENT IMAGING SYSTEM USING A CHOLESTERIC TO NEMATIC PHASE TRANSITION...	Dec. 26, 1972
3,707,323	(U.S.)	LIQUID CRYSTAL DEVICES AND SYSTEMS FOR ULTRASONIC IMAGING...	Dec. 26, 1972
1,302,482	(British)	VISUAL DISPLAY DEVICE...	Jan. 10, 1973
3,711,181	(U.S.)	OPTICAL NOTCH FILTER...	Jan. 16, 1973
3,711,713	(U.S.)	ELECTRICALLY CONTROLLED THERMAL IMAGING SYSTEM USING A CHOLESTERIC TO NEMATIC PHASE TRANSITION...	Jan. 16, 1973
3,712,047	(U.S.)	TIME DISPLAY DEVICE FOR TIMEPIECES...	Jan. 23, 1973
3,713,156	(U.S.)	SURFACE AND SUBSURFACE DETECTION DEVICE...	Jan. 23, 1973
1,303,947	(British)	OPTOELECTRICAL DISPLAY ELEMENT...	Jan. 24, 1973
1,304,268	(British)	IMAGE DISPLAY DEVICE...	Jan. 24, 1973
1,304,554	(British)	LIQUID CRYSTAL DISPLAY DEVICE...	Jan. 24, 1973
3,716,289	(U.S.)	ELECTRO-OPTICAL DISPLAY DEVICE USING SMECTIC-NEMATIC LIQUID CRYSTAL MIXTURES...	Feb. 13, 1973
3,716,290	(U.S.)	LIQUID-CRYSTAL DISPLAY DEVICE...	Feb. 13, 1973
3,716,658	(U.S.)	LIQUID-CRYSTAL TELEVISION SYSTEM...	Feb. 13, 1973
1,306,912	(British)	LIQUID CRYSTAL CELLS...	Feb. 14, 1973
1,307,809	(British)	CAPRONYLHYDROXY-BENZOIC ACID ESTERSM THEIR USE AS NEMATO-GENIC COMPOUNDS AND NEMATOGENIC MIXTURES CONTAINING THEM...	Feb. 21, 1973
3,718,380	(U.S.)	IMAGING SYSTEM IN WHICH EITHER A LIQUID CRYSTALLINE MATERIAL OR AN ELECTRODE IS SHAPED IN AN IMAGE CONFIGURATION...	Feb. 27, 1973
3,718,381	(U.S.)	LIQUID CRYSTAL ELECTRO-OPTICAL MODULATORS...	Feb. 27, 1973
3,718,382	(U.S.)	LIQUID CRYSTAL IMAGING SYSTEM IN WHICH AN ELECTRICAL FIELD IS CREATED BY AN X-Y ADDRESS SYSTEM...	Feb. 27, 1973
3,718,842	(U.S.)	LIQUID CRYSTAL MOUNTING STRUCTURE...	Feb. 27, 1973
1,308,208	(British)	DISPLAY SYSTEM...	Feb. 28, 1973
1,308,237	(British)	IMINES AND THEIR USE IN ELECTRO-OPTIC COMPOSITIONS AND DEVICES...	Feb. 28, 1973

3,720,456	(U.S.)	METHOD FOR NARROWING THE BANDWIDTH OF AN OPTICAL SIGNAL... Mar. 13, 1973
1,309,558	(British)	METHOD OF DETECTING ELECTROMAGNETIC RADIATION AND HOMOGENEOUS COMPOSITIONS FOR USE IN PERFORMING THE METHOD... Mar. 14, 1973
3,722,998	(U.S.)	LIQUID CRYSTAL APPARATUS FOR REDUCING CONTRAST... Mar. 27, 1973
3,723,346	(U.S.)	TEMPERATURE INDICATOR USING THE SMECTIC 'A' PHASE OF A LIQUID CRYSTAL... Mar. 27, 1973
3,723,651	(U.S.)	OPTICALLY-SCANNED LIQUID-CRYSTAL PROJECTION DISPLAY... Mar. 27, 1973
3,725,899	(U.S.)	DATA EXHIBITING SCREEN DEVICE WITH A LIQUID-CRYSTAL LAYER, AND METHOD OF MANUFACTURE... Apr. 3, 1973
3,726,584	(U.S.)	LIGHT MODULATOR SYSTEM... Apr. 10, 1973
3,727,527	(U.S.)	PHOTOGRAPHIC APPARATUS WITH LIQUID CRYSTAL VOLTAGE INDICATOR... Apr. 17, 1973
3,728,007	(U.S.)	REFLECTIVE TYPE LIQUID CRYSTAL DISPLAY DEVICE HAVING IMPROVED OPTICAL CONTRAST... Apr. 17, 1973
3,728,008	(U.S.)	LIQUID CRYSTAL DISPLAY... Apr. 17, 1973
3,730,607	(U.S.)	INDICATOR SCREEN WITH CONTROLLED VOLTAGE TO MATRIX CROSSPOINTS THEREOF... May 1, 1973
3,731,986	(U.S.)	DISPLAY DEVICES UTILIZING LIQUID CRYSTAL LIGHT MODULATION... May 8, 1973
3,732,119	(U.S.)	TEMPERATURE SENSITIVE VISUAL DISPLAY DEVICE... May 8, 1973
3,732,429	(U.S.)	LIQUID CRYSTAL DEVICE... May 8, 1973
1,316,213	(British)	REAR PROJECTION SYSTEM... May 9, 1973
1,316,497	(British)	METHODS AND APPARATUS FOR VISUAL REPRESENTATION, IN A MODEL, OF A SURVEYED MEDIUM... May 9, 1973
3,733,485	(U.S.)	EXPOSURE METER FOR THERMAL IMAGING DEVICES... May 15, 1973
3,734,597	(U.S.)	PROCESS FOR PRODUCING A COLOR STATE IN A DISPLAY DEVICE... May 22, 1973
3,734,598	(U.S.)	LIQUID CRYSTAL DISPLAY DEVICE HAVING AN INCLINED REAR REFLECTOR... May 22, 1973
1,318,007	(British)	ARRANGEMENT FOR LOCALLY ALTERING THE BRIGHTNESS AND/OR COLOR OF AN INDICATOR MEDIUM... May 23, 1973
1,318,011	(British)	NEMATIC LIQUID CRYSTAL MATERIAL... May 23, 1973
1,3.8,012	(British)	NEMATIC LIQUID CRYSTAL COMPOUNDS AND MIXTURES... May 23, 1973
3,736,047	(U.S.)	LIQUID CRYSTAL DISPLAY DEVICE WITH INTERNAL ANTI-REFLECTION CASTING... May 29, 1973
3,737,567	(U.S.)	STEREOSCOPIC APPARATUS HAVING LIQUID CRYSTAL FILTER VIEWER... Jun. 5, 1973
3,740,717	(U.S.)	LIQUID CRYSTAL DISPLAY... Jun. 19, 1973

"""""""""""""

INVENTOR INDEX

Abbott, C.P.,III.,	3,718,842		Becker, J.H., (Cont.)	3,707,322	3,711,713
Adams, J.E.,	3,642,348	3,652,148		3,718,380	3,718,382
	3,655,971	3,666,947	Bleha, W.P.Jr.,	3,732,429	
	3,666,948	3,669,525	Borden, H.C.Jr.,	3,702,723	
	3,679,290	3,680,950	Borel, J.,	3,716,290	
	3,687,515	3,697,152	Borowski, K.,	3,727,527	
	3,704,056	3,707,322	Braunstein, M.,	3,732,429	
	3,711,181	3,711,713	Brenden, B.B.,	1,194,544	
	3,718,380	3,718,382	Caplan, S.,	3,612,654	3,614,210
	3,720,456	3,726,584		3,647,280	3,689,131
Aiken, W.R.,	3,734,598		Cartmell, J.V.,	3,578,844	3,585,381
Allan, F.V.,	3,728,008			3,600,060	3,674,338
Allen, P.J.,	3,604,930			3,697,297	3,700,306
Anderson, T.G.,	3,733,485			3,732,119	3,734,597
Assouline, G.,	3,663,086	3,718,381	Castellano, J.A.,	3,540,796	3,597,044
Augustine, C.F.,	3,693,084			3,674,342	3,703,329
Baker, C.E.,	3,650,608		Churchill, D.,	3,578,844	3,585,381
Baltzer, D.H.,	3,620,889			3,600,060	3,674,338
Becker, J.H.,	3,652,148	3,655,971		3,697,297	3,700,306
				3,732,119	3,734,597

Name	Patent No.	Patent No.
Cindrich, I.,	3,572,907	
Conners, G.H.,	3,592,527	
Creagh, L.T.,	3,655,270	3,716,289
Curtin, H.R.,	1,194,544	
Dailey, J.L.,	3,711,181	
Davis, F.,	3,576,761	3,619,254
	3,647,279	
DeKoster, H.A.,	3,524,726	
Dreyer, J.F.,	3,592,526	3,597,043
Fergason, J.L.,	3,529,156	3,580,864
	3,627,408	3,627,699
	3,663,390	3,723,346
	3,731,986	
Flannery, J.B.Jr.,	3,666,947	3,680,950
	3,687,515	
Fontanel, A.,	1,316,497	
Freiser, M.J.,	3,625,591	
Friel, R.N.,	3,674,342	
Garfein, A.,	3,667,039	
Gaynor, J.,	3,733,485	
Gelber, R.M.,	3,736,047	
Girard, P.,	3,691,755	3,712,047
Goldberg, N.N.,	3,529,156	3,580,864
	3,627,699	3,663,390
Goldmacher, J.E.,	3,499,702	3,540,796
	3,650,603	3,703,331
Gorog, I.,	3,723,651	
Grabmaier, J.,	3,730,607	
Grau, G.,	1,316,497	
Greubel, W.,	3,725,899	
Haas, W.E.L.,	3,652,148	3,655,971
	3,666,947	3,666,948
	3,669,525	3,679,290
	3,680,950	3,687,515
	3,707,322	3,711,713
	3,718,380	3,718,382
	3,720,456	3,726,584
Haller, I.,	3,625,591	3,656,834
Hanlon, T.F.,	3,569,614	3,700,805
Hareng, M.,	3,718,381	
Hedman, C.L.Jr.,	3,674,341	3,728,007
Heilmeier, G.H.,	3,499,112	3,499,702
	3,503,673	3,519,330
	3,551,026	3,575,491
	3,575,493	3,600,061
	3,650,603	3,655,269
	3,703,331	
Hodson, T.L.,	3,585,381	3,617,374
Hofstein, S.R.,	3,505,804	
Hsieh, P.Y.,	3,728,008	
Huener, R.C.,	3,740,717	
Huggins, H.A.,	3,656,834	
Jacobs, J.W.,	3,648,280	
Jankowitz, G.,	3,527,945	
Johnson, R.N.,	3,628,268	
Jones, D.,	3,690,745	3,716,289
Jones, J.W.,	3,585,381	3,617,374
Kahn, F.J.,	3,684,053	
Kessler, C.W.,	3,637,291	
Kessler, L.W.,	3,707,323	
Klein, R.I.,	3,612,654	3,647,280
	3,689,131	
Koopman, D.E.,	3,700,306	
Kratomi, S.,	3,737,567	
Krueger, H.,	3,730,607	
Kubitzek, A.,	3,727,527	
Lechner, B.J.,	3,532,813	3,575,492
Leder, L.B.,	3,679,290	3,697,152
Leiba, E.,	3,663,086	3,718,381
Lewis, B.L.,	3,675,989	
Lu, S.,	3,690,745	3,716,289
Madrid, R.W.,	3,622,224	3,642,348
	3,652,148	3,704,056
	3,707,322	3,711,713
	3,718,380	3,718,382
Marlow, F.J.,	3,503,672	3,654,606
Mash, D.H.,	1,304,268	
Matthies, D.L.,	3,622,226	3,661,444
Mauer, P.B.,	3,592,527	
Mechlowitz, B.,	3,666,947	3,666,948
Miller, A.,	3,623,795	3,697,297
Miller, R.E.,	3,732,119	
Mitsui, H.,	3,668,861	
Morgan, D.K.,	3,740,717	
Morse, J.E.,	3,722,998	
Myrenne, K.-D.S.,	3,674,341	3,728,007
Nao, R.A.,	3,653,745	
Nester, E.O.,	3,575,492	3,654,606
Ngo, D.-T.,	3,645,604	
Nicastro, L.J.,	3,588,225	
Niemiec, S.J.,	3,740,717	
Pietsch, R.,	3,675,989	
Pothier, R.G.,	3,713,156	
Rackman, M.I.,	3,716,658	
Rafuse, M.J.,	3,675,987	
Reilly, J.N.,	3,718,842	
Rindner, W.,	3,667,039	
Robert, J.,	3,716,290	
Rubin, D.C.,	3,667,039	
Sawyer, S.P.,	3,707,323	
Segawa, H.,	3,704,625	
Seto, H.,	3,704,625	
Sharpless, E.N.,	3,647,279	
Shaw, H.E., Jr.,	3,590,371	
Small, E.A., Jr.,	3,736,047	
Soref, R.,	3,675,988	
Spitz, E.,	3,663,086	
Stein, R.J.,	3,666,881	
Taylor, G.W.,	3,623,795	
Taylor, T.R.,	3,723,346	
Trzaskas, T.T.,	3,637,291	
Ueda, M.,	3,704,625	
Uleman, T.,	1,306,912	
Walton, R.S.,	3,613,351	
Wank, M.R.,	3,569,709	
Whitaker, J.G.,	3,617,374	
Wild, P.,	3,705,310	1,318,007
Wysocki, J.J.,	3,622,224	3,642,348
	3,652,148	3,655,971
	3,697,150	3,704,056
	3,707,322	3,711,713
	3,718,380	3,718,382
Zanoni, L.A.,	3,499,112	3,503,673
	3,600,061	

"""""""""""""

PATENT HOLDER INDEX

AGFA-Gevaert A.G., 3,727,527
American Micro-Systems, Inc., 3,702,723
Barnes Engineering Co., 3,527,945
Battelle Development Corp., The, 1,194,544
Bell & Howell Co., 3,733,485
Bell Telephone Laboratories, Inc., 3,645,604 3,694,053
Bendix Corp., 3,693,084
Brown, Boveri & Co., Ltd., 3,705,310 1,318,007
Chain Lakes Research Corp., 3,572,907
Commissariat a l'Energie Atomique, 3,716,290
Display Technology Corp., 3,734,598
Eastman Kodak Co., 3,722,998
General Motors Corp., 3,648,280
General Time Corp., 3,524,726
Hamilton Watch Comp., 3,613,351
Hughes Aircraft Co., 3,732,429
Institut Francais du Petrol des Carburants et Lubrifiants, 1,316,497
International Business Machines Corp., 3,625,591 3,656,834
International Liquid Crystal Co., 3,723,346 3,731,986
Kabushiki Kaisha Suwa Seikosha, 3,668,861
Kabushiki Kaisha Saini Seikosha and Kabushiki Kaisha Hattori Tokeitsen, 1,318,011 1,318,012
Liquid Crystal Industries, Inc., 3,576,761 3,619,254 3,647,279
Longines, S.A., 1,303,947
National Cash Register Company, The, 1,201,230 3,578,844 3,585,381 3,600,06 3,617,374
                                      3,637,291 3,674,338 3,697,297 3,700,306 1,302,482
                                      3,732,119 1,316,213 3,734,594
Olivetti & Co., 3,728,008
Optical Coating Laboratory, Inc., 3,569,709 3,736,047
Polacoat, Inc., 3,502,526 3,597,043
PPG Industries, Inc., 3,590,371 1,306,912
Radiation Incorporated, 3,675,989
RCA Corporation, 3,499,112 3,499,702 3,503,672 3,503,673 3,505,804 3,519,330
                  3,532,813 3,540,796 3,551,026 1,219,840 1,220,169 1,227,616
                  1,228,606 3,575,491 3,575,492 3,575,493 3,588,225 3,597,044
                  3,600,061 1,246,847 3,612,654 3,614,210 3,622,226 3,623,795
                  1,263,277 1,263,278 3,647,280 3,650,603 3,653,745 3,654,606
                  3,655,269 3,655,971 3,661,444 1,273,779 1,276,523 3,674,342
                  3,689,131 3,703,329 3,703,331 1,304,554 1,308,208 1,308,237
                  3,723,651 3,740,717
Rolex S.A., 3,691,755 3,712,047
Sankyo Keiryoki Kabishiki Kaisha, 3,704,625
SCM Corp., 3,674,341 3,728,007
Siemens A.G., 3,725,899 3,730,607
Sperry Rand Corp., 3,675,987 3,675,988
Standard Telephones and Cables Ltd., 1,304,268
Texas Instruments, Inc., 3,650,608 3,690,745 3,716,289 3,718,842
Thomson-CSF, 1,251,790 1,258,739 3,663,086 3,718,381
U.S.A., NASA, 3,667,039
U.S.A., Secretary of the Army, 3,628,268
U.S.A., Secretary of the Navy, 3,604,930
Vari-Light Corp., 3,620,889
Western Electric Corp., 3,663,390
Westinghouse Electric Corp., 3,529,156 3,580,864 3,627,408 3,627,699 1,309,558
Xerox Corp., 3,622,224 3,642,348 3,652,148 3,666,947 3,666,948 3,669,525 3,679,290
              3,680,950 3,687,515 3,697,150 3,697,152 3,704,056 3,707,322 3,711,181
              3,711,713 3,718,380 3,718,382 3,720,456 3,726,584
Zenith Radio Corp., 3,707,323

BIBLIOGRAPHY

Books and Conference Reports

Berkowitz, G. (ed.), NOVEL AUDIO-VISUAL IMAGING SYSTEMS, Proceedings of the SPSE Seminar in New York City on Sept. 23-24, 1971 (1971)

Brown, G.H., Doane, J.W., Fishel, D.L., Neff, V., Drauglis, E., LIQUID CRYSTALS, AD-725949 (AFML-TR-71-20), April, 1971.

Brown, G.H., Doane, J.W., Neff, V.D., REVIEW OF THE STRUCTURE AND PHYSICAL PROPERTIES OF LIQUID CRYSTALS, Butterworth, London and also CRC Press, Cleveland, Ohio, (1971)

Brown, G.H., Labes, M.M., LIQUID CRYSTALS 3 (pts. 1 & 2), Gordon & Breach Science Publishers, New York, London, (1972)

FARADAY SYMPOSIUM ON LIQUID CRYSTALS, Faraday Society, London, (1971)

Hoseman, R., (ed.), Abstracts, "THIRD INTERNATIONAL LIQUID CRYSTAL CONFERENCE," Berlin, West Germany, Aug. 24-29, 1970.

Johnson, J.F., Porter, R.S., (eds.), LIQUID CRYSTALS AND ORDERED FLUIDS, (Proc. Am. Chem. Soc. Symp., New York City, Sept. 10-12, 1969, Plenum Press, New York-London, (1970)

Kallard, T., (ed.), LIQUID CRYSTALS AND THEIR APPLICATIONS, Optosonic Press, New York, (1970)

Longley-Cook, M.T., THERMAL CONDUCTIVITY OF P-P'-AZOXYANISOLE, Univ. Arizona, Tucson, thesis, 1972, University Microfilms, Ann Arbor, Michigan, Order No. 72-11975.

Martins, Assis Farinha, CONTRIBUTION A L'ETUDE DE LA DYNAMIQUE MOLECULAIRE DANS LES PHASES NEMATIQUE ET ISOTROPE DES CRISTAUX LIQUIDS, thesis, Published in Portugaliae Physica v8 n1-2 (1972)

Parker, J.H. Jr., ANOMALOUS ALIGNMENT AND DOMAINS IN A NEMATIC LIQUID CRYSTAL, Univ. Maine, Orono, thesis, University Microfilms, Ann Arbor, Michigan, Order No. 72-15649. (1971)

Shih, C.-S., THEORY OF ORIENTATION CORRELATIONS IN NEMATIC LIQUID CRYSTALS, Yale Univ., New Haven, thesis, 1971, University Microfilms, Ann Arbor, Michigan, Order No. 71-31010.

Wulf, A., SOME STATISTICAL MECHANICAL MODELS FOR SYSTEMS OF LONG LINEAR MOLECULES: APPLICATION TO THE ISOTROPIC-NEMATIC LIQUID CRYSTAL TRANSITION, Univ. Maryland, thesis, 1970, University Microfilms, Ann Arbor, Mich. Order No. 71-4102.

Articles

1. GLASSY STATES OF LIQUID, LIQUID CRYSTAL AND OF CRYSTAL STUDIED BY CALORIMETRIC AND OTHER METHODS, Adachi, K., Suga, H., Seki, S., p30: Progr & Abstr 2nd Int'l Conf Calorimetry and Thermodynamics, Orono, Maine, 12-14 July, '71.

2. GLASSY STATE OF LIQUID, LIQUID CRYSTAL, AND OF CRYSTAL (Abstr.), Adachi, K., Haida, O., Matsuo, T., Sorai, M., Suga, H., Seki, S., Acta Crystallogr A (Denmark) vA28 pt4 suppl p5129 (15 Jly '72)

3. THE EFFECTIVE ROTARY POWER OF CHOLESTEROL, Adams, J.E., Leder, L., Chem Phys Lett (Holland) v6 n2 p90 (15 Jly '70)

4. CHARACTERIZATION OF MOLECULAR ROLE IN PITCH DETERMINATION IN LIQUID CRYSTAL MIXTURES, Adams, J.E., Haas, W.E.L., p118: Abstr 3rd Int'l Liquid Crystal Conf., Berlin, W. Germany, (Aug '70)

5. DISPERSIVE REFLECTION IN CHOLESTERICS, Adams, J.E., Haas, W.E.L., Mol Cryst & Liq Cryst v11 n3 p229 (Oct '70)

6. THE EFFERCTIVE ROTARY POWER OF THE FATTY ESTERS OF CHOLESTEROL, Adams, J.E., Haas, W.E.L., Wysocki, J.J., p463: "Liquid Crystals and Ordered Fluids," Plenum Press (1970)

7. CHOLESTERIC FILMS AS OPTICAL FILTERS, Adams, J.E., Haas, W.E.L., Dailey, J., J Appl Phys v42 n10 p4096 (Sep '71)

8. CHARACTERIZATION OF MOLECULAR ROLE IN PITCH DETERMINATION IN LIQUID CRYSTAL MIXTURES, Adams, J.E., Haas, W.E.L., Mol Cryst & Liq Cryst v15 n1 p27 (Oct '71)

9. SENSITIVITY OF CHOLESTERIC FILMS TO ULTRAVIOLET EXPOSURE, Adams, J.E., Haas, W.E.L., J Electrochem Soc v118 n12 p2026 (Dec '71)

10. THE RELATIONSHIP BETWEEN PITCH CHANGE AND STIMULUS IN CHOLESTERICS, Adams, J.E., Haas, W.E.L., Mol Cryst & Liq Cryst v16 n1&2 p33 (Feb '72)

11. EQUATIONS OF MOTION IN LIQUID CRYSTALS OF THE NEMATIC TYPE, Aero, E.L., Bulygin, A.N., Prikl Mat & Mekh (USSR) v35 n5 p879 (1971)

12. LINEAR MECHANICS OF LIQUID CRYSTAL MEDIA, Aero, E.L., Bulygin, A.N., Fiz Tver Tela (USSR) v13 n6 p1701 (Jne '71)

13. THE KINEMATICS OF NEMATIC LIQUID CRYSTALS, Aero, E.L., Bulygin, A.N., Prikl Mekh (USSR) p97 (Mch '72)

14. VISCOSITY AND THERMAL CONDUCTIVITY OF LIQUID CRYSTALS IN A MAGNETIC FIELD, Aero, E.L., Bulygin, A.N., Soviet Phys - Tech Phys v17 n4 p694 (Oct '72)

15. OPTICAL WAVE PROPAGATION IN CHOLESTERIC LIQUID CRYSTALS WITH FINITE THICKNESS, Aihara, M., Inaba, H., Opt Commun v3 n2 p77 (Apr '71)

16. INTERPRETATION OF THERMODYNAMIC DERIVATIVES NEAR THE LIQUID CRYSTAL PHASE TRANSITION IN P-AZOXYANISOLE, Alben, R., Mol Cryst & Liq Cryst v10 n1-2 p21 (May '70)

17. PRETRANSITION EFFECTS IN NEMATIC LIQUID CRYSTALS: MODEL CALCULATIONS, Alben, R., Mol Cryst & Liq Cryst v13 n3 p193 (Jne '71)

18. THE CHARACTERIZATION OF ORDER IN NEMATIC LIQUID CRYSTALS, Alben, R., Solid State Commun v11 n8 p1081 (15 Oct '72)

19. ROOM-TEMPERATURE NEMATIC LIQUID CRYSTAL MIXTURES WITH POSITIVE DIELECTRIC ANISOTROPY, Alder, C.J., Raynes, E.P., J Phys D: Appl Phys v6 n5 pL33 (20 Mch '73)

20. A METHOD FOR REDUCING THE DECAY TIME OF A LIQUID CRYSTAL, Alimonda, A., Meyer, V., IEEE Trans Electron Devices vED-20 n3 p332 (Mch '73)

21. LIVING SYSTEMS, Ambrose, E.J., p17: Abstr 3rd Int'l Liquid Crystal Conf., Berlin, West Germany (Aug '70)

22. LIQUID CRYSTALLINE PHENOMENA AT THE CELL SURFACE, Ambrose, E.J., Faraday Symp on Liquid Crustals, Faraday Soc., London (1971)

23. FABRICATING PATTERNS ON CONDUCTIVE GLASS, Ambrosia, A., Sambucetti, C.J., IBM Tech Discl Bull v13 n5 p1320 (Oct '70)

24. RAMAN SCATTERING FROM NEMATIC LIQUID CRYSTALLINE AZOXYBENZENES, Amer, N.M., Shen, Y.-R., J Chem Phys v56 n6 p2654 (15 Mch '72)

25. LOW FREQUENCY RAMAN MODE IN SMECTIC LIQUID CRYSTALS NEAR THE PHASE TRANSITIONS, Amer, N.M., Shen, Y.-R., Solid State Commun v12 n4 p263 (15 Feb '73)

26. A LATTICE MODEL FOR CRYSTAL-NEMATIC AND NEMATIC-ISOTROPIC LIQUID TRANSITIONS, Andrews, J.T.S., p122: Abstr 3rd Int'l Liquid Crystal Conf., Berlin, W. Germany (Aug '70)

27. ABOLISHING LASER-DISPLAY 'SPECKLE', Anon, Electronic Design v19 n16 p24 (5 Aug '71)

27A. GUIDANCE IN LIQUID CRYSTAL FILM IS REPORTED IN INTEGRATED-OPTICS SESSION AT 'OSA' MEETING, Anon, Laser Focus v9 n7 p20 (Jly '73)

28. TWO LIQUID CRYSTAL PHASES WITH NEMATIC MORPHOLOGY IN LATERALLY SUBSTITUTED PHENYLENDIAMINE DERIVATIVES, Arora, S.L., Fergason, J.L., Saupe, A., Mol Cryst & Liq Cryst v10 n3 p243 (Jne '70)

29. MOLECULAR STRUCTURE AND MESOMORPHIC PROPERTIES OF PHENYLENEALKOXYBENZOATES, Arora, S.L., Fergason, J.L., Taylor, T.R., p86: Abstr 3rd Int'l Liquid Crystal Conf., Berlin, W. Germany (Aug '70)

30. POLYMORPHISM OF SMECTIC PHASES WITH SMECTIC 'A' MORPHOLOGY, Arora, S.L., Taylor, T.R., Fergason, J.L., p321: "Liquid Crystals and Ordered Fluids," Plenum Press (1970)

31. EFFECTS OF MOLECULAR GEOMETRY ON THE FORMATION OF SMECTIC PHASES, Arora, S.L., Fergason, J.L., Faraday Symp on Liquid Crystals, Faraday Soc., London (1971)

32. ELECTRO-OPTIC PERFORMANCE OF A NEW ROOM-TEMPERATURE NEMATIC LIQUID CRYSTAL, Ashford, A., Constant, J., Kirton, J., Raynes, E.P., Electronics Letters v9 n5 p118 (8 Mch '73)

33. CONTINUUM MECHANICS OF NEMATIC LIQUIDS, Aslaksen, E.W., Phys Kond Mater (Germany) v14 n1 p80 (1971)

34. TWO-DIMENSIONAL ORDER IN A NEMATIC LIQUID CRYSTAL NEAR THRESHOLD FOR DYNAMIC SCATTERING, Aslaksen, E.W., Ineichen, B., J Appl Phys v42 n2 p882 (Feb '71)

35. STEADY STATE OF A NEMATIC LIQUID CRYSTAL ABOVE THE THRESHOLD FOR DYNAMIC SCATTERING, Aslaksen, E.W., Mol Cryst & Liq Cryst v15 n2 p121 (Nov '71)

36. COMMENT ON THE ELECTROHYDRODYNAMIC INSTABILITY IN NEMATIC CRYSTALS, Aslaksen, E.W., J Appl Phys v43 n3 p776 (Mch '72)

37. LIQUID CRYSTALS, Assouline, G., Leiba, E., Rev Tech Thomson-CSF v1 n4 p483 (Dec '69)

38. DIFFRACTION D'UN FAISCEAU LASER PAR UN CRISTAL LIQUIDE NEMATIQUE SOUMIS A UN CHAMP ELECTRIQUE, Assouline, G., Dmitrieff, A., Hareng, M., Leiba, E., Comptes Rendus (Ser.B) v271 n16 p857 (19 Oct '70)

39. SCATTERING OF LIGHT BY NEMATIC LIQUID CRYSTALS, Assouline, G., Dmitrieff, A., Hareng, M., Leiba, E., J Appl Phys v42 n6 p2567 (May '71)

40. AN ATTENUATOR UTILIZING LIQUID CRYSTALS, Assouline, G., Hareng, M., Leiba, E.,

Rev Tech Thomson-CSF (France) v3 n3 p473 (Sep '71)

41. LIQUID CRYSTAL AND PHOTOCONDUCTOR IMAGE CONVERTER, Assouline, G., Hareng, M., Leiba, E., IEEE Proc v59 n9 p155 (Sep '71)

42. APPLICATIONS OF CHOLESTERIC LIQUID CRYSTALS TO NONDESTRUCTIVE TESTING, Assouline, G., Hareng, M., Leiba, E., Electron & Microelectron Ind (France) n147 p35 (Oct '71)

43. ELECTRICAL MEASUREMENT IN NEMATIC LIQUID CRYSTALS, Assouline G., Hareng, M., Leiba, E., IEEE Trans Electron Dev vED-18 n10 p959 (Oct '71)

44. TWO-COLOR LIQUID CRYSTAL DISPLAY, Assouline, G., Hareng, M., Leiba, E., Electron Lett (GB) v7 n23 p699 (18 Nov '71)

45. A LIQUID CRYSTAL DISPLAY WITH ELECTRICALLY CONTROLLED BIREFRINGENCE, Assouline, G., Hareng, M., Leiba, E., Roncillat, M., Electron Lett (GB) v8 n2 p45 (27 Jan '72)

46. IMAGE TRANSFORMATION BY A PHOTOCONDUCTING LAYER ON A NEMATIC LIQUID CRYSTAL, Assouline, G., Hareng, M., Leiba, E., Comptes Rendus (Ser. B) v274 n10 p692 (6 Mch '72)

47. LIQUID CRYSTALS, Assouline, G., Hareng, M., Leiba, E., Rev Electrotec (Spain) v17 n194 p29 (Jne '72)

48. 31-NORCYCLOARTANOL FATTY ESTERS: CHOLESTERIC LIQUID CRYSTALS FROM A TRITERPENE OF PLANT ORIGIN, Atallah, A.M., Nicholas, H.J., Mol Cryst & Liq Cryst v19 n1 p1 (May '72)

49. LIQUID CRYSTALLINE PROPERTIES OF FATTY ACID ESTERS OF LOPHENOL, A CHOLESTEROL BIOSYNTHETIC INTERMEDIATE, Atallah, A.M., Nicholas, H.J., Mol Cryst & Liq Cryst v18 n3&4 p321 (Oct '72)

50. INFLUENCE OF THE POSITION OF RING UNSATURATION IN STEROIDS AND TRITERPENES ON THE TYPE AND FORMATION OF MESOPHASES. I: INFLUENCE OF THE DELTA 8 DOUBLE BOND, Atallah, A.M., Nicholas, H.J., Mol Cryst & Liq Cryst v18 n3&4 p339 (Oct '72)

51. INFLUENCE OF THE POSITION OF RING UNSATURATION IN STEROIDS AND TRITERPENES ON THE TYPE AND FORMATION OF MESOPHASES. II: INFLUENCE OF THE DELTA 7 DOUBLE BOND, Atallah, A.M., Nicholas, H.J., Mol Cryst & Liq Cryst v19 n3&4 p217 (Jan '73)

52. EFFECT OF BOUNDARY CONDITIONS ON THE PERFORMANCE OF NEMATIC LIQUID CRYSTAL DISPLAYS, Attergut, S., Cole, H.S., p92: 1972 SID Symp Dig of Tech Papers, San Frabcisco, (6-8 Jne '72)

53. DIGITAL HOLOGRAPHIC MEMORY WITH ACOUSTO-OPTIC DEFLECTION ACCESS, Auffret, R., Lacroix, R., Sapriel, J., Onde Elec (France) v51 n9 p787 (Oct '71)

54. LIQUID CRYSTAL APPLICATIONS, Auffret, R., Treheux, M., Echo Rech (France) n67 p16 (Jan '72)

55. APPLICATION DES CRISTAUX LIQUIDES AU TRAITEMENT OPTIQUE DE L'INFORMATION, Auvergne, M., Roddier, F., Comptes Rendus (Ser. B) v273 p1088 (20 Dec '71)

56. COLOR DISPLAY ELEMENT USING NEMATIC LIQUID CRYSTALS, Aviram, A., Haller, I., IBM Tech Discl Bull v14 n11 p3536 (Apr '72)

57. PACKAGING OF LIQUID CRYSTAL DISPLAYS, Aviram, A., IBM Tech Discl Bull v15 n2 p580 (Jly '72)

58. SURFACE REORIENTATION LIQUID CRYSTAL DISPLAY DEVICE, Aviram, A., Gladstone, G.L., IBM Tech Discl Bull v15 n6 p1812 (Nov '72)

59. MULTICOLOR DISPLAY, Aviram, A., Freiser, M.J., IBM Tech Discl Bull v15 n8 p2538 (Jan '73)

60. SIMPLIFIED APPROACH TO THE PROPAGATION OF POLARIZED LIGHT IN ANISOTROPIC MEDIA. - APPLICATION TO LIQUID CRYSTALS, Azzam, R.M.A., Bashara, N.M., Opt Soc Am J v62 n11 p1252 (Nov '72)

61. TRAJECTORIES DESCRIBING THE EVOLUTION OF POLARIZED LIGHT IN HOMOGENEOUS ANISOTROPIC MEDIA AND LIQUID CRYSTALS, Azzam, R.M.A., Merrill, B.E., Bashara, N.M., Appl Opt v12 n4 p764 (Apr '73)

62. NEMATIC LIQUIDS AS REACTION MEDIA FOR THE CLAISEN REARRANGEMENT, Bacon, W.E., Brown, G.H., p84: Abstr 3rd Int'l Liquid Crystal Conf., Berlin, W. Germany (Aug '70)

63. NEMATIC SOLVENTS AS MEDIA FOR THE CLAISEN REARRANGEMENT, Bacon, W.E., Brown, G.H., Mol Cryst & Liq Cryst v12 n3 p229 (Feb '71)

64. ABSORPTION AND DISPERSION MEASUREMENTS OF LONGITUDINAL ULTRASONIC WAVES OF 1GHz IN THREE BODIES POSSESSIONG A CHOLESTERIC STATE, Bacri, J.-C., Comptes Rendus (Ser.B) v270 n25 p1589 (22 Jan '70)

65. HELICAL TWISTING POWER OF STEROIDAL SOLUTES IN CHOLESTERIC MESOPHASES, Baessler, H., Labes, M.M., J Chem Phys v52 n2 p631 (15 Jan '70)

66. THE ABSENCE OF HELICAL INVERSION IN SINGLE COMPONENT CHOLESTERIC LIQUID CRYSTALS, Baessler, H., Malya, P.A.G., Nes, W.R., Labes, M.M., Mol Cryst & Liq Cryst v6 n3&4 p329 (Feb '70)

67. DETERMINATION OF THE PITCH OF A CHOLESTERIC LIQUID CRYSTAL BY INFRARED TRANSMISSION MEASUREMENTS, Baessler, H., Labes, M.M., Mol Cryst & Liq Cryst v6 n3&4 p419 (Feb '70)

68. DIPOLE RELAXATION IN A LIQUID CRYSTAL, Baessler, H., Beard, R.B., Labes, M.M., J Chem Phys v52 n5 p2292 (1 Mch '70)

69. DIPOLE RELAXATION IN A LIQUID CRYSTAL, Baessler, H., Beard, R.B., Labes, M.M., p103: Abstr 3rd Int'l Liq Cryst Conf., Berlin, W. Germany (Aug '70)

70. EFFECTS OF SOLUTES ON THE HELICAL TWIST OF CHOLESTERIC LIQUID CRYSTALS, Baessler, H., Teucher, I., Labes, M.M., p119: Abstr 3rd Int'l Liq Cryst Conf., Berlin, W. Germany (Aug '70)

71. EFFECTS OF ORGANIC ADDITIVES ON PARAFFIN CHAIN ELECTROLYTE SOLUTIONS, Bain, R.M., Hyde, A.J., Faraday Symposium on Liquid Crystals, Faraday Society, London (1971)

72. EFFECT OF CHARGE-TRANSFER ACCEPTORS ON DYNAMIC SCATTERING IN A NEMATIC LIQUID CRYSTAL, Baise, A.I., Teucher, I., Labes, M.M., Appl Phys Lett v21 n4 p142 (15 Aug '72)

73. X-RAY DIFFRACTION STUDIES ON ORIENTED LYOTROPIC MESOMORPHIC PHASES, Balmbra, R.R., Clunie, J.S., p78: Abstr 3rd Int'l Liq Cryst Conf., Berlin, W. Germany (Aug '70)

74. MESOMORPHIC PHASE STRUCTURE IN THE POTASSIUM OLEATE & WATER SYSTEM, Balmbra, R.R., Bucknall, D.A.B., Clunie, J.S., Mol Cryst & Liq Cryst v11 n2 p173 (Sep '70)

75. NEMATIC CRYSTALS COME OF AGE, Baltzer, D.H., Electro-Optical Systems Design v2 n1 p72 (Jan '70)

76. TEMPERATURE DEPENDENCE OF BIREFRINGENCE OF BIREFRINGENCE IN LIQUID CRYSTALS, Balzarini, D.A., Phys Rev Lett v25 n14 p914 (5 Oct '70)

77. NEMATIC-PHASE N.M.R. INVESTIGATION OF ROTATIONAL ISOMERISM: CONFORMATIONS OF 2-FURANALDEHYDE, Barili, P.L., Lunazzi, L., Veracini, L., Mol Phys (GB) v24 n3 p673 (Sep '72)

78. EFFECT OF SUBSTITUENT LOCATION ON THE THERMODYNAMIC PROPERTIES OF CHOLESTERYL HALOBENZOATE HALOCINNAMATES, Barrall, E.M., II., Bredfeldt, K.E., Mol Cryst & Liq Cryst v18 n3&4 p195 (Oct '72)

79. PHASE MODULATION OF COHERENT LIGHT WITH THE AID OF LIQUID CRYSTALS, Basov, N.G., Berezin, P.D., Blinov, L.M., Kompanetz, I.N., Morozov, V.N., Nikitin, V.V., JETP Lett v15 n4 p138 (20 Feb '72)

80. THE LIQUID CRYSTALLINE STATE COPOLYMERIZATION, Baturin, A.A., Amerik, Y.B., Krentsel, B.A., p140: Abstr 3rd Int'l Liq Cryst Conf., Berlin, W. Germany (Aug '70)

81. THE LIQUID CRYSTALLINE STATE COPOLYMERIZATION, Baturin, A.A., Amerik, Y.B., Krentsel, B.A., Mol Cryst & Liq Cryst v16 n1&2 p117 (Feb '72)

82. QUENCHING OF FLUORESCENCE AT THE NEMATIC-ISOTROPIC PHONE TRANSITION, Baur, G., Stieb, A., Meier, G., J Appl Phys v44 n4 p1905 (Apr '73)

83. PHOTOCONDUCTOR LIGHT GATED LIQUID CRYSTALS USED FOR OPTICAL DATA, Beard, T.D., p18: 1971 Annual Meeting Opt Soc Am, Ottawa, Canada, 5-8 Oct '71.

84. PHOTOACTIVATED LIQUID CRYSTAL LIGHT VALVES (EMPLOYING PHOTOCONDUCTIVE FILMS), Beard, T.D., Bleha, W.P., Braunstein, M., Jacobson, A.D., Margerum, J.D., Wong, S.-Y., p34: 17th Int'l Electron Devices Meeting (Abstr.), IEEE, Washington, D.C., 11-13 Oct '71.

85. PHOTOCONDUCTOR + LIGHT-GATED LIQUID CRYSTALS USED FOR OPTICAL DATA (Abstr.), Beard, T.D., Opt Soc Am J v61 n11 p1559 (Nov '71)

86. 'AC' LIQUID-CRYSTAL LIGHT VALVE, Beard, T.D., Bleha, W.P., Wong, S.-Y., Appl Phys Lett v22 n3 p90 (1 Feb '73)

87. CERENKOV STRUCTURE RADIATION IN CHOLESTERIC LIQUID CRYSTALS, Belyakov, V.A., Orlov, V.P., Phys Lett A (Holland) v42A n1 p3 (6 Nov '72)

88. HOW TO ELIMINATE REFLECTIONS IN CHOLESTERIC LIQUID CRYSTAL PHOTOGRAPHY, Benjamin, A., Photog Applic Sci Tech Medicine p30 (Mch '71)

89. AQUEOUS DYE AGGREGATES AS LIQUID CRYSTALS, Berg, R.A., Haxby, B.A., Mol Cryst & Liq Cryst v12 n1 p93 (Dec '70)

90. BRAGG REFLECTION OF OBLIQUELY INCIDENT LIGHT BY THIN FILMS OF CHOLESTERIC LIQUID CRYSTALS, Berreman, D.W., Scheffer, T.J., p23: 1970 Spring Meeting Opt Soc Am, Philadelphia, Pa. (7-10 Apr. '70)

91. BRAGG REFLECTION OF OBLIQUELY INCIDENT LIGHT BY THIN FILMS OF CHOLESTERIC CRYSTALS (Abstr.), Berreman, D.W., Scheffer, T.J., Opt Soc Am J v60 n5 p725 (May '70)

92. REFLECTION AND TRANSMISSION BY PERFECTLY ORDERED CHOLESTERIC LIQUID CRYSTAL FILMS: THEORY AND VERIFICATION, Berreman, D.W., Scheffer, T.J., p39: Abstr 3rd Int'l Liq Cryst Conf., Berlin, W. Germany (Aug '70)

93. BRAGG REFLECTION OF LIGHT FROM SINGLE-DOMAIN CHOLESTERIC LIQUID-CRYSTAL FILMS, Berreman, D.W., Scheffer, T.J., Phys Rev Lett v25 n9 p577 (31 Aug '70)

94. REFLECTION AND TRANSMISSION BY SINGLE-DOMAIN CHOLESTERIC LIQUID CRYSTAL FILMS: THEORY AND VERIFICATION, Berreman, D.W., Scheffer, T.J., Mol Cryst & Liq Cryst v11 n4 p395 (Nov '70)

95. OPTICAL EFFECTS OF BOUNDARY PROXIMITY IN CHOLESTERIC LIQUID CRYSTAL FILMS, Berreman, D.W., Scheffer, T.J., p38: Abstr 1971 Spring Meeting Opt Soc Am, Tucson, Arizona, 5-8 Apr '71.

96. OPTICAL EFFECTS OF BOUNDARY PROXIMITY IN CHOLESTERIC LIQUID CRYSTAL FILMS (Abstr.), Berreman, D.W., Scheffer, T.J., Opt Soc Am J v61 n5 p679 (May '71)

97. ORDER VERSUS TEMPERATURE IN CHOLESTERIC LIQUID FROM REFLECTANCE SPECTRA, Berreman, D.W., Scheffer, T.J., Phys Rev (A) v5 n3 p1397 (Mch '72)

98. SOLID SURFACE SHAPE AND THE ALIGNMENT OF AN ADJACENT NEMATIC LIQUID CRYSTAL, Berreman, D.W., Phys Rev Lett v28 n26 p1683 (26 Jne '72)

99. SPATIAL DISTRIBUTION OF LIGHT SCATTERED BY 'PAA' IN APPLIED ELECTRIC FIELD, Bertolotti, M., Daino, B., Scudieri, F., Sette, D., p30: Abstr 3rd Int'l Liq Cryst Conf., Berlin, W. Germany (Aug '70)

100. SPATIAL DISTRIBUTION OF LIGHT SCATTERED BY P-AZOXYANISOLE IN APPLIED ELECTRIC FIELD, Bertolotti, M., Daino, B., Scudieri, F., Sette, D., Mol Cryst & Liq Cryst v15 n2 p133 (Nov '71)

101. LIGHT SCATTERING BY FLUCTUATIONS INDUCED BY AN APPLIED ELECTRIC FIELD IN A NEMATIC LIQUID CRYSTAL (APAPA), Bertolotti, M., Daino, B., DiPorto, P., Scudieri, F., Sette, D., J Physique (Colloque No.1) suppl to v33 n2-3 pC1-63 (Feb-Mch '72)

102. ACOUSTIC MODULATION OF LIGHT BY NEMATIC LIQUID CRYSTALS, Bertolotti, M., Martellucci, S., Scudieri, F., Sette, D., Appl Phys Lett v21 n2 p74 (15 Jly '72)

103. ACOUSTIC MODULATION OF LIGHT BY NEMATIC LIQUID CRYSTALS, Bertolotti, M., Martellucci, S., Scudieri, F., Sette, D., Appl Phys Lett v21 n2 p74 (15 Jly '72)

104. ELECTRIC AND OPTIC BEHAVIOR OF ALIGNED SAMPLES OF NEMATIC LIQUID CRYSTALS, Bertolotti, M., Scudieri, F., Sette, D., J Appl Phys v43 n10 p3914 (Oct '72)

105. LOW FREQUENCY RAMAN SPECTRA OF P-METHOXY BENZYLIDENE P-N-BUTYLANILINE (solid, nematic and isotropic states), Billard, J., Delhaye, M., Merlin, J.C., Vergotten, G., Comptes Rendus (Ser. B) v273 n25 p1105 (20 Dec '71)

106. FAST MICRODIAGNOSTICS OF THE TWIST DIRECTION OF MESOMORPHIC PHASES, Billard, J., Comptes Rendus (Ser. B) v274 n5 p333 (31 Jan '72)

107. NEUTRON SCATTERING STUDY OF SELF-DIFFUSION IN LIQUID CRYSTALS, Blinc, R., Dimic, V., Phys Lett A (Holland) v31A n10 p531 (18 May '70)

108. SELF-DIFFUSION SPIN RELAXATION AND LONG RANGE ORDER IN LIQUID CRYSTALS, Blinc, R., Pirs, J., Vilfan, M., Zupancic, I., Dimic, V., p62: Abstr 3rd Int'l Liq Cryst Conf., Berlin (Aug '70)

109. SELF-DIFFUSION AND MOLECULAR ORDER IN LYOTROPIC LIQUID CRYSTALS, Blinc, R., Easwaran, K., Pirs, J., Vilfan, M., Zupancic, I., Phys Rev Lett v25 n19 p1327 (9 Nov '70)

110. SELF-DIFFUSION IN LIQUID CRYSTALS, Blinc, R., Dimic, V., Pirs, J., Vilfan, M., Zupancic, I., Mol Cryst & Liq Cryst v14 n1-2 p97 (Aug '71)

111. MEASUREMENT OF SELF-DIFFUSION IN LIQUID CRYSTALS BY A MULTIPLE-PULSE N.M.R. METHOD, Blinc, R., Pirs, J., Zupancic, I., Phys Rev Lett v30 n12 p546 (19 Mch '73)

112. POLYMERIZATION OF METHACRYLOYLOXY BENZOIC ACID WITHIN LIQUID CRYSTALLINE MEDIA, Blumstein, A., Kitagawa, N., Blumstein, R., p142: Abstr 3rd Int'l Liq Cryst Conf., Berlin, W. Germany (Aug '70)

113. POLYMERIZATION OF P-METHACRYLOLOXY BENZOIC ACID WITHIN LIQUID CRYSTALLINE MEDIA, Blumstein, A., Kitagawa, N., Blumstein, R., Mol Cryst & Liq Cryst v12 n3 p215 (Feb '71)

114. BIREFRINGENCE OF CHOLESTERIC LIQUID CRYSTAL FILMS, Boettcher, B., Graber, G., p37: Abstr 3rd Int'l Liq Cryst Conf., Berlin, W. Germany (Aug '70)

115. DOPPELBRECHUNG AN DUENNEN SCHICHTEN CHOLESTERISCHER FLUESSIGKEITEN, Boettcher, B., Graber, G., Mol Cryst & Liq Cryst v14 n1-2 p1 (Aug '71)

116. LOW ELECTROOPTIC THRESHOLD IN NEW LIQUID CRYSTALS, Boller, A., Scherrer, H., Schadt, M., Wild, P., IEEE Proc v60 n8 p1002 (Aug '72)

117. P/MOS CHIP DRIVES LIQUID CRYSTAL DISPLAY FOR DIGITAL ALARM CLOCK, Borden, H., Mingione, J., Nance, P., Electronics v45 n3 p66 (31 Jan '72)

118. CRYSTAL TO LIQUID CRYSTAL TRANSITION STUDIED BY RAMAN SCATTERING, Borer, W.J., Mitra, S.S., Brown, C.W., Phys Rev Lett v27 n7 p379 (16 Aug '71)

119. LIQUID CRYSTAL DISPLAYS - A MARKET READY TO BLOOM, Bosso, A.J., Electro-optical Systems Design v4 n1 p28 (Jan '72)

120. MOLECULAR ORDER AT THE FREE SURFACE OF A NEMATIC LIQUID CRYSTAL FROM LIGHT REFLECTIVITY MEASUREMENTS, Bouchiat, M.A., Langevin-Cruchon, D., Phys Lett v34A n6 p331 (5 Apr '71)

121. HELICAL DISINCLINATION PAIRS IN CHOLESTERICS, Bouligand, Y., Kleman, M., J Phys (France) v31 n11 p1041 (Nov-Dec '70)

122. RECHERCHES SUR LES TEXTURES DES ETATS MESOMORPHES. I. LES ARRANGEMENTS FOCAUX DANS LES SMECTIQUES: RAPPELS ET CONSIDERATIONS THEORETIQUES, Bouligand, Y., J Phys (France) v33 n5-6 p525 (May-Jne '72)

123. RECHERCHES SUR LES TEXTURES DES ETATS MESOMORPHES. 2: LES CHAMPS POLYGONAUX DANS LES

CHOLESTERIQUES, Bouligand, Y., J Phys (France) v33 n7 p715 (Jly '72)

124. IONIC RESIDUAL CONDUCTION IN THE ISOTROPIC PHASE OF A NEMATIC LIQUID CRYSTAL, Briere, G., Gaspard, F., Herino, R., Chem Phys Lett (Holland) v9 n4 p285 (15 May '71)

125. CORRELATION BETWEEN CHEMICAL AND ELECTROCHEMICAL REACTIVITY AND D.C. CONDUCTION IN THE ISOTROPIC PHASE OF A LIQUID CRYSTALLINE P-METHOXYBENZYLIDENE-P-N-BUTYLANILINE (MBBA), Briere, G., Herino, R., Mondon, F., Mol Cryst & Liq Cryst v19 n2 p157 (Dec '72)

126. THEORY OF MAGNETIC SUSPENSIONS IN LIQUID CRYSTALS, Brochard, F., De Gennes, P.G., J Phys (France) v31 n7 p691 (Jly '70)

127. IMPEDOMETRIC MEASUREMENTS OF VISCOSITY IN CHOLESTERIC LIQUID CRYSTALS, Brochard, F., J Phys (France) v32 n8-9 p685 (Aug-Sep '71)

128. MOUVEMENTS DE PAROIS DANS UNE LAME MINCE NEMATIQUE, Brochard, F., J Phys (France) v33 n5-6 p607 (May-Jne '72)

129. DYNAMICS OF THE ORIENTATION OF A NEMATIC-LIQUID-CRYSTAL FILM IN A VARIABLE MAGNETIC FIELD, Brochard, F., Pieranski, P., Guyon, E., Phys Rev Lett v28 n26 p1681 (26 Jne '72)

130. BIREFRINGENCE D'ECOULEMENT ET ATTENUATION DU SON AU VOISINAGE DE LA TRANSITION SMECTIC 'A'-SMECTIC 'C', Brochard, F., Comptes Rendus (Ser. B) v276 n2 p87 (8 Jan '73)

131. THERMAL FLUCTUATIONS IN THE NEMATIC MESOPHASE, Brooks, S.A., Luckhurst, G.R., Pedulli, G.F., Chem Phys Lett (Holland) v11 n2 p159 (1 Oct '71)

132. LIQUID CRYSTALS AND THEIR ROLES IN INANIMATE AND ANIMATE SYSTEMS, Brown, G.H., Am Sci v60 n1 p64 (Jan-Feb '72)

133. PROPERTIES OF LIQUID CRYSTALS, Brown, G.H., p34: Program, 84th Meeting Ac Soc Am, Miami Beach, Fla., 28 Nov - 1 Dec, '72.

134. STRUCTURE, PROPERTIES, AND SOME APPLICATIONS OF LIQUID CRYSTALS (Abstr.), Brown, G.H., Opt Soc Am J v63 n4 p504 (Apr '73)

135. INDICES DES MELANGES DE PARA-AZOXYANISOLE ET DE PARA-AZOXYPHENETOLE DANS L'ETAT NEMATIQUE. INTERPRETATION DES RESULTATS A L'AIDE DE LA THEORIE DE MAIER ET SAUPE, Brunet-Germain, M., Mol Cryst & Liq Cryst v11 n3 p289 (Oct '70)

136. MEASUREMENTS OF SPECIFIC ROTARY POWER AND PITCH OF MIXTURES OF CHOLESTEROL AND P-AZOXY-ANISOLE. COMPARISON WITH RESULTS OF THE MAUGUIN-deVRIES THEORY, Brunet-Germain, M., Acta Cryst (Internat) v26a Pt.6 p595 (1 Nov '70)

137. INDICES OF P-METHOXYBENZYLIDENE P-N-BUTYLANILINE (MBBA), Brunet-Germain, M., Comptes Rendus (Ser. B) v271 n21 p1075 (23 Nov '70)

138. INTERPRETATION OF GRANDJEAN DISCONTINUITIES IN A CHOLESTERIC STRUCTURE USING deVRIES THEORY, Brunet-Germain, M., Comptes Rendus (Ser. B) v274 n17 p1036 (24 Apr '72)

139. N.M.R. SPECTRUM OF 3,3'-BIS-ISOXALOZE IN NEMATIC PHASE, Bucci, P., Franchini, P.F., Serra, A.M., Veracini, C.A., Chem Phys Lett (Holland) v8 n5 p421 (1 Mch '71)

140. CONFORMATION STUDIES OF MOLECULES PARTIALLY ORIENTED IN THE NEMATIC PHASES: N.M.R. SPECTRA OF 5,5'-BISISOXAZOLE, Bucci, P., Veracini, C.A., J Chem Phys v56 n3 p1290 (1 Feb '72)

141. CONFORMATIONAL STUDIES OF MOLECULES PARTIALLY ORIENTED IN THE NEMATIC PHASES: FURAN-2,5-DIALDEHYDE, Bucci, P., Veracini, C.A., Longeri, M., Chem Phys Lett (Holland) v15 n3 p396 (16 Aug '72)

142. INFRARED SPECTROSCOPIC MEASUREMENTS ON THE CRYSTAL-NEMATIC TRANSITION, Bulkin, B.J., Grunbaum, D., p303: "Liquid Crystals and Ordered Fluids," Plenum Press, 1970.

143. VIBRATIONAL SPECTRA OF LIQUID CRYSTALS. II. THE RAMAN SPECTRUM OF P-AZOXYANISOLE IN CRYSTAL, NEMATIC AND ISOTROPIC PHASES, 10-100 cm^{-1} REGION. Bulkin, B.J., Prochaska, F.T.,

J Chem Phys v54 n2 p635 (15 Jan '71)

144. VIBRATIONAL SPECTRA OF LIQUID CRYSTALS. IV. INFRARED AND RAMAN SPECTRA OF PHOSPHOLIPID-WATER MIXTURES, Bulkin, B.J., Krishnamachari, N., J Am Chem Soc v94 n4 p1109 (23 Feb '72)

145. VIBRATIONAL SPECTRA OF LIQUID CRYSTALS. VI. RELATIVE INTENSITIES OF RAMAN SPECTRA OF CHOLESTERIC AND NEMATIC SOLUTIONS, Bulkin, B.J., Lephards, J.O., Krishnan, K., Mol Cryst & Liq Cryst v19 n3-4 p295 (Jan '73)

146. HOLOGRAPHY RECORDING WITH AN ELECTROOPTIC LIQUID CRYSTAL CELL, Burchardt, C.B., Schadt, M., Helfrich, W., Appl Opt v10 n9 p2196 (Sep '71)

147. AN N.M.R. STUDY OF THE BARRIER TO INTERNAL ROTATION, THE MOLECULAR STRUCTURE AND THE INDIRECT SPIN-SPIN COUPLING CONSTANTS OF O-XYLENE PARTIALLY ORIENTED IN A NEMATIC PHASE, Burnell, E.E., Diehl, P., Mol Phys (GB) v24 n3 p489 (Sep '72)

148. DRIVER CIRCUITRY FOR LIQUID CRYSTAL DISPLAYS, Buser, M.S., Eurocon 71 Digest, Lausanne, Switzerland, 18-22 Oct '71.

149. LIQUID CRYSTAL DISPLAY DEVICE, Bush, R.F., Seiden, P.E., IBM Tech Discl Bull v14 n1 p223 (Jne '71)

150. ANTIBIOTIC POLYMERS, Butcher, B.T., Stanfield, M.K., Stewart, G.T., Wagle, S.S., Zemelman, R., p95: Abstr 3rd Int'l Liq Cryst Conf., Berlin, W. Germany (Aug '70)

151. DISPLAY TECHNOLOGY WITH LIQUID CRYSTALS, Byatt, D.W.G., Mess & Pruef (Germany) v8 n2 p85 (Feb '72)

152. NUCLEAR QUADRUPOLAR RELAXATION OF N^{14} IN A NEMATIC LIQUID, Cabane, B., Clark, G., p57: Abstr 3rd Int'l Liq Cryst Conf., Berlin, W. Germany (Aug '70)

153. EFFECT OF THE DOUBLE SCATTERING OF X-RAYS ON THE INTENSITY DIFFRACTED BY A NEMATIC PHASE OF P-AZOXYPHENETOLE, Cabos, C., Malet, G., Delord, P., Comptes Rendus (Ser. B) v273 n4 p199 (26 Jly '71)

154. DIFFUSION DES RAYONS X PAR UNE PHASE NEMATIQUE ORIENTEE DE PARA-AZOXYPHENETOLE. DISTRIBUTIONS CYLINDRIQUES D'ATOMES, Cabos, C., Delord, P., Comptes Rendus (Ser. B) v275 n11 p387 (11 Sep '72)

155. CALCULATIONS OF THE NUCLEAR RELAXATION TIME T_1 IN A TWO-DIMENSIONAL QUASI-NEMATIC SYSTEM, Caille, A., Solid State Commun v10 n6 p571 (15 Mch '72)

156. N.M.R. STUDIES OF SELECTED CHOLESTERIC COMPOUNDS, Cameron, L.M., Callender, R.E., p121: Abstr 3rd Int'l Liq Cryst Conf., Berlin, W. Germany (Aug '70)

157. N.M.R. STUDIES OF SELECTED CHOLESTERIC COMPOUNDS, Cameron, L.M., Callender, R.E., Kramer, A.J., Mol Cryst & Liq Cryst v16 n1-2 p75 (Feb '72)

158. ULTRASONIC ABSORPTION MEASUREMENTS IN LIQUID CRYSTALS, Candau, S., Martinoty, P., p123: Abstr 3rd Int'l Liq Cryst Conf., Berlin, W. Germany (Aug '70)

159. ANOMALOUS ALIGNMENT IN THE SMECTIC PHASE OF A LIQUID CRYSTAL OWING TO AN ELECTRIC FIELD, Carr, E.F., Phys Rev Lett v24 n15 p807 (13 Apr '70)

160. ORDERING IN THE SMECTIC PHASE OWING TO ELECTRIC FIELDS, Carr, E.F., p99: Abstr 3rd Int'l Liq Cryst Conf., Berlin, W. Germany (Aug '70)

161. EFFECTS OF ELECTRIC FIELDS ON MIXTURES OF NEMATIC AND CHOLESTERIC LIQUID CRYSTALS, Carr, E.F., Parker, J.H., McLemore, D.P., p201: "Liquid Crystals and Ordered Fluids," Plenum Press, 1970.

162. ORDERING IN THE SMECTIC PHASE OWING TO ELECTRIC FIELDS, Carr, E.F., Mol Cryst & Liq Cryst v13 n1 p27 (Apr '71)

163. ANGULAR DEPENDENT STUDIES IN LIQUID CRYSTALS INVOLVING PERMEABILITY AND DIELECTRIC ANISOTROPIES, Carr, E.F., Murty, C.R.K., Mol Cryst & Liq Cryst v18 n3&4 p369 (Oct '72)

164. ANGULAR DEPENDENT STUDIES IN LIQUID CRYSTALS INVOLVING PERMEABILITY AND DIELECTRIC ANISOTROPIES (errata for above) Mol Cryst & Liq Cryst v20 n2 p191 (Mch '73)

165. LIQUID CRYSTAL DIFFRACTION GRATING, Carroll, T.O., J Appl Phys v43 n3 p767 (Mch '72)

166. DEPENDENCE OF CONDUCTION-INDUCED ALIGNMENT OF NEMATIC LIQUID CRYSTALS UPON VOLTAGE ABOVE THRESHOLD, Carroll, T.O., J Appl Phys v43 n4 p1342 (Apr '72)

167. NOW THAT THE HEAT IS OFF, LIQUID CRYSTALS CAN SHOW THEIR COLORS, Castellano, J.A., Electronics v43 n14 p64 (6 Jly '70)

168. LIQUID CRYSTALS V. ROOM TEMPERATURE NEMATIC MATERIALS DERIVED FROM P-ALKYLCARBONATO-P'-ALKOXYPHENYL BENZOATES, Castellano, J.A., McCaffrey, M.T., Goldmacher, J.E., p87: Abstr 3rd Int'l Liq Cryst Conf., Berlin, W. Germany (Aug '70)

169. LIQUID CRYSTALS IV. ELECTRO-OPTIC EFFECTS IN P-ALKOXYBENZYLIDENE-P'-AMINOALKYLPHENONES AND RELATED COMPOUNDS, Castellano, J.A., McCaffrey, M.T., p293: "Liquid Crystals and Ordered Fluids," Plenum Press, 1970.

170. LIQUID CRYSTALS V.: NEMATIC MATERIALS DERIVED FROM P-ALKYLCARBONATO-P'-ALKOXYPHENYL BENZOATES, Castellano, J.A., McCaffrey, M.T., Goldmacher, J.E., Mol Cryst & Liq Cryst v12 n4 p345 (Mch '71)

171. ELECTRO-OPTIC PROPERTIES OF LIQUID CRYSTALS, Castellano, J.A., p153: Proc 21st Electronic Components Conf., Washington, D.C., 10-12 May '71.

172. LIQUID CRYSTAL DISPLAYS, Castellano, J.A., p95: Proceedings "Novel Audio-visual imaging systems," SPSE Seminar, N.Y.C. 23/24 Sep.'71.

173. MESOMORPHIC MATERIALS FOR ELECTROOPTICAL APPLICATION, Castellano, J.A., Ferroelectrics (GB) v3 n1 p29 (Nov '71)

174. OPTICAL CONTRAST ENHANCEMENT IN LIQUID CRYSTAL DEVICES BY SPATIAL FILTERING, Caulfield, H.J., Soref, R.A., Appl Phys Lett v18 n1 p5 (1 Jan '71)

175. A LOW TEMPERATURE LIQUID CRYSTAL EXHIBITING SMECTIC MORPHOLOGY, Champa, R.A., Mol Cryst & Liq Cryst v16 n1&2 p175 (Feb '72)

176. LOW MELTING LIQUID CRYSTALLINE HETEROCYCLIC ANILS, Champa, R.A., Mol Cryst & Liq Cryst v19 n3&4 p233 (Jan '73)

177. INTERFEROMETRIC STUDY OF LIQUID CRYSTALLINE SURFACES, Chandrasekhar, S., Madhusudana, N.V., Acta Cryst (International) v26A Pt.1 p153 (Jan '70)

178. THEORY OF MELTING OF MOLECULAR CRYSTALS: THE LIQUID CRYSTALLINE PHASE, Chandrasekhar, S., Shashidhar, R., Tara, N., Mol Cryst & Liq Cryst v10 n4 p337 (Jly '70)

179. OPTICAL ROTATORY DISPERSION OF CHOLESTERIC LIQUID CRYSTALS, Chandrasekhar, S., Prasad, J.S., p38: Abstr 3rd Int'l Liq Cryst Conf., Berlin, W. Germany (Aug '70)

180. MOLECULAR THEORY OF NEMATIC LIQUID CRYSTALS, Chandrasekhar, S., Madhusudana, N.V., p42: Abstr 3rd Int'l Liq Cryst Conf., Berlin, W. Germany (Aug '70)

181. THEORY OF MELTING OF MOLECULAR CRYSTALS, II.: SOLID-SOLID AND MELTING TRANSITIONS, Chandrasekhar, S., Shashidhar, R., Tara, N., Mol Cryst & Liq Cryst v12 n3 p245 (Feb '71)

182. MOLECULAR STATISTICAL THEORY OF NEMATIC LIQUID CRYSTALS, Chandrasekhar, S., Madhusudana, N.V., Acta Crystallogr A (Denmark) v27a Pt.4 p303 (1 Jly '71)

183. THEORY OF ROTATORY DISPERSION OF CHOLESTERIC LIQUID CRYSTALS, Chandrasekhar, S., Prasad, J.S., Mol Cryst & Liq Cryst v14 n1&2 p115 (Aug '71)

184. THEORY OF MELTING OF MOLECULAR CRYSTALS: THE NEMATIC LIQUID CRYSTALLINE PHASE, Chandrasekhar, S., Shashidhar, R., Indian J Pure Appl Phys v9 n11 p975 (Nov '71)

185. RELATIONSHIP BETWEEN ELASTICITY AND ORIENTATIONAL ORDER IN NEMATIC LIQUID CRYSTALS, Chandrasekhar, S., Madhusudana, N.V., Shubha, K., Faraday Symposium of Liquid Crystals, London, Faraday Society (1971)

186. MOLECULAR STATISTICAL THEORY OF NEMATIC LIQUID CRYSTALS. II: Relation between elasticity and orientational order, Chandrasekhar, S., Madhusudana, N.V., Shubha, K., Acta Crystallogr A (Denmark) vA28 Pt.1 p28 (1 Jan '72)

187. THEORY OF MELTING OF MOLECULAR CRYSTALS. III: THE EFFECT OF SHORT RANGE ORIENTATIONAL ORDER ON LIQUID CRYSTALLINE TRANSITIONS, Chandrasekhar, S., Shashidhar, R., Mol Cryst & Liq Cryst v16 n1&2 p21 (Feb '72)

188. MOLECULAR THEORY OF NEMATIC LIQUID CRYSTALS, Chandrasekhar, S., Madhusudana, N.V., Mol Cryst & Liq Crist v17 n1 p37 (May '72)

189. CHARACTERIZATION OF LIQUID CRYSTALS: THE LOW FREQUENCY RAMAN SCATTERING OF CHOLESTERYL PROPIONATE NONANOATE AND PALMITATE, Chang, R., Mol Cryst & Liq Cryst v12 n2 p105 (Jan '71)

190. APPLICATION OF POLARIMETRY AND INTERFEROMETRY TO LIQUID CRYSTAL FILM RESEARCH, Chang, R., Mater Res Bull v7 n4 p267 (Apr '72)

191. "SECONDARY" DYNAMIC SCATTERING IN NEGATIVE NEMATIC LIQUID CRYSTALS FILMS, Chang, R., J Appl Phys v44 n4 p1885 (Apr '73)

192. OPTICAL WAVEGUIDE MODULATION USING NEMATIC LIQUID CRYSTAL, Channin, D.J., Appl Phys Lett v22 n8 p365 (15 Apr '73)

193. LIQUID CRYSTALLINE PROPERTIES OF PHOSPHOLIPIDS AND BIOLOGICAL MEMBRANES, Chapman, D., Faraday Symposium on Liquid Crystals, London, Faraday Society (1971)

194. N.M.R. STUDY OF MOLECULAR MOTIONS IN THE MESOPHASES OF POTASSIUM LAURATE - D_2O SYSTEM, Charvolin, J., Rigny, P., p68: Abstr 3rd Int'l Liq Cryst Conf., Berlin, (Aug '70)

195. N.M.R. STUDY OF MOLECULAR MOTIONS IN THE MESOPHASES OF POTASSIUM LAURATE - D_2O SYSTEM, Charvolin, J., Rigny, P., Mol Cryst & Liq Cryst v15 n3 p211 (Dec '71)

196. PROTON RELAXATION STUDY OF PARAFFIN CHAIN MOTIONS IN A LYOTROPIC LIQUID CRYSTAL, Charvolin, J., Rigny, P., J Chem Phys v58 n9 p3999 (1 May '73)

197. PRETRANSITIONAL ANOMALY IN THE BEND ELASTIC CONSTANT FOR A NEMATIC TO SMECTIC 'A' TRANSITION, Cheung, L., Meyer, R.B., Phys Lett A v43A n3 p261 (12 Mch '73)

198. INFLUENCE OF FLOW ON THE ORIENTATION OF NEMATIC P-AZOXYANISOLE, Chistyakov, I.G., Sukharev, S.K., Kristallografiya v16 n5 p1052 (1971)

199. STRUCTURE OF THE NEMATIC MESOPHASE OF P-ANISALAMINOAZOBENZENE, Chistyakov, I.G., Sukharev, S.K., Soviet Phys - Crystallography v17 n6 p1108 (May-Jne '73)

200. EFFECTS OF MAGNETIC FIELD ON THE OPTICAL TRANSMISSION IN CHOLESTERIC LIQUID CRYSTALS, Chou, S.C., Cheung, L., Meyer, R.B., Solid State Commun v11 n8 p977 (15 Oct '72)

201. GAS-LIQUID CHROMATOGRAPHY DETERMINATION OF THE DEGREE OF ORDER IN A NEMATIC MESOPHASE, Chow, L.C., Martire, D.E., p131: Abstr 3rd Int'l Liq Cryst Conf., Berlin (Aug '70)

202. THERMODYNAMICS OF SOLUTIONS WITH LIQUID CRYSTAL SOLVENTS. III: MOLECULAR INTERPRETATION OF SOLUBILITY IN NEMATOGENIC SOLVENTS, Chow, L.C., Martire, D.E., J Phys Chem v75 n13 p2005 (24 Jne '71)

203. THERMODYNAMICS OF SOLUTIONS WITH LIQUID CRYSTAL SOLVENTS. IV: 'GLC' DETERMINATION OF THE DEGREE OF ORDER IN A NEMATIC MESOPHASE, Chow, L.C., Martire, D.E., Mol Cryst & Liq Cryst v14 n3&4 p293 (Sep '71)

204. PROTON MAGNETIC RESONANCE LINE IN LIQUID CRYSTALLINE ANISYLIDENE p-AMINOPHENYLACETATE, (APAPA), Christman, S.B., Moses, H.A., Cohen, P.S., Smith, E.P., Fink, L.J., J Chem Phys v53 n1 p456 (1 Jly '70)

205. COHERENCE LENGTH IN THE ISOTROPIC PHASE OF A NEMATIC LIQUID CRYSTAL: MBBA, Chu, B., Bak, C.S., Lin, F.L., J Chem Phys v56 n7 p3717 (1 Apr '72)

206. COHERENCE LENGTH IN THE ISOTROPIC PHASE OF A ROOM-TEMPERATURE NEMATIC CRYSTAL, Chu, B., Bak, C.S., Lin, F.L., Phys Rev Lett v28 n17 p1111 (24 Apr '72)

207. ORIENTATION OSCILLATIONS OF THE DOMAIN STRUCTURE OF LIQUID CRYSTALS. MECHANISM OF FORMATION OF HEXAGONAL DOMAIN STRUCTURES IN CONSTANT ELECTRIC FIELDS, Chuvyrov, A.N., Trofimov, A.N., Soviet Phys - Crystallography v17 n6 p1056 (May-Jne '73)

208. SHEAR EFFECTS ON CHOLESTERIC LIQUID CRYSTALS, Ciliberti, D.F., Dixon, G.D., Scala, L.C., Mol Cryst & Liq CRyst v20 n1 p27 (Feb '73)

209. BINARY MIXTURES OF 'MBBA' AND RODLIKE MOLECULES, Cladis, P.E., Rault, J., Burger, J.-P., p97: Abstr 3rd Int'l Liq Cryst Conf., Berlin, W. Germany (Aug '70)

210. BINARY MIXTURES OF ROD-LIKE MOLECULES WITH P-METHOXYBENZYLIDENE-P'-N-BUTYLANAILINE, Cladis, P.E., Rault, J., Burger, J.-P., Mol Cryst & Liq Cryst v13 n1 p1 (Apr '71)

211. ON A NEW METHOD FOR DECORATION OF THE MESOMORPHIC PHASE OF 'MBBA', Cladis, P.E., Kleman, M., Pieranski, P., Comptes Rendus (Ser. B) v273 n6 p275 (9 Sep '71)

212. THE CHOLESTERIC DOMAIN TEXTURE, Cladis, P.E., Kleman, M., Mol Cryst & Liq Cryst v16 n1&2 p1 (Feb '72)

213. NEW METHOD FOR MEASURING THE TWIST ELASTIC CONSTANT AND THE SHEAR VISCOSITY FOR NEMATICS, Cladis, P.E., Phys Rev Lett v28 n25 p1629 (19 Jne '72)

214. NON-SINGULAR DISCLINATIONS OF STRENGTH S=+1 IN NEMATICS, Cladis, P.E., Kleman, M., J Physique (France) v33 n5-6 p591 (May-Jne '72)

215. LIGHT SCATTERING BY DEFORMATION OF THE PLANE TEXTURE OF SMECTIC AND CHOLESTERIC LIQUID CRYSTALS, Clark, N.A., Pershan, P.S., Phys Rev Lett v30 n1 p3 (1 Jan '73)

216. STRAIN-INDUCED INSTABILITY OF MONODOMAIN SMECTIC 'A' AND CHOLESTERIC LIQUID CRYSTALS, Clark, N.A., Meyer, R.B., Appl Phys Lett v22 n10 p493 (15 May '73)

217. LIQUID CRYSTALS FOR NUMERIC ALPHA-NUMERIC DISPLAYS, Clary, R.M., IEEE Intercon '73, Session 43/1, New York, 26-30 Mch '73.

218. OSCILLATION MODES OF NEMATIC LIQUID CRYSTALS, Conners, G.H., Paxton, K.B., J Appl Phys v43 n7 p2959 (Jly '72)

219. ELECTRO-OPTIC EFFECTS IN NEMATIC LIQUID CRYSTALS, Cook, B.D., Nemec, J., p312: 21st Annual Southwestern IEEE Conf., Houston, Texas, 28-30 Apr '71.

220. MAPPING ULTRASONIC FIELDS WITH CHOLESTERIC LIQUID CRYSTALS, Cook, B.D., Werchan, R.E., Ultrasonics (GB) p101 (Apr '71)

221. ON THE TRANSIENT SCATTERING OF LIGHT BY PULSED LIQUID CRYSTAL CELLS, Cosentino, L.S., IEEE Trans Electron Devices vED-18 n12 p1192 (Dec '71)

222. LIQUID CRYSTAL PROPERTIES OF METHYL SUBSTITUTED STILBENES, Cox, R.J., Mol Cryst & Liq Cryst v19 n2 p111 (Dec '72)

223. LIQUID CRYSTAL DISPLAYS, Creagh, L.T., Kmetz, A.R., Reynolds, R.A., p630: 1971 IEEE Int'l Conv Digest, New York, 22-25 Mch '71.

224. PERFORMANCE CHARACTERISTICS OF NEMATIC LIQUID CRYSTAL DISPLAY DEVICES, Creagh, L.T., Kmetz, A.R., Reynolds, R.A., IEEE Trans Electron Devices vED-18 n9 p672 (Sep '71)

225. PERFORMANCE ADVANTAGES OF LIQUID CRYSTAL DISPLAYS WITH SURFACTANT PRODUCED HOMOGENEOUS ALIGNMENT, Creagh, L.T., Kmetz, A.R., p90: 1972 SID Symp, San Francisco, 6-8 Jne '72.

226. PROPAGATING PLANE DISINCLINATION SURFACES IN NEMATIC LIQUID CRYSTALS, Currie, P.K., Mol Cryst & Liq Cryst v19 n3-4 p249 (Jan '73)

227. IMPLICATIONS OF PARODI'S RELATION FOR WAVES IN NEMATICS, Currie, P.K., Solid State Communic v12 n1 p31 (1 Jan '73)

228. ON FORM BIREFRINGENCE OF SOME SMECTIC LIQUID CRYSTALS, Cvikl, B., Moroi, D., Franklin, W., Mol Cryst & Liq Cryst v12 n3 p267 (Feb '71)

229. A LIQUID CRYSTAL DEVICE FOR RELATIVE POROSITY MEASUREMENTS, Dalby, E.D., Rev Sci Instrum v42 n10 p1540 (Oct '71)

230. MESOMORPHIC BEHAVIOR OF CHOLESTERYL ESTERS II: TRANS-P-N-ALKOXYCINNAMATES OF CHOLESTEROL, Dave, J.S., Vora, R.A., p94: Abstr 3rd Int'l Liq Cryst Conf., Berlin (Aug '70)

231. MESOMORPHIC BEHAVIOR OF SCHIFF'S BASE COMPOUNDS - I. N-N'-DI(4-N-ALKOXY-1-NAPHTHYLIDENE)-BENZIDINES, Dave, J.S., Prajapati, A.P., Vora, R.A., p137A: Abstr 3rd Int'l Liq Cryst Conf., Berlin, W. Germany (Aug '70)

232. MESOMORPHIC BEHAVIOR OF THE CHOLESTERYL ESTERS - I: P-N-ALKOXYBENZOATES OF CHOLESTEROL, Dave, J.S., Vora, R.A., p477: "Liquid Crystals and Ordered Fluids," Plenum Press, 1970.

233. MESOMORPHIC BEHAVIOR OF SCHIFF'S BASE COMPOUNDS - I. N,N'-DI(4-N-ALKOXY-1-NAPHTHYLIDENE)-BENZIDINES, Dave, J.S., Kurian, G., Prajapati, A.P., Vora, R.A., Mol Cryst & Liq Cryst v14 n3-4 p307 (Sep '71)

234. MESOMORPHIC BEHAVIOR OF CHOLESTERYL ESTERS - II. TRANS-P-N-ALKOXY CINNAMATES OF CHOLESTEROL, Dave, J.S., Vora, R.A., Mol Cryst & Liq Cryst v14 n3-4 p319 (Sep '71)

235. SOME SOLUBILITY CHARACTERISTICS OF CHOLESTERYL ESTERS, Davis, G.J., Porter, R.S., Mol Cryst & Liq Cryst v6 n3-4 p377 (Feb '70)

236. THERMAL TRANSITIONS OF CHOLESTERYL ESTERS OF C_{18} ALIPHATIC ACIDS, Davis, G.J., Porter, R.S., Steiner, J.W., Small, D.M., Mol Cryst & Liq Cryst v10 n3 p331 (Jne '70)

237. AN INTERCOMPARISON OF TEMPERATURES AND HEATS OF TRANSITION FOR ESTERS OF CHOLESTEROL, Davis, G.J., Porter, R.S., Barrall, E.M.II., Mol Cryst & Liq Cryst v11 n4 p319 (Nov '70)

238. OPTICAL SCATTERING IN A NEMATIC LIQUID CRYSTAL INDUCED BY ACOUSTIC SURFACE WAVES, Davis, L.E., Chambers, J., Electron Lett v7 n11 p287 (3 Jne '71)

239. LIQUID CRYSTALS IN ULTRASONIC, ELECTRIC, AND OPTICAL FIELDS, Davis, L.E., IEEE Trans Sonics & Ultrasonics vSU-19 n3 p390 (Jly '72)

240. DOMAIN STRUCTURES OF A NEMATIC LIQUID CRYSTAL IN A MAGNETIC FIELD, De Gennes, P.G., Solid State Commun v8 n3 p213 (1 Feb '70)

241. SOME PROBLEMS OF LIQUID CRYSTAL MICROSTRUCTURE, De Gennes, P.G., p18: Abstr 3rd Int'l Liq Cryst Conf., Berlin, W. Germany (Aug '70)

242. ELECTROHYDRODYNAMIC EFFECTS IN NEMATICS. II., De Gennes, P.G., Comments, Solid State Phys (GB) v3 n5 p148 (Dec'70 - Jan '71)

243. SHORT RANGE ORDER EFFECTS IN THE ISOTROPIC PHASE OF NEMATICS & CHOLESTERICS, De Gennes, P.G., Mol Cryst & Liq Cryst v12 n3 p193 (Feb '71)

244. WALL MOTION IN A NEMATIC SLAB UNDER ROTATING FIELDS, De Gennes, P.G., J Physique (France) v32 n10 p789 (Oct '71)

245. POSSIBLE EXPERIMENTS ON TWO-DIMENSIONAL NEMATICS, De Gennes, P.G., Faraday Symposium on Liquid Crystals, London, Faraday Society (1971).

246. X-RAY SCATTERING BY NEMATIC FLUIDS, De Gennes, P.-G., Comptes Rendus (Ser. B) v274 n2 p142 (10 Jan '72)

247. TENTATIVE MODEL FOR THE SMECTIC 'B'-PHASE, De Gennes, P.G., Phys Lett (Holland) v38A n4 p219 (14 Feb '72)

248. ON THE SMECTIC 'A' TO SMECTIC 'B' REVERSIBLE TRANSITION, De Gennes, P.G., Comptes Rendus (Ser. B) v274 n11 p758 (13 Mch '72)

249. AN ANALOGY BETWEEN SUPERCONDUCTORS AND SMECTICS, De Gennes, P.- G., Solid State Commun v10 n9 p753 (1 May '72)

250. SUR LA STRUCTURE DU COEUR DES CONIQUES FOCALES DANS LES SMECTIQUES 'A', De Gennes, P.-G., Comptes Rendus (Ser. B) v275 n15 p549 (9 Oct '72)

251. MONOCHROMATIC EFFECT IN NEMATICS, De Gennes, P.-G., Phys Lett (Holland) v41A n5 p479 (23 Oct '72)

252. PROPRIETES ACOUSTIQUES DES CRISTAUX LIQUIDES, De Gennes, P.-G., J Physique (France) colloque C6, suppl au no. 11-12 v33 p80 (Nov-Dec '72)

253. ELECTROHYDRODYNAMIC INSTABILITIES IN NEMATIC LIQUID CRYSTALS, De Jeu, W.H., Gerritsma, C.J., Van Boxtel, A.M., Phys Lett A (Holland) v34a n4 p203 (8 Mch '71)

254. INSTABILITIES OF NEMATIC LIQUID CRYSTALS IN PULSATING ELECTRIC FIELDS, De Jeu, W.H., Phys Lett A (Holland) v37a n5 p365 (20 Dec '71)

255. ON THE RELATION BETWEEN MOLECULAR STRUCTURE AND LIQUID CRYSTALLINE BEHAVIOR, De Jeu, W.H., Van der Veen, J., Philips Res Rep (Holland) v27 n2 p172 (Apr '72)

256. ELECTROHYDRODYNAMIC INSTABILITIES IN SOME NEMATIC AZOXY COMPOUNDS WITH DIELECTRIC ANISOTROPIES OF DIFFERENT SIGN, De Jeu, W.H., Gerritsma, C.J., J Chem Phys v56 n10 p4752 (15 May '72)

257. RELAXATION OF THE DIELECTRIC CONSTANT AND ELECTROHYDRODYNAMIC INSTABILITIES IN A LIQUID CRYSTAL, De Jeu, W.H., Gerritsma, C.J., Van Zanten, P., Goossens, W.J.A., Phys Lett A v39a n5 p355 (5 Jne '72)

258. INSTABILITIES IN ELECTRIC FIELDS OF NEMATIC LIQUID CRYATALS WITH POSITIVE DIELECTRIC ANISOTROPY: DOMAINS, LOOP DOMAINS AND REORIENTATION, De Jeu, W.H., Gerritsma, C.J., Lathouwers, Th.W., Chem Phys Lett v14 n4 p503 (15 Jne '72)

258A. INSTABILITIES IN ELECTRIC FIELDS OF A NEMATIC LIQUID CRYSTAL WITH LARGE NEGATIVE DIELECTRIC ANISOTROPY, De Jeu, W.H., Van der Veen, J., Phys Lett A v44a n4 p277 (18 Jne '73)

259. BUCKLING INSTABILITY OF THE LAYERS IN A SMECTIC-A LIQUID CRYSTAL, Delaye, M., Ribotta, R., Durand, G., Phys Lett A v44a n2 p139 (21 May '73)

260. EFFECT OF PRESSURE ON THE MESOMORPHIC IN P-AZOXYANISOLE (PAA), Deloche, B., Cabane, B., Jerome, D., p134: Abstr 3rd Int'l Liq Cryst Conf., Berlin, W. Germany (Aug '70)

261. EFFECT OF PRESSURE ON THE MESOMORPHIC TRANSITION IN PARA-AZOXYANISOLE (PAA), Deloche, B., Cabane, B., Jerome, D., Mol Cryst & Liq Cryst v15 n3 p197 (Dec '71)

262. COUPLING OF HYDROGEN BONDING TO ORIENTATIONAL FLUCTUATION MODES IN THE LIQUID CRYSTAL 'PHBA', Deloche, B., Cabane, B., Mol Cryst & Liq Cryst v19 n1 p25 (Nov '72)

263. DIFFRACTION EFFECT ON A CYLINDRICAL DISTRIBUTION OF ATOMS; APPLICATION TO X-RAY SCATTERING BY A NEMATIC PHASE OF PARA-AZOXYANISOLE, Delord, P., Malet, G., Comptes Rendus (Ser. B) v270 n17 p1107 (27 Apr '70)

264. INVESTIGATION OF A SMECTIC TETRAMORPHOUS SUBSTANCE, Demus, D., Diele, S., Klapperstueck, M., Link, V., Zaschke, H., p76: Abstr 3rd Int'l Liq Cryst Conf., Berlin (Aug '70)

265. INVESTIGATION OF A SMECTIC TETRAMORPHOUS SUBSTANCE, Demus, D., Diele, S., Klapperstueck, M., Link, V., Zaschke, H., Mol Cryst & Liq Cryst v15 n2 p161 (Nov '71)

266. A NEW VARIANT OF POLYMORPHISM IN THE LIQUID CRYSTAL STATE: A SUBSTANCE HAVING SMECTIC-B AND NEMATIC MODIFICATIONS, Demus, D., Klapperstueck, M., Rurainski, R., Marzotko, D., Z Phys Chem Leipzig v246 n5/6 p385 (1971)

267. ANOMALOUS DENSITIES IN LIQUID CRYSTALLINE BIS-(4-N-ALKOXYBENZAL)-1,4-PHENYLENDIAMINES, Demus, D., Rurianski, R., Mol Cryst & Liq Cryst v16 n1-2 p171 (Feb '72)

268. CRYSTALS WITH AN INCOMPLETELY ORDERED STRUCTURE, Denoyer, F., Lambert, M., Levelut, A.M., Guinier, O., Kristallografiya (USSR) v16 n6 p1140 (1971)

269. THE STATISTICAL MECHANICS FOR LONG SEMI-FLEXIBLE MOLECULES: A THEORY OF THE NEMATIC MESO-PHASE, De Rocca, A.G., Wulf, A., p45: Abstr 3rd Int'l Liq Cryst Conf., Berlin (Aug '70)

270. DIELECTRIC PROPERTIES OF A NEMATIC LIQUID CRYSTAL WITH ELLIPSOIDAL MOLECULES, Derzhanski, A.I., Petrov, A.G., C.R. Acad Bulg Sci (Bulgaria) v24 n5 p569 (1971)

271. A POSSIBLE RELATIONSHIP BETWEEN THE DIELECTRIC PERMEABILITY AND THE PIEZOELECTRIC PROPERTIES OF NEMATIC LIQUID CRYSTALS, Derzhanski, A., Petrov, A.G., Phys Lett A v34a n7 p427 (19Apr71)

272. INVERSE CURRENTS AND CONTACT BEHAVIOR OF SOME NEMATIC LIQUID CRYSTALS, Derzhanski, A., Petrov, A.G., Phys Lett A v36a n4 p307 (13 Sep '71)

273. A MOLECULAR-STATISTICAL APPROACH TO THE PIEZOELECTRIC PROPERTIES OF NEMATIC LIQUID CRYSTALS, Derzhanski, A., Petrov, A.G., Phys Lett A v36a n6 p483 (11 Oct '71)

274. 'AC' CONDUCTIVITY OF LIQUID CRYSTAL IN ELECTRIC AND MAGNETIC FIELDS, Derzhanski, A., Grigorov, L., Tenchov, B., C.R. Acad Bulg (Bulgaria) v25 n2 p163 (1972)

275. A MOLECULAR-STATISTICAL APPROACH TO THE PIEZOELECTRIC PROPERTIES OF NEMATIC LIQUID CRYSTALS - ONE DIMENSIONAL MODEL, Derzhanski, A., Petrov, A.G., C.R. Acad Bulg (Bulgaria) v25 n2 p167 (1972)

276. A MOESSBAUER COMPARISON OF THE RECOILFREE FRACTION OF A SUPERCOOLED SMECTIC LIQUID CRYSTAL WITH ITS SOLID STATE, Detjen, R.E., Uhrich, D.L., Sheley, C.F., Phys Lett A v42a n7 p522 (29 Jan '73)

277. DEFORMATION OF NEMATIC LIQUID CRYSTALS IN AN ELECTRIC FIELD, Deuling, H.J., Mol Cryst & Liq Cryst v19 n2 p123 (Dec '72)

278. EVIDENCE FOR THE EXISTENCE OF MORE THAN ONE TYPE OF NEMATIC PHASE, De Vries, A., Mol Cryst & Liq Cryst v10 n1-2 p31 (May '70)

279. X-RAY STUDIES OF SOME SMECTIC, NEMATIC AND ISOTROPIC PHASES, De Vries, A., p71: Abstr 3rd Int'l Liq Cryst Conf., Berlin, W. GErmany (Aug '70)

280. X-RAY PHOTOGRAPHIC STUDIES OF LIQUID CRYSTALS. II. APPARENT MOLECULAR LENGTH AND THICKNESS IN THREE PHASES OF ETHYL-P-ETHOXYBENZAL-P-AMINOBENZOATE, De Vries, A., Mol Cryst & Liq Cryst v11 n4 p361 (Nov '70)

281. X-RAY PHOTOGRAPHIC STUDIES OF LIQUID CRYSTALS. III. STRUCTURE DETERMINATION OF THE SMECTIC PHASE OF 4-BUTYLOXYBENZAL-4'-ETHYLANILINE, De Vries, A., Fishel, D.L., Mol Cryst & Liq Cryst v16 n4 p311 (Apr '72)

282. ON THE CALCULATION OF THE MOLECULAR CYLINDRICAL DISTRIBUTION FUNCTION AND THE ORDER PARAMETER FROM X-RAY DIFFRACTION DATA OF LIQUID CRYSTALS, De Vries, A., J Chem Phys v56 n9 p4489 (1 May '72)

283. SOME COMMENTS ON CYLINDRICAL DISTRIBUTION FUNCTIONS, WITH SPECIAL REFERENCE TO X-RAY STUDIES ON LIQUID CRYSTALS, De Vries, A., Acta Crystallogr A (Denmark) v28A Pt.6 p659 (1 Nov '72)

284. X-RAY PHOTOGRAPHIC STUDIES OF LIQUID CRYSTALS. IV. THE ISOTROPIC, NEMATIC AND SMECTIC 'A' PHASES OF SOME 4-ALKOXYBENZAL-4'-ETHYLANILINES, De Vries, A., Mol Cryst & Liq Cryst v20 n2 p119 (Mch '73)

285. CONTROL CIRCUIT FOR LIQUID CRYSTAL CELLS, Dickson, L.D., Sensenbaugh, J.D., IBM Tech Discl Bull v13 n11 p3517 (Apr '71)

286. X-RAY DIFFRACTION AND POLYMORPHISM OF SMECTIC LIQUID CRYSTALS, Diele, S., Sackman, H., Brand, P., p75: Abstr 3rd Int'l Liq Cryst Conf., Berlin, W. Germany (Aug '70)

287. X-RAY DIFFRACTION AND POLYMORPHISM OF SMECTIC LIQUID CRYSTALS. I. A-, B-, AND C-MODIFICATIONS, Diele, S., Brand, P., Sackman, H., Mol Cryst & Liq Cryst v16 n1-2 p105 (Feb '72)

288. X-RAY DIFFRACTION AND POLYMORPHISM OF SMECTIC LIQUID CRYSTALS. II. D- and E-MODIFICATIONS, Diele, S., Brand, P., Sackman, H., Mol Cryst & Liq Cryst v17 n2 p163 (Jne '72)

289. MESOMORPHIC BEHAVIOR OF COMPOUNDS OF LOW THERMAL PHASE STABILITY, Dietrich, H.J. Steiger, E.L., Mol Cryst & Liq Cryst v16 n3 p263 (Mch '72)

290. ANISOTROPY OF THE DIELECTRIC CONSTANT AND CONDUCTIVITY OF P-METHOXY-BENZYLIDENE P-N-BUTYLANILINE IN THE NEMATIC PHASE, Diguet, D., Rondelez, F. Durand, G., Comptes Rendus (Ser. B) v271 n18 p954 (2 Nov '70)

291. STUDY OF MOLECULAR MOTIONS IN LIQUID CRYSTALLINE MBBA BY COLD NEUTRON SCATTERING, Dimic, V., Barbic, L., Blinc, R., Phys Status Solidi (Germany) v54 n1 p121 (1 Nov '72)

292. THE LOW FREQUENCY MODE IN LIQUID CRYSTALS, Dimic, V., Osredkar, M., Mol Cryst & Liq Cryst v19 n3-4 p189 (Jan '73)

293. THERMAL HYSTERESIS IN CHOLESTERIC COLOR RESPONSES, Dixon, G.D., Scala, L.C., Mol Cryst & Liq Cryst v10 n3 p317 (Jne '70)

294. THERMAL DECOMPOSITION OF CHOLESTERYL OLEYL CARBONATE, Dixon, G.D., Scala, L.C., Mol Cryst & Liq Cryst v10 n3 p327 (Jne '70)

295. PIEZOELECTRIC DOMAINS IN LIQUID CRYSTALS, Dmitriev, S.G., Zh Eksp & Teor Fiz (USSR) v61 n5 p2049 (1971)

296. PIEZOELECTRIC DOMAINS IN LIQUID CRYSTALS, Dmitriev, S.G., Soviet Phys - JETP v34 n5 p1093 (May '72)

297. SPIN RELAXATION IN NEMATIC LIQUID CRYSTALS AND IN SOLUTES IN LIQUID CRYSTALS, Doane, J.W., Murphy, J.A., Visintainer, J.J., p55: Abstr 3rd Liq Cryst Conf., Berlin (Aug '70)

298. SPIN-LATTICE RELAXATION IN THE NEMATIC LIQUID CRYSTAL PHASE, Doane, J.W., Johnson, D.L., Chem Phys Lett (Holland) v6 n4 p291 (15 Aug '70)

299. THE ANGULAR DEPENDENCE OF T_1 IN LIQUID CRYSTALS, Doane, J.W., Moroi, D.S., Chem Phys Lett (Holland) v11 n3 p339 (15 Oct '71)

300. MEASUREMENT OF THE SELF DIFFUSION CONSTANT IN SMECTIC-A LIQUID CRYSTALS, Doane, J.W., Parker, R.S., p60: 17th Congr Ampere N.M.R. and Related Phenomena, Turku, Finland (Aug '72)

301. POSSIBLE SECOND-ORDER NEMATIC-SMECTIC 'A' PHASE TRANSITION, Doane, J.W., Parker, R.S., Cvikl, B., Johnson, D.L., Fishel, D.L., Phys Rev Lett v28 n26 p1694 (26 Jne '72)

302. LOW TEMEPERATURE CHIRAL NEMATIC LIQUID CRYSTAL S DERIVED FROM BETA-METHYLBUTYLANILINE, Dolphin, D., Mulsiani, Cheng, J., Meyer, R.B., J Chem Phys v58 n2 p413 (15 Jan '73)

303. CORRELATED N.M.R. AND E.P.R. STUDIES IN A NEMATIC LIQUID CRYSTAL, Dong, R.Y., Marusic, M., Schwerdtfeger, C.F., Solid State Commun v8 n19 p1577 (1 Oct '70)

304. EVIDENCE FOR NONEXPONENTIAL TIME CORRELATION FUNCTIONS IN A LIQUID CRYSTAL, Dong, R.Y., Forbes, W.F., Pintar, M.M., Solid State Commun v9 n2 p151 (15 Jan '71)

305. ON THE ANISOTROPY OF NUCLEAR SPIN RELAXATION IN A NEMATIC LIQUID CRYSTAL, Dong, R.Y., Chem Phys Lett (Holland) v9 n6 p600 (15 Jne '71)

306. PROTON SPIN RELAXATION IN THE LIQUID CRYSTAL 'PAA', Dong, R.Y., J Chem Phys v55 n1 p145 (1 Jly '71)

307. PROTON SPIN RELAXATION STUDY OF THE CHOLESTERIC-TO-NEMATIC PHASE TRANSITION, Dong, R.Y., Pintar, M.M., Forbes, W.F., J Chem Phys v55 n5 p2449 (1 Sep '71)

308. PROTON SPIN-LATTICE RELAXATION IN THE NEMATIC LIQUID CRYSTAL 'MBBA', Dong, R.Y., Forbes, W.F., Pintar, M.M., Mol Cryst & Liq Cryst v16 n3 p213 (Mch '72)

309. PROTON SPIN-LATTICE RELAXATION IN THE NEMATIC LIQUID CRYSTAL PHASE, Dong, R.Y., J Magn Resonance v7 n1 p60 (May '72)

310. STUDY OF LIQUID-CRYSTALLINE STRUCTURES OF POLYSTYRENE-POLYBUTADIENE BLOCK COPOLYMERS BY SMALL-ANGLE X-RAY SCATTERING AND ELECTRON MICROSCOPY, Douy, A., Gallot, B.R., Mol Cryst & Liq Cryst v14 n3-4 p191 (Sep '71)

311. SELECTIVE REFLECTION OF CHOLESTERIC LIQUID CRYSTALS, Dreher, R., Meier, G., Saupe, A., p34: Abstr 3rd Int'l Liq Cryst Conf., Berlin, W. Germany (Aug '70)

312. SELECTIVE REFLECTION BY CHOLESTERIC LIQUID CRYSTALS, Dreher, R., Meier, G., Saupe, A., Mol Cryst & Liq Cryst v13 n1 p17 (Apr '71)

313. REFLECTION PROPERTIES OF DISTORTED CHOLESTERIC LIQUID CRYSTALS, Dreher, R., Solid State Commun v12 n6 p519 (15 Mch '73)

314. PLANE PROBLEMS IN ELASTICITY THEORY OF NEMATIC LIQUID CRYSTALS, Dreizin, Yu.A., Dykhne, A.M., Zh Eksp & Teor Fiz (USSR) v61 n5 p2140 (1971)

315. PLANE PROBLEMS IN ELASTICITY THEORY OF NEMATIC LIQUID CRYSTALS, Dreizin, Yu.A., Dykhne, A.M., Soviet Phys - JETP v34 n5 p1140 (May '72)

316. EPITAXY OF NEMATIC LIQUID CRYSTALS, Dreyer J.F., p24: Abstr 3rd Int'l Liq Cryst Conf., Berlin, W. Germany (Aug '70)

317. A LIQUID CRYSTAL DEVICE FOR ROTATING THE PLANE OF POLARIZED LIGHT, Dreyer, J.F., p25: Abstr 3rd Int'l Liq Cryst Conf., Berlin, W. Germany (Aug '70)

318. THE ALIGNMENT OF MOLECULES IN THE NEMATIC LIQUID CRYSTAL STATE, Dreyer, J.F., p311: "Liquid Crystals and Ordered Fluids," Plenum Press, 1970.

319. A PIEZO OPTICAL TRANSDUCER, Dreyer, J.F., Opt Spectra v5 n11 p28 (Nov '71)

320. THE CARBONACEOUS MESOPHASE FORMED IN THE PYROLISIS OF GRAPHITIZABLE ORGANIC MATERIALS, Dubois, J., Agache, C., White, J.L., Metallogr v3 n3 p337 (Sep '70)

321. HYDRODYNAMIC INSTABILITIES OF NEMATIC LIQUID CRYSTALS UNDER 'AC' ELECTRIC FIELDS, Dubois-Violette, E., De Gennes, P.-G., Parodi, O., J Phys (France) v32 n4 p305 (Apr '71)

322. HYDRODYNAMIC INSTABILITIES OF A NEMATIC LIQUID CRYSTAL UNDER A THERMAL GRADIENT, Dubois-Violette, E., Comptes Rendus (Ser. B) v273 n21 p923 (22 Nov '71)

323. THEORY OF INSTABILITIES OF NEMATICS UNDER 'AC' ELECTRIC FIELDS: SPECIAL EFFECTS NEAR THE CUT OFF FREQUENCY, Dubois-Violette, E., J Phys (France) v33 n1 p95 (Jan '72)

324. HYDRODYNAMIC INSTABILITIES OF CHOLESTERICS UNDER A THERMAL GRADIENT, Dubois-Violette, E., J Physique (France) v34 n1 p107 (Jan '73)

325. LOCAL FIELD EFFECTS IN ORIENTED NEMATIC LIQUID CRYSTALS, Dunmur, D.A., Chem Phys Lett (Holland) v10 n1 p49 (1 Jly '71)

326. EFFET ELECTROHYDRODYNAMIQUE DANS UN CRISTAL LIQUIDE NEMATIQUE, Durand, G., Veyssie, M., Rondelez, F., Leger, L., Comptes Rendus (Ser.B) v270 p97 (5 Jan '70)

327. SUR LA PORTEE DES ONDULATIONS DE COUCHES DANS LES SMECTIQUES 'A', Durand, G., Comptes Rendus (Ser. B) v275 n17 p629 (23 Oct '72)

328. UNE NOUVELLE METHODE DE DETERMINATION DES COEFFICIENTS DE DISSUSION DIFFERENTIELS ISO-THERMES BASEE SUR L'UTILISATION DE L'INTERFEROMETRIE HOLOGRAPHIQUE HOLOGRAPHIQUE ET LA

SIMULATION PAR ORDINATEUR, Durou, C., Giraudou, J.-C., Mahenc, J., Comptes Rendus (Ser. C) v275C n15 p761 (9 Oct '72)

329. STRUCTURE OF ORGANIC IONIC MELT MESOPHASES, Duruz, J.J., Ubbelohde, A.R., Proc Roy Soc A (GB) v330 n1580 p1 (1972)

330. N.M.R. RELAXATION IN CHOLESTEROL AND CHOLESTERIC LIQUID CRYSTALS, Dybowski, C.R., Wade, C.G., J Chem Phys v55 n4 p1576 (15 Aug '71)

331. VISCOELASTIC PROPERTIES OF LIQUID CRYSTALLINE BIOLOGICAL MATERIALS IN THE VICINITY OF ISOTROPIC-CHOLESTERIC PHASE TRANSITION, Dyro, J.F., Edmonds, P.D., Barberian, J.G., Silage, D., IEEE Trans Sonics & Ultrasonics vSU-19 n3 p403 (Jly '72)

332. THEORY OF DISCLINATIONS IN LIQUID CRYSTALS, Dzyaloshinski, I.E., Zh Eksp & Teor Fiz (USSR) v58 n4 p1443 (1970)

333. ULTRASONIC INVESTIGATION OF THE NEMATIC-ISOTROPIC PHASE TRANSITION IN 'MBBA', Eden, D., Garland, C.W., Williamson, R.C., J Che, Phys v58 n5 p1861 (1 Mch '73)

334. LIQUID CRYSTAL WATCH, Edmonds, H.D., Hill, C.P., IBM Tech Discl Bull v15 n10 p3041 (Mch '73)

335. ELECTRON MICROSCOPY OF MESOMORPHIC STRUCTURES OF AQUEOUS LIPID PHASES. I. THE SYSTEM POTASSIUM OLEATE/WATER, Eins, S., Mol Cryst & Liq Cryst v11 n2 p119 (Sep '70)

335A. STOP BANDS FOR OPTICAL WAVE PROPAGATION IN CHOLESTERIC LIQUID CRYSTALS, Elachi, C., Yeh, C., Opt Soc Am J v63 n7 p840 (Jly '73)

336. THE ODD-EVEN EFFECT IN STERYL w-PHENYLALKANOATES, Elser, W., Pohlmann, J.L.W., Boyd, P.R., p89: Abstr 3rd Int'l Liq Cryst Conf., Berlin, W. Germany (Aug '70)

337. STRUCTURE DEPENDENCE OF CHOLESTERIC MESOPHASES, Elser, W., Pohlmann, J.L.W., p91: Abstr 3rd Int'l Liq Cryst Conf., Berlin, W. Germany (Aug '70)

338. S-CHOLESTERYL ALKANETHIOATES, Elser, W., Pohlmann, J.L.W., Boyd, P.R., Mol Cryst & Liq Cryst v11 n3 p279 (Oct '70)

339. STRUCTURE DEPENDENCE OF CHOLESTERIC MESOPHASES III. MINOR CHANGES WITHIN THE 17BETA-SIDE CHAIN OF CHOLESTEROL, Elser, W., Pohlmann, J.L.W., Boyd, P.R., Mol Cryst & Liq Cryst v13 n3 p255 (Jne '71)

340. THE ODD-EVEN EFFECT IN STERYL w-PHENYLALKANOATES, Elser, W., Pohlmann, J.L.W., Boyd, P.R., Mol Cryst & Liq Cryst v15 n2 p175 (Nov '71)

341. CHOLESTERYL N-ALKYL CARBONATES, Elser, W., Pohlmann, J.L.W., Boyd, P.R., Mol Cryst & Liq Cryst v20 n1 p77 (Feb '73)

342. THE SELECTIVE LIGHT REFLECTION BY PLANE TEXTURES, Ennulat, R.D., p33: Abstr 3rd Int'l Liq Cryst Conf., Berlin, W. Germany (Aug '70)

343. MESOMORPHISM OF HOMOLOGOUS SERIES. II. ODD-EVEN BEHAVIOR, Ennulat, R.D., Brown, A.J., p93: Abstr 3rd Int'l Liq Cryst Conf., Berlin, W. Germany (Aug '70)

344. MESOMORPHISM OF HOMOLOGOUS SERIES. II. ODD-EVEN EFFECT, Ennulat, R.D., Brown, A.J., Mol Cryst & Liq Cryst v12 n4 p367 (Mch '71)

345. THERMAL RADIOGRAPHY UTILIZING LIQUID CRYSTALS, Ennulat, R.D., Fergason, J.L., Mol Cryst & Liq Cryst v13 n2 p148 (May '71)

346. THE SELECTIVE LIGHT REFLECTION BY PLANE TEXTURES, Ennulat, R.D., Mol Cryst & Liq Cryst v13 n4 p337 (Jly '71)

347. NORMAL STRESS EFFECTS IN CHOLESTERIC MESOPHASES, Erhardt, P.F., Pochan, J.M., Richards, W.C., J Chem Phys v57 n9 p3596 (1 Nov '72)

348. SINGULAR SOLUTIONS IN LIQUID CRYSTAL THEORY, Ericksen, J.L.,

p181: "Liquid Crystals and Ordered Fluids," Plenum Press, 1970.

349. ISOTROPIC-NEMATIC PHASE TRANSITION IN LIQUID CRYSTALS, Fan, C., Stephen, M.J., Phys Rev Lett v25 n8 p500 (24 Aug '70)

350. FLUCTUATION AND LIGHT SCATTERING IN CHOLESTERIC LIQUID CRYSTALS, Fan, C., Kramer, L., Stephen, M.J., Phys Rev (A) v2 n6 p2482 (1 Dec '70)

351. COMMENTS ON PIEZOELECTRIC EFFECT IN LIQUID CRYSTALS, Fan, C., Mol Cryst & Liq Cryst v13 n1 p9 (Apr '71)

352. DISCLINATION LINES IN LIQUID CRYSTALS, Fan, C., Phys Lett (A) v34a n6 p335 (5 Apr '71)

353. LIQUID CRYSTAL NUMERICAL DISPLAY, Fan, G.J.-Y., IBM Tech Discl Bull v15 n3 p769 (Aug '72)

354. THERMAL FLUCTUATIONS AND PROTON SPIN-LATTICE RELAXATION IN NEMATIC LIQUID CRYSTALS, Farinha-Martins, A., p56: Abstr 3rd Int'l Liq Cryst Conf., Berlin, W. Germany (Aug '70)

355. LIGHT ACTUATED LIQUID CRYSTAL LIGHT VALVES, Feldman, M., White, D.L., p42: Int'l electron devices meeting (abstr.), Washington, D.C., 28-30 Oct '70.

356. COMMENTS ON A "SIMPLIFIED ELECTROHYDRODYNAMIC TREATMENT OF THRESHOLD EFFECTS IN NEMATIC LIQUID CRYSTALS," Felici, N.J., Tobazeon, R., Heilmeier, G.H., IEEE Proc v60 n2 p241 (Feb '72)

357. EXPERIMENTS WITH CHOLESTERIC LIQUID CRYSTALS, Fergason, J.L., Am J Phys v38 n4 p425 (Apr '70)

358. APPLICATION OF LIQUID CRYSTALS, Fergason, J.L., p20: Abstr 3rd Int'l Liq Cryst Conf., Berlin, W. Germany (Aug '70)

359. RADIOGRAPHY UTILIZING CHOLESTERIC LIQUID CRYSTALS, Fergason, J.L., Ennulat, R.D., p32: Abstr 3rd Int'l Liq Cryst Conf., Berlin, W. Germany (Aug '70)

360. SUR LA BIREFRINGENCE ELECTRIQUE DU P-METHOXY-BENZYLIDENE P-N-BUTYLANILINE (MBBA) EN PHASE ISOTROPE, Filippini, J.-C., Comptes Rendus (Ser. B) v275 n10 p349 (4 Sep '72)

361. PROTON MAGNETIC LINES IN LIQUID CRYSTALLINE 4,4'-DIPROPOXYAZOXYBENZENE, 4,4'-DIBUTOXY-AZOXYBENZENE, AND DIETHYL 4,4'- AZOXYDIBENZOATE, Fink, J.J., Moses, H.A., Cohen, P.S., J Chem Phys v56 n12 p6198 (15 Jne '72)

362. THERMOTROPIC LIQUID CRYSTALS. II. TRANSITION TEMPERATURES AND MESOPHASE IDENTIFICATIONS FOR SOME ANILS, Fishel, D.L., Patel, P.R., Mol Cryst & Liq Cryst v17 n2 p139 (Jne '72)

363. TEMPERATURE DEPENDENCE OF P-AZOXYANISOLE IN CRYSTAL AND NEMATIC PHASES, Ford, W.G.F., J Chem Phys v56 n12 p6270 (15 Jne '72)

364. HYDRODYNAMICS OF LIQUID CRYSTALS, Forster, D., Lubenski, T.C., Martin, P.C., Swift, J., Pershan, P.S., Phys Rev Lett v26 n17 p1016 (26 Apr '71)

365. THEORY OF DIFFUSION IN LIQUID CRYSTAL MESOPHASES, Franklin, W.M., p125: Abstr 3rd Int'l Liq Cryst Conf., Berlin, W. Germany (Aug '70)

366. DIFFUSION THEORY IN LIQUID CRYSTALS, Franklin, W.M., Mol Cryst & Liq Cryst v14 n3-4 p227 (Sep '71)

367. ORDERED STATES OF A NEMATIC LIQUID, Freiser, M.J., Phys Rev Lett v24 n19 p1041 (11 May '70)

368. SUCCESSIVE TRANSITIONS IN A NEMATIC LIQUID, Freiser, M.J., p44: Abstr 3rd Int'l Liq Cryst Conf., Berlin, W. Germany (Aug '70)

369. SUCCESSIVE TRANSITIONS IN A NEMATIC LIQUID, Freiser, M.J., Mol Cryst & Liq Cryst v14 n1-2 p165 (Aug '71)

370. LIQUID CRYSTAL POLYCHROMATIC DISPLAY DEVICE, Freiser, M.J., IBM Tech Discl Bull v15 n2 p700 (Jly '72)

371. LIGHT SCANNER EMPLOYING A NEMATIC LIQUID CRYSTAL, Freiser, M.J., IBM Tech Discl Bull v15 n8 p2540 (Jan '73)

372. LIQUID CRYSTALS AND EMULSION PROPERTIES, Friberg, S., p146: Abstr 3rd Int'l Liq Cryst Conf., Berlin, W. Germany (Aug '70)

373. LIQUID CRYSTALLINE PHASES IN EMULSIONS, Friberg, S., J Colloid & Interface Sci v37 n2 p291 (Oct '71)

374. APPLICATION OF DISLOCATION THEORY TO LIQUID CRYSTALS, Friedel, J., Kleman, M., p607: Proc Conf fundamental aspect of dislocation theory, Washington, D.C., 21-25 Apr '71.

375. ELECTRON PARAMAGNETIC RESONANCE STUDY OF TWO SMECTIC 'A' LIQUID CRYSTALS, Fryburg, G.C., Gelerinter, E., Fishel, D.L., Mol Cryst & Liq Cryst v16 n1-2 p39 (Feb '72)

376. PROTON N.M.R. IN A LIQUID CRYSTAL SOLUTION, Fung, B.M., Gerace, M.J., J Chem Phys v53 n3 p1171 (1 Aug '70)

377. MEASUREMENT OF DEUTERIUM NUCLEAR QUADRUPOLE COUPLING CONSTANT IN LIQUID CRYSTAL SOLUTIONS, Fung, B.M., Wei, I.Y., p60: Abstr 3rd Int'l Liq Cryst Conf., Berlin, W. Germany (Aug '70)

378. THE VISCOSITY COEFFICIENTS OF A ROOM-TEMPERATURE LIQUID CRYSTAL (MBBA), Gahwiller, Ch., Phys Lett (A) v36a n4 p311 (13 Sep '71)

379. TEMPERATURE DEPENDENCE OF FLOW ALIGNMENT IN NEMATIC LIQUID CRYSTALS, Gahwiller, Ch., Phys Rev Lett v28 n24 p1554 (12 Jne '72)

380. ELECTROHYDRODYNAMIC INSTABILITY IN A NEMATIC LIQUID CRYSTAL: EFFECT OF AN ADDITIONAL STABILIZING 'AC' ELECTRIC FIELD ON THE SPATIAL PERIOD OF "CHEVRONS," Galerne, Y., Durand, G., Veyssie, M., Pontikis, V., Phys Lett (A) v38a n6 p449 (13 Mch '72)

381. SPATIAL PERIODS OF BEND OSCILLATIONS IN THE DIELECTRIC ELECTROHYDRODYNAMICAL INSTABILITY OF A NEMATIC LIQUID CRYSTAL, Phys Rev (A) v6 n1 p484 (Jly '72)

382. QUASIELECTRIC RAYLEIGH SCATTERING IN A SMECTIC 'C' LIQUID CRYSTAL, Galerne, Y., Martinand, J.L., Durand, G., Veyssie, M., Phys Rev Lett v29 n9 p562 (28 Aug '72)

383. LIQUID CRYSTALLINE STRUCTURES FOR BINARY SYSTEMS α-ω SOAPS/WATER, ELECTRIC INTERACTIONS IN THESE SYSTEMS, Gallot, B., p77: Abstr 3rd Int'l Liq Cryst Conf., Berlin, W. Germany (Aug'70)

384. STUDY OF LIQUID-CRYSTALLINE STRUCTURES OF 'AB' AND 'ABA' POLYBUTADIENE - POLYSTYRENE BLOCK COPOLYMERS, BY SMALL ANGLE X-RAY SCATTERING AND ELECTRON MICROSCOPY, Gallot, B., Douy, A., p80: Abstr 3rd Int'l Liq Cryst Conf., Berlin, W. Germany (Aug '70)

385. PHASE DIAGRAM OF SYSTEMS: BLOCK-COPOLYMER/PREFERENTIAL SOLVENT OF ONE BLOCK, Gallot, B., Gervais, M., Douy, A., p81: Abstr 3rd Int'l Liq Cryst Conf., Berlin, W. Germany (Aug '70)

386. LIQUID CRYSTALLINE STRUCTURES OF BINARY SYSTEMS α-ω SOAP/WATER, Gallot, B.R., Mol Cryst & Liq Cryst v13 n4 p323 (Jly '71)

387. NONDESTRUCTIVE LIQUID CRYSTAL METHOD OF LOCATING 'FET' GATE OXIDE SHORTS, Garbarino, P.L., Sandison, R.D., IBM Tech Discl Bull v15 n6 p1738 (Nov '72)

388. MAGNETIC PROPERTIES OF NEMATIC SUBSTANCES. STUDY OF PARAOXYANISOLE AND P-(P'-METHOXY-BENZYLIDENE) AMINO-BUTYL-BENZENE, Gasparoux, H., Regaya, B., Prost, J., Comptes Rendus (Ser. B) v272 n20 p1168 (17 May '71)

389. DETERMINATION DIRECTE DE L'ANISOTROPIE MAGNETIQUE DE CRISTAUX LIQUIDES NEMATIQUES, Gasparoux, H., Prost, J., J Physique (France) v32 n11-12 p953 (Nov-Dec '71)

390. AN E.P.R. STUDY OF A SMECTIC 'C' LIQUID CRYSTAL: A NEW METHOD FOR DETERMINING THE TILT ANGLE, Gelerinter, E., Fryburg, G.C., Appl Phys Lett v18 n3 p84 (1 Feb '71)

391. N.M.R. SPECTRA OF 1,2-DIFLUOROBENZENE IN NEMATIC SOLVENTS. ANISOTROPY OF INDIRECT FLUORINE COUPLINGS AND MOLECULAR GEOMETRY, Gerritsen, J., MacLean, C., p59: Abstr 3rd Int'l Liq Cryst Conf., Berlin, W. Germany (Aug '70)

392. N.M.R. SPECTRA OF 1,2-DIFLUOROBENZENE AND 1,1-DIFLUOROETHENE IN NEMATIC SOLVENTS. ANISOTROPY OF INDIRECT FLUORINE COUPLINGS AND MOLECULAR GEOMETRY, Gerritsen, J., MacLean, C., Mol Cryst & Liq Cryst v12 n2 p97 (Jan '71)

393. A MICROWAVE ANALOG OF OPTICAL ROTATION IN CHOLESTERIC LIQUID CRYSTALS, Gerritsen, H.J., Yamaguchi, R.T., Am J Phys v39 n8 p920 (Aug '71)

394. STRUCTURAL EFFECTS IN CHOLESTERIC LIQUID CRYSTALS, Gerritsma, C.J., Van Zanten, P., p102: Abstr 3rd Int'l Liq Cryst Conf., Berlin, W. Germany (Aug '70)

395. THE HELICAL TWIST IN A CHOLESTERIC GRANDJEAN-CANO PATTERN, Gerritsma, C.J., Goossens, W.J.A., Niessen, A.K., Phys Lett (A) v34a n7 p354 (19 Apr '71)

396. DISTORTION OF A TWISTED NEMATIC LIQUID CRYSTAL BY A MAGNETIC FIELD, Gerritsma, C.J., De Jeu, W.H., Van Zanten, P., Phys Lett (A) v36a n5 p389 (27 Sep '71)

397. PERIODIC PERTURBATION IN THE CHOLESTERIC PLANE TEXTURE, Gerritsma, C.J., Van Zanten, P., Phys Lett (A) v37a n1 p47 (25 Oct '71)

398. ELECTRIC-FIELD-INDUCED TEXTURE TRANSFORMATION AND PITCH CONTRACTION IN A CHOLESTERIC LIQUID CRYSTAL, Gerritsma, C.J., Van Zanten, P., Mol Cryst & Liq Cryst v15 n3 p257 (Dec '71)

399. THE DEPENDENCE OF THE ELECTRIC-FIELD-INDUCED CHOLESTERIC-NEMATIC TRANSITION ON THE DIELECTRIC ANISOTROPY, Gerritsma, C.J., Van Zanten, P., Phys Lett (A) v42a n2 p127 (20 Nov '72)

400. MAGNETIC-FIELD-INDUCED MOTION OF DISCLINATIONS IN A TWISTED NEMATIC LAYER, Gerritsma, C,J., Geurst, J.A., Spruitt, A.M.J., Phys Lett (A) v43a n4 p356 (26 Mch '73)

401. PHASE DIAGRAM OF SYSTEMS: BLOCK COPOLYMER PREFERENTIAL SOLVENT OF ONE BLOCK, Gervais, M., Douy, A., Gallot, B., Mol Cryst & Liq Cryst v13 n4 p289 (Jly '71)

402. - 404. deleted...

405. THEORY OF ELECTRICALLY INDUCED HYDRODYNAMIC INSTABILITIES IN SMECTIC LIQUID CRYSTALS, Geurstm J.A., Goossens, W.J.A., Phys Lett (A) v41a n4 p369 (9 Oct '72)

406. DYNAMICAL BEHAVIOR OF A NEMATIC LIQUID CRYSTAL JUST ABOVE THE NEMATIC-ISOTROPIC TRANSITION FROM SPIN-LATTICE RELAXATION, Ghosh, S.K., Tettamanti, E., Indovina, P.L., Phys Rev Lett v29 n10 p638 (4 Sep '72)

407. SELF-DIFFUSION OF NEMATIC LIQUID CRYSTALS IN THE ISOTROPIC PHASE NEAR THE NEMATIC-ISOTROPIC TRANSITION, Ghosh, S.K., Tettamanti, E., Phys Lett (A) v43a n4 p361 (26 Mch '73)

408. LYOTROPIC LIQUID CRYSTALS OF COMMERCIAL SURFECTANTS, Giua, P.E., p147: Abstr 3rd Int'l Liq Cryst Conf., Berlin, W. Germany (Aug '70)

409. LIQUID CRYSTAL DISPLAY DEVICES, Gladstone, G.L., Sadagopan, V., IBM Tech Discl Bull v14 n5 p1472 (Oct '71)

410. NONLINEAR BEHAVIOR OF THE ELECTROHYDRODYNAMIC INSTABILITY THRESHOLD IN NEMATIC LIQUID CRYSTALS, Gladstone, G.L., Phys Lett (A) v37a n4 p325 (6 Dec '71)

411. LIQUID CRYSTAL DISPLAY DEVICE CONFIGURATION, Gladstone, G.L., Sadagopan, V., IBM Tech Discl Bull v15 n2 p437 (Jly '72)

412. OPTICAL HARMONIC GENERATION IN LIQUID CRYSTALS, Goldberg, L.S., Schnur, J.M., Radio Electronic Engr (GB) v39 n5 p279 (May '70)

413. LIQUID CRYSTALS III. NEMATIC MESOMORPHISM IN BENZYLIDENE ANILS CONTAINING A TERMINAL ALCOHOL GROUP, Goldmacher, J.E., McCaffrey, M.T., p375: "Liquid Crystals and Ordered

Fluids," Plenum Press, 1970.

414. DYNAMIC SCATTERING IN THE HOMEOTROPIC AND HOMOGENEOUS TEXTURES OF A NEMATIC LIQUID CRYSTAL, Gooch, C.H., Tarry, H.A., J Phys D: Appl Phys v5 n4 pL25 (Apr '72)

415. MATRIX-ADDRESSED LIQUID CRYSTAL DISPLAYS, Gooch, C.H., Low, J.J., J Phys D: Appl Phys v5 n7 p1218 (Jly '72)

416. THE USE OF LIQUID CRYSTALS TO MAKE PROGRAMMABLE GRATICULES IN OPTICAL SYSTEMS, Gooch, C.H., Bottomley, R., Low, J.J., Tarry, H.A., J Phys E: Sci Instrum v6 n5 p485 (May '73)

417. A MOLECULAR THEORY OF THE CHOLESTERIC PHASE, Goossens, W.J.A., Phys Lett (A) v31a n8 p413 (20 Apr '70)

418. A MOLECULAR THEORY OF THE CHOLESTERIC PHASE, Goossens, W.J.A., p52: Abstr 3rd Int'l Liq Cryst Conf., Berlin, W. Germany (Aug '70)

419. A MOLECULAR THEORY OF THE CHOLESTERIC PHASE AND OF THE TWISTING POWER OF OPTICALLY ACTIVE MOLECULES IN A NEMATIC LIQUID CRYSTAL, Goossens, W.J.A., Mol Cryst & Liq Cryst v12 n3 p237 (Feb '71)

420. THE CONDUCTION REGIME IN A NEMATIC LIQUID CRYSTAL WITH NEGATIVE DIELECTRIC ANISOTROPY AT FREQUENCIES ABOVE THE DIELECTRIC RELAXATION FREQUENCY, Goossens, W.J.A., Phys Lett (A) v40a n2 p95 (3 Jly '72)

421. INFLUENCE OF ELECTRIC AND MAGNETIC FIELDS ON THE DIELECTRIC CONSTANT OF CHOLESTERIC LIQUID CRYSTALS, Gopalakrishna, C.V.S.S.V., Avadhanlu, M.N., Murty, C.R.K., Indian J Pure & Appl Phys v8 n3 178 (Mch '70)

422. APPLICATION OF THE LANDAU THEORY OF PHASE TRANSITION TO LIQUID-LIQUID CRYSTALS TRANSITIONS, Goshen, S., Mukamel, D., Shtrikman, S., Solid State Commun v9 n10 p649 (15 May '71)

423. LIQUID CRYSTAL MATRIX DISPLAYS USING ADDITIONAL SOLID LAYERS FOR SUPPRESSION OF PARASITIC CURRENTS, Grabmeier, J.G., Greubel, W.F., Krueger, H.H., p31: Abstr 3rd Int'l Liq Cryst Conf., Berlin, W. Germany (Aug '70)

424. LIQUID CRYSTAL MATRIX DISPLAYS USING ADDITIONAL SOLID LAYERS FOR SUPPRESSION OF PARASITIC CURRENTS, Grabmeier, J.G., Greubel, W.F., Krueger, H.H., Mol Cryst & Liq Cryst v15 n2 p95 (Apr '71)

425. SMALL ANGLE X-RAY STUDIES OF LIQUID CRYSTAL PHASE TRANSITIONS. II. SURFACE IMPURITY AND ELECTRIC FIELD EFFECTS, Gravatt, C.C., Brady, G.W., p455: "Liquid Crystals and Ordered Fluids," Plenum Press, 1970.

426. THE LIQUID CRYSTALLINE BEHAVIOR OF ALKYL, ARYL, AND ARYLALKYL 4-P-SUBSTITUTED BENZYLIDENE-AMINOCINNAMATES AND ALPHA-METHYLCINNAMATES, Gray, G.W., Harrison, K.J., p82: Abstr 3rd Int'l Liq Cryst Conf., Berlin, W. Germany (Aug '70)

427. THE LIQUID CRYSTALLINE PROPERTIES OF ALKYL, ARYL, AND ARYLALKYL 4-P-SUBSTITUTED-BENZYLIDENE-AMINOCINNAMATES AND ALPHA-METHYLCINNAMATES, Gray, G.W., Harrison, K.J., Mol Cryst & Liq Cryst v13 n1 p37 (Apr '71)

428. SOME EFFECTS OF MOLECULAR STRUCTURAL CHANGE ON LIQUID CRYSTALLINE PROPERTIES, Gray, G.W., Harrison, K.J., Faraday Symposium on Liquid Crystals, Faraday Society, London (1971)

429. NEW FAMILY OF NEMATIC LIQUID CRYSTALS FOR DISPLAYS, Gray, G.W., Harrison, K.J., Nash, J.A., Electronics Lett v9 n6 p130 (22 Mch '73)

430. NEMATIC HETEROCYCLIC DIESTERS, Green, D.C., Young, W.R., IBM Tech Discl Bull v15 n8 p2467 (Jan '73)

431. VISUALIZATION OF INTERNAL STRUCTURE BY MICROWAVE HOLOGRAPHY, Gregoris, L.G., Iizuka, K., IEEE Proc v58 n5 p791 (May '70)

432. ELECTRICALLY CONTROLLABLE DOMAINS IN NEMATIC LIQUID CRYSTALS, Greubel, W., Wolff, U., Appl Phys Lett v19 n7 p213 (1 Oct '71)

433. CORRELATION BETWEEN ELECTRICAL PROPERTIES AND OPTICAL BEHAVIOR OF NEMATIC LIQUID CRYSTALS, Gruler, H., Meier, G., p106: Abstr 3rd Int'l Liq Cryst Conf., Berlin, W. Germany (Aug '70)

434. CORRELATION BETWEEN ELECTRICAL PROPERTIES AND OPTICAL BEHAVIOR OF NEMATIC LIQUID CRYSTALS, Gruler, H., Meier, G., Mol Cryst & Liq Cryst v12 n4 p289 (Mch '71)

435. ELECTRIC-FIELD-INDUCED DEFORMATION IN ORIENTED LIQUID CRYSTALS OF THE NEMATIC TYPE, Gruler, H., Meier, G., Mol Cryst & Liq Cryst v16 n4 p299 (Apr '72)

436. ELASTIC CONSTANTS OF NEMATIC LIQUID CRYSTALS. I. THEORY OF THE NORMAL DEFORMATION, Gruler, H., Scheffer, T.J., Meier, G., Z Naturforsch v27a n6 p966 (Jne '72)

437. LIQUID CRYSTAL DISPLAYS, Gurtler, R.W., Maze, C., IEEE Spectrum v9 n11 p25 (Nov '72)

438. THERMAL CONDUCTIVITY OF A NEMATIC FILM UNDER MAGNETIC FIELD, Guyon, E., Pieranski, P., Brochard, F., Comptes Rendus (Ser. B) v273 n12 p486 (20 Sep '71)

439. EXPERIMENTAL STUDY OF CONVECTION IN A NEMATIC LIQUID CRYSTAL FILM, Guyon, E., Pieranski, R., Comptes Rendus (Ser. B) v274 n9 p656 (28 Feb '72)

440. A.C.-FIELD-INDUCED GRANDJEAN PLANE TEXTURE IN MIXTURES OF ROOM-TEMPERATURE NEMATICS AND CHOLESTERICS, Haas, W., Adams, J., Flannery, J.B., Phys Rev Lett v24 n11 p577 (16 Mch '70)

441. ELECTRO-OPTIC EFFECT IN A ROOM-TEMPERATURE NEMATIC LIQUID CRYSTAL, Haas, W., Adams, J., Flannery, J.B., p109: Abstr 3rd Int'l Liq Cryst Conf., Berlin, W. Germany (Aug '70)

442. NEW ELECTRO-OPTIC EFFECT IN A ROOM-TEMPERATURE NEMATIC LIQUID CRYSTAL, Haas, W., Adams, J., Flannery, J.B., Phys Rev Lett v25 n19 p1326 (9 Nov '70)

443. ELECTRIC FIELD EFFECTS ON THE SYSTEM OLEYL CHOLESTERYL CARBONATE - CHOLESTERYL CHLORIDE, Haas, W., Adams, J., J Electrochem Soc v118 n8 p1372 (Aug '71)

444. OPTICAL STORAGE EFFECTS IN LIQUID CRYSTALS, Haas, W., Adams, J., Dir, G., Chem Phys Lett (Holland) v14 n1 p95 (1 May '72)

445. HELICAL TWISTING POWER OF A DIASTEREOMERIC PAIR OF BRANCHED-CHAIN ESTERS OF CHOLESTEROL, Hakemi, H., Labes, M.M., J Chem Phys v58 n4 p1318 (15 Feb '73)

446. LIGHT SCATTERING SPECTRUM OF A NEMATIC LIQUID CRYSTAL, Haller, I., Litster, J.D., p26: Abstr 3rd Int'l Liq Cryst Conf., Berlin, W. Germany (Aug '70)

447. ACHROMATIC POLARIZATION ROTATOR, Haller, I., Freiser, M.J., IBM Tech Discl Bull v13 n5 p1211 (Oct '70)

448. TEMPERATURE DEPENDENCE OF NORMAL MODES IN A NEMATIC LIQUID CRYSTAL, Haller, I., Litster, J.D., Phys Rev Lett v25 n22 p1550 (30 Nov '70)

449. EFFECT OF END-CHAIN POLARITY ON THE MESOPHASE STABILITY OF SOME SUBSTITUTED SCHIFF-BASES, Haller, I., Cox, R.J., p393: "Liquid Crystals and Ordered Fluids," Plenum Press, 1970.

450. LIGHT SCATTERING SPECTRUM OF A NEMATIC LIQUID, Haller, I., Litster, J.D., Mol Cryst & Liq Cryst v12 n3 p277 (Feb '71)

451. ALIGNING NEMATIC LIQUID CRYSTALS, Haller, I., Huggins, H.A., IBM Tech Discl Bull v13 n11 p3237 (Apr '71)

452. ON THE MEASUREMENT OF INDICES OF REFRACTION OF NEMATIC LIQUIDS, Haller, I., Huggins, H.A., Freiser, M.J., Mol Cryst & Liq Cryst v16 n1-2 p53 (Feb '72)

453. ELASTIC CONSTANTS OF THE NEMATIC LIQUID CRYSTALLINE PHASE OF P-METHOXY BENZYLIDENE-P-N-BUTYLANILINE ('MBBA'), Haller, I., J Chem Phys v57 n4 p1400 (15 Aug '72)

454. LIQUID CRYSTALS. PROPERTIES AND APPLICATIONS TO ELECTRONICS, Hampel, B., Elektro-Anzeiger (Germany) v25 n7 p106 (22 Mch '72)

455. NEW DISPLAY USING LIQUID CRYSTAL, Hareng, M., Revue Tech Thomson-CSF v3 n4 p782 (Dec '71)

456. LIQUID CRYSTAL MATRIX DISPLAY BY ELECTRICALLY CONTROLLED BIREFRINGENCE, Hareng, M., Assouline, G., Leiba, E., IEEE Proc v60 n7 p (Jly '72)

457. EFFET DU CHAMP ELECTRIQUE SUR LA BIREFRINGENCE DE CRISTAUX LIQUIDES NEMATIQUES, Hareng, M., Leiba, E., Assouline, G., Mol Cryst & Liq Cryst v17 n3-4 p361 (Jly '72)

458. LA BIREFRINGENCE ELECTRIQUEMENT CONTROLEE DANS LES CRISTAUX LIQUIDES NEMATIQUES, Hareng, M., Assouline, G., Leiba, E., Appl Opt v11 n12 p2920 (Dec '72)

459. LIQUID CRYSTAL DISPLAYS, Harrison, L., Electron (GB) n13 p17 (12 Oct '72)

460. A STUDY OF A LIQUID CRYSTAL SYSTEM, Hartshorne, N.H., Microscope (GB) v19 n4 p424 (Oct '71)

461. SWITCHING OF NEMATIC LIQUID CRYSTALS BY SURFACE WAVES, Heidrich, P.F., Pole, R.V., Powell, C.G., Tseng, C.-C., Yu, H.N., IBM Tech Discl Bull v15 n1 p165 (Jne '72)

462. FURTHER STUDIES OF THE DYNAMIC SCATTERING MODE IN NEMATIC LIQUID CRYSTALS, Heilmeier, G.H., Zanoni, L.A., Barton, L.A., IEEE Trans Electron Devices vED-17 n1 p22 (Jan '70)

463. ORIENTATIONAL OSCILLATIONS IN NEMATIC LIQUID CRYSTALS, Heilmeier, G.H., Helfrich, W., Appl Phys Lett v16 n4 p155 (15 Feb '70)

464. LIQUID CRYSTAL DISPLAY DEVICE, Heilmeier, G.H., Sci Am v222 n4 p100 (Apr '70)

465. A REVIEW OF ELECTRIC AND MAGNETIC FIELD EFFECTS IN LIQUID CRYSTALS, Heilmeier, G.H., p14: Abstr 3rd Int'l Liq Cryst Conf., Berlin, W. Germany (Aug '70)

466. SOME EXPERIMENTS ON ELECTRIC FIELD INDUCED STRUCTURAL CHANGES IN A MIXED LIQUID CRYSTAL SYSTEM, Heilmeier, G.H., Zanoni, L.A., Goldmacher, J.E., p215: "Liquid Crystals and Ordered Fluids," Plenum Press, 1970.

467. A SIMPLIFIED ELECTROHYDRODYNAMIC TREATMENT OF THRESHOLD EFFECTS IN NEMATIC LIQUID CRYSTALS, Heilmeier, G.H., IEEE Proc v59 n3 p422 (Mch '71)

468. EFFECT OF ELECTRIC FIELDS ON THE TEMPERATURE OF PHASE TRANSITIONS OF LIQUID CRYSTALS, Helfrich, W., Phys Rev Lett v24 n5 p201 (2 Feb '70)

469. TORQUES IN SHEARED NEMATIC LIQUID CRYSTALS: A SIMPLE MODEL IN TERMS OF THE THEORY OF DENSE FLUIDS, Helfrich, W., p124: Abstr 3rd Int'l Liq Cryst Conf., Berlin (Aug '70)

470. TORQUES IN SHEARED NEMATIC LIQUID CRYSTALS: A SIMPLE MODEL IN TERMS OF THE THEORY OF DENSE FLUIDS, Helfrich, W., J Chem Phys v53 n6 p2267 (15 Sep '70)

471. DEFORMATION OF CHOLESTERIC LIQUID CRYSTALS WITH LOW THRESHOLD VOLTAGE, Helfrich, W., Appl Phys Lett v17 n12 p531 (15 Dec '70)

472. CAPILLARY VISCOMETRY OF CHOLESTERIC LIQUID CRYSTALS, Helfrich, W., p405: "Liquid Crystals and Ordered Fluids," Plenum Press, 1970.

473. THE STRENGTH OF PIEZOELECTRICITY IN LIQUID CRYSTALS, Helfrich, W., Z Naturforsch A v26a n5 p833 (May '71)

474. A SIMPLE METHOD TO OBSERVE THE PIEZOELECTRICITY OF LIQUID CRYSTALS, Helfrich, W., Phys Lett (A) v35a n6 p393 (12 Jly '71)

475. ELECTROHYDRODYNAMIC AND DIELECTRIC INSTABILITIES OF CHOLESTERIC LIQUID CRYSTALS, Helfrich, W., J Chem Phys v55 n2 p839 (15 Jly '71)

476. BIREFRINGENCE OF NEMATOGENIC LIQUIDS CAUSED BY ELECTRICAL CONDUCTION, Helfrich, W., Schadt, M., Phys Rev Lett v27 n9 p561 (30 Aug '71)

477. OPTICALLY ACTIVE SMECTIVE LIQUID CRYSTAL, Helfrich, W., Oh, C.S., Mol Cryst & Liq Cryst v14 n3-4 p289 (Sep '71)

478. FLOW ALIGNMENT OF WEAKLY ORDERED NEMATIC LIQUID CRYSTALS, Helfrich, W., J Chem Phys v56 n6 p3187 (15 Mch '72)

479. FLOW ALIGNMENT OF NEMATIC LIQUID CRYSTALS, Helfrich, W., Helv Phys Acta (Switzerland) v45 n1 p35 (30 Jne '72)

480. ORIENTING ACTION OF SOUND ON NEMATIC LIQUID CRYSTALS, Helfrich, W., Phys Rev Lett v29 n24 p1583 (11 Dec '72)

481. FIELD INDUCED SURFACE CHARGES IN NEMATIC ELECTROLYTE SYSTEMS, Hepke, G., Schneider, F., Chem Phys Lett v13 n6 p548 (15 Apr '72)

482. KINETICS OF ALIGNMENT OF A NEMATIC LIQUID CRYSTAL IN MAGNETIC FIELDS, Heppke, G., Schneider, F., Z Naturforsch v27a n6 p976 (Jne '72)

483. NOISE MEASUREMENTS IN NEMATIC LIQUID CRYSTALS, Herczfeld, P.R., Hartman, Mol Cryst & Liq Cryst v18 n2 p157 (Sep '72)

484. ELECTRICAL AND OPTICAL PROPERTIES OF NEMATIC LIQUID CRYSTALS, Hisamitsu, S., Yoshino, K., Inuishi, Y., Technol Rpt Osaka Univ (Japan) v22 ns 1027-1052 p201 (Feb '72)

485. POLYMORPHISME D'UN COMPOSE NEMATIQUE DERIVE DE L'ACIDE CINNAMIQUE, Hochapfel, A., Berchet, D., Perron, R., Petit, J., Mol Cryst & Liq Cryst v13 n2 p165 (May '71)

486. NENDORECS IN LIQUID CRYSTALS, Holland, W.P., p46: Abstr 3rd Int'l Liq Cryst Conf., Berlin, W. Germany (Aug '70)

487. DYNAMIC SCATTERING BY A NEMATIC LIQUID CRYSTAL WHERE THE APPLIED ELECTRIC FIELD IS NORMAL TO THE INCIDENT LIGHT, Holloway, W.W.Jr., Rafuse, M.J., J Appl Phys v42 n13 15395 (Dec '71)

488. CIRCULAR DICHROISM AND ROTATORY DISPERSION NEAR ABSORPTION BANDS OF CHOLESTERIC LIQUID CRYSTAL, Holzwarth, G., Holzwarth, N.A.W., Opt Soc Am J v63 n3 p324 (Mch '73)

489. INFRARED CIRCULAR DICHROISM AND LINEAR DICHROISM OF LIQUID CRYSTALS, Holzwarth, G., Chabay, I., Holzwarth, N.A.W., J Chem Phys v58 n11 p4816 (1 Jne '73)

490. LIQUID CRYSTAL DISPLAY, Hornberger, W.P., IBM Tech Discl Bull v12 n10 p1697 (Mch '70)

491. PARACRYSTALLINE LATTICES IN MESOPHASES AND REAL STRUCTURES, Hosemann, R., Mueller, B., Mol Cryst & Liq Cryst v10 n3 p273 (Jne '70)

492. PHASE DIAGRAMS OF BINARY NEMATIC MESOPHASE SYSTEMS, Hsu, E. C.-H., Johnson, J.F., Mol Cryst & Liq Cryst v20 n2 p177 (Mch '73)

493. HYDRODYNAMICS OF NEMATIC LIQUID CRYSTALS, Huang, H.-W., Phys Rev Lett v26 n25 p1525 (21 Jne '71)

494. PARAMETRIC STUDY OF THE OPTICAL STORAGE EFFECT IN MIXED LIQUID CRYSTAL SYSTEMS, Hulin, J.P., Appl Phys Lett v21 n10 p455 (15 Nov '72)

495. A MOLECULAR FIELD TREATMENT OF LIQUID CRYSTALLINE MIXTURES, Humphries, R.L., James, P.G., Luckhurst, G.R., Faraday Symposium of Liquid Crystals, Faraday Society, London, 1971.

496. MOLECULAR FIELD TREATMENT OF NEMATIC LIQUID CRYSTALS, Humphries, R.L., James, P.G., Luckhurst, G.R., JCS Faraday Trans (II) v68 n6 p1031 (1972)

497. EXPERIMENTS ON MAGNETIC ORIENTATION OF LIQUID CRYSTALS OF POLY-GAMMA-BENZYL-L-GLUTAMATE, Iizuka, E., Go, Y., J Phys Soc Japan v31 n4 p1205 (Oct '71)

498. A LIQUID CRYSTAL FILM USED FOR MAPPING AN ACOUSTIC FIELD, Iizuka, K., IEEE Proc v58 n2 p288 (Feb '70)

499. TEMPERATURE DEPENDENCE OF THE VISCOSITY COEFFICIENTS OF LIQUID CRYSTALS, Imura, H., Okano, K., Japanese J Appl Phys v11 n10 p1440 (Oct '72)

500. FRICTION COEFFICIENT FOR A MOVING DISINCLINATION IN A NEMATIC LIQUID CRYSTAL, Imura, H., Okano, K., Phys Lett (A) v42a n6 p403 (15 Jan '73)

501. INTERACTION BETWEEN DISINCLINATIONS IN NEMATIC LIQUID CRYSTALS, Imura, H., Okano, K., Phys Lett (A) v42a n6 p405 (15 Jan '73)

502. OPTICAL PATTERN PROCESSING UTILIZING NEMATIC LIQUID CRYSTALS, Inokuchi, S., Morita, Y., Sakurai, Y., Appl Opt v11 n10 p2223 (Oct '72)

503. PRESENT STATUS OF LIQUID CRYSTAL DISPLAYS IN JAPAN, Inoue, T., JEE (Japan) n66 p24 (May '72)

504. DISPLAY OF 'HF' ACOUSTIC HOLOGRAMS UTILIZING LIQUID CRYSTALS, Intlekofer, M.J., Auth, D.C., Fourney, M.E., p83: Imaging Techniques for Testing and Inspection, SPIE Seminar Proc., Vol. 29 (Feb '72)

505. LIQUID CRYSTAL HOLOGRAPHY OF HIGH-FREQUENCY ACOUSTIC WAVES, Intlekofer, M.J., Auth, D.C., Appl Phys Lett v20 n4 p151 (15 Feb '72)

506. PHASE TRANSITION IN TWO-DIMENSIONAL LIQUID CRYSTALS, Isihara, A., Wadati, M., J Chem Phys v55 n9 p4678 (1 Nov '71)

507. TEMPERATURE RESOLUTION OF THE HUMAN EYE BY MEANS OF CHOLESTERIC LIQUID CRYSTAL FILMS, Ittner, G., Boettcher, B., p35: Abstr 3rd Int'l Liq Cryst Conf., Berlin (Aug '70)

508. LIVING ORGANISMS AND LIQUID CRYSTALS, Iwayanagis, S., Sugira, Y., Oyo Buturi v40 n5 p539 (May '71)

509. PHOTOACTIVATED LIQUID CRYSTAL LIGHT VALVES, Jacobson, A.D., p70: 1972 SID symp digest of tech papers, San Francisco, 6-8 Jne '72.

510. HYDRODYNAMICS OF LIQUID CRYSTALS, Jahnig, F., Schmidt, H., Faraday Symposium of Liquid Crystals, Faraday Society, London, 1971.

511. HYDRODYNAMICS OF LIQUID CRYSTALS, Jahnig, F., Schmidt, H., Ann Phys v71 n1 p129 (May '72)

512. ELECTRO-OPTIC RESPONSE TIMES IN LIQUID CRYSTALS, Jakeman, E., Raynes, E.P., Phys Lett (A) v39a n1 p69 (10 Apr '72)

513. THE ANISOTROPIC PSEUDO-POTENTIAL FOR NEMATIC LIQUID CRYSTALS, James, P.G., Luckhurst, G.R., Mol Phys (GB) v19 n4 p489 (Oct '70)

514. REPRESENTATION OF THE ANISOTROPIC PSEUDO-POTENTIAL FOR NEMATOGENS, James, P.G., Luckhurst, G.R., Mol Phys (GB) v20 n4 p761 (Apr '71)

515. ANISOTROPY OF SELF-DIFFUSION IN A LIQUID CRYSTAL STUDIED BY NEUTRON QUASIELASTIC SCATTERING, Janik, J.A., Janik, J.M., Otnes, K., Riste, T., p128: Abstr 3rd Int'l Liq Cryst Conf., Berlin, W. Germany (Aug '70)

516. ANISOTROPY OF SELF-DIFFUSION IN A LIQUID CRYSTAL STUDIED BY NEUTRON QUASIELASTIC SCATTERING, Janik, J.A., Janik, J.M., Otnes, K., Riste, T., Mol Cryst & Liq Cryst v15 n3 p189 (Dec '71)

517. THIN FILM SURFACE ORIENTATION FOR LIQUID CRYSTALS, Janning, J.L., Appl Phys Lett v21 n4 p173 (15 Aug '72)

518. CHOLESTERIC ENERGIES, Jenkins, J.T., J Fluid Mech (GB) v45 Pt13 p465 (15 Feb '71)

519. ON A MATERIAL COEFFICIENT IN CHOLESTERIC LIQUID CRYSTALS, Jenkins, J.T., Mol Cryst & Liq Cryst v18 n3-4 p309 (Oct '72)

520. LIQUID CRYSTALS AND NEWTON'S RINGS, Jeppesen, M.A., Hughes, W.T., Am J Phys v38 n2 p199 (Feb '70)

521. DUTERIUM AND PROTON MAGNETIC RESONANCE INVESTIGATION OF LAMELLAR LYOTROPIC LIQUID CRYSTALS, Johansson, A., Drakenberg, T., p66: Abstr 3rd Int'l Liq Cryst Conf., Berlin (Aug '70)

522. PROTON AND DEUTERON MAGNETIC RESONANCE STUDIES OF LAMELLAR LYOTROPIC MESOPHASES, Johansson, A., Drakenberg, T., Mol Cryst & Liq Cryst v14 n1-2 p23 (Aug '71)

523. DYNAMIC SCATTERING IN A ROOM-TEMPERATURE NEMATIC LIQUID CRYSTAL, Jones, D., Creagh, L., Lu, S., Appl Phys Lett v16 n2 p61 (15 Jan '70)

524. DESIGN OF LIQUID-CRYSTAL DISPLAYS FOR LOW-POWER ELECTRONIC POCKET CALCULATORS, Jones, D., Lu, S., p58: 17th Int'l Electron Devices Meeting (Abstr.) Washington, 11-13 Oct '71)

525. FIELD-EFFECT LIQUID CRYSTAL DEVICES, Jones, D., Lu, S., p100: 1972 SID Symp, Digest of Tech Papers, 6-8 Jne '72, San Francisco, Calif.

526. ELECTRIC-FIELD-INDUCED COLOR CHANGES AND PITCH DILATION IN CHOLESTERIC LIQUID CRYSTAL, Kahn, F.J., Phys Rev Lett v24 n5 p209 (2 Feb '70)

527. CHOLESTERIC LIQUID CRYSTALS FOR OPTICAL APPLICATIONS, Kahn, F.J., Appl Phys Lett v18 n6 p231 (15 Mch '71)

528. LIGHT SCATTERING IN NEMATIC LIQUID CRYSTALS AND APPLICATIONS TO SCHLIEREN LIGHT VALVES, Kahn, F.J., White, D.L., p37: Abstr 1971 Spring Meeting of the Opt Soc Am, Tucson, Arizona (5-8 Apr '71)

529. LIGHT SCATTERING IN NEMATIC LIQUID CRYSTALS AND APPLICATIONS TO SCHLIEREN LIGHT VALVES, Kahn, F.J., White, D.L., Opt Soc Am J v61 n5 p678 (May '71)

530. THE CHOLOPHOR: A PASSIVE POLARIZATION-SWITCHED LIQUID CRYSTAL SCREEN FOR MULTICOLOR LASER DISPLAYS, Kahn, F.J., LaMacchia, J.T., IEEE Trans Electron Devices vED-18 n9 p733 (Sep '71)

531. ELECTRIC-FIELD-INDUCED ORIENTATIONAL DEFORMATION OF NEMATIC LIQUID CRYSTALS: TUNABLE BIREFRINGENCE, Kahn, J.F., Appl Phys Lett v20 n5 p199 (1 Mch '72)

532. 'I.R.'-LASER-ADDRESSED THERMO-OPTIC SMACTIC LIQUID-CRYSTAL STORAGE DISPLAYS, Kahn, F.J., Appl Phys Lett v22 n3 p111 (1 Feb '73)

533. ORIENTATION OF LIQUID CRYSTALS BY SURFACE COUPLING AGENTS, Kahn, F.J., Appl Phys Lett v22 n8 p386 (15 Apr '73)

534. PHOTO VOLTAIC EFFECT IN THE NEMATIC LIQUID CRYSTAL, Kamei, H., Katayama, Y., Ozawa, T., Japanese J Appl Phys v11 n9 p1385 (Sep '72)

535. MOSSBAUER EFFECT IN THE SMECTIC MESOPHASE, Kaplan, J.I., Glasser, M.L., Mol Cryst & Liq Cryst v11 n1 p103 (Aug '70)

536. ON THE STATISTICAL THEORY OF THE NEMATIC MESOPHASE, Kaplan, J.I., Drauglis, E., Chem Phys Lett (Holland) v9 n6 p645 (15 Jne '71)

537. DEPENDENCE OF ISOTROPIC-NEMATIC TRANSITION TEMPERATURE ON ALKYL CHAIN LENGTH IN HOMOLOGOUS SERIES, Kaplan, J.I., J Chem Phys v57 n7 p3015 (1 Oct '72)

538. ELECTRO-OPTICAL PHENOMENA IN THIN LAYERS OF LIQUID CRYSTALS, Kapustin, A.P., Chumakova, S.P., Kristallografiya (USSR) v15 n5 p1091 (Sep '70)

539. INFLUENCE OF AN ALTERNATING ELECTRIC FIELD ON THE TRANSPARENCY OF A THIN LAYER OF THE SMECTIC PHASE OF P-N-OCTYLOXYBENZOIC ACID, Kapustin, A.P., Kovatov, Z.Kh., Mamaeva, L.S., Trofimov, A.N., Izv Vuz Fiz (USSR) n9 p22 (1970)

540. THE TEMPERATURE DEPENDENCE OF ULTRASONIC VELOCITY IN LIQUID CRYSTALS IN THE PHASE TRANSITION REGION, Kapustin, A.P., Mart'yanova, L.I., Kristallografiya (USSR) v16 n3 p648 (1971)

541. PHOTOELASTIC EFFECTS IN LIQUID CRYSTALS, Kapustin, A.P., Kuvatov, Z.Kh., Trofimov, A.N., Izv VUZ Fiz (USSR) n4 p150 (1971)

542. PERIODIC ORIENTED OSCILLATIONS OF THE DOMAIN STRUCTURE OF LIQUID CRYSTALS IN ALTERNATING ELECTRIC FIELDS, Kapustin, A.P., Trofimov, A.N., Chuvyrov, A.N., Soviet Phys - Crystallography v17 n1 p157 (Jly-Aug '72)

543. void (duplicate)

544. VISCOUS BEHAVIOR OF CHOLESTERIC LIQUID CRYSTALS, Kartha, C.G., Agarwal, R.K.L.P., J Phys Soc Japan v28 n2 p470 (Feb '70)

545. ULTRASONIC ABSORPTION IN CHOLESTERIC LIQUID CRYSTALS, Kartha, C.G., Padmini, A.R.K.L., Sastry, G.S., J Phys Soc Japan v31 n2 p617 (Aug '71)

546. ULTRASONIC VELOCITY STUDIES IN CHOLESTERIC LIQUID CRYSTALS, Kartha, C.G., Padmini, R.K.L., Indian J Pure Appl Phys v9 n9 p725 (Sep '71)

547. ULTRASONIC AND VISCOUS BEHAVIOR OF POLYMESOMORPHIC LIQUID CRYSTAL, Kartha, C.G., Padmini, A.R.K.L., J Phys Soc Japan v31 n3 p904 (Sep '71)

548. ELECTROHYDRODYNAMIC INSTABILITIES IN A HIGH-PURITY NEMATIC LIQUID CRYSTAL, Kashnow, R.A., Cole, H.S., J Appl Phys v42 n5 p2134 (Apr '71)

549. A NEMATIC LIQUID CRYSTAL AS A PHASE GRATING OF ELECTRICALLY VARIABLE SPATIAL FREQUENCY, Kashnow, R.A., Bigelow, J.E., p96: 1972 SID Symp Dig of Tech Papers, San Francisco, Calif. 6-8 Jne '72)

550. THICKNESS MEASUREMENTS OF NEMATIC LIQUID CRYSTAL LAYERS, Kashnow, R.A., Rev Sci Instrum v43 n12 p1837 (Dec '72)

551. OPTICAL PROPERTIES OF CHOLESTERIC LIQUID CRYSTALS, Kats, E.I., Zh Eksp Teor Fiz (USSR) v59 n5 p1854 (1970)

552. CHERENKOV RADIATION IN CHOLESTERIC LIQUID CRYSTALS, Kats, E.I., Zh Eksp & Teor Fiz (USSR) v61 n4 p1686 (1971)

553. CERENKOV RADIATION IN CHOLESTERIC LIQUID CRYSTALS, Kats, E.I., Soviet Phys - JETP v34 n4 p899 (Apr '72)

554. LIQUID-CRYSTAL IMAGE INTENSIFIER AND RECORDER, Kazan, B., IBM Tech Discl Bull v12 n6 p864 (Nov '69)

555. LIQUID CRYSTAL DISPLAY WITH LONG LIFE, Kazan, B., IBM Tech Discl Bull v14 n10 p3018 (Mch '72)

556. LIQUID CRYSTALS: MATERIAL WITH A HOT FUTURE, Kaye, D., Electronic Design n19 p76 (13Sep'70)

557. INFRARED INTERFEROMETRY WITH A CO_2 LASER SOURCE AND LIQUID CRYSTAL DETECTION, Keilman, F., Appl Opt v9 n6 p1319 (Jne '70)

558. A LIQUID-CRYSTALLINE (NEMATIC) PHASE WITH A PARTICULARLY LOW SOLIFICATION POINT, Kelker, H., Scheurle, B., Angew Chem v8 p884 (1969)

559. NEW APPLICATIONS OF NEMATIC PHASES IN GAS-LIQUID-CHROMATOGRAPHY, Kelker, H., Jainz, J., Sabel, J., Winterscheidt, H., p132: Abstr 3rd Int'l Liq Cryst Conf., Berlin (Aug '70)

560. EINIGE NEUE ANWENDUNGEN NEMATISCHER PHASEN IN DER GASCHROMATOGRAPHIE, Kelker, H., Scheurle, B., Sabel, J. Jainz, J., Winterscheidt, H., Mol Cryst & Liq Cryst v12 n2 p113 (Jan '71)

561. KONZENTRATIONSBEDINGTE GRANDJEAN-STUFEN CHOLESTERINISCHER PHASEN, Kelker, H., Mol Cryst & Liq Cryst v15 n4 p347 (Jan '72)

562. ULTRASONIC DETERMINATION OF ANISOTROPIC SHEAR AND BULK VISCOSITIES IN NEMATIC LIQUID CRYSTALS, Kemp, K.A., Letcher, S.V., Phys Rev Lett v27 n24 p1634 (13 Dec '71)

563. ULTRASONIC ABSORPTION AND VISCOSITY ANISOTROPY IN ORIENTED NEMATIC LIQUID CRYSTALS, Kemp, K.A., Letcher, S.V., IEEE Trans Sonics & Ultrasonics vSU-19 n3 p408 (Jly '72)

564. LIQUID CRYSTAL DEVICE FOR THE VISUALIZATION OF ELECTRIC FIELDS, Kerllenevich, B., Chapunov, E., Coche, A., Rev Sci Instrum v42 n10 p1545 (Oct '71)

565. EFFECTS OF THE ADDITION OF CHOLESTERICS ON NEMATIC LIQUID CRYSTAL PROPERTIES, Kerllenevich, B., Coche, A., J Appl Phys v42 n13 p5313 (Dec '71)

566. HEAT TRANSPORT IN LIQUID CRYSTALS, Kessler, J.O., Longley-Cook, M.T., p129: Abstr 3rd Int'l Liq Cryst Conf., Berlin, W. Germany (Aug '70)

567. MAGNETIC ALIGNMENT OF NEMATIC LIQUID CRYSTALS, Kessler, J.O., p361: "Liquid Crystals and Ordered Fluids," Plenum Press, 1970.

568. ULTRASONIC STIMULATION OF OPTICAL SCATTERING IN NEMATIC LIQUID CRYSTALS, Kessler, L.W., Sawyer, S.P., Appl Phys Lett v17 n10 p440 (15 Nov '70)

569. LIQUID CRYSTAL ELECTROPHOTOGRAPHIC DISPLAY, Keyes, R.W., IBM Tech Discl Bull v12 n12 p2135 (May '70)

570. N.M.R. INVESTIGATIONS ON (BENZENE) CHROMIUM TRICARBONYL ORIENTED IN A NEMATIC PHASE, Khetrapal, C.L., Kunwar, A.C., Kanekar, C.R., Diehl, P., Mol Cryst & Liq Cryst v12 n2 p179 (Jan '71)

571. N.M.R. SPECTRA OF BICYCLIC COMPOUNDS ORIENTED IN THE NEMATIC PHASE. PART I: THE SPECTRUM OF QUINOXALINE, Khetrapal, C.L., Kunwar, A.C., Mol Cryst & Liq Cryst v15 n4 p363 (Jan '72)

572. THE STRUCTURE OF 2,1,3-BENZOTHIADIAZOLE FROM THE N.M.R. SPECTRUM IN THE LIQUID CRYSTALLINE PHASE, Khetrapal, C.L., Patankar, A.V., Mol Cryst & Liq Cryst v15 n4 p367 (Jan '72)

573. N.M.R. SPECTRA OF BICYCLIC COMPOUNDS ORIENTED IN THE NEMATIC PHASE. Pt.II: THE SPECTRUM OF PHTHALAZINE, Khetrapal, C.L., Saupe, A., Kunwar, A.C., Mol Cryst & Liq Cryst v17 n2 p121 (Jne '72)

574. PROTON MAGNETIC RESONANCE STUDIES OF METHYLMERCURIC HALIDES IN ISOTROPIC AND NEMATIC LIQUID CRYSTAL SOLUTIONS, Khetrapal, C.L., Saupe, A., Mol Cryst & Liq Cryst v19 n3-4 p195 (Jan '73)

575. ELECTRICALLY CONTROLLABLE HOLOGRAMS USING NEMATIC LIQUID CRYSTALS (Abstr.), Kiemle, H., Wolff, U., Opt Soc Am J v60 n11 p1563 (Nov '70)

576. APPLICATION DE SRISTAUX LIQUIDES EN HOLOGRAPHIE OPTIQUE, Kiemle, H., Wolff, U., Opt Commun v3 n1 p26 (Mch '71)

577. THEORY OF OPTICAL ACTIVITY IN NEMATIC LIQUID CRYSTALS, Kimura, H., J Phys Soc Japan v30 n5 p1273 (May '71)

578. THE POTENTIAL BARRIERS OF THE 'PAA' MOLECULES IN CRYSTAL, NEMATIC AND ISOTROPIC PHASES (I.R. SPECTRA), Kirov, N., Simova, P., Phys Lett (A) v37a n1 p51 (25 Oct '71)

579. ELECTRO-OPTIC EFFECTS IN LIQUID CRYSTALS, Kirton, J., Raynes, E.P., Endeavour (GB) v32 n116 p71 (May '73)

580. APPARATUS FOR THE CALIBRATION OF SHEAR SENSITIVE LIQUID CRYSTALS, Klein, E.J., Margozzi, A.P., Rev Sci Instrum v41 n2 p238 (Feb '70)

581. DEFECT DENSITIES IN DIRECTIONAL MEDIA, MAINLY LIQUID CRYSTALS, Kleman, M., Philosophical Mag v27 n5 p1057 (May '73)

582. COMPARISON OF THE RADIOLYTICALLY INDUCED POLYMER FORMATION IN ALKONYL CARBONATES IN THE MESOMORPHIC STATE AND THE CRYSTALLINE STATE, Klingen, T. J., Wright, J.R., p138: Abstr 3rd Int'l Liq Cryst Conf., Berlin, W. Germany (Aug '70)

583. RADIOLYTICALLY INDUCED POLYMER FORMATION IN ALKENYL CARBORANES: PHASE EFFECTS, Klingen, T.J., Wright, J.R., Mol Cryst & Liq Cryst v13 n2 p173 (May '71)

584. EXPERIMENTAL COMPARISON OF MULTIPLEXING TECHNIQUES FOR LIQUID CRYSTAL DISPLAYS, Kmetz, A.R., p66: 1972 SID Symp Digest of Tech Papers, San Francisco, Calif. 6-8 Jne '72.

585. ESTIMATION OF NEMATIC-ISOTROPIC POINTS OF NEMATIC LIQUID CRYSTALS, Knaak, L.E., Rosenberg, H.M., Serve, M.P., Mol Cryst & Liq Cryst v17 n2 p171 (Jne '72)

586. CYCLOARTENYL PALMITATE: A NATURALLY OCCURRING ESTER THAT FORMS A CHOLESTERIC MESOPHASE, Knapp, F.F., Nicholas, H.J., Mol Cryst & Liq Cryst v6 n3-4 p319 (Feb '70)

587. STRUCTURAL STUDIES OF THE CHOLESTERIC MESOPHASE, Knapp, F.F., Nicholas, H.J., p147: "Liquid Crystals and Ordered Fluids," Plenum Press, 1970.

588. A DIGITAL CLOCK WITH LIQUID CRYSTAL INDICATORS, Knauer, R., Siemens Electron Components Bull (Germany) v6 n4 p87 (Oct '71)

589. ON THE THEORY OF TRANSLATIONAL AND ORIENTATIONAL MELTING WITH APPLICATION TO LIQUID CRYSTALS, Kobayashi, K.K., Phys Lett (A) v31a n3 p125 (Feb '70)

590. THEORY OF TRANSLATIONAL AND ORIENTATIONAL MELTING WITH APPLICATION TO LIQUID CRYSTALS, Kobayashi, K.K., J Phys Soc Japan v29 n1 p101 (Jly '70)

591. THEORY OF TRANSLATIONAL AND ORIENTATIONAL MELTING WITH APPLICATION TO LIQUID CRYSTALS, Kobayashi, K.K., p130: Abstr 3rd Int'l Liq Cryst Conf., Berlin, W. Germany (Aug '70)

592. THEORY OF TRANSLATIONAL AND ORIENTATIONAL MELTING WITH APPLICATION TO LIQUID CRYSTALS, Kobayashi, K.K., Mol Cryst & Liq Cryst v13 n2 p137 (May '71)

593. STATISTICAL DYNAMIC THEORY OF THE LIQUID CRYSTALS, Kobayashi, K.K., Oyo Buturi v40 n5 p532 (May '71)

594. void (duplicate)

595. LIQUID CRYSTALS AND LASERS, Kobayashi, K.K., Kagaku (Japan) v41 n8 p442 (1971)

596. PREPARATION OF ALPHANUMERIC INDICATORS WITH LIQUID CRYSTALS, Kobayashi, K.K., Shimojo, T., Kasano, K., Tsunda, I., p68: 1972 SID Symp Digest of Tech Papers, San Francisco, Calif., 6-8 Jne '72.

597. ON THE COLLECTIVE MODE IN LIQUID CRYSTALS, Kobayashi, K.K., Franklin, W.M., Moroi, D.S., Phys Lett (A) v42a n6 p449 (15 Jan '73)

598. MOLECULAR THEORY OF COLLECTIVE MODES IN NEMATIC LIQUID CRYSTALS, Kobayashi, K.K., Franklin, W.M., Moroi, D.S., Phys Rev (A) v7a n5 p1781 (May '73)

599. REAL-TIME DETECTION OF METALLIC OBJECTS USING LIQUID CRYSTAL MICROWAVE HOLOGRAMS, Kock, W.E., IEEE Proc v60 n9 p1105 (Sep '72)

600. EASTMAN LIQUID CRYSTAL PRODUCTS, Kodak, a Kodak Publication (No. JJ-14)

601. ELECTROHYDRODYNAMIC FLOW IN NEMATIC LIQUID CRYSTALS, Koelmans, H., Van Boxtel, A.M., Phys Lett (A) v32a n1 p32 (1 Jne '70)

602. ELECTROHYDRODYNAMIC FLOW IN NEMATIC LIQUID CRYSTALS, Koelmans, H., Van Boxtel, A.M., p113: Abstr 3rd Int'l Liq Cryst Conf., Berlin, W. Germany (Aug '70)

603. ELECTROHYDRODYNAMIC FLOW IN NEMATIC LIQUID CRYSTALS, Koelmans, H., Van Boxtel, A.M., Mol Cryst & Liq Cryst v12 n2 p185 (Jan '71)

604. THE ELECTRIC FIELD EFFECT OF NEMATIC ETHYL P-ANISILDENE-P-AMINOCINNAMATE, Kogure, O., Murase, K., Japan J Appl Phys v9 n10 p1280 (Oct '70)

605. PRODUCTION OF THIN LAYERS OF LIQUID CRYSTALS, Kolenko, E., Lopatina, L., Borodzulja, V., Proc 5th Czechoslovakia Conf Electronics & Vacuum Phys, Brno, 16-19 Oct '72.

606. CONTROLLABLE LIQUID CRYSTAL TRANSPARENCY FOR RECORDING OF HOLOGRAMS, Kompanets, I.N., Morozov, V.N., Nikitin, V.V., Blinov, L.M., Soviet J Quantum Electr v2 n3 p260 (Nov/Dec'72)

607. LIQUID CRYSTALS AND THEIR USE IN THE TESTING OF MATERIALS, Kopp, W.-U., Prakt Metallogr (Germany) v9 n7 p370 (Jly '72)

608. TRIPLET E.S.R.-SPECTRA OF CHARGE TRANSFER COMPLEXES IN LIQUID AND SINGLE CRYSTALS, Krebs, P., Sackmann, E., Schwartz, J., Chem Phys Lett v8 n5 p417 (1 Mch '71)

609. ORIENTATION DISTRIBUTION FUNCTION OF AROMATIC MOLECULES IN FROZEN LIQUID CRYSTALS FROM FROM THEIR TRIPLET E.S.R.-SPECTRA, Krebs, P., Sackmann, E., Mol Phys v23 n2 p437 (Feb '72)

610. AN X-RAY DIFFRACTION STUDY OF THREE MESOPHASES SHOWING SMECTIC 'A' MORPHOLOGY, Krigbaum, W.R., Poirier, J.C., Costello, M.J., Mol Cryst & Liq Cryst v20 n2 p133 (Mch '73)

611. SCATTERING OF LIGHT BY LIQUID CRYSTALLINE P-AZOXYANISOLE, Krishnamurty, D., Subramhanyam, H.S., p27: Abstr 3rd Int'l Liq Cryst Conf., Berlin, W. Germany (Aug '70)

612. SCATTERING OF LIGHT BY LIQUID CRYSTALLINE P-AZOXYANISOLE, Krishnamurty, D., Subramhanyam, H.S., Mol Cryst & Liq Cryst v14 n3-4 p209 (Sep '71)

613. PROTON MAGNETIC RESONANCE ABSORPTION IN N-BUTYL STEREATE, Krishnamurty, K.S., Krishnamurty, D., Mol Cryst & Liq Cryst v6 n3-4 p407 (Feb '70)

614. PRIFICATION OF LIQUID CRYSTALS, Kusabayashi, N., Oyo Buturi v40 n5 p523 (May '71)

615. INDICES DE REFRACTION DU P-METHOXY-P'-DIPHENYLACETYLENE EN PHASE NEMATIQUE, Labrunie, G., Bresse, M., Comptes Rendus (Ser. B) v276 n15 p647 (9 Apr '73)

616. BEHAVIOR OF A DEIONIZED LIQUID CRYSTAL SUBJECTED TO UNIPOLAR INJECTION IN NEMATIC AND ISOTROPIC PHASE, Lacroix, J.-C., Tobazeon, R., Appl Phys Lett v20 n7 p251 (1 Apr '72)

617. LIQUID CRYSTALS AND ELECTRONICS, Lafonta, P., Electron & Microelectron Ind (France) n159 p3 (15 Jne '72)

618. THE STRANGE WORLD OF LIQUID CRYSTALS, Laisk, E., Electron Aust (Australia) v32 n11 p12 (Feb '71)

619. REPLY TO PHASE TRANSITIONS IN TWO-DIMENSIONAL LIQUID CRYSTALS, Lakatos, K., J Chem Phys v55 n9 p4679 (1 Nov '71), see ibid v55 n9 p4678 (1971)

620. LIQUID CRYSTAL DISPLAYS, Lancaster, D., Radio-Electron v43 n2 p33 (Feb '72)

621. SPECTRE DES FLUCTUATIONS THERMIQUES A LA SURFACE LIBRE D'UN CRISTAL LIQUIDE NEMATIQUE, Langevin, D., Bouchiat, M.A., J Physique (France) v33 n1 p101 (Jan '72)

622. ANALYSE SPECTRALE DE LA LUMIERE DIFFUSEE PAR LA SURFACE LIBRE D'UN CRISTAL LIQUIDE NEMATIQUE. MESURE DE LA TENSION SUPERFICIELLE ET DES COEFFICIENTS DE VISCOSITE, Langevin, D., J Physique (France) v33 n2-3 p249 (Feb-Mch '72)

623. LIGHT SCATTERING FROM THE FREE SURFACE OF A NEMATIC LIQUID CRYSTAL, Langevin, D., Bouchiat, M.A., J Physique (France) colloque no. 1 suppl to v33 n2-3 pC1-77 (Feb-Mch '72)

624. FILAMENTOUS BACTERIAL VIRUSES VIII. LIQUID CRYSTALS OF FD, Lapointe, J., Marvin, D.A., Mol Cryst & Liq Cryst v19 n3-4 p269 (Jan '73)

625. NEMATIC ORDERING OF HARD RODS DERIVED FROM A SCALED PARTICLE TREATMENT, Lasher, G., J Chem Phys v53 n11 p4141 (1 Dec '70)

626. MONTE CARLO RESULTS FOR A DISCRETE-LATTICE MODEL OF NEMATIC ORDERING, Lasher, G., Phys Rev (A) v5 n3 p1350 (Mch '72)

627. NEMATIC LIQUID CRYSTAL ORDER - A MONTE CARLO CALCULATION, Lebwohl, P.A., Lasher, G., Phys Rev (A) v6a n1 p426 (Jly '72)

628. LIQUID CRYSTAL MATRIX DISPLAYS, Lechner, B.J., Marlowe, F.J., Nester, E.O., Tults, J., IEEE Proc v59 n11 p1566 (Nov '71)

629. RECHERCHES SUR LES SUBSTANCES MESOMORPHES. II. ARYLIDENE AMINOCINNAMATES SUBSTITUES, Leclerc, M., Billard, J., Jacques, J., Mol Cryst & Liq Cryst v10 n4 p429 (Jly '70)

630. ON SOME LIQUID CRYSTAL PROPERTIES OF CHOLESTERYL 2-(2-ETHOXYETHOXY)ETHYL CARBONATE AND RELATED COMPOUNDS, Leder, L.B., Chem Phys Lett v6 n4 p285 (15 Aug '70)

631. LIQUID CRYSTAL PROPERTIES OF SOME STERYL CHLORIDES, Leder, L.B., J Chem Phys v54 n11 p4671 (1 Jne '71)

632. HALFWIDTH OF THE CHOLESTERIC LIQUID CRYSTAL REFLECTIONS BAND, Leder, L.B., Olechna, D., Opt Commun v3 n5 p295 (Jly '71)

633. ROTATORY SENSE AND PITCH OF CHOLESTERIC LIQUID CRYSTALS, Leder, L.B., J Chem Phys v55 n6 p2649 (15 Sep '71)

634. RIGHT-ROTATORY CHOLESTERIC LIQUID CRYSTALS: DERIVATIVES OF $\Delta^{8(14)}$ - CHOLESTANOL, Leder, L.B., J Chem Phys v58 n3 p1118 (1 Feb '73)

635. WAVE PROPAGATION IN NEMATIC LIQUID CRYSTALS, Lee, J.D., Eringen, A.C., J Chem Phys v54 n12 p5027 (15 Jne '71)

636. ALIGNMENT OF NEMATIC LIQUID CRYSTALS, Lee, J.D., Eringen, A.C., J Chem Phys v55 n9 p4504 (1 Nov '71)

637. BOUNDARY EFFECTS OF ORIENTATION OF NEMATIC LIQUID CRYSTALS, Lee, J.D., Eringen, A.C., J Chem Phys v55 n9 p4509 (1 Nov '71)

638. CONTINUUM THEORY OS SMECTIC LIQUID CRYSTALS, Lee, J.D., Eringen, A.C., J Chem Phys v58 n10 p4203 (15 May '73)

639. BIREFRINGENCE OF TWO UNIFORM TEXTURES OF DI-(4-N-DECYLOXYBENZAL)-2-CHLORO 1,4-PHENYLEN-DIAMINE IN THE SMECTIC PHASE, Lefevre, M., Martinand, J.L., Durand, G., Veyssie, M., Comptes Rendus (Ser. B) v273 n11 p403 (13 Sep '71)

640. OBSERVATION OF WALL MOTIONS IN NEMATICS, Leger, L., Solid State Commun v10 n8 p697 (15 Apr '72)

641. N.M.R. STUDY OF THE ORDER PARAMETER IN A NEMATIC PHASE AT ROOM TEMPERATURE, Le Pesant, J.P., Papon, P., p61: Abstr 3rd Int'l Liq Cryst Conf., Berlin (Aug '70)

642. MAGNETIC FIELD EFFECT ON THE STRUCTURE OF NEMATIC LIQUID CRYSTAL DROPLETS IN EMULSION IN AN ISOTROPIC LIQUID, Le Roy, P., Debeauvais, F., Candau, S., Comptes Rendus (Ser. B) v274 n6 p419 (7 Feb '72)

643. GAIN AND VISUALIZATION OF THE MODES OF A THERMALLY STABILIZED H.C.N. LASER (LIQUID CRYSTAL CAMERA), Lesieur, J.P., Sexton, M.C., Veron, D., J Phys (D) v5 n7 p1212 (Jly '72)

644. SOME MAGNETO-HYDROSTATIC EFFECTS IN NEMATIC LIQUID CRYSTALS, Leslie, F.M., J Phys (D): Appl Phys v3 n6 p889 (Jne '70)

645. DISTORTION OF ORIENTATION PATTERNS IN LIQUID CRYSTALS BY A MAGNETIC FIELD, Leslie, F.M., p104: Abstr 3rd Int'l Liq Cryst Conf., Berlin, W. Germany (Aug '70)

646. DISTORTION OF TWISTED ORIENTATION PATTERNS IN LIQUID CRYSTALS BY MAGNETIC FIELDS, Leslie, F.M., Mol Cryst & Liq Cryst v12 n1 p57 (Dec '70)

647. THE THEORY OF LIQUID CRYSTALS, Leslie, F.M.,
p50: Conf on theor rheology, Aberystwyth, Wales, UK (Univ College Wales), 13-16 Apr '71.

648. THERMO-MECHANICAL COUPLING IN CHOLESTERIC LIQUID CRYSTALS, Leslie, F.M.,
Faraday Symposium on Liquid Crystals, Faraday Society, London, 1971.

649. MAGNETOHYDRODYNAMIC EFFECTS IN THE NEMATIC MESOPHASE, Leslie, F.M.,
Chem Phys Lett v13 n4 p368 (15 Mch '72)

650. DYNAMIC SHEAR PROPERTIES OF SOME SMECTIC LIQUID CRYSTALS, Letcher, S.V., Barlow, A.J.,
Phys Rev Lett v26 n4 p172 (25 Jan '71)

651. ULTRASONIC MEASUREMENTS IN LIQUID CRYSTALS, Letcher, S.V.,
p35: Program 84th Meeting Acoust Soc Am, Miami Beach, Fla., 28 Nov - 1 Dec '72)

652. STRUCTURE OF SMECTIC 'B' LIQUID CRYSTALS, Levelut, A.M.,
Comptes Rendus *Ser. B) v272 n17 p1018 (26 Apr '71)

653. LINEAR LOCAL ORDER IN MOLECULAR CRYSTALS AND LIQUID CRYSTALS (Abstr.), Levelut, A.M.,
Lambert, M., Acta Crystallogr (A) (Denmark) v28a Pt.4 suppl p5130 (15 Jly '72)

654. BRILLOUIN SCATTERING FROM SMECTIC LIQUID CRYSTALS, Liao, Y., Clark, N.A., Pershan, P.S.,
Phys Rev Lett v30 n14 p639 (2 Apr '73)

655. ANISOTROPIC ULTRASONIC WAVE PROPAGATION IN A NEMATIC LIQUID CRYSTAL PLACED IN A MAGNETIC
FIELD, Lieberman, E.D., Lee, J.D., Moon, F.C., Appl Phys Lett v18 n7 p280 (1 Apr '71)

656. COPOLYMERISATION DES MONOMERES MESOMORPHES DANS LA PHASE NEMATIQUE EN PRESANCE D'UN CHAMP
MAGNETIQUE, Liebert, L., Strzelecki, L., Comptes Rendus (Ser. C) v276 n8 p647 (19 Feb '73)

657. COUNTER-ION BINDING IS SOME LYOTROPIC LIQUID CRYSTALLINE PHASES STUDIED BY N.M.R.,
Lindblom, G., Lindman, B., p65: Abstr 3rd Int'l Liq Cryst Conf., Berlin (Aug '70)

658. ION BINDING IN LIQUID CRYSTALS STUDIED BY N.M.R. I. THE CETYLTRIMETHYLAMMONIUM BROMIDE/
HEXANOL/WATER SYSTEM, Lindblom, G., Lindman, B., Mol Cryst & Liq Cryst v14 n1-2 p49 (Aug
'71)

659. CRITICAL SLOWING OF FLUCTUATIONS IN A NEMATIC LIQUID CRYSTAL, Litster, J.D., Stinson, T.W.,
J Appl Phys v41 n3 p996 (1 Mch '70)

660. CRITICAL POINTS AND ALMOST CRITICAL POINTS (NEMATIC CRYSTALS AND FERROMAGNETS), Litster,
J.D., Conf on Dynamical Aspects of Critical Phenomena, New York, 9-11 Jne '70; Gordon and
Breach, London, 1972, p152.

661. THE ELECTROCHEMISTRY OF THE LIQUID CRYSTAL 'MBBA': THE ROLE OF ELECTRODE REACTION IN
DYNAMIC SCATTERING, Lomax, A., Hirasawa, R., Bard, A.J.,
J Electrochem Soc v119 n12 p1619 (Dec '72)

662. ORIENTATION OF P-DITHIIN IN A LYOTROPIC LIQUID CRYSTAL, Long, R.C.Jr., Goldstein, J.H.,
J Mol Spectrosc v40 n3 p632 (Dec '71)

663. HEAT TRANSPORT IN LIQUID CRYSTALS, Longley-Cook, M., Kessler, J.O.,
Mol Cryst & Liq Cryst v12 n4 p315 (Mch '71)

664. ANISOTROPIC ULTRASONIC PROPERTIES OF A NEMATIC LIQUID CRYSTAL, Lord, A.E.Jr., Labes, M.M.,
Phys Rev Lett v25 n9 p570 (31 Aug '7o)

665. ULTRASONIC INVESTIGATION OF THE CHOLESTERIC-NEMATIC TRANSITION, Lord, A.E.,Jr.,
Mol Cryst & Liq Cryst v18 n3-4 p313 (Oct '72)

666. ELECTRICAL FIELD DISTRIBUTION ASSOCIATED WITH DYNAMIC SCATTERING IN NEMATIC LIQUID
CRYSTALS, Lu, S., Jones, D., Appl Phys Lett v16 n12 p484 (15 Jne '70)

667. LIGHT DIFFARCTION PHENOMENA IN AN A.C.-EXCITED NEMATIC LIQUID CRYSTAL SAMPLE, Lu, S.,
Jones, D., J Appl Phys v42 n5 p2138 (Apr '71)

668. LIQUID CRYSTAL DISPLAYS FOR ELECTRONIC TIME-KEEPING, Lu, S., Jones, D., Halberstram, M., Eurocon 71 Digest, Lausanne, Switzerland, 18-22 Oct '71.

669. A CALCULATION OF THE ELASTIC CONSTANT K_{11} FOR A NEMATIC LIQUID CRYSTAL, Lubensky, T.C., Phys Lett (A) v33a n4 p202 (2 Nov '70)

670. MOLECULAR DESCRIPTION OF NEMATIC LIQUID CRYSTALS, Lubensky, T.C., Phys Rev (A) v2 n6 p2497 (1 Dec '70)

671. HYDRODYNAMICS OF CHOLESTERIC LIQUID CRYSTALS, Lubensky, T.C., Phys Rev (A) v6a n1 p452 (Jly '72)

672. LOW-TEMPERATURE PHASE OF INFINITE CHOLESTERICS, Lubensky, T.C., Phys Rev Lett v29 n4 p206 (24 Jly '72)

673. LIQUID CRYSTALS AND BROKEN SYMMETRY HYDRODYNAMICS, Lubensky, T.C., p34: Program, 84th Meeting Ac Soc Am, Miami Beach, Fla., 28 Nov - 1 Dec '72)

674. CHOLESTERIC-NEMATIC PHASE TRANSITIONS, Luckhurst, G.R., Smith, H.J., p116: Abstr 3rd Int'l Liq Cryst Conf., Berlin, W. Germany (Aug '70)

675. THE ALIGNMENT OF CYBOTACTIC GROUPS IN A NEMATIC MESOPHASE BY AN ELECTRIC FIELD, Luckhurst, G.R., Chem Phys Lett v9 n4 p289 (15 May '71)

676. MOLECULAR ORGANIZATION IN THE SMECTIC MESOPHASE OF ETHYL 4-AZOXYBENZOATE, Luckhurst, G.R., Sanson, A., Mol Cryst & Liq Cryst v16 n1-2 p179 (Feb '72)

677. LIQUID CRYSTALS, Luckhurst, G.R., Phys Bull (GB) v23 p279 (May '72)

678. ANGULAR DEPENDENT LINEWIDTHS FOR A SPIN PROBE DISSOLVED IN A LIQUID CRYSTAL, Luckhurst, G.R., Sanson, A., Mol Phys v24 n6 p1297 (Dec '72)

679. MOLECULAR ORGANIZATION IN THE NEMATIC AND SMECTIC 'A' MESOPHASES OF 4-N-BUTYLOXYBENZYLIDENE-4'-ACETOANILINE, Luckhurst, G.R., Setaka, M., Mol Cryst & Liq Cryst v19 n2 p179 (Dec'72)

680. SOLUTE-SOLVENT INTERACTIONS WITHIN THE NEMATIC MESOPHASE, Luckhurst, G.R., Setaka, M., Mol Cryst & Liq Cryst v19 n3-4 p279 (Jan '73)

681. LIQUID CRYSTAL CELL HAVING NONUNIFORM THICKNESS, Ludeman, C.P., Strnad, R.J., IBM Tech Discl Bull v15 n4 p1349 (Sep '72)

682. AN X-RAY STUDY OF THE MESOPHASES OF CHOLESTERYL STEARATE, Lydon, J.E., p72: Abstr 3rd Int'l Liq Cryst Conf., Berlin, W. Germany (Aug '70)

683. LIQUID CRYSTAL DISPLAYS FOR MATCHED FILTERING, MacAnally, R.B., Appl Phys Lett v18 n2 p54 (15 Jan '71)

684. ORIENTATIONAL ORDER IN ANISALDAZINE IN THE NEMATIC PHASE, Madhusudana, N.V., Shashidhar, R., Chandrasekhar, S., p29: Abstr 3rd Int'l Liq Cryst Conf., Berlin, W. Germany (Aug '70)

685. ORIENTATIONAL ORDER IN ANISALDAZINE IN THE NEMATIC PHASE, Madhusudana, N.V., Shashidhar, R., Scahndrasekhar, S., Mol Cryst & Liq Cryst v13 n1 p61 (Apr '71)

686. CHARACTERISTICS DATA OF LIQUID CRYSTAL FIELD DETECTORS, Magura, K., Nachrichtech Z (NTZ, Germany) v23 n9 p440 (Sep '70)

687. EFFECT OF ULTRASOUND ON A NEMATIC LIQUID CRYSTAL, Mailer, H., Likins, K.L., Taylor, T.R., Fergason, J.L., Appl Phys Lett v18 n4 p105 (15 Feb '71)

688. COTTON EFFECT INDUCED BY THE EFFECT OF A CHOLESTERIC MESOPHASE ON ACHIRAL MOLECULES, Mainusch, K.-J., Stegemeyer, H., Z Phys Chem (Frankfurt) v77 n1-6 p210 (1972)

689. ABSORPTION AND FLUORESCENCE MEASUREMENTS IN COMPENSATED CHOLESTERIC MESOPHASES, Pt.I: ORIENTATION OF CHROMOPHORES IN LIQUID CRYSTAL SOLVENTS, Mainusch, K.-J., Mueller, U.,

Pollmann, P., Stegemeyer, H., Z Naturforsch v27a n11 p1677 (Nov '72)

690. ZIRKULARPOLARISATION DER FLUORESZENZ ACHIRALER MOLEKUELE IN CHOLESTERISCHEN MESOPHASEN, Mainusch, K.-J., Pollmann, P., Stegemeyer, H., Naturwiss v60 n1 p48 (Jan '73)

691. APPLICATION DES CRISTAUX LIQUIDES AU CONTROLE NONDESTRUCTIF, Manaranche, J.C., J Phys D: Appl Phys v5 n6 p1120 (Jne '72)

692. PROPAGATION OF POLARIZED LIGHT IN A CHOLESTERIC STRUCTURE, Marathay, A.S., Opt Commun (Holland) v3 n6 p369 (Aug '71)

693. MATRIX-OPERATOR DESCRIPTION OF THE PROPAGATION OF POLARIZED LIGHT THROUGH CHOLESTERIC LIQUID CRYSTALS, Marathay, A.S., Opt Soc Am J v61 n10 p1363 (Oct '71)

694. REVERSIBLE ULTRAVIOLET IMAGING WITH LIQUID CRYSTALS, Margerum, J.D., Nimoy, J., Wong, S.-Y., Appl Phys Lett v17 n2 p51 (15 Jly '70)

695. TRANSPARENT PHASE IMAGES IN PHOTOACTIVATED LIQUID CRYSTALS, Margerum, J.D., Beard, T.D., Bleha, W.P., Wong, S.-Y., Appl Phys Lett v19 n7 p216 (1 Oct '71)

696. MECHANISM OF SHEAR INDUCED STRUCTURAL CHANGE IN LIQUID CRYSTALS. CHOLESTERIC-NEMATIC MIXTURES, Marsh, D., Pochan, J., J Chem Phys v58 n7 p2835 (1 Apr '73)

697. EXTENDED DEBYE THEORY FOR DIELECTRIC RELAXATIONS IN NEMATIC LIQUID CRYSTALS, Martin, A.S., Meier, G., Saupe, A., Faraday Symp on Liquid Crystals, Faraday Society, London, 1971.

698. A NEW ELASTIC-HYDRODYNAMIC THEORY OF LIQUID CRYSTALS, Martin, P.C., Swift, J., Pershan, P.S., p54: Abstr 3rd Int'l Liq Cryst Conf., Berlin, W. Germany (Aug '70)

699. NEW ELASTIC-HYDRODYNAMIC THEORY OF LIQUID CRYSTALS, Martin, P.C., Pershan, P.S., Swift, J., Phys Rev Lett v25 n13 p844 (28 Sep '70)

700. UNIFIED HYDRODYNAMIC THEORY FOR CRYSTALS, LIQUID CRYSTALS AND NORMAL FLUIDS, Martin, P.C., Parodi, O., Pershan, P.S., Phys Rev (A) v6 n6 p2401 (Dec '72)

701. ELECTRIC FIELD QUENCHING OF THERMAL FLUCTUATIONS OF ORIENTATION IN A NEMATIC LIQUID CRYSTAL, Martinand, J.L., Durand, G., Solid State Commun v10 n9 p815 (1 May '72)

702. ULTRASONIC RELAXATION IN A NEMATIC LIQUID CRYSTAL, Martinoty, P., Candau, S., Comptes Rendus (Ser. B) v271 n2 p107 (15 Jly '70)

703. VISCOSITY MEASUREMENTS OF NEMATIC LIQUID CRYSTALS BY A SHEAR WAVE REFLECTANCE TECHNIQUE, Martinoty, P., Candau, S., p127: Abstr 3rd Int'l Liq Cryst Conf., Berlin (Aug '70)

704. DETERMINATION OF VISCOSITY COEFFICIENTS OF A NEMATIC LIQUID CRYSTAL USING A SHEAR WAVES REFLECTANCE TECHNIQUE, Martinoty, P., Candau, S., Mol Cryst & Liq Cryst v14 n3-4 p243 (Sep '71)

705. DYNAMIC PROPERTIES NEAR THE NEMATIC-ISOTROPIC TRANSITION OF A LIQUID CRYSTAL, Martinoty, P., Candau, S., Debeauvais, F., Phys Rev Lett v27 n17 p1123 (25 Oct '71)

706. ULTRASONIC IMPEDOMETRIC STUDIES IN CHOLESTERIC LIQUID CRYSTALS, Martinoty, P., Candau, S., Phys Rev Lett v28 n21 p1361 (22 May '72)

707. MESURE DE L'IMPEDANCE MECANIQUE DE CISAILLEMENT DES MESOPHASES NEMATIQUE ET CHOLESTERIQUE AUX FREQUENCES ULTRASONORES, Martinoty, P., Candau, S., J Physique Colloque C6, supplement n11-12 v33 p81 (Nov-Dec '72)

708. THERMAL FLUCTUATIONS AND PROTON SPIN-LATTICE RELAXATION IN NEMATIC LIQUID CRYSTALS, Martins, A.F., Mol Cryst & Liq Cryst v14 n1-2 p85 (Aug '71)

709. A SIMPLE MODEL FOR SELF-DIFFUSION IN THE ISOTROPIC PHASE OF NEMATIC LIQUID CRYSTALS, Martins, A.F., Phys Lett (A) v38a n3 p211 (31 Jan '72)

710. TRANSITION BETWEEN CONDUCTION AND DIELECTRIC REGIMES OF THE ELECTROHYDRODYNAMIC INSTABILI-

...TIES IN A NEMATIC LIQUID CRYSTAL, Martunand, J.J., Michalovcz, A., Salin, D., Phys Lett (A) v39a n3 p181 (8 May '72)

711. EVIDENCE OF THE EXISTENCE OF TWO SOLID PHASES OF 'MBBA' IN SPECIFIC HEAT MEASUREMENT, Mayer, J., Waluga, T., Janik, J.A., Phys Lett (A) v41a n2 p102 (11 Sep '72)

712. LIQUID CRYSTALS VII. THE MESOMORPHIC BEHAVIOR OF HOMOLOGOUS P-ALKOXY-P'-ACYLOXYAZOXY-BENZENES, McCaffrey, M.T., Castellano, J.A., Mol Cryst & Liq Cryst v18 n3-4 p209 (Oct '72)

713. TEMPERATURE DEPENDENCE OF ORIENTATIONAL ORDER IN A NEMATIC LIQUID CRYSTAL AT CONSTANT MOLAR VOLUME, McColl, J.R., Shih, C.S., Phys Rev Lett v29 n2 p85 (10 Jly '72)

714. THE FELIX, A FAST LIQUID CRYSTAL, IS EASY ON POWER, EASY ON THE EYE, McDermott, J., Electron Design n20 p44 (28 Sep '72)

715. ANOMALOUS EFFECTS OWING TO ELECTRIC FIELDS IN THE NEMATIC PHASE OF 4,4'-DI-N-HEPTYLOXY-AZOXY BENZENE, McLemore, D.P., Carr, E.F., J Chem Phys v57 n8 p3245 (15 Oct '72)

716. SIMPLE MOLECULAR MODEL FOR THE SMECTIC 'A' PHASE OF LIQUID CRYSTALS, McMillan, W.L., Phys Rev (A) v4 n3 p1238 (Sep '71)

717. X-RAY SCATTERING FROM LIQUID CRYSTALS. I. CHOLESTERYL NONANOATE AND MYRISTATE, McMillan, W.L., Phys Rev (A) v6 n3 p936 (Sep '72)

718. MEASUREMENT OF SMECTIC-A-PHASE ORDER-PARAMETER FLUCTUATIONS IN THE NEMATIC PHASE OF P-N-OCTYLOXY BENZYLIDENE-P'-TOLUIDINE, McMillan, W.L., Phys Rev (A) v7a n5 p1673 (May '73)

719. EFFECT OF INTERMOLECULAR FORCES ON THE ANNIHILATION OF POSITRONS IN A NEMATIC LIQUID CRYSTAL, McNutt, J.D., Kinnison, W.W., Phys Rev (B) v5 n3 p826 (1 Feb '72)

720. MEASUREMENT OF THE ROTATIONAL VISCOSITY COEFFICIENT AND THE SHEAR-ALIGNMENT ANGLE IN NEMATIC LIQUID CRYSTALS, Meiboom, S., Hewitt, R.C., Phys Rev Lett v30 n7 p261 (12 Feb '73)

721. TYPE CHARACTER IMPRESSION COMPARISON GAUGE, Meier, H.E., IBM Tech Discl Bull v12 n12 p2290 (May '70)

722. ELECTRIC FIELD HYSTERESIS EFFECTS IN CHOLESTERIC LIQUID CRYSTALS, Melamed, L., Rubin, D., Appl Phys Lett v16 n4 p149 (15 Feb '70)

723. ELECTRIC-FIELD HYSTERESIS EFFECTS ON THE TRANSMITTANCE OF LIQUID CRYSTAL FILMS, Melamed, L., Rubin, D., p15: 1970 Spring Meeting of the Opt Soc Am, Philadelphia, Pa., 7-10 Apr '70)

724. ELECTRIC-FIELD HYSTERESIS EFFECTS ON THE TRANSMITTANCE OF LIQUID CRYSTAL FILMS (Abstr.), Melamed, L., Rubin, D., Opt Soc Am J v60 n5 p717 (May '70)

725. ELECTRIC-FIELD HYSTERESIS EFFECTS IN THE TRANSMISSION PROPERTIES OF LIQUID CRYSTALS, Melamed, L., Rubin, D., Report NASA-TM-X-2044 (Jly '70)

726. SELECTED OPTICAL PROPERTIES OF CHOLESTERIC LIQUID CRYSTALS, Melamed, L., Rubin, D., Appl Opt v10 n5 p1103 (May '71)

727. STRESS-OPTIC EFFECTS IN A LIQUID CRYSTAL VELOCIMETER, Melamed, L., p11: 1971 Annual Meeting Opt Soc Am, Ottawa, Canada, 5-8 Oct '71.

728. STRESS-OPTIC EFFECT IN A LIQUID CRYSTAL VELOCIMETER (Abstr.), Melamed, L., Opt Soc Am J v61 n11 p1551 (Nov '71)

729. THERMALLY ADDRESSED ELECTRICALLY ERASED HIGH-RESOLUTION LIQUID CRYSTAL LIGHT VALVES, Melchior, H., Kahn, F.J., Maydan, D., Fraser, D.B., Appl Phys Lett v21 n8 p392 (15 Oct '72)

730. THE POLYMORPHIC AND MESOMORPHIC BEHAVIOR OF FOUR ESTERS OF CHOLESTEROL, Merritt, W.G., Cole, G.D., Walker, W.W., Mol Cryst & Liq Cryst v15 n2 p105 (Nov '71)

731. POINT DISCLINATIONS AT A NEMATIC-ISOTROPIC LIQUID INTERFACE, Meyer, R.B., p112: Abstr 3rd Int'l Liq Cryst Conf., Berlin, W. Germany (Aug '70)

732. POINT DISCLINATION AT A NEMATIC-ISOTROPIC LIQUID INTERFACE, Meyer, R.B., Mol Cryst & Liq Cryst v16 n4 p355 (Apr '72)

733. ON THE EXISTENCE OF EVEN INDEXED DISCLINATIONS IN NEMATIC LIQUID CRYSTALS, Meyer, R.B., Philosophical Mag v27 n2 p405 (Feb '73)

734. THE INTERACTION BETWEEN A DISCLINATION IN A NEMATIC LIQUID CRYSTAL AND A RUBBED SURFACE, Meyer, R.B., Solid State Communic v12 n7 p585 (1 Apr '73)

735. ELECTROHYDRODYNAMIC INSTABILITIES IN NEMATIC LIQUID CRYSTALS IN LOW-FREQUENCY FIELDS, Meyerhofer, D., Sussman, A., Appl Phys Lett v20 n9 p337 (1 May '72)

736. ELECTRO-OPTIC AND HYDRODYNAMIC PROPERTIES OF NEMATIC LIQUID CRYSTAL FILMS WITH FREE SURFACES, Meyerhofer, D., Sussman, A., Williams, R., J Appl Opt v43 n9 p3685 (Sep '72)

737. N.M.R. DETERMINATION OF SOME DEUTERIUM QUADRUPOLE COUPLING CONSTANTS IN NEMATIC SOLUTIONS, Millett, F.S., Dailey, B.P., J Chem Phys v56 n7 p3249 (1 Apr '72)

738. ROLE OF CHARGE-FLUCTUATION FORCES IN ADLINEATION OF SIMILAR MOLECULES, Mishra, R.K., Mol Cryst & Liq Cryst v10 n1-2 p85 (May '70)

739. PRINCIPLES UNDERLYING "FLUCTUATING CRYSTALLINITY" IN CELL MEMBRANES AND OTHER BIOLOGICAL STRUCTURES AND ITS IMPLICATIONS, Mishra, R.K., DeVries, A., Brown, G.H., Petscher, E.J., Barenberg, S., Geil, P.H., Falor, W.H., p96: Abstr 3rd Int'l Liq Cryst Conf., Berlin (Aug '70)

740. LIQUID CRYSTALS: A NEW TECHNIQUE FOR THERMAL MAPPING OF ELECTRONIC COMPONENTS, Mizell, L., p35: 4th Int'l Microelectronics Conf., Munich, Germany, 9-11 Nov '70.

741. APPLICATION OF NEMATIC LIQUID CRYSTALS TO CONTROL THE COHERENCE OF LASER BEAMS, Mizuno, H., Tanaka, S., Opt Communic v3 n5 p320 (Jly '71)

742. LIQUID CRYSTALS IN ELECTRONICS, Moravec, F., Lescinsky, M., Hoff, F., Slaboproudy Obsor (Czechoslovakia) v33 n5 p195 (1972)

743. LIQUID CRYSTAL LIGHT VALVE (in Czech), Moravec, F., Czechoslovak J Phys vA22 n3 p320 (1972)

744. EQUIVALENT PERMITTIVITY TENSOR FOR PARALLEL ANISOTROPIC HOMOGENEOUS STRATIFIED MEDIA, Moroi, D.S., Mol Cryst & Liq Cryst v18 n3-4 p327 (Oct '72)

745. BIREFRINGENCE OF COMPENSATED CHOLESTERIC LIQUID CRYSTALS, Mueller, W.U., Stegemeyer, H., Berichte der Bunsen Gesellschaft v77 n1 p20 (Jan '73)

746. SOUND VELOCITY IN A NEMATIC LIQUID CRYSTAL, Mullen, M.E., Luthi, B., Stephen, M.J., Phys Rev Lett v28 n13 p799 (27 Mch '72)

747. SYNTHESIS AND LIQUID CRYSTALLINE PROPERTIES OF TERMINALLY SUBSTITUTED BENZYLIDENEANILINE, Murase, K., Elec Commun Lab Tech J (Japan) v21 n6 p1159 (1972)

748. AN N.M.R. MEASUREMENT OF THE DIFFUSION ANISOTROPY IN A NEMATIC LIQUID CRYSTAL, Murphy, J.A., Doane, J.W., Mol Cryst & Liq Cryst v13 n1 p93 (Apr '71)

749. HELICAL TWISTING POWER IN MIXTURES OF NEMATIC AND CHOLESTERIC LIQUID CRYSTALS, Nakagiri, T., Kodama, H., Kobayashi, K.K., Phys Rev Lett v27 n9 p564 (3 Aug '71)

750. A HELICAL TWISTING POWER IN MIXTURES OF NEMATIC AND CHOLESTERIC LIQUID CRYSTALS, Nakagiri,T., Phys Lett (A) v36a n5 p427 (27 Sep '71)

751. NEMATIC LIQUID CRYSTALS AND THEIR APPLICATIONS, Nannichi, Y., Ohnishi, Y., J Inst Electronics Commun Engrs Japan v53 n3 p358 (Mch '70)

752. STIMULATED RAMAN SCATTERING IN A NEMATIC LIQUID CRYSTAL, Narasimha Rao, D.V.L.G., Agrawal, D.K., Phys Lett (A) v37a n5 p383 (20 Dec '71)

753. CONDUCTIVE SPACERS FOR DISPLAY DEVICES, Narasimham, M.A.,
IBM Tech Discl Bull v15 n8 p2566 (Jan '73)

754. LIQUID CRYSTAL LINE FLARING CHARACTER FORMAT, Nassimbene, E.G.,
IBM Tech Discl Bull v15 n5 p1544 (Oct '72)

755. TEMPERATURE DEPENDENCE OF ANISOTROPIC-ULTRASONIC PROPAGATION IN A NEMATIC LIQUID CRYSTAL,
Natale, G.G., Phys Rev Lett v28 n22 p1439 (29 May '72)

756. ANISOTROPIES OF THE ^{19}F CHEMICAL SHIFTS IN FLUOROBENZENE COMPOUNDS FROM N.M.R. IN LIQUID CRYSTALS, Nehring, J., Saupe, A., J Chem Phys v52 n3 p1307 (1 Feb '70)

757. EXPERIMENTAL AND THEORETICAL STUDIES OF ALIGNMENT SINGULARITIES IN LIQUID CRYSTALS,
Nehring, J., p51: Abstr 3rd Int'l Liq Cryst Conf., Berlin (Aug '70)

758. ON THE ELASTIC THEORY OF UNIAXIAL LIQUID CRYSTALS, Nehring, J., Saupe, A.,
J Chem Phys v54 n1 p337 (1 Jan 71)

759. CALCULATION OF THE ELASTIC CONSTANTS OF NEMATIC LIQUID CRYSTALS, Nehring, J., Saupe, A.,
J Chem Phys v56 n11 p5527 (1 June '72)

760. THE FORMATION OF THREADS IN THE DYNAMIC SCATTERING MODE OF NEMATIC LIQUID CRYSTALS,
Nehring, J., Petty, M.S., Phys Lett (A) v40a n4 p307 (31 Jly '72)

761. CALCULATION OF THE STRUCTURE AND ENERGY OF NEMATIC THREADS, Nehring, J.,
Phys Rev (A) v7a n5 p1737 (May '73)

762. EFFECTS OF ELECTRIC FIELDS ON NEMATIC LIQUID CRYSTALS, Nemec, J., Cook, B.D.,
p105: Abstr 3rd Int'l Liq Cryst Conf., Berlin, W. Germany (Aug '70)

763. PHASE AND TEMPERATURE DEPENDENCE OF POSITION ANNIHILATION IN LIQUID CRYSTALS,
Nicholas, J.B., Ache, H.J., J Chem Phys v57 n4 p1597 (15 Aug '72)

764. SPIN RELAXATION IN NEMATIC SOLVENTS, Nordio, P.L., Rigatti, G., Segre, U.,
J Chem Phys v56 n5 p2117 (1 Mch '72)

765. A QUANTUM THEORY OF LIQUID CRYSTALS, Novakovic, L., Shukla, G.C.,
Fizika (Zagreb) v4 n1 p29 (Jan '72)

766. THE EFFECT OF HETEROCYCLIC NITROGEN ON THE MESOMORPHIC BEHAVIOR OF 4-ALKOXYBENZYLIDENE-
-2-ALKOXY-5'-AMINOPYRIDINES, Oh, C.S., Mol Cryst & Liq Cryst v19 n2 p95 (Dec '72)

767. MAGNETIC LIQUID CRYSTALLINE COMPOSITIONS, O'Hern, R.J., Voorhis, A.M.K.,
IBM Tech Discl Bull v13 n8 p2265 (Jan '71)

768. ELECTRO-OPTICAL PROPERTIES OF NEMATIC LIQUID CRYSTAL FILMS, Ohtsuka, T., Tsukamoto, M.,
Japanese J Appl Phys v10 n8 p1046 (Aug '71)

769. 'AC' ELECTRIC-FIELD-INDUCED CHOLESTERIC-NEMATIC PHASE TRANSITION IN MIXED LIQUID CRYSTAL
FILMS, Ohtsuka, T., Tsukamoto, M., Japanese J Appl Phys v12 n1 p20 (Jan '73)

770. LIQUID CRYSTAL MATRIX DISPLAY, Ohtsuka, T., Tsukamoto, M., Tsuchiya, M.,
Japanese J Appl Phys v12 n3 p371 (Mch '73)

771. ELECTRO-OPTIC EFFECT IN CELLS OF A ROOM-TEMPERATURE NEMATIC LIQUID CRYSTAL:'MBBA',
Onnagawa, H., Ootake, T., Miyashita, K., Oyo Buturi v40 n5 p510 (May '71)

772. A.C.-D.C. TECHNIQUE FOR RAPID CONICAL-HELICAL PERTURBATION IN CHOLESTERIC LIQUID CRYSTALS,
Oron, N., Labes, M.M., Appl Phys Lett v21 n5 p243 (1 Sep '72)

773. A.C.-D.C. REGIMES IN THE ELECTROHYDRODYNAMICAL UNSTABILITIES OF A NEMATIC LIQUID CRYSTAL,
Orsay Liquid Crystal Group, p98: Abstr 3rd Int'l Liq Cryst Conf., Berlin, (Aug '70)

774. HYDRODYNAMIC INSTABILITIES IN NEMATIC LIQUIDS UNDER A.C. ELECTRIC FIELDS,
Orsay Liquid Crystal Group, Phys Rev Lett v25 n24 p1642 (14 Dec '70)

775. THEORY OF LIGHT SCATTERING BY NEMATICS, Orsay Liquid Crystal Group, p195: "Liquid Crystals and Ordered Fluids," Plenum Press, 1970.

776. RECENT EXPERIMENTAL INVESTIGATIONS IN NEMATIC AND CHOLESTERIC MESOPHASES, Orsay Liquid Crystal Group, p447: "Liquid Crystals and Ordered Fluids," Plenum Press, 1970.

777. A.C. AND D.C. REGIMES OF THE ELECTROHYDRODYNAMIC INSTABILITIES IN NEMATIC LIQUID CRYSTALS, Orsay Liquid Crystal Group, Mol Cryst & Liq Cryst v12 n3 p251 (Feb '71)

778. LES CRISTAUX LIQUIDES, Orsay Liquid Crystal Group, La Recherche v2 n12 p433 (May '71)

779. VISCOSITY MEASUREMENTS BY QUASI ELASTIC LIGHT SCATTERING IN P-AZOXYANISOL, Orsay Liquid Crystal Group, Mol Cryst & Liq Cryst v13 n2 p187 (May '71)

780. SIMPLIFIED ELASTIC THEORY FOR SMECTICS 'C', Orsay Liquid Crystal Group, Solid State Commun v9 n10 p653 (15 May '71)

781. QUASI ELASTIC RAYLEIGH SCATTERING IN AN SMECTIC LIQUID CRYSTAL, Orsay Liquid Crystl Group, J Physique (France) Colloque No.1 suppl to v33 n2-3 pC1-76 (Feb-Mch '72)

782. TEMPERATURE DEPENDENCE OF THE SELF-DIFFUSION TENSOR OR PARA-AZOXYANISOLE, Otnes, K., Pynn, R., Janik, J.A., Janik, J.M., Phys Lett (A) v38a n5 p335 (28 Feb '72)

783. POLYMERIZATION OF A NEMATIC LIQUID CRYSTAL MONOMER, Paleos, C.M., Labes, M.M., p141: Abstr 3rd Int'l Liq Cryst Conf., Berlin, W. Germany (Aug '70)

784. POLYMERIZATION OF A NEMATIC LIQUID CRYSTAL MONOMER, Paleos, C.M., Labes, M.M., Mol Cryst & Liq Cryst v11 n4 p385 (Nov '70)

785. CONDITIONS OF STABILITY FOR LIQUID CRYSTALLINE PHOSPHOLIPID MEMBRANES, Papahadjopoulos, D., Ohki, S., p13: "Liquid Crystals and Ordered Fluids," Plenum Press, 1970.

786. A STATISTICAL MODEL FOR TRANSITIONS IN NEMATIC LIQUID CRYSTALS, Papon, P., LePesant, J.P., Chem Phys Lett v12 n2 p331 (15 Dec '71)

787. STRUCTURAL RELAXATION AND SOUND PROPAGATION IN CHOLESTERIC LIQUID CRYSTALS, Papoular, M., Phys Lett (A) v31a n2 p65 (26 Jan '70)

788. ONDES DE SURFACE DANS UN CRISTAL LIQUIDE NEMATIQUE, Papoular, M., Rapini, A., J Physique (France) C1 suppl n4 v31 pC1-27 (Apr '70)

789. MAGNETIC STABILIZATION OF NEMATIC TURBULENCE, Papoular, M., Phys Lett (A) v38a n3 p173 (31 Jan '72)

790. ANOMALOUS ALIGNMENT AND DOMAINS IN A NEMATIC LIQUID CRYSTAL, Parker, J.H., Carr, E.F., J Chem Phys v55 n4 p1846 (15 Aug '71)

791. TRANSIENT SURFACE TEMPERATURE RESPONSE OF LIQUID CRYSTAL FILMS, Parker, R., Mol Cryst & Liq Cryst v20 n2 p99 (Mch '73)

792. STRESS TENSOR FOR A NEMATIC LIQUID CRYSTAL, Parodi, O., J Physique (France) v31 n7 p581 (Jly '70)

793. CHOLESTERIC LIQUID CRYSTALS IN OSCILLATING MAGNETIC FIELDS, Parsons, J.D., Hayes, C.F., Solid State Communic v12 n5 p299 (1 Mch '73)

794. INTERFACIAL AND TEMPERATURE GRADIENT EFFECTS ON THERMAL CONDUCTIVITY OF A LIQUID CRYSTAL, Patharkar, M.N., Rajan, V.S.V., Picot, J.J.C., Mol Cryst & Liq Cryst v15 n3 p225 (Dec '71)

795. BIREFRINGENCE AND POLYMORPHISM OF LIQUID CRYSTALS, Pelzl, G., Sackmann, H., Faraday Symposium on Liquid Crystals, Faraday Society, London, 1971.

796. BIREFRINGENCE OF SMECTIC MODIFICATIONS OF THE HOMOLOGOUS THALLIUM SOAPS, Pelzl, G., Sackmann, H., Mol Cryst & Liq Cryst v15 n1 p75 (Oct '71)

797. VOLTAGE-INDUCED VORTICITY AND OPTICAL FOCUSING IN LIQUID CRYSTALS, Penz, P.A., Phys Rev Lett v24 n25 p1405 (22 Jne '70)

798. ELECTRIC FIELD INDUCED VORTICITY IN P-AZOXYANISOLE, Penz, P.A., p111: Abstr 3rd Int'l Liq Cryst Conf., Berlin, W. Germany (Aug '70)

799. ORDER PARAMETER DISTRIBUTION FOR THE ELECTROHYDRODYNAMIC MODE OF A NEMATIC LIQUID CRYSTAL, Penz, P.A., Mol Cryst & Liq Cryst v15 n2 p141 (Nov '71)

800. ELECTROHYDRODYNAMIC SOLUTIONS FOR NEMATIC LIQUID CRYSTALS, Penz, P.A., Ford, G.W., Appl Phys Lett v20 n11 p415 (1 Jnu '72)

801. ELECTROMAGNETIC HYDRODYNAMICS OF LIQUID CRYSTALS, Penz, P.A., Ford, G.W., Phys Rev (A) v6 n1 p414 (Jly '72)

802. ELECTROHYDRODYNAMIC SOLUTIONS FOR THE HOMEOTROPIC NEMATIC-LIQUID CRYSTAL GEOMETRY, Penz, P.A., Phys Rev (A) v6 n4 p1676 (Oct '72)

803. ASPECTS OF PHYSICAL PROPERTIES AND APPLICATIONS OF LIQUID CRYSTALS, Petrie, S.E.B., Buecher, H.K., Klingbiel, R.T., Rose, P.I., Eastman Organic Chemical Bull, Kodak Publication No. JJ-60-731, v45 n2 p1 (1973)

804. MOLECULAR PARAMETERS AND DIELECTRIC ANISOTROPY OF LIQUID CRYSTAL P-AZOXYANISOLE, Petrov, A.G., C.R. Acad Bulg Sci (Bulgaria) v24 n5 p573 (1971)

805. RAMAN SPECTRA OF LIQUID AND CRYSTALLINE 'ClCN' AND 'BrCN', Pezolet, M., Savioe, R., J Chem Phys v54 n12 p5267 (15 Jne '71)

806. CHOLESTERIC LIQUID CRYSTALS: AMBIENT TEMPERATURE EFFECTS, Pick, P.G., Fabijanic, J., Mol Cryst & Liq Cryst v15 n4 p371 (Jan '72)

807. THE EFFECT OF AMBIENT LIGHT ON LIQUID CRYSTAL TAPES, Pick, P.G., Fabijanic, J., Stewart, A., Mol Cryst & Liq Cryst v20 n1 p47 (Feb '73)

808. EFFECT OF OPTICAL ROTATION ON LOW-ANGLE LIGHT-SCATTERING PATTERNS, Picot, C., Stein, R.S., J Polymer Sci A-2 v8 Pt.A-2 n9 p1491 (Sep '70)

809. STATIC AND DYNAMIC BEHAVIOR OF A NEMATIC LIQUID CRYSTAL IN A MAGNETIC FIELD. PART I: STATIC RESULTS, Pieranski, P., Brochard, F., Guyon, E., J Physique (France) v33 n7 p681 (Jly '72)

810. STATIC AND DYNAMIC BEHAVIOR OF A NEMATIC LIQUID CRYSTAL IN A MAGNETIC FIELD. PART II: DYNAMICS, Pieranski, P., Brochard, F., Guyon, E. J Physique (France) v34 n1 p35 (Jan '73)

811. HEAT CONVECTION IN LIQUID CRYSTALS HEATED FROM ABOVE, Pieranski, P., Dubois-Violette, E., Guyon, E., Phys Rev Lett v30 n16 p736 (16 Apr '73)

812. HIGH FREQUENCY ELECTROHYDRODYNAMIC EFFECT IN LIQUID CRYSTALS, Pikin, S.A., Zh Eksp & Teor Fiz (USSR) v61 n5 p2133 (1971)

813. HIGH-FREQUENCY ELECTRO-HYDRODYNAMICAL EFFECT IN LIQUID CRYSTALS, Pikin, S.A., Soviet Phys - JETP v34 n5 p1137 (May '72)

814. MAGNETIC PROPERTIES OF LIQUID CRYSTALS, Pincus, P., J Appl Phys v41 n3 p974 (1 Mch '70)

815. MESURE DE L'ANISOTROPIE DIAMAGNETIQUE D'UNE CONFIGURATION NEMATIQUE ORIENTEE PAR UN CHAMP MAGNETIQUE, Poggi, Y., Aleonard, R., Comptes Rendus (Ser.B) v276 n15 p643 (9 Apr '73)

816. NUCLEATION STUDIES OF SUPERCOOLED CHOLESTERIC LIQUID CRYSTALS, Pochan, J.M., Gibson, H.W., Am Chem Soc J v93 n5 p1279 (10 Mch '71)

817. SHEAR-INDUCED TEXTURE CHANGES IN CHOLESTERIC LIQUID CRYSTAL MIXTURES, Pochan, J.M., Erhardt, P.F., Phys Rev Lett v27 n12 p790 (20 Sep '71)

818. MECHANISM OF SHEAR-INDUCED STRUCTURAL CHANGES IN LIQUID CRYSTALS - CHOLESTERIC MIXTURES,

Pochan, J.M., Marsh, D.G., J Chem Phys v57 n3 p1193 (1 Aug '72)

819. CRYSTAL NUCLEATION STUDIES IN SUPERCOOLED MESOMORPHIC PHASES OF CHOLESTERYL DERIVATIVES, Pochan, J.M., Gibson, H.W., Am Chem Soc J v94 n16 p5573 (9 Aug '72)

820. ON THE EXISTENCE OF NEMATIC SECONDARY STRUCTURES, Pohl, L., Steinstraesser, R., p70: Abstr 3rd Int'l Liq Cryst Conf., Berlin (Aug '70)

821. STRUCTURE DEPENDENCE OF CHOLESTERIC MESOPHASES, Pohlmann, J.L.W., Elser, W., p90: Abstr 3rd Int'l Liq Cryst Conf., Berlin (Aug '70)

822. STRUCTURE DEPENDENCE OF CHOLESTERIC MESOPHASES, Pohlmann, J.L.W., Elser, W., Boyd, P.R., p92: Abstr 3rd Int'l Liq Cryst Conf., Berlin (Aug '70)

823. STRUCTURE DEPENDENCE OF CHOLESTERIC MESOPHASES II., Pohlmann, J.L.W., Elser, W., Boyd, P.R., Mol Cryst & Liq Cryst v13 n3 p243 (Jne '71)

824. STRUCTURE DEPENDENCE OF CHOLESTERIC MESOPHASES IV. 17BETA-ALKYL SUBSTITUTED ANDROST-5-EN-3BETA-OLS, Pohlmann, J.L.W., Elser, W., Boyd, P.R., Mol Cryst & Liq Cryst v13 n3 p271 (Jne '71)

825. CHOLESTERYL w-PHENYLALKANOATE, Pohlmann, J.L.W., Elser, W., Boyd, P.R., Mol Cryst & Liq Cryst v20 n1 p87 (Feb '73)

826. NEMATIC LIQUID CRYSTAL USED AS AN INSTANTANEOUS HOLOGRAPHIC MEDIUM, Poisson, F., Opt Communic v6 n1 p43 (Sep '72)

827. FLOW BEHAVIOR OF COMPENSATED CHOLESTERIC LIQUID CRYSTAL MIXTURES, Pollmann, P., Z Naturforsch (A) v27a n4 p719 (Apr '72)

828. THERMODYNAMICS OF LIQUID CRYSTALS, Porter, R.S., Johnson, J.F., p10: Nat Acad Sci Autumn Meeting, Hanover, N.H., 13-15 Oct '69)

829. THE INFLUENCE OF CONTACT MATERIALS AND WORKING CONDITIONS ON THE SCATTERING PROPERTIES OF NEMATIC LIQUID CRYSTALS (Abstr.), Pospisil, M., Proc 5th Czechoslovakia Conf Electronics & Vacuum Phys, Brno, Czechoslovakia, 16-19 Oct '72.

830. OBSERVATION OF ALIGNMENT IN A QUENCHED LIQUID CRYSTAL WITH THE MOSSBAUER EFFECT, Potasek, M.J. Munck, E., Groves, J.L., Debrunner, P.G., Che, Phys Lett v15 n1 p55 (15 Jly '72)

831. THE AGGREGATION OF POLY-GAMMA-BENZYL-L-GLUTAMATE IN MIXED SOLVENT SYSTEMS, Powers, J.C.Jr., p365: "Liquid Crystals and Ordered Fluids", Plenum Press, 1970.

832. THERMALLY ACTIVATED LIQUID CRYSTAL DISPLAY, Powers, J.V., IBM Tech Discl Bull v15 n6 p1811 (Nov '72)

833. TRANSPARENCY CHARACTERISTICS OF SEVERAL CHOLESTERYL ESTERS, Poziomek, E.J., Novak, T.J., Mackay, R.A., Mol Cryst & Liq Cryst v15 n4 p283 (Jan '72)

834. TRANSITIONS IN MESOPHASE FORMING SYSTEMS. I. TRANSFORMATION KINETICS AND PRETRANSITION EFFECTS IN CHOLESTERYL MYRISTATE, Price, F.P., Wendorff, J.H., J Phys Chem v75 n18 p2839 (2 Sep '71)

835. TRANSITIONS IN MESOPHASES FORMING SYSTEMS, II. TRANSFORMATION KINETICS AND PROPERTIES OF CHOLESTERYL ACETATE, Price, F.P., Wendorff, J.H., J Phys Chem v75 n18 p2849 (2 Sep '71)

836. COMMENTS ON THE LATTICE MODEL OF LIQUID CRYSTALS, Priest, R.G., Phys Rev Lett v26 n8 p423 (22 Feb '71)

837. A CALCULATION OF THE ELASTIC CONSTANTS OF A NEMATIC LIQUID CRYSTAL, Priest, R.G., Mol Cryst & Liq Cryst v17 n2 p129 (Jne '72)

838. THEORY OF THE FRANK ELASTIC CONSTANTS OF NEMATIC LIQUID CRYSTALS, Priest, R.G., Phys Rev (A) v7a n2 p720 (Feb '73)

839. DETERMINATION OF TWIST VISCOSITY COEFFICIENT IN THE NEMATIC MESOPHASE, Prost, J., Gasparoux, H., Phys Lett (A) v36a n3 p245 (30 Aug '71)

840. MODIFICATION UNDER THE ACTION OF A MAGNETIC FIELD OF THE ORIENTATION OF MOLECULES IN A DROP OF A NEMATIC SUBSTANCE, Prost, J., Gasparoux, H., Comptes Rendus (Ser. B) v273 n9 p335 (30 Aug '71)

841. PERIODIC STRUCTURE INDUCED IN A NEMATIC PHASE BY A ROTATING MAGNETIC FIELD, Prost, J. Canet, Q., Comptes Rendus *Ser. B) v274 n1 p54 (3 Jan '72)

842. COMMENTS ON THERMOMECHANICAL COUPLING IN CHOLESTERIC LIQUID CRYSTALS, Prost, J., Solid State Communic v11 n1 p183 (1 Jly '72)

843. ELECTROMAGNETIC MICROWAVES VISUALIZATION WITH LIQUID CRYSTALS, Puyhaubert, J., Onde Elec (France) v52 n5 p213 (May '72)

844. CARBONATE SCHIFF BASE NEMATIC LIQUID CRYSTALS: SYNTHESIS AND ELECTRO-OPTIC PROPERTIES, Rafuse, M.J., Soref, R.A., Mol Cryst & Liq Cryst v18 n2 p95 (Sep '72)

845. ON THE ORIENTATIONAL TRANSITION IN LIQUID CRYSTALS, Raich, J.C., Etters, R.D., Chem Phys Lett v6 n5 p491 (1 Sep '70)

846. AN ANALYSIS OF THE INTERFACIAL EFFECT ON THE HEAT CONDUCTION IN A NEMATIC MESOPHASE, Rajan, V.S.V., Picot, J.J.C., Mol Cryst & Liq Cryst v17 n2 p109 (Jne '72)

847. ANISOTROPIC THERMAL DIFFUSION IN NEMATIC 'APAPA', Rajan, V.S.V., Patharkar, M.N., Picot, J.J.C., Mol Cryst & Liq Cryst v18 n3-4 p279 (Oct '72)

848. THERMAL TRANSPORT PHENOMENA IN NEMATIC LIQUID CRYSTALS: A REVIEW, Rajan, V.S.V., Picot, J.J.C., Mol Cryst & Liq Cryst v20 n1 p47 (Feb '73)

849. NON-FOURIER HEAT CONDUCTION IN A NEMATIC MESOPHASE, Rajan, V.S.V., Picot, J.J.C., Mol Cryst & Liq Cryst v20 n1 p69 (Feb '73)

850. STIMULATED BRILLOUIN SCATTERING IN A LIQUID CRYSTAL, Rao, D.V.G.L.N., Phys Lett (A) v32A n7 p533 (7 Sep '70)

851. INSTABILITES MQGNETIQUES D'UN SMECTIQUE 'C', Rapini, A., J Physique (France) v33 n2-3 p237 (Feb-Mch '72)

852. POUVOIR ROTATOIRE ACOUSTIQUE D'UN VERRE CHOLESTERIQUE, Rapini, A., Comptes Rendus (Ser. B) v275 n19 p701 (6 Nov '72)

853. FERRONEMATICS, Rault, J., Cladis, P.E., Burger, J.P., Phys Lett (A) v32a n3 p199 (29 Jne'70)

854. CHOLESTERIC TEXTURE NEAR T_c AND IN PRESENCE OF A MAGNETIC FIELD, Rault, J., Cladis, P.E., p120: Abstr 3rd Int'l Liq Cryst Conf., Berlin (Aug '70)

855. ON A NEW METHOD OF STUDYING THE MOLECULAR ORIENTATION AT THE SURFACE OF A CHOLESTERIC CRYSTAL, Rault, J., Comptes Rendus (Ser. B) v272 n22 p1275 (2 Jne '71)

856. CHOLESTERIC TEXTURE NEAR T_c AND IN PRESENCE OF A MAGNETIC FIELD, Rault, J., Cladis, P.E., Mol Cryst & Liq Cryst v15 n1 p1 (Oct '71)

857. HELICAL DISLOCATION LINES IN CHOLESTERICS, Rault, J., Solid State Commun v9 n22 p1965 (15 Nov '71)

858. CREATION DE LIGNES DE DISLOCATION DANS UN CHOLESTERIQUE PAR L'APPLICATION D'UN CHAMP MAGNETIQUE, Rault, J., Mol Cryst & Liq Cryst v16 n1-2 p143 (Feb '72)

859. IRREGULARITIES ON DISLOCATION LINES IN NEMATIC AND CHOLESTERIC CRYSTALS, Rault, J., J Physique (France) v33 n4 p383 (Apr '72)

860. TWISTED NEMATIC LIQUID CRYSTAL ELECTRO-OPTIC DEVICES WITH AREAS OF REVERSE TWIST, Raynes, E.P., Electronics Lett (GB) v9 n5 p101 (8 Mch '73)

861. A STUDY OF ETHYLENE CARBONATE AND ETHYLENE MONOTHIOCARBONATE AS SOLUTES IN A THERMOTROPIC NEMATIC PHASE, Raza, M.A., Reeves, L.W., J Magn Resonance (USA) v8 n3 p222 (Nov '72)

862. APPLICATION DES MESURES DE SUSCEPTIBILITE MAGNETIQUE A L'ETUDE DE LA TRANSITION CHOLESTERIQUE - NEMATIQUE, Regaya, B., Gasparoux, H., Prost, J., Rev Phys Applique (France) v7 n2 p83 (Jne '72)

863. LIGHT SCATTERING STUDIES OF ORIENTATION CORRELATION IN CHOLESTERYL ESTERS, Rhodes, M.B., Porter, R.S., Chu, W., Stein, R.S., Mol Cryst & Liq Cryst v10 n3 p295 (Jne '70)

864. HOLOMICROSCOPY OF LIQUID CRYSTALS, Rhodes, M.B., p73: Abstr 3rd Int'l Liq Cryst Conf., Berlin (Aug '70)

865. RAYLEIGH SCATTERING INDUCED BY STATIC BENDS OF LAYERS IN A SMECTIC 'A' LIQUID CRYSTAL, Ribotta, R. Durand, G., Litster, J.D., Solid State Commun v12 n1 p27 (1 Jan '73)

866. CHOLESTERIC LIQUID CRYSTALS FOR OPTICAL PROCESSING, Richard, S.P., Marathay, A.S., p19: 1971 Annual Meeting, Opt Soc Am, Ottawa, Canada, 5-8 Oct '71.

867. CHOLESTERIC LIQUID CRYSTALS FOR OPTICAL PROCESSING (Abstr.), Richard, S.P., Marathay, A.S., Opt Soc Am J v61 n11 p1559 (Nov '71)

868. ORIENTATION OF SOLUTE MOLECULES IN NEMATIC LIQUID CRYSTAL SOLVENT, Robertson, J.C., Yim, C.T., Gilson, D.F.R., Canadian J Chem v49 n13 p2345 (1 Jly '71)

869. DEFORMATION OF THE PLANAR TEXTURE OF A CHOLESTERIC CRYSTAL UNDER THE ACTION OF AN ELECTRIC FIELD, Rondelez, F., Arnould, H., Comptes Rendus (Ser. B) v273 n13 p549 (27 Sep '71)

870. DIELECTRIC AND RESISTIVITY MEASUREMENTS ON ROOM TEMPERATURE NEMATIC 'MBBA', Rondelez, F., Diguet, D., Durand, G., Mol Cryst & Liq Cryst v15 n2 p183 (Nov '71)

871. ELECTROHYDRODYNAMIC EFFECTS IN CHOLESTERIC LIQUID CRYSTALS UNDER A.C. ELECTRIC FIELD, Rondelez, F., Arnould, H., Phys Rev Lett v28 n12 p735 (20 Mch '72)

872. DISTORTIONS OF A PLANAR CHOLESTERIC STRUCTURE INDUCED BY A MAGNETIC FIELD, Rondelez, F., Hulin, J.P., Solid State Commun v10 n11 p1009 (1 Jne '72)

873. INCOHERENT CROSS-SECTION FOR NEUTRON QUASI-ELASTIC SCATTERING IN LIQUID CRYSTALS, Rosciszewski, K., Acta Phys Pol (A) (Poland) v41a n5 p549 (May '72)

874. BRILLOUIN SCATTERING IN A CHOLESTERIC LIQUID CRYSTAL NEAR THE CHOLESTERIC-ISOTROPIC TRANSITION, Rosen, H., Shen, Y.R., Mol Cryst & Liq Cryst v18 n3-4 p285 (Oct '72)

875. THE EFFECT ON THERMAL NEMATIC STABILITY OF SCHIFF'S BASES UPON REVERSAL OF TERMINAL SUBSTITUENTS, Rosenberg, H.M., Champa, R.A., Mol Cryst & Liq Cryst v11 n2 p191 (Sep '70)

876. FLUID PHASES OF RIGID MOLECULES OF HIGH ASYMMETRY, Runnels, L.K., Colvin, C., p48: Abstr 3rd Int'l Liq Cryst Conf., Berlin, W. Germany (Aug '70)

877. NATURE OF THE RIGID-ROD MESOPHASE, Runnels, L.K., Colvin, C., J Chem Phys v53 n11 p4219 (1 Dec '70)

878. FLUID PHASES OF HIGHLY ASYMETRIC MOLECULES: PLATE-SHAPED MOLECULES, Runnels, L.K., Colvin, C., Mol Cryst & Liq Cryst v12 n4 p299 (Mch '71)

879. FLUORESCENCE POLARIZATION MEASUREMENTS ON MOLECULES ORIENTED IN LIQUID CRYSTALS, Sackmann, E., Rehm, D., Chem Phys Lett (Holland) v4 n9 p537 (15 Jan '70)

880. BIREGRINGENCE OF THE SMECTIC MODIFICATION OF THALLIUM SOAPS, Sackmann, E., Pelzl, G., p40: Abstr 3rd Int'l Liq Cryst Conf., Berlin (Aug '70)

881. RELATION BETWEEN THE PRINCIPAL POLARIZABILITIES OF A MOLECULE AND ITS AVERAGE ORIENTATION IN NEMATIC LIQUID CRYSTALS, Sackmann, E., Mohwald, H., Chem Phys Lett v12 n3 p467 (1 Jan '72)

882. CIRCULAR DICHROISM OF HELICALLY ARRANGED MOLECULES IN CHOLESTERIC PHASES, Sackmann, E.,

Voss, J., Chem Phys Lett (Holland) v14 n4 p528 (15 Jne '72)

883. ON OPTICAL POLARIZATION MEASUREMENTS IN LIQUID CRYSTALS, Sackmann, E., Moehwald, H., J Chem Phys v58 n12 p5407 (15 Jne '73)

884. A NOVEL METHOD FOR DETERMINING THE EXISTENCE AND CHIRALITY OF CHOLESTERIC MESOPHASES, Saeva, F.D., Mol Cryst & Liq Cryst v18 n3-4 p375 (Oct '72)

885. MICROSCOPIC STUDY ON THE MOLECULAR ARRANGEMENTS IN THE SMECTIC 'A', 'B' and 'C' MODIFICATIONS, Sakagami, S., Takase, A., Nakamizo, M., Kakiyama, H., Mol Cryst & Liq Cryst v19 n3-4 p303 (Jan '73)

886. TEMPERATURE DEPENDENCE OF THE LINEAR GROWTH-RATE OF LIQUID-CRYSTAL PHASES OF PARA-AZOXY-ANISOLE AND PARA-AZOXYPHENETOL, Sakevich, N.M., Kristallografiya (USSR) v16 n3 p650 (1971)

887. INFRARED HOLOGRAPHY WITH LIQUID CRYSTALS, Sakusabe, T., Kobayashi, S., Japanese J Appl Phys v10 n6 p758 (Jne '71)

888. CHOLESTERIC AND NEMATIC STRUCTURES OF POLY-GAMMA-BENZYL-L-GLUTAMATE, Samulski, E.T., Tobolsky, A.V., p111: "Liquid Crystals and Ordered Fluids," Plenum Press, 1970.

889. BROWNIAN-MOTION CONTRIBUTIONS TO RELAXATION IN NEMATIC LIQUID CRYSTALS, Samulski, E.T., Dybowski, C.R., Wade, C.G., Chem Phys Lett (Holland) v11 n1 p113 (15 Sep '71)

890. INTERMOLECULAR AND INTRAMOLECULAR CONTRIBUTIONS TO PROTON RELAXATION IN LIQUID CRYSTALS, Samulski, E.T., Dybowski, C.R., Wade, C.G., Phys Rev Lett v29 n6 p340 (7 Aug '72)

891. NON LINEAR RESPONSE OF TWO DIMENSIONAL NEMATICS TO AN APPLIED FIELD, Sarma, G., Solid State Commun v10 n11 p1049 (1 Jne '72)

892. ANOMALOUS DIELECTRIC ABSORPTION IN 'MBBA' AT THE NEMATIC-ISOTROPIC PHASE TRANSITION POINT, Sasabe, H., Ooizumi, K., Japan J Appl Phys v11 n11 p1750 (Nov '72)

893. PRESSURE DEPENDENCE OF THE CRYSTAL-NEMATIC TRANSITION TEMPERATURE IN 'MBBA', Sasabe, H., Ooizumi, K., Japan J Appl Phys v11 n11 p1751 (Nov '72)

894. CHOLESTERIC-NEMATIC PHASE TRANSITIONS IN COMPENSATED LIQUID CRYSTALS, Sato, S., Wada, M., Japanese J Appl Phys v10 n8 p1106 (Aug '71)

895. MOLECULAR ORIENTATION EFFECTS IN COMPENSATED LIQUID CRYSTALS, Sato, S., Wada, M., Japanese J Appl Phys v11 n10 p1566 (Oct '72)

896. CHOLESTERIC-NEMATIC PHASE TRANSITIONS IN MIXTURES OF CHOLESTERYL CHLORIDE AND NEMATIC LIQUID CRYSTALS, Sato, S., Wada, M., Japanese J Appl Phys v11 n11 p1752 (Nov '72)

897. SPECTROSCOPY IN LIQUID CRYSTALS AND LIQUID CRYSTAL SOLUTIONS, Saupe, A., p16: Abstr 3rd Int'l Liq Cryst Conf., Berlin, W. Germany (Aug '70)

898. ULTRAVIOLET, INFRARED AND MAGNETIC RESONANCE SPECTROSCOPY ON LIQUID CRYSTALS, Saupe, A., Mol Cryst & Liq Cryst v16 n1-2 p87 (Feb '72)

899. LONG TERM STABILITY OF CHOLESTERIC LIQUID CRYSTAL SYSTEMS, II. Scala, L.C., Dixon, G.D., Mol Cryst & Liq Cryst v10 n4 p411 (Jly '70)

900. VOLTAGE-DEPENDENT OPTICAL ACTIVITY OF A TWISTED NEMATIC LIQUID CRYSTAL, Schadt, M., Helfrich, W., Appl Phys Lett v18 n4 p127 (15 Feb '71)

901. DIELECTRIC PROPERTIES OF SOME NEMATIC LIQUID CRYSTALS WITH STRONG POSITIVE DIELECTRIC ANISOTROPY, Schadt, M., J Chem Phys v56 n4 p1494 (15 Feb '72)

902. KERR EFFECT IN THE ISOTROPIC PHASE OF SOME NEMATOGENIC COMPOUND, Schadt, M., Helfrich, W., Mol Cryst & Liq Cryst v17 n3-4 p355 (Jly '72)

903. ELECTRIC FIELD ORDERING OF PARA-AZOXYANISOLE LIQUID CRYSTAL MOLECULES STUDIED BY ELECTRON SPIN RESONANCE, Schara, M., Sentjurc, M., Solid State Commun v8 n8 p593 (15 Apr '70)

904. STRUCTURES AND ENERGIES OF GRANDJEAN-CANO LIQUID CRYSTAL DISCLINATIONS, Scheffer, T.J., Phys Rev (A) v5 n3 p1327 (Mch '72)

905. ELECTRIC AND MAGNETIC FIELD INVESTIGATIONS OF THE PERIODIC GRIDLIKE DEFORMATION OF A CHOLESTERIC LIQUID CRYSTAL, Scheffer, T.J.,

906. INTRODUCTION TO THE CHEMISTRY OF LIQUID CRYSTALS FOR DISPLAYS, Scherrer, H., Eurocon 71 Digest, Lausanne, Switzerland, 18-22 Oct '71.

907. DEFORMATION OF NEMATIC LIQUID CRYSTALS WITH VERTICAL ORIENTATION IN ELECTRIC FIELDS, Schiekel, M.F., Fahrenschon, K., Appl Phys Lett v19 n10 p391 (15 Nov '71)

908. LIQUID CRYSTALS FOR MODERN INDICATION TECHNIQUES, Schiekel, M., Radio Elektron Schau (Austria) n2 p92 (1972)

909. MONOCHROMATIC AND MULTI-COLORED INDICATING ELEMENTS WITH NEMATIC LIQUID CRYSTALS ON THE BASIS OF THE 'DAP' EFFECT, Schiekel, M., Tech Mitt AEG-Telefunken (Germany) v62 n3 p113 ('72)

910. MULTICOLOR MATRIX-DISPLAY BASED ON THE DEFORMATION OF VERTICALLY-ALIGNED NEMATIC LIQUID CRYSTAL PHASES, Schiekel, M.F., Fahrenschon, K., p98: 1972 SID Symposium, Digest of Technical Papers, San Francisco, Calif., 6-8 Jne '72.

911. LIQUID CRYSTALLINE CURVATURE ELECTRICITY: THE BENDING MODE OF 'MBBA', Schmidt, D., Schadt, M., Helfrich, W., Z Naturforsch v27a n2 p277 (Feb '72)

912. RAMAN SPECTRA OF VARIOUS PHASES OF THE LIQUID CRYSTALLINE MATERIAL, P-AZOXYANISOLE, Schnur, J.M., Hass, M., Adair, W.L., Phys Lett (A) v41a n4 p326 (9 Oct '72)

913. ON THE VALIDITY OF THE MAIER-SAUPE THEORY OF THE NEMATIC TRANSITION, Schultz, T.D., p43: Abstr 3rd Int'l Liq Cryst Conf., Berlin, W. Germany (Aug '70)

914. ON THE VALIDITY OF THE MAIER-SAUPE THEORY OF THE NEMATIC TRANSITION, Schultz, T.D., Mol Cryst & Liq Cryst v14 n1-2 p147 (Aug '71)

915. E.P.R. STUDY OF THE TEMPERATURE DEPENDENCE OF MOLECULAR ROTATION IN A NEMATIC LIQUID CRYSTAL, Schwerdfeger, C.F., Marusic, M., p64: Abstr 3rd Int'l Liq Cryst Conf., Berlin (Aug '70)

916. E.P.R. STUDY OF THE TEMPERATURE DEPENDENCE OF MOLECULAR ROTATION IN NEMATIC LIQUID CRYSTALS, Schwerdfeger, C.F., Marusic, M., Mackay, A., Dong, R.Y., Mol Cryst & Liq CRyst v12 n4 p395 (Mch '71)

917. SPIN LABEL STUDIES OF ORIENTED SMECTIC LIQUID CRYSTALS (A MODEL SYSTEM FOR BILAYER MEMBRANES) Seelig, J., Am Chem Soc J v92 n13 p3881 (Jly '70)

918. MOLECULAR ARCHITECTURE OF LIQUID CRYSTALLINE BILAYERS, Seelig, J., Limacher, H., Bader, P., Am Chem Soc J v94 n18 p6364 (6 Sep '72)

919. LIQUID CRYSTAL DISPLAY, Selzo, C.A., IBM Tech Discl Bull v13 n5 p1387 (Oct '70)

920. LIQUID CRYSTAL ORDERING IN THE MAGNETIC AND ELECTRIC FIELDS AS STUDIED IN 4,4'-BIS HEPTYL-OXYAZOXYBENZENE BY ELECTRON PARAMAGNETIC RESONANCE, Sentjurc, M., Schara, M., p63: Abstr 3rd Int'l Liq Cryst Conf., Berlin, W. Germany (Aug '70)

921. LIQUID CRYSTAL ORDERING IN THE LAGNETIC AND ELECTRIC FIELDS STUDIED IN 4,4'-DI-N-HEPTYL-OXYAZOXYBENZENE BY E.R.P., Sentjurc, M., Schara, M., Mol Cryst & Liq Cryst v12 n2 p133 (Jan '71)

922. VISUAL OBSERVATION OF 'RF' MAGNETIC FIELDS USING CHOLESTERIC LIQUID CRYSTALS, Sethares, J.C., Gulaya, S., Appl Opt v9 n12 p2795 (Dec '70)

923. ELECTRICAL CONDUCTIVITY IN CHOLESTERIC LIQUID CRYSTALS, Shaw, D.G., Kaufmann, J.W., J Chem Phys v54 n6 p2424 (15 Mch '71)

924. DIELECTRIC PROPERTIES OF A CHOLESTERIC CRYSTAL, Shaw, D.G., Kaufmann, J.W., Phys Status Solidi A (Germany) v12 n2 p637 (1972)

925. PHASE-MATCHED THIRD-HARMONIC GENERATION IN CHOLESTERIC LIQUID CRYSTALS, Shelton, J.W., Shen, Y.R., Phys Rev Lett v25 n1 p23 (6 Jly '70)

926. UMKLAPP OPTICAL THIRD-HARMONIC GENERATION IN CHOLESTERIC LIQUID CRYSTALS, Shelton, J.W., Shen, Y.R., Phys Rev Lett v26 n10 p538 (8 Mch '71)

927. ELECTRO-OPTIC SWITCHING IN LOW-LOSS LIQUID CRYSTAL WAVEGUIDES, Sheridan, J.P., Schnur, J.M., Giallorenzi, T.G., Appl Phys Lett v22 n11 p560 (1 Jne '73)

928. LATTICE MODEL FOR BIAXIAL LIQUID CRYSTALS, Shih, C.-S., Alben, R., J Chem Phys v57 n8 p3055 (15 Oct '72)

929. MAGNETIC FIELD DEPENDENCE OF THE CAPACITY OF A TWISTED NEMATIC LIQUID CRYSTAL CELL, Shtrikman, S., Wohlfarth, E.P., Wand, Y., Phys Lett (A) v37a n5 p369 (20 Dec '71)

930. REAL-TIME VISUAL RECONSTRUCTION OF INFRARED HOLOGRAMS, Simpson, W.A., Deeds, W.E., Appl Opt v9 n2 p499 (Feb '70)

931. MESOMORPHIC POLYMERS, Skoulios, A., p23: Abstr 3rd Int'l Liq Cryst Conf., Berlin (Aug '70)

932. PHASE TRANSITIONS IN MESOMORPHIC BENZYLIDENEANILINES, Smith, G.W., Gardlund, Z.G., Curtis, R.J., Mol Cryst & Liq Cryst v19 n3-4 p327 (Jan '73)

933. ELECTRONIC NUMBERS, Sobel, A., Sci Am v228 n6 p65 (Jne '73)

934. INDUCED CIRCULAR DICHROISM IN CHOLESTERIC LIQUID CRYSTALS, Soeva, F.D., Wysocki, J.J., Am Chem Soc J v93 n22 p5928 (3 Nov '71)

935. ELECTRONICALLY SCANNED ANALOG LIQUID CRYSTAL DISPLAYS, Soref, R.A., Appl Opt v9 n6 p1323 (Jne '70)

936. THERMO-OPTIC EFFECTS IN NEMATIC-CHOLESTERIC MIXTURES, Soref, R.A., J Appl Phys v41 n7 p3022 (Jne '70)

937. COLOR CONTRAST CROSS-GRID DISPLAYS USING UNDISTURBED CHOLESTERICS, Soref, R.A., IEEE Proc v58 n7 p1163 (Jly '70)

938. SOLID FACTS ABOUT LIQUID CRYSTALS, Soref, R.A., Laser Focus v6 n9 p45 (Sep '70)

939. THE PHYSICAL BASIS OF DISPLAY EFFECTS IN LIQUID CRYSTALS, Soref, R.A., p12: NEREM Record (1970)

940. LIQUID CRYSTAL LIGHT CONTROL EXPERIMENTS, Soref, R.A., A chapter in "Physics of Opto-Electronic Materials," W.A. Albers, Jr., editor, Plenum Press, New York, 1971.

941. THE PHYSICS AND CHEMISTRY OF LIQUID CRYSTALS, Soref, R.A., SID Digest of Technical Papers, v11 p122 (1971)

942. ELECTRICALLY CONTROLLED BIREFRINGENCE OF THIN NEMATIC FILMS, Soref, R.A., Rafuse, M.J., J Appl Phys v43 n5 p2029 (May '72)

943. LIQUID CRYSTAL DISPLAY PHENOMENA, Soref, R.A., Proc Soc Information Display v13 p95 (1972)

944. TRANSVERSE DIELD EFFECTS IN NEMATIC LIQUID CRYSTALS, Soref, R.A., Appl Phys Lett v22 n4 p165 (15 Feb '73)

945. LIQUID CRYSTAL BAR-GRAPH DISPLAYS, Soref, R.A., IEEE Proc v61 n3 p384 (Mch '73)

946. INTERDIGITAL FIELD-EFFECT LIQUID CRYSTAL DISPLAYS, Soref, R.A., 1973 SID Symposium, Digest of Technical Papers (to be published in 1973)

947. THE N.M.R. SPECTRA OF ORIENTED 1,1-DIFLUOROETHYLENE, AND TETRAFLUOROETHYLENE, Spiesecke, H., Saupe, A., Mol Cryst & Liq Cryst v6 n3-4 p287 (Feb '70)

948. GEOMETRICAL STRUCTURE DATA OF 1-^{13}C-ETHYLENE, Spiesecke, H., p58: Abstr 3rd Int'l Liq Cryst Conf., Berlin (Aug '70)

949. ELECTRICALLY ORIENTED X-RAY DIFFRACTION PATTERNS OF LIQUID CRYSTALLINE SOLUTIONS OF POLY-GAMMA-BENZYL-L-GLUTAMATE, Stamatoff, J.B., Mol Cryst & Liq Cryst v16 n1-2 p137 (Feb '72)

950. OPTICAL ROTATORY POWER OF LIQUID CRYSTAL MIXTURES, Stegemeyer, H., Mainusch, K.-J., Chem Phys Lett v6 n1 p5 (1 Jly '70)

951. OPTICAL PROPERTIES OF LIQUID CRYSTAL MIXTURES, Stegemeyer, H., Mainusch, K.-J., Stranski, I.N., p36: Abstr 3rd Int'l Liq Cryst Conf., Berlin (Aug '70)

952. OPTICAL ROTATORY POWER OF LIQUID CRYSTAL MIXTURES. II. MIXTURE OF A NEMATOGENIC AND A NON-MESOMORPHIC CHIRALIC COMPOUND, Stegemeyer, H., Mainusch, K.-J., Steigner, E., Chem Phys Lett v8 n5 p425 (1 Mch '71)

953. SIMULTANEOUS EXISTENCE OF TWO LIQUID CRYSTAL PHASES OF HELICAL STRUCTURE, Stegemeyer, H., Muller, U., Naturwissenschaften (Germany) v58 n12 p621 (Dec '71)

954. INDUCTION OF OPTICAL ACTIVITY IN A NEMATIC MESOPHASE BY l-MENTHOL. OPTICAL PROPERTIES OF LIQUID CRYSTAL MIXTURES, Stegemeyer, H., Mainusch, K.-J., Chem Phys Lett v16 n1 p38 (15 Sep '72)

955. MESOMORPHIC SCHIFF'S BASES DERIVED FROM P-ACETOXY-, P-N-BUTYL, AND P-N-BUTOXY-ANILINES, Steiger, E.L., Dietrich, H.J., Mol Cryst & Liq Cryst v16 n3 p279 (Mch '72)

956. A TWO-FREQUENCY COINCIDENCE ADDRESSING SCHEME FOR NEMATIC LIQUID CRYSTAL DISPLAYS, Stein, C.R., Kashnov, R.A., p112: 17th Int'l Electron Devices Meeting (Abstr.), Washington, D.C., 11-13 Oct '71.

957. A TWO-FREQUENCY COINCIDENCE ADDRESSING SCHEME FOR NEMATIC-LIQUID-CRYSTAL DISPLAYS, Stein, C.R., Kashnov, R.A., Appl Phys Lett v19 n9 p343 (1 Nov '71)

958. RECENT ADVANCES IN FREQUENCY COINCIDENCE MATRIX ADDRESSING OF LIQUID CRYSTAL DISPLAYS, Stein, C.R., Kashnov, R.A., p64: 1972 SID Symp Digest of Tech Papers, San Francisco, Calif., 6-8 Jne '72)

959. LIQUID CRYSTALS: PRINCIPLES AND APPLICATIONS IN ELECTRO-OPTICS, Steinstrasser, R., Elektron Ind (Germany) v3 n11 pEP-97 (Nov '72)

960. REDUCTION OF THE ISOTROPIC-NEMATIC TRANSITION TEMPERATURE OF LIQUID CRYSTALS DUE TO THERMAL FLUCTUATIONS OF THE CONFORMATION OF LONG MOLECULES, Stenschke, H., Solid State Commun v10 n7 p653 (1 Apr '72)

961. HYDRODYNAMICS OF LIQUID CRYSTALS, Stephen, M.J., Phys Rev (A) v2 n4 p1558 (Oct '70)

962. LIQUID CRYSTALS: PERSPECTIVES, PROSPECTS, AND PRODUCTS, Stepke, E., Electro-optical Systems Design p20 (Feb '72)

963. PRETRANSITIONAL PHENOMENA IN THE ISOTROPIC PHASE OF A NEMATIC LIQUID CRYSTAL, Stinson, T.W., Litster, J.D., Phys Rev Lett v25 n8 p503 (24 Aug '70)

964. STATIC AND DYNAMIC BEHAVIOR NEAR THE ORDER DISORDER TRANSITION OF NEMATIC LIQUID CRYSTALS, Stinson, T.W., Litster, J.D., Clark, N.A., J Phys (France) colloque no.1 suppl to v33 n2-3 pC1-69 (Feb-Mch '72)

965. CORRELATION RANGE OF FLUCTUATIONS OF SHORT-RANGE ORDER IN THE ISOTROPIC PHASE OF A LIQUID CRYSTAL, Stinson, T.W., Litster, J.D., Phys Rev Lett v30 n15 p688 (9 Apr '73)

966. LIQUID CRYSTALS IN TWO DIMENSIONS, Straley, J.P., Phys Rev (A) v4 n2 p675 (Aug '71)

967. ZWANZIG MODEL FOR LIQUID CRYSTALS, Straley, J.P., J Chem Phys v57 n9 p3694 (1 Nov '72)

968. OPTICAL REFLECTION FROM CHOLESTERIC LIQUID CRYSTAL FILMS, Subramanyam, S.V.,

Appl Opt v10 n2 p317 (Feb '71)

969. NUCLEAR MAGNETIC RESONANCE IN LIQUID CRYSTALS, Sung, C.C.,
Chem Phys Lett v10 n1 p35 (1 Jly '71)

970. IONIC EQUILIBRIUM AND IONIC CONDUCTANCE IN THE SYSTEM TETRA-ISOPENTYL AMMONIUM NITRATE - P-AZOXYANISOLE, Sussman, A., p114: Abstr 3rd Int'l Liq Cryst Conf., Berlin (Aug '70)

971. IONIC EQUILIBRIUM AND IONIC CONDUCTANCE IN THE SYSTEM TETRA-ISO-PENTYL AMMONIUM NITRATE-P--AZOXYANISOLE, Sussman, A., Mol Cryst & Liq Cryst v14 n1-2 p183 (Aug '71)

972. DYNAMIC SCATTERING LIFE IN THE NEMATIC COMPOUND P-METHOXYBENZYLIDENE-P'-AMINO PHENYL ACETATE AS INFLUENCED BY CURRENT DENSITY, Sussman, A., Appl Phys Lett v21 n4 p126 (15 Aug '72)

973. SECONDARY HYDRODYNAMIC STRUCTURE IN DYNAMIC SCATTERING, Sussman, A.,
Appl Phys Lett v21 n6 p269 (15 Sep '72)

974. ON PYROELECTRICAL EFFECTS IN LIQUID CRYSTALS, Szymanski, A.,
p110: Abstr 3rd Int'l Liq Cryst Conf., Berlin (Aug '70)

975. E.S.R. MEASUREMENTS OF THE DEGREE OF ORDER IN NEMATIC PHASE, Sy, D., Sanson, A., Ptak, M.,
Solid State Commun v10 n10 p985 (15 May '72)

976. ELECTRET THERMAL ANALYSIS OF LIQUID CRYSTALS, Takamatsu, T., Ootake, H., Fukada, E.,
Rep Inst Phys & Chem Res (Japan) v47 n4 p116 (1971)

977. CONSIDERATIONS ON THS SURFACE TENSION AND VISCOSITY ANOMALIES OF LIQUID CRYSTALS, Tamamushi, B.-I., p126: Abstr 3rd Int'l Liq Cryst Conf., Berlin (Aug '70)

978. NOVEL ELECTRO-OPTICAL STORAGE EFFECT IN A CERTAIN SMECTIC LIQUID CRYSTAL, Tani, C.,
Appl Phys Lett v19 n7 p241 (1 Oct '71)

979. ANISOTROPIC NUCLEAR SPIN-LATTICE RELAXATION IN A NEMATIC LIQUID CRYSTAL, Tarr, C.E., Nickerson, M.A., Smith, C.W., Appl Phys Lett v17 n8 p318 (15 Oct '70)

980. NUCLEAR SPIN-LATTICE RELAXATION IN A LIQUID CRYSTAL ORDERED IN A.C. ELECTRIC FIELDS,
Tarr, C.A., Fuller, A.M., Nickerson, M.A., Appl Phys Lett v19 n6 p179 (15 Sep '71)

981. IMPROVED ELECTRONIC WATCHES, Taylor, G.W., Lefkowitz, I., IEEE Proc v61 n4 p487 (Apr '73)

982. BIAXIAL LIQUID CRYSTALS, Taylor, T.R., Fergason, J.L., Arora, S.L.,
Phys Rev Lett v24 n8 p359 (23 Feb '70)

983. OPTICAL AND STRUCTURAL EFFECT IN TYPE 'C' SMECTIC PHASES, Taylor, T.R., Arora, S.L., Fergason, J.L., p41: Abstr 3rd Int'l Liq Cryst Conf., Berlin (Aug '70)

984. TEMEPERATURE-DEPENDENT TILT ANGLE IN THE SMECTIC 'C' PHASE OF A LIQUID CRYSTAL, Taylor, T.R., Arora, S.L., Fergason, J.L., Phys Rev Lett v25 n11 p722 (14 Sep '70)

985. CURRENT- AND MAGNETIC-FIELD-INDUCED ORDER AND DISORDER IN ORDERED NEMATIC LIQUID CRYSTALS, Teaney, D.T., Migliori, A., J Appl Phys v41 n3 p998 (1 Mch '70)

986. SHARP ORDER BREAKDOWN FOR PARALLEL FLOW IN NEMATICS, Teaney, D.T.,
IBM Tech Discl Bull v12 n12 p2244 (May '70)

987. SMALL-ANGLE X-RAY SCATTERING NEAR NEMATIC-ISOTROPIC TRANSITION TEMPERATURE, Terauchi, H., Nakatsu, K., Kusabayashi, S., Japan J Appl Phys v11 n5 p763 (May '72)

988. X-RAY SCATTERING FROM A SMECTIC LIQUID CRYSTAL, Terauchi, H., Takeuchi, T., Kusabayashi, S.,
Japan J Appl Phys v11 n12 p1862 (Dec '72)

989. PROPERTIES OF STRUCTURALLY STABILIZED ANIL-TYPE NEMATIC LIQUID CRYSTALS, Teucher, I., Paleos, C.M., Labes, M.M., Mol Cryst & Liq Cryst v11 n2 p187 (Sep '70)

990. DIFFUSION THROUGH NEMATIC LIQUID CRYSTALS, Teucher, I., Baessler, H., Labes, M.M.,

Nat Phys Sci (GB) v299 n1 p25 (4 Jan '71)

991. MAGNETIC FIELD EFFECTS ON THE DYNAMIC SCATTERING THRESHOLD IN A NEMATIC LIQUID CRYSTAL, Teucher, I., Labes, M.M., J Chem Phys v54 n9 p4130 (1 May '71)

992. BIREFRINGENCE AND OPTICAL ROTATORY DISPERSION OF A COMPENSATED CHOLESTERIC LIQUID CRYSTAL, Teucher, I., Ko, K., Labes, M.M., J Chem Phys v56 n7 p3308 (1 Apr '72)

993. BIREFRINGENCE AND OPTICAL ROTATORY DISPERSION OF A COMPENSATED CHOLESTERIC LIQUID CRYSTAL, Teucher, I., Ko, K., Labes, M.M., J Chem Phys v56 n7 p3308 (1972)

994. APPLICATION OF NEMATIC LIQUID CRYSTALS FOR THE INVESTIGATION OF P-N JUNCTIONS AND INSULATING LAYERS, Thiessen, K., Tuyen, L.T., Phys Status Solidi (A) v13 n1 p73 (16 Sep '72)

995. PROTON SPIN THERMOMETRY AT LOW FIELDS IN THE SOLID PHASE OF LIQUID CRYSTALS, Thompson, R.T., Pintar, M.M., p60: 17th Congress Ampere NMR and related phenomena, Turku, Finland, 21-26 Aug '72.

996. VARIABLE FREQUENCY PULSED N.M.R. STUDY OF LYOTROPIC LIQUID CRYSTALS, Tiddy, G.J.T., Nat Phys Sci (GB) v230 n14 p136 (5 Apr '71)

997. FLUOROCARBON SURFACTANT/WATER MESOMORPHIC PHASES, Tiddy, G.J.T., Faraday Symposium on Liquid Crystals, Faraday Society, London, 1971.

998. EFFECT OF PRESSURE ON THE PHASE TRANSITIONS IN NEMATIC LIQUID CRYSTALS, Tikhomirova, N.A., Vistin, L.K., Nosov, V.N., Soviet Phys - Crystallography v17 n5 p878 (Mch/Apr '73)

999. SCALED PARTICLE THEORY OF TWO DIMENSIONAL ANISOTROPIC FILMS, Timling, K., p50: Abstr 3rd Int'l Liq Cryst Conf., Berlin, 1970 (Aug '70)

1000. THE PLACE FOR A KNOWLEDGE OF THE IMPORTANCE OF THE CRYSTALLINE-LIQUID STATE IN THE CHEMISTRY AND TECHNOLOGY OF HIGH POLYMERS, Tinius, K., Plaste + Kautschuk (Germany) v18 n6 p408 (1971)

1001. ELECTROHYDRODYNAMIC PHENOMENA IN CONNECTION WITH UNIPOLAR INJECTION OF CHARGE CARRIERS IN 'MBBA' DEIONIZED BY ELECTRODIALYSIS. ALSO: NEMATIC LIQUID CRYSTAL PURIFICATION BY ELECTRO-DIALYSIS BEHAVIOR OF 'MBBA' IN ELECTRIC FIELD, Tobazeon, N.N., Filippini, N.N., Borel, N.N., Robert, N.N., Poggi, N.N., p117: Abstr 3rd Int'l Liq Cryst Conf., Berlin (Aug '70)

1002. USE OF LIQUID CRYSTALS TO MAKE INFRARED RADIATION VISIBLE, Tolmachev, A.V., Kuz'michev, V.M., JETP Letters v14 n4 p144 (20 Aug '71)

1002[a]. VISUALIZATION OF RADIATION IN THE MILLIMETER AND SUBMILLIMETER RANGE BY MEANS OF LIQUID CRYSTALS, Tolmachev, A.V., Govorun, E.Ya., Kuz'michev, V.M., Soviet Phys - JETP v36 n2 p309 (Feb '73)

1003. LIQUID CRYSTAL VIEWING SCREEN, Tompkins, E.N., p37: Abstr 1971 Spring Meeting, Opt Soc Am, Tucson, Arizona, 5-8 Apr '71.

1004. A MIXED LIQUID CRYSTAL WITH NEW ELECTROOPTIC EFFECT, Toriyama, K., Aoyagi, T., Nomura, S., Japan J Appl Phys v9 n5 p584 (May '70)

1006. OPTICAL TRANSIENT BEHAVIOR OF NEMATIC LIQUID CRYSTALS IN AN ELECTRIC FIELD, Toriyama, K., Japan J Appl Phys v9 n9 p1190 (Sep '70)

1007. LIQUID CRYSTALS AS INDICATOR ELEMENTS, Toriyama, K., Aoyagi, T., Oyo Buturi v40 n5 p560 (May '71)

1008. APPLICATION OF THE CONTINUUM THEORY TO NEMATIC LIQUID CRYSTALS, Tseng, H.C., Silver, D.L., Finlayson, B.A., Phys Fluids v15 n7 p1213 (Jly '72)

1009. EVIDENCE OF NEMATIC ORDER FROM THE RAMAN SPECTRA OF VITRIFIED LIQUID CRYSTALS OF N-ANISYLI-DENE -4(PHENYLAZO)-1-NAPHTYLAMINE (APN), Tsenter, M.Ya., Bobovich, Ya.S., Belyaevskaya, N.M., Opt & Spectrosc v29 n2 p27 (Aug '70)

1010. A FACSIMILE PRINTER UTILIZING AN ARRAY OF LIQUID CRYSTAL CELLS, Tults, J., Proc Soc Inf Display v12 n4 p199 (1971)

1011. LIQUID CRYSTALS IN MOLTEN SALT SYSTEMS, Ubbelohde, A.R., Michels, H.J., Duroz, I.J., Nature v228 n5266 p50 (3 Oct '70)

1012. RELAXATION DYNAMOMETER FOR STUDYING MOLTEN SALT MESOPHASES, Ubbelohde, A.R. Michels, H.J., Duroz, J.J., J Phys E: Sci Instr v5 n3 p283 (Mch '72)

1013. MOLECULAR ARRANGEMENT OF NEMATIC LIQUID CRYSTALS, Uchida, T., Watanabe, H., Wada, M., Japanese J Appl Phys v11 n10 p1559 (Oct '72)

1014. SYNTHESIS OF LIQUID CRYSTALS, Uemo, S., Oyo Buturi v40 n5 p523 (May '71)

1015. DETECTION OF DEFECTS IN PRINTED CIRCUITS, Uhls, D.L., IBM Tech Discl Bull v15 n5 p1670 (Oct '72)

1016. MOESSBAUER INVESTIGATION OF THE SMECTIC LIQUID CRYSTALLINE STATE, Uhrich, D.L., Wilson, J.M., Resch, W.A., Phys Rev Lett v24 n8 p355 (23 Feb '70)

1017. CYLINDRICAL FUNCTION OF INTERATOMIC DISTANCES FOR LIQUID CRYSTALS, Vainshtein, B.K., Kosterin, E.A., Chistyakov, I.G., Soviet Phys - Doklady v16 n7 p504 (Jan '72)

1018. THE STRUCTURE OF LIQUID CRYSTALS (abstr.), Vainshtein, B.K., Chistyakov, I.G., Acta Crystallogr (Denmark) vA28 Pt.4 suppl p5134 (15 Jly '72)

1019. THE CONFORMATION OF AROMATIC SCHIFF BASES IN CONNECTION WITH LIQUID CRYSTALLINE PROPERTIES, Van Der Veen, J., p85: Abstr 3rd Int'l Liq Cryst Conf., Berlin, (Aug '70)

1020. THE CONFORMATION OF AROMATIC SCHIFF BASES IN CONNECTION WITH LIQUID CRYSTALLINE PROPERTIES, Van Der Veen, J., Grobben, A.H., Mol Cryst & Liq Cryst v15 n3 p239 (Dec '71)

1021. LOW MELTING LIQUID CRYSTALLINE P,P'-DI-N-ALKYLAZOXY- AND AZOBENZENES, Van Der Veen, J., De Jeu, W.H., Grobben, A.H., Boven, J., Mol Cryst & Liq Cryst v17 n3-4 p291 (Jly '72)

1022. ON THE MAGNETIC THRESHOLD FOR THE ALIGNMENT OF A TWISTED NEMATIC CRYSTAL, Van Doorn, C.Z., Phys Lett (A) v42a n7 p537 (29 Jan '73)

1023. CHEMISTRY OF LIQUID CRYSTALS, Van Meter, J.P., Eastman Organic Chemical Bulletin v45 n1 p1 (1973) Kodak publication no. JJ60-723.

1024. LIQUID CRYSTALS AND THEIR APPLICATIONS, Vanner, K.C., ERA J (GB) p13 (Autumn, 1971)

1025. REDUCTION OF THE SWITCHING TIME OF A LIQUID CRYSTAL OPTICAL TRANSPARENCY, Vasil'ev, A.A., Kompanets, I.N., Nikitin, V.V., Soviet J Quantum Electr v2 n3 p263 (Nov-Dec '72)

1026. A COMPUTER PROGRAM FOR LIQUID CRYSTAL SEARCHING, Verbit, L., Tuggey, R.L., p150: Abstr 3rd Int'l Liq Cryst Conf., Berlin, W. Germany (Aug '70)

1027. A LINE NOTATION FOR THERMOTROPIC LIQUID CRYSTAL PHASE TRANSITIONS, Verbit, L., Mol Cryst & Liq Cryst v15 n1 p89 (Oct '71)

1028. SYNTHESIS AND LIQUID CRYSTAL PROPERTIES OF SOME ACETYLENE DERIVATIVES, Verbit, L., Tuggey, R.L., Mol Cryst & Liq Cryst v17 n1 p49 (May '72)

1029. OBSERVATION OF THE MODES OF A CONTINUOUS HYDROGEN CYANADE LASER USING LIQUID CRYSTALS, Veron, D., Comptes Rendus (Ser. B) v274 n17 p1013 (24 Apr '72)

1029A. INFLUENCE OF MAGNETIC FIELD ON LIGHT SCATTERING IN NEMATIC LIQUID CRYSTALS, Veselago, V.G., Korobkin, Yu.V., Leonov, Yu.S., JETP Lett v17 n10 p397 (20 May '73)

1030. CHROMATOGRAPHIC AND THERMOOPTICAL INVESTIGATION OF THE PHASE TRANSITIONS IN LIQUID CRYSTAL SYATEMS, Vigalok, R.V., Palikhov, N.A., Vigdergauz, M.S., Soviet Phys - Crystallogr v17 n4 p731 (Jan-Feb '73)

1031. MECHANISMS FOR SPIN-LATTICE RELAXATION IN NEMATIC LIQUID CRYSTALS, Vilfan, M., Blinc, R., Doane, J.W., Solid State Commun v11 n8 p1073 (15 Oct '72)

1032. QUADRUPOLE AND PROTON SPIN-LATTICE RELAXATION IN THE NEMATIC LIQUID CRYSTALLINE PHASE, Visintainer, J.J., Doane, J.W., Fishel, D.L., Mol Cryst & Liq Cryst v13 n1 p69 (Apr '71)

1033. ELECTROSTRUCTURAL EFFECT AND OPTICAL PROPERTIES OF A CERTAIN CLASS OF LIQUID CRYSTALS AND THEIR BINARY MIXTURES, Vistin, L.K., Kristallografiya (USSR) v15 n3 p594 (May '70)

1034. A NEW ELECTROSTRUCTURAL EFFECT IN LIQUID CRYSTALS OF NEMATIC TYPE, Vistin, L.K., Soviet Phys - Doklady v15 n10 p908 (Apr '71)

1035. DIELECTRIC HYSTERESIS IN LIQUID SINGLE CRYSTALS, Vistin, L.K., Soviet Phys - Crystallogr v17 n4 p735 (Jan-Feb '73)

1036. POLYMORPHISM IN CHOLESTERYL ESTERS: CHOLESTERYL PALMITATE, Vogel, M.J., Rarral, E.M.II., Mignosa, C.P., p135: Abstr 3rd Int'l Liq Cryst Conf., Berlin (Aug '70)

1037. EFFECT OF SOLVENT TYPE ON THE THERMODYNAMIC PROPERTIES OF NORMAL ALIPHATIC CHOLESTERYL ESTERS, Vogel, M.J., Barrall, E.M.II., Mignosa, C.P., p333: "Liquid Crystals and Ordered Fluids," Plenum Press, 1970.

1038. POLYMORPHISM IN CHOLESTERYL ESTERS: CHOLESTERYL PALMITATE, Vogel, M.J., Barrall, E.M.II., Mignosa, C.P., IBM J Res/Dev v15 n1 p52 (Jan '71)

1039. POLYMORPHISM IN CHOLESTERYL ESTERS: CHOLESTERYL PALMITATE, Vogel, M.J., Barrall, E.M.II., Mol Cryst & Liq Cryst v15 n1 p49 (Oct '71)

1040. MODULATED SPIN ECHO TRAINS FROM LIQUID CRYSTALS, Vold, R.L., Chan, S.O., J Chem Phys v53 n1 p449 (1 Jly '70)

1041. LIQUID CRYSTALS, Von Planta, C., Microtecnic (Switzerland) v26 n6 p359 (Aug '72)

1042. THEORY OF LIQUID CRYSTALS, Wadati, M., Isihara, A., Mol Cryst & Liq Cryst v17 n2 p95 (Jne '72)

1043. SUBSTRATE TREATMENT FOR LIQUID CRYSTAL DISPLAY, Wagner, P.R., IBM Tech Discl Bull v13 n10 p2961 (Mch '71)

1044. A NEW OPTICAL METHOD FOR STUDYING THE VISCOELASTIC BEHAVIOR OF NEMATIC LIQUID CRYSTALS, Wahl, J., Fischer, F., Opt Commun v5 n5 p341 (Aug '72)

1045. VOLTAGE EFFECT ON CHOLESTERIC LIQUID CRYSTALS, Wako, T., Nakamura, K., Mol Cryst & Liq Cryst v19 n2 p141 (Dec '72)

1046. VARIATION OF D.C. DOMAIN THRESHOLD IN A NEMATIC LIQUID CRYSTAL UNDER CONTINUAL DYNAMIC SCATTERING, Wargocki, F.E., Lord, A.E.,Jr., J Appl Phys v44 n1 p531 (Jan '73)

1047. NUCLEAR SPIN RELAXATION IN NEMATIC LIQUID CRYSTALS, Watkins, C.L., Johnson, C.S.Jr., J Phys Chem v75 n16 p2452 (5 Aug '71)

1048. EFFECTS OF MAGNETIC FIELD ON ATTENUATION OF ULTRASONIC WAVES IN A NEMATIC LIQUID CRYSTAL, Wetsel, G.C.Jr., Speer, R.S., Lowry, B.A., Woodard, W.R., J Appl Phys v43 n4 p1495 (Apr '72)

1049. MAGNETIC LIQUID CRYSTAL COMPOSITION, Wheelock, C.E., IBM Tech Discl Bull v13 n8 p2266 (Jan '71)

1050. LIQUID CRYSTAL LIGHT VALVES, White, D.L., Feldman, M., Electron Lett v6 n26 p837 (31 Dec '70)

1051. CONTRAST ENHANCEMENT IN MATRIX ADDRESSED LIQUID CRYSTAL LIGHT VALVE ARRAYS BY ELECTRONIC QUENCHING, Wild, P.J., Nehring, J., Eurocon 71 Digest, Lausanne, Switzerland, 18-22 Oct '71.

1052. TURN-ON TIME REDUCTION AND CONTRAST ENHANCEMENT IN MATRIX-ADDRESSED LIQUID CRYSTAL LIGHT VALVES, Wild, P.J., Nehring, J., Appl Phys Lett v19 n9 p335 (1 Nov '71)

1053. MATRIX-ADDRESSED LIQUID CRYSTAL PROJECTION DISPLAY, Wild, P.J., p62: 1972 SID Symposium, Digest of Tech Papers, San Francisco, Calif., 6-8 June '72.

1054. THERMODYNAMICS OF SOLUTIONS WITH LIQUID CRYSTAL SOLVENTS. VI. THE EFFECT OF ORGANIC SOLUTES ON THE SELECTIVE REFLECTION OF VISIBLE LIGHT BY CHOLESTERIC MIXTURES, Willey, D.G., Martire, D.E., Mol Cryst & Liq Cryst v18 n1 p55 (Aug '72)

1055. ON THE MEASUREMENT OF THE ELASTIC CONSTANTS OF TWIST AND BEND FOR THE NEMATIC LIQUID CRYSTAL P-N-METHOXY BENZYLIDENE-P-BUTYLANILINE (MBBA), Williams, C., Cladis, P.E., Solid State Commun v10 n4 p357 (15 Feb '72)

1056. NONSINGULAR S=+1 SCREW DISCLINATION LINES IN NEMATICS, Williams, C., Pieranski, P., Cladis, P.E., Phys Rev Lett v29 n2 p90 (10 Jly '72)

1057. SURFACE DISCLINATION LINES IN 'MBBA', Williams, C., Vitek, V., Kleman, M., Solid State Commun v12 n6 p581 (15 Mch '73)

1058. LIQUID CRYSTAL DOMAINS IN A LONGITUDINAL ELECTRIC FIELD, Williams, R., J Chem Phys v56 n1 p147 (1 Jan '72)

1059. THEORY OF MOESSBAUER SPECTRAL ASYMETRY OF QUADRUPOLE SPLIT LINES IN LIQUID CRYSTALS, Wilson, J.M., Uhrich, D.L., Mol Cryst & Liq Cryst v13 n1 p85 (Apr '71)

1060. COMPOSITION AND TEMPERATURE VS LIQUID CRYSTALLINITY IN AMPHIPHILIC SYSTEMS, Winsor, P.A., p21: Abstr 3rd Int'l Liq Cryst Conf., Berlin, W. Germany (Aug '70)

1061. LIQUID CRYSTALLINITY IN RELATION TO COMPOSITION AND TEMPERATURE IN AMPHIPHILIC SYSTEMS, Winsor, P.A., Mol Cryst & Liq Cryst v12 n2 p141 (Jan '71)

1062. ACIEVEMENT OF A PRONOUNCED THRESHOLD IN THE DYNAMIC SCATTERING OF NEMATIC LIQUID CRYSTALS, Wolff, U., p108: Abstr 3rd Int'l Liq Cryst Conf., Berlin, W. Germany (Aug '70)

1063. OPTICAL-FIELD-INDUCED ORDERING IN THE ISOTROPIC PHASE OF A NEMATIC LIQUID CRYSTAL, Wong, G.K.L., Shen, Y.R., Phys Rev Lett v30 n19 p895 (7 May '73)

1064. INTEGRAL EQUATION APPROACH TO LIQUID CRYSTALS, Workman, H., Fixman, M., J Chem Phys v58 n11 p5024 (1 Jne '73)

1065. ELECTRIC FIELD INDUCED DOMAINS IN TWISTED NEMATIC LIQUID CRYSTALS, Wright, J.J., Dawson, J.F., Phys Lett (A) v43a n2 p145 (26 Feb '73)

1066. STATISTICAL MECHANICS FOR LONG SEMIFLEXIBLE MOLECULES: A MODEL FOR THE NEMATIC MESOPHASE, Wulf, A., DeRocco, A.G., J Chem Phys v55 n1 p12 (1 Jly '71)

1067. DISTRIBUTION FUNCTION THEORY OF NEMATIC LIQUID CRYSTALS, Wulf, A., J Chem Phys v55 n9 p4512 (1 Nov '71)

1068. CONTINUED KINETIC STUDY OF THE CHOLESTERIC-NEMATIC AND NEMATIC-CHOLESTERIC TRANSITION IN A LIQUID CRYSTAL FILM, Wysocki, J.J., p100: Abstr 3rd Int'l Liq Cryst Conf., Berlin (Aug '70)

1069. KINETIC STUDY OF THE ELECTRIC-FIELD INDUCED CHOLESTERIC-NEMATIC TRANSITION IN LIQUID CRYSTAL FILMS: I. RELAXATION TO THE CHOLESTERIC STATE, Wysocki, J.J., Adams, J.E., Olechna, D.J., p419: "Liquid Crystals and Ordered Fluids," Plenum Press, 1970.

1070. CONTINUED KINETIC STUDY OF THE CHOLESTERIC-NEMATIC TRANSITION IN A LIQUID CRYSTAL FILM, Wysocki, J.J., Mol Cryst & Liq Cryst v14 n1-2 p71 (Aug '71)

1071. LIQUID CRYSTAL DISPLAY DEVICES, Yamada, T., Kawashi, T., J Inst Electronics Commun Engrs Japan v53 n4 p469 (Apr '70)

1072. NON-NEWTONIAN VISCOSITY OF LIQUID CRYSTALS FORMED BY CHOLESTERYL OLEYL CARBONATE,

Yamada, T., Fukada, E., Japanese J Appl Phys v12 n1 p68 (Jan '73)

1073. LIGHT SCATTERING STUDY OF THE DYNAMICAL BEHAVIOR OF ORDERING JUST ABOVE THE PHASE TRANSITION TO A CHOLESTERIC LIQUID CRYSTAL, Yang, C.C.,
Phys Rev Lett v28 n15 p952 (10 Apr '72)

1074. ON MEASURING NUCLEAR MAGNETIC SHIELDING ANISOTROPIES IN LIQUID CRYSTAL SOLVENTS, Yannoni, C.S., IBM J Res/Dev v15 n1 p59 (Jan '71)

1075. STUDIES OF CHEMICAL SHIFT ANISOTROPY IN LIQUID CRYSTAL SOLVENTS, IV. RESULTS FOR SOME FLUORINE COMPOUNDS, Yannoni, C.S., Dailey, B.P., Ceasar, G.P.,
J Chem Phys v54 n9 p4020 (1 May '71)

1076. DIELECTRIC ANISOTROPY OF THE LIQUID CRYSTAL PHASE OF P-AZOXYANISOLE PRODUCED BY MAGNETIC FIELD, Yano, S., Kasatori, T., Kuwahara, M., Aoki, K., J Chem Phys v57 n1 p571 (1 Jly '72)

1077. NEMATIC ORDER AND STRUCTURE IN AMORPHOUS POLYMERS, Yeh, G.S.Y.,
p79: Abstr 3rd Int'l Liq Cryst Conf., Berlin, W. Germany (Aug '70)

1078. THE PREPARATION AND THERMODYNAMICS OF SOME HOMOLOGOUS MITRONES, A NEW GROUP OF LIQUID CRYSTALS, Young, W.R., Haller, I., Aviram, A., p136: Abstr 3rd Int'l Liq Cryst Conf., Berlin, W. Germany (Aug '70)

1079. MESOMORPHIC SUBSTANCES CONTAINING SOME GROUP IV ELEMENTS, Young, W.R., Haller, I., Green, D.C., p137: Abstr 3rd Int'l Liq Cryst Conf., Berlin, W. Germany (Aug '70)

1080. MESOMORPHIC PROPERTIES OF THE HETEROCYCLIC ANALOGS OF BENZYLIDENE-4-AMINO-4'-METHOXY-BIPHENYL, Young, W.R., Haller, I., Williams, L.,
p383: "Liquid Crystals and Ordered Fluids," Plenum Press, 1970.

1081. PREPARATION AND THERMODYNAMICS OF SOME HOMOLOGOUS NITRONES, A NEW GROUP OF LIQUID CRYSTALS, Young, W.R., Haller, I., Aviram, A., IBM J Res/Dev v15 n1 p41 (Jan '71)

1082. MESOMORPHIC SUBSTANCES CONTAINING SOME GROUP IV ELEMENTS, Young, W.R., Haller, I., Green, D.C., Mol Cryst & Liq Cryst v13 n4 p305 (Jly '71)

1083. PREPARATION OF THERMODYNAMICS OF SOME HOMOLOGOUS NITRONES, A NEW GROUP OF LIQUID CRYSTALS, Young, W.R., Young, W.R., Haller, I., Aviram, A.,
Mol Cryst & Liq Cryst v13 n4 p357 (Jly '71)

1084. MESOMORPHISM IN THE 4,4'-DIALKOXY-TRANS-STILBENES, Young, W.R., Haller, I., Aviram, A., Mol Cryst & Liq Cryst v15 n4 p311 (Jan '72)

1085. COMBINATION REFLECTIVE/TRANSMISSIVE LIQUID CRYSTAL DISPLAY, Young, W.R.,
IBM Tech Discl Bull v15 n8 p2435 (Jan '73)

1086. ANISOTROPIC MASS DIFFUSION IN LIQUID CRYSTALS, Yun, C.-K., Fredrickson, A.G.,
Mol Cryst & Liq Cryst v12 n1 p73 (Dec '70)

1087. HEAT GENERATION IN NEMATIC MESOPHASES SUBJECTED TO MAGNETIC FIELDS, Yun, C.-K., Fredrickson, A.G., p239: "Liquid Crystals and Ordered Fluids," Plenum Press, 1970.

1088. THERMAL CONDUCTION NEAR AN INTERFACE IN A NEMATIC LIQUID CRYSTAL SUBJECTED TO A MAGNETIC FIELD, Yun, C.-K., Picot, J.J.C., Fredrickson, A.G., J Appl Phys v42 n12 p4764 (Nov '71)

1089. INDUCED VISCOSITY CHANGES IN A POLARIZABLE LIQUID, Yun, C.-K., Fredrickson, A.G.,
Phys Fluids v16 n1 p1 (Jan '73)

1090. ON A FRICTION COEFFICIENT OF NEMATIC LIQUID CRYSTALS, Yun, C.-K., Phys Lett (A) v43a n4 p369 (26 Mch '73)

1091. RAMAN SCATTERING IN LIQUID CRYSTALS WITH AN ARGON LASER, Zhdanova, A.S., Gorelik, V.S., Shushinskii, M.M., Opt & Spectrosc v31 n6 p490 (Dec '71)

1092. ON THE POSSIBILITIES OF PERTURBED INTERNAL REFLECTION SPECTROPHOTOMETRY IN THE STUDY OF THE LIQUID CRYSTALLINE STATE OF A SUBSTANCE, Zolotarev, V.M., Belyaevskaya, N.M., Bobovich, Ya.S., Opt & Spectrosc v28 n1 p104 (Jan '70)

" " " " " " " " " " "

Please insert the following...

402. CONTINUUM THEORY AND FOCAL CONIC TEXTURE FOR LIQUID CRYSTALS OF THE SMECTIC MESOPHASE, Geurst, J.A., Phys Lett (A) v34a n5 p283 (22 Mch '71)

403. GENERALIZED VORTICITY IN THE THEORY OF LIQUID CRYSTALS, Geurst, J.A., Phys Lett (A) v36a n1 p63 (2 Aug '71)

404. CONTINUUM THEORY OF TYPE-A SMECTIC LIQUID CRYSTALS, Geurst, J.A., Phys Lett (A) v37a n4 p279 (6 Dec '71)

" " " " " " " " " " "

The following material was received past the deadline for indexing...

THEORY OF THE CHANGE IN CHOLESTERIC PITCH NEAR CHOLESTERIC-SMECTIC PHASE TRANSITIONS, Alben, R., Mol Cryst & Liq Cryst v20 n3-4 p231 (Apr '73)

G.E. EXHIBITS TELETYPE WITH LIQUID CRYSTAL DISPLAY, Anon, Electro-optical Systems Design v5 n7 p4 (Jly '73)

LIQUID CRYSTALS - A VIABLE NEW MEDIUM, Astle, B., Opt Spectra v7 n7 p35 (Jly '73)

PROTON MAGNETIC RESONANCE IN NEMATIC LIQUID CRYSTALS: ACETOXYBENZAL-P-ANISIDINE, ACETOXYBENZAL--P-AMINOAZOBENZENE, AND ANISAL-P-AMINOAZOBENZENE, Avadhanlu, M.N., Murty, C.R.K., Mol Cryst & Liq Cryst v20 n3-4 p221 (Apr '73)

DYNAMIQUE DES FLUCTUATIONS PRES D'UNE TRANSITION SMECTIQUE 'A' - NEMATIQUE DE 2° ORDRE, Brochard, F., J Phys (France) v34 n5-6 p411 (May-Jne '73)

ORIENTATIONAL ORDER IN THE VICINITY OF A SECOND ORDER SMECTIC 'A' TO NEMATIC PHASE TRANSITION, Cabane, B., Clark, W.G., Solid State Commun v13 n2 p129 (15 Jly '73)

MOLECULAR ALIGNMENT AND CONDUCTIVITY ANISOTROPY IN A NEMATIC LIQUID CRYSTAL, Carr, E.F., Chou, L.S., J Appl Phys v44 n7 p3365 (Jly '73)

ELECTRON-BEAM ADDRESSABLE LIQUID CRYSTAL DISPLAY WITH STORAGE CAPABILITY, Chang, I.F., IBM Tech Discl Bull v16 n1 p353 (Jne '73)

SOME MICROSTRUCTURAL OBSERVATIONS OF 'MBBA' LIQUID CRYSTAL FILMS, Chang, R., Mol Cryst & Liq Cryst v20 n3-4 p267 (Apr '73)

NEMATIC LIQUID CRYSTAL MATERIALS FOR DISPLAYS, Creagh, L.T., IEEE Proc v61 n7 p814 (Jly '73)

HELIXINVERSION IN EINEM BINAEREN MISCHSYSTEM NEMATISCH/CHOLESTERISCH, Finkelmann, H., Stegemeyer, H., Z Naturforsch v28a n5 p799 (May '73)

DIRECT DETERMINATION OF THE FIVE INDEPENDENT VISCOSITY COEFFICIENTS OF NEMATIC LIQUID CRYSTALS, Gaehwiller, Ch., Mol Cryst & Liq Cryst v20 n3-4 p301 (Apr '73)

A HYBRID LIQUID CRYSTAL DISPLAY WITH A SMALL NUMBER OF INTERCONNECTIONS, Gerritsma, C.J., Lorteye, J.H.J., IEEE Proc v61 n7 p829 (Jly '73)

REFLECTIVE LANDS FOR USE IN MULTILAYER CERAMIC LIQUID CRYSTAL WATCH DIALS, Haddad, M.M., Kaiser, H.D., Schmeckenbecher, A.F., IBM Tech Discl Bull v16 n1 p42 (Jne '73)

RETARDING CRYSTALLIZATION OF SUPERCOOLED NEMATIC LIQUID CRYSTALS, Haller, I., Young, W.R., IBM Tech Discl Bull v16 n1 p119 (Jne '73)

PROPERTIES OF A HOMOLOGOUS SERIES OF O-HYDROXY SUBSTITUTED ANILS AND SOME BINARY MIXTURES, Hirata, H., Waxman, S.N., Teucher, I., Labes, M.M., Mol Cryst & Liq Cryst v20 n3-4 p343 (Apr '73)

LIQUID CRYSTALS IN INTEGRATED OPTICS, Hu, C., Whinnery, J.R., IEEE J Quantum Electron vQE-9 n6 p684 (Jne '73)

KERR RESPONSE OF NEMATIC LIQUIDS, Johnston, A.R., J Appl Phys v44 n7 p2971 (Jly '73)

SURFACE-PRODUCED ALIGNMENT OF LIQUID CRYSTALS, Kahn, F.C., Taylor, G.N., Schonhorn, H., IEEE Proc v61 n7 p823 (Jly '73)

THE STRUCTURE OF A CHOLESTERIC MESOPHASE PERTURBED BY A MAGNETIC FIELD, Luckhurst, G.R., Smith, H.J., Mol Cryst & Liq Cryst v20 n3-4 p319 (Apr '73)

NUCLEAR MAGNETIC RESONANCE STUDIES OF SMECTIC LIQUID CRYSTALS, Luz, Z., Meiboom, S., J Chem Phys v59 n1 p275 (1 Jly '73)

LASER-ADDRESSED LIGHT VALVES USING LIQUID CRYSTALS, Maydan, D., IEEE J Quantum Electron vQE-9 n6 p707 (Jne '73)

INFRARED LASER ADDRESSING OF MEDIA FOR RECORDING AND DISPLAYING OF HIGH-RESOLUTION GRAPHIC INFORMATION, Maydan, D., IEEE Proc v61 n7 p1007 (Jly '73)

MEASUREMENT OF SMECTIC-PHASE ORDER-PARAMETER FLUCTUATIONS IN THE NEMATIC PHASE OF HEPTYLOXY-AZOXYBENZENE, McMillan, W.L., Phys Rev (A) v8 n1 p328 (Jly '73)

LIGHT SCATTERING CHARACTERISTICS IN LIQUID CRYSTAL STORAGE MATERIALS, Meyerhofer, D., Pasierb, E.F., Mol Cryst & Liq Cryst v20 n3-4 p279 (Apr '73)

EFFECT OF BULK IMPURITIES ON THE TRANSPARENCY OF CHOLESTERYL NONANOATE, Novak, T.J., Poziomek, E.J., Mackay, R.A., Mol Cryst & Liq Cryst v20 n3-4 p203 (Apr '73)

FLUORESCENCE OF PYRENE AND PHENANTRENE IN CHOLESTERYL NONANOATE AS A FUNCTION OF TEMPERATURE, Novak, T.J., Mackay, R.A., Poziomek, E.J., Mol Cryst & Liq Cryst v20 n3-4 p213 (Apr '73)

FREQUENCY DEPENDENCE OF THE PROTON SPIN RELAXATION OF THE DIPOLAR ENERGY IN A LIQUID CRYSTAL, Sharp, A.R., Forbes, W.F., Pintar, M.M., J Chem Phys v59 n1 p460 (1 Jly '73)

ELECTRON PARAMAGNETIC RESONANCE STUDIES OF A VISCOUS NEMATIC LIQUID CRYSTAL. II. EVIDENCE COUNTER TO A SECOND-ORDER PHASE CHANGE, Shutt, W.E., Gelerinter, E., Fryburg, G.C., Sheley, C.F., J Chem Phys v59 n1 p143 (1 Jly '73)

RESPONSE OF NEMATIC LIQUID CRYSTALS TO VAN DER WAALS FORCES, Smith, E.R., Ninham, B.W., Physica (Holland) v66 n1 p111 (15 May '73)

A MOESSBAUER STUDY OF A Sn-119 BEARING SOLUTE IN AN ORDERED SMECTIC LIQUID CRYSTAL, AT 77°K, Uhrich, D.L., Hsu, Y.Y., Fishel, D.L., Wilson, J.M., Mol Cryst & Liq Cryst v20 n3-4 p349 (Apr '73)

CAN TWO-DIMENSIONAL MESOPHASES SOLIDIFY?, Yorke, E.D., DeRocco, A.G., J Chem Phys v59 n1 p92 (1 Jly '73)

" " " " " " " " " " " "

AUTHOR INDEX

Ache, H.J., 763
Adachi, K., 1, 2
Adair, W.L., 912
Adams, J., 440, 441, 442, 443, 444
Adams, J.E., 3, 4, 5, 6, 7, 8, 9, 10, 1069
Aero, E.L., 11, 12, 13, 14
Agache, C., 320
Agarwal, R.K.L.P., 544
Agrawl, D.K., 752
Aihara, M., 15
Alben, R., 16, 17, 18, 928
Alder, C.J., 19
Aleonard, R., 815
Alimonda, A., 20
Ambrose, E.J., 21, 22
Ambrosia, A., 23
Amer, N.M., 24, 25
Amerik, Y.B., 80, 81
Andrews, J.T.S., 26
Anon, 27, 27A
Aoki, K., 1076
Arnould, H., 869, 871
Arora, S.L., 28, 29, 30, 31, 982, 983, 984
Ashford, A., 32
Aslaksen, E.W., 33, 34, 35, 36
Assouline, G., 37, 38, 39, 40, 41, 42, 43,
   44, 45, 46, 47, 456, 457, 458
Atallah, A.M., 48, 49, 50, 51
Attergut, S., 52
Auffret, R., 53, 54
Auth, D.C., 504, 505
Auvergne, M., 55
Avadhanlu, M.N., 421
Aviram, A., 56, 57, 58, 59, 1078, 1081, 1083,
   1084
Azzam, R.M.A., 60, 61
Bacon, W.E., 62, 63
Bacri, J.-C., 64
Bader, P., 918
Baessler, H., 65, 66, 67, 68, 69, 70, 990
Bain, R.M., 71
Baise, A.I., 72
Bak, C.S., 205, 206
Balmbra, R.R., 73, 74
Baltzer, D.H., 75
Balzarini, D.A., 76
Barberian, J.G., 331
Barbic, L., 291
Bard, A.J., 661
Barenberg, S., 739
Barili, P.L., 77
Barlow, A.J., 650
Barrall, E.M.II., 78, 237, 1036, 1037, 1038, 1039
Barton, L.A., 462
Bashara, N.M., 60, 61
Basov, N.G., 79
Baturin, A.A., 80, 81
Baur, G., 82
Beard, R.B., 68, 69
Beard, T.D., 83, 84, 85, 86, 695
Belyaevskaya, N.M., 1009, 1092
Belyakov, V.A., 87
Benjamin, A., 88
Berchet, D., 485
Berezin, P.D., 79
Berg, R.A., 89
Berreman, D.W., 90, 91, 92, 93, 94, 95, 96,
   97, 98
Bertolotti, M., 99, 100, 101, 102, 103, 104
Bigelow, J.E., 549
Billard, J., 105, 106, 629
Bleha, W.P., 84, 86, 695
Blinc, R., 107, 108, 109, 110, 111, 291, 1031
Blinov, L.M., 79, 606
Blumstein, A., 112, 113
Blumstein, R., 112, 113
Bobovich, Ya. S., 1009, 1092
Boettcher, B., 114, 115, 507
Boller, A., 116
Borden, H., 117
Borel, N.N., 1001
Borer, W.J., 118
Borodzulja, V., 605
Bosso, A.J., 119
Bottomley, R., 416
Bouchiat, M.A., 120, 621, 623
Bouligand, Y., 121, 122, 123
Boven, J., 1021
Boyd, P.R., 336, 338, 339, 340, 341, 822,
   823, 824, 825
Brady, G.W., 425
Brand, P., 286, 287, 288
Braunstein, M., 84
Bredfeldt, K.E., 78
Briere, G., 124, 125
Brochard, F., 126, 127, 128, 129, 130, 438,
   809, 810
Brooks, S.A., 131
Brown, A.J., 343, 344
Brown, C.W., 118
Brown, G.H., 62, 63, 132, 133, 134, 739
Brunet-Germain, M., 135, 136, 137, 138
Bucci, P., 139, 140, 141
Bucknall, A.B., 74
Buecher, H.K., 803

Bulkin, B.J., 142, 143, 144, 145
Bulygin, A.N., 11, 12, 13, 14
Burckhardt, C.B., 146
Burger, J.P., 209, 210, 853
Burnell, E.E., 147
Buser, M.S., 148
Bush, R.F., 149
Butcher, B.T., 150
Byatt, D.W.G., 151
Cabane, B., 152, 260, 261, 262
Cabos, C., 153, 154
Caille, A., 155
Callender, R.E., 156, 157
Cameron, L.M., 156, 157
Candau, S., 158, 702, 703, 704, 705
Canet, Q., 841
Carr, E.F., 159, 160, 161, 162, 163, 164, 715, 790
Carroll, T.O., 165, 166
Castellano, J.A., 167, 168, 169, 170, 171, 172, 173, 712
Caulfield, H.J., 174
Ceasar, G.P., 1075
Chabay, I., 489
Chambers, J., 238
Champa, R.A., 175, 176
Chan, S.O., 1040
Chandrasekhar, S., 177, 178, 179, 180, 181, 182, 183, 184, 185, 186, 187, 188, 684, 685
Chang, R., 189, 190, 191
Channin, D.J., 192
Chapman, D., 193
Chapunov, E., 564
Charvolin, J., 194, 195, 196
Cheng, J., 302
Cheung, L., 197, 200
Chistyakov, I.G., 198, 199, 1017, 1018
Chou, S.C., 200
Chow, L.C., 201, 202, 203
Christman, S.B., 204
Chu, B., 205, 206
Chu, W., 863
Chumakova, S.P., 538
Chuvyrov, A.N., 207, 542
Ciliberti, D.F., 208
Cladis, P.E., 209, 210, 211, 212, 213, 214, 853, 854, 856, 1055, 1056
Clark, G., 152
Clark, N.A., 215, 216, 654, 964
Clary, R.M., 217
Clunie, J.S., 73, 74
Coche, A., 564, 565
Cohen, P.S., 204, 361
Cole, G.D., 730
Cole, H.S., 52, 548
Colvin, C., 876, 877, 878
Conners, G.H., 218
Constant, J., 32,
Cook, B.D., 219, 220, 762
Cosentino, L.S., 221
Costello, M.J., 610
Cox, R.J., 222, 449
Creagh, L.T., 223, 224, 225, 523

Currie, P.K., 226, 227
Curtis, R.J., 932
Cvikl, B., 228, 301
Dailey, B.P., 737, 1075
Dailey, J., 7
Daino, B., 99, 100, 101
Dalby, E.D., 229
Dave, J.S., 230, 231, 232, 233, 234
Davis, G.J., 235, 236, 237
Davis, L.E., 238, 239
Dawson, J.F., 1065
Debeauvais, F., 705, 706, 707
Debrunner, P.G., 830
Deeds, W.E., 930
DeGennes, P.-G., 126, 240, 241, 242, 243, 244, 245, 246, 247, 248, 249, 250, 251, 252, 321
De Jeu, W.H., 253, 254, 255, 256, 257, 258, 258A, 396, 1021
Delaye, M., 259
Delhaye, M., 105
Deloche, B., 260, 261, 262
Delord, P., 153, 154, 263
Demus, D., 264, 265, 266, 267
Denoyer, F., 268
De Rocco, A.G., 269, 1066
Derzhanski, A.I., 270, 271, 272, 273, 274, 275
Detjen, R.E., 276
Deuling, H.J., 277
De Vries, A., 278, 279, 280, 281, 282, 283, 284, 739
Dickson, L.D., 285
Diehl, P., 147, 570
Diele, S., 264, 265, 286, 287, 288
Dietrich, H.J., 289, 955
Diguet, D., 290, 870
Dimic, V., 107, 108, 110, 291, 292
Di Porto, P., 101
Dir, G., 444
Dixon, G.D., 208, 293, 294, 899
Dmitrieff, A., 38, 29
Dmitriev, S.G., 295, 296
Doane, J.W., 297, 298, 299, 300, 301, 748, 1031, 1032
Dolphin, D., 302
Dong, R.Y., 303, 304, 305, 306, 307, 308, 309, 916
Douy, A., 310, 384, 385, 401
Drakenberg, T., 521, 522
Drauglis, E., 536
Dreher, R., 311, 312, 313
Dreizin, Yu. A., 314, 315
Dreyer, J.F., 316, 317, 318, 319
Dubois, J., 320
Dubois-Violette, E., 321, 322, 323, 324, 811
Dunmur, D.A., 325
Durand, G., 259, 290, 326, 327, 380, 381, 382, 639, 701, 865, 870
Durou, C., 328
Duroz, I.J., 1011, 1012
Duruz, J.J., 329
Dybowski, C.R., 330, 889, 890
Dykhane, A.M., 314, 315

Dyro, J.F., 331
Dzyaloshinsky, I.E., 332
Easwaran, K., 109
Eden, D., 333
Edmonds, H.D., 334
Edmonds, P.D., 331
Eins, S., 335
Elachi, C., 335A
Elser, W., 336, 337, 338, 339, 340, 341, 821, 822, 823, 824, 825
Ennulat, R.D., 342, 343, 344, 345, 346, 359
Erhardt, P.F., 347, 817
Ericksen, J.L., 348
Eringen, A.C., 635, 636, 637, 638
Etters, R.D., 845
Fabijanic, J., 806
Fahrenschon, K., 907, 910
Falor, W.H., 739
Fan, C., 349, 350, 351, 352
Fan, G.J.-Y., 353
Farinha-Martins, A., 354
Feldman, M., 355, 1050
Felici, N.J., 356
Fergason, J.L., 28, 29, 30, 31, 345, 357, 358, 359, 687, 982, 983, 984
Filippini, J.-C., 360
Filippini, N.N., 1001
Fink, J.J., 361
Fink, L.J., 204
Finlayson, B.A., 1008
Fischer, F., 1044
Fishel, D.L., 281, 301, 362, 375, 1032
Fixman, M., 1064
Flannery, J.B., 440, 441, 442
Forbes, W.F., 304, 306, 307, 308
Ford, G.W., 800, 801
Ford, W.G.F., 363
Forster, D., 364
Fourney, M.E., 504
Franchini, P.F., 139
Franklin, W., 228
Franklin, W.M., 365, 366, 597, 598
Fraser, D.B., 729
Fredrickson, A.G., 1086, 1087, 1088, 1089
Freiser, M.J., 59, 367, 368, 369, 370, 371, 447, 452
Friberg, S., 372, 373
Friedel, J., 374
Fryburg, G.C., 375, 390
Fukada, E., 976, 1072
Fuller, A.M., 980
Fung, B.M., 376, 377
Gahwiller, Ch., 378, 379
Galerne, Y., 380, 381, 382
Gallot, B., 383, 384, 385, 386, 401
Gallot, B.R., 310
Garbarino, P.L., 387
Gardlund, Z.G., 932
Garland, C.W., 333
Gaspard, F., 124
Gasparoux, H., 388, 389, 839, 840, 862
Geil, P.H., 739
Gelerinter, E., 375, 390
Gerace, M.J., 376
Gerritsen, J., 391, 392, 393

Gerritsma, C.J., 253, 256, 257, 258, 394, 395, 396, 397, 398, 399, 400
Gervais, M., 385, 401
Geurst, J.A., 400, 402, 403, 404, 405
Ghosh, S.K., 406, 407
Giallorenzi, T.G., 927
Gibson, H.W., 816, 819
Gilson, D.F.R., 868
Giraudou, J.-C., 328
Giua, P.E., 408
Gladstone, G.L., 58, 409, 410, 411
Glasser, M.L., 535
Go, Y., 497
Goldberg, L.S., 412
Goldmacher, J.E., 168, 170, 413, 466
Goldstein, J.H., 662
Gooch, C.H., 414, 415, 416
Goossens, W.J.A., 257, 395, 405, 417, 418, 419, 420
Gopalakrishna, C.V.S.S.V., 421
Gorelik, V.S., 1091
Goshen, S., 422
Govorun, E.Ya., 1002A
Graber, G., 114, 115
Grabmaier, J.G., 423, 424
Gravatt, C.C., 425
Gray, G.W., 426, 427, 428, 429
Green, D.C., 430, 1079, 1082
Gregoris, L.G., 431
Greubel, W., 432
Greubel, W.F., 423, 424
Grigorov, L., 274
Grobben, A.H., 1020, 1021
Groves, J.L., 830
Gruler, H., 433, 434, 435, 436
Grunbaum, D., 142
Guinier, O., 268
Gulaya, S., 922
Gurtler, R.W., 437
Guyon, E., 129, 438, 439, 809, 810, 811
Haas, W., 440, 441, 442, 443, 444
Haas, W.E.L., 4, 5, 6, 7, 8, 9, 10
Haida, O., 2
Hakemi, H., 445
Halbertstam, M., 668
Haller, I., 56, 446, 447, 448, 449, 450, 451, 452, 453, 1078, 1079, 1080, 1081, 1082, 1083, 1084
Hampel, B., 454
Hareng, M., 38, 39, 40, 41, 42, 43, 44, 45, 46, 47, 456, 457, 458
Harrison, K.J., 426, 427, 428, 429
Harrison, L., 459
Hartman, 483
Hartshorne, N.H., 460
Hass, M., 912
Haxby, B.A., 89
Hayes, C.F., 793
Heidrich, P.F., 461
Heilmeier, G.H., 356, 462, 463, 464, 465, 466, 467
Helfrich, W., 146, 463, 468, 469, 470, 471, 472, 473, 474, 475, 476, 477, 478, 479, 480, 900, 901, 902, 911
Heppke, G., 481, 482

Herczfeld, P.R., 483
Herino, R., 124, 125
Hewitt, R.C., 720
Hill, C.P., 334
Hirasawa, R., 661
Hisamitsu, S., 484
Hochapfel, A., 485
Hoff, F., 742
Holland, W.P., 486
Holloway, W.W., Jr., 487
Holzwarth, G., 488, 489
Holzwarth, N.A.W., 488, 489
Hornberger, W.P., 490
Hosemann, R., 491
Hsu, E.C.-H., 492
Huang, H.-W., 493
Huggins, H.A., 451, 452
Hughes, W.T., 520
Hulin, J.P., 494, 872
Humphries, R.L., 495, 496
Hyde, A.J., 71
Iizuka, E., 497
Iizuka, K., 431, 498
Imura, H., 499, 500, 501
Inaba, H., 15
Indovina, P.L., 406
Ineichen, B., 34
Inokuchi, S., 502
Inoue, T., 503
Intlekofer, M.J., 504, 505
Inuishi, Y., 484
Isihara, A., 506, 1042
Ittner, G., 507
Iwayanagi, S., 508
Jacobson, A.D., 84, 509
Jacques, J., 629
Jahnig, F., 510, 511
Jaintz, J., 559, 560
Jakeman, E., 512
James, P.G., 495, 496, 513, 514
Janik, J.A., 515, 516, 782
Janik, J.M., 515, 516, 782
Janning, J.L., 517
Jenkins, J.T., 518, 519
Jeppesen, M.A., 520
Jerome, D., 260, 261
Johansson, A., 512, 522
Johnson, C.S., Jr., 1047
Johnson, D.L., 298, 301
Johnson, J.F., 492, 828
Jones, D., 523, 524, 525, 666, 667, 668
Kahn, F.J., 526, 527, 528, 529, 530, 531, 532, 533, 729
Kakiyama, H., 885
Kamei, H., 534
Kanekar, C.R. 570
Kaplan, J.I., 535, 536, 537
Kapustin, A.P., 538, 539, 540, 541, 542
Kartha, C.G., 544, 545, 546, 547
Kasano, K., 596
Kasatori, T., 1076
Kashnov, R.A., 548, 549, 550, 956, 957, 958
Katayama, Y., 534
Kats, E.I., 551, 552, 553

Kaufmann, J.W., 923, 924
Kawashi, T., 1071
Kazan, B., 554, 555
Kaye, D., 556
Keilman, F., 557
Kelker, H., 558, 559, 560, 561
Kemp, K.A., 562, 563
Kerllenevich, B., 564, 565
Kessler, J.O., 566, 567, 663
Kessler, L.W., 568
Keyes, R.W., 569
Khetrapal, C.L., 570, 571, 572, 573, 574
Kiemle, H., 575, 576
Kimura, H., 577
Kinnison, W.W., 719
Kirov, N., 578
Kirton, J., 32, 579
Kitagawa, N., 112, 113
Klapperstueck, M., 264, 265, 266
Klein, E.J., 580
Kleman, M., 121, 211, 212, 214, 374, 581, 1057
Klingbiel, R.T., 803
Klingen, T.J., 582, 583
Knetz, A.R., 223, 224, 225, 584
Knaak, L.E., 585
Knapp, F.F., 586, 587
Knauer, R., 588
Ko, K., 992, 993
Kobayashi, K.K., 589, 590, 591, 592, 593, 594, 595, 596, 597, 598, 749
Kobayashi, S., 887
Kock, W.E., 599
Kodak, 600
Kodama, H., 749
Koelmans, H., 601, 602, 603
Kogure, O., 604
Kolenko, E., 605
Kompanets, I.N., 79, 606, 1025
Kunwar, A.C., 570, 571, 573
Kopp, W.-U., 607
Korobkin, Yu.V., 1029A
Kosterin, E.A., 1017
Kramer, A.J., 157
Kramer, L., 350
Krebs, P., 608, 609
Krentsel, B.A., 80, 81
Krigbaum, W.R., 610
Krishnamachari, N., 144
Krishnamurty, D., 611, 612, 613, 614
Krishnamurty, K.S., 614
Krishnan, K., 145
Krueger, H.H., 423, 424
Kurian, G., 233
Kusabayashi, N., 614
Kusabayashi, S., 987, 988
Kuvatov, Z.Kh., 539, 541
Kuwahara, M., 1076
Kuz'michev, V.M., 1002, 1002A
Labes, M.M., 65, 66, 67, 68, 69, 70, 72, 445, 664, 772, 783, 784, 989, 990, 991, 992, 993
Labrunie, G., 615
Lacroix, J.-C., 616

Lacroix, R., 53
Lafonta, P., 617
Laisk, E., 618
Lakatos, K., 619
LaMacchia, J.T., 530
Lambert, M., 268, 653
Lancaster, D., 620
Langevin, D., 621, 622, 623
Langevin-Cruchon, D., 120
Lapointe, J., 624
Lasher, G., 625, 626, 627
Lathouwers, Th.W., 258
Lebwohl, P.A., 627
Lechner, B.J., 628
Leclerq, M., 629
Leder, L., 3
Leder, L.B., 630, 631, 632, 633, 634
Lee, J.D., 635, 636, 637, 638, 655
Lefevre, M., 639
Lefkowitz, I., 981
Leger, L., 326, 640
Leiba, E., 37, 38, 39, 40, 41, 42, 43, 44, 45, 46, 47, 456, 457, 458
Leonov, Yu.S., 1029A
LePesant, J.P., 640, 786
Lephardt, J.O., 145
Le Roy, P., 642
Lescinsky, M., 742
Lesieur, J.P., 643
Leslie, F.M., 644, 645, 646, 647, 648, 649
Letcher, S.V., 562, 563, 650, 651
Levelut, A.M., 268, 652, 653
Liao, Y., 654
Lieberman, E.D., 655
Liebert, L., 656
Likins, K.L., 687
Limacher, H., 918
Lin, F.L., 205, 206
Lindblom, G., 657, 658
Lindman, B., 657, 658
Link, V., 264, 265
Litster, J.D., 446, 448, 450, 659, 660, 865, 963, 964, 965
Lomax, A., 661
Long, R.C., Jr., 662
Longeri, M., 141
Longley-Cook, M.T., 566, 663
Lopatina, L., 605
Lord, A.E., Jr., 664, 665, 1046
Low, J.J., 415, 416
Lowry, B.A., 1048
Lu, S., 523, 524, 525, 666, 667, 668
Lubensky, T.-C., 364, 669, 670, 671, 672, 673
Luckhurst, G.R., 131, 495, 496, 513, 514, 674, 675, 676, 677, 678, 679, 680
Ludeman, C.P., 681
Lunazzi, L., 77
Luthi, B., 746
Lydon, J.E., 682
MacAnally, R.B., 683
Mackay, A., 916
Mackay, R.A., 833
MacLean, C., 391, 392
Madhusudana, N.V., 177, 180, 182, 185, 186, 188, 649, 684

Magura, K., 686
Mahenc, J., 328
Mailer, H., 687
Mainusch, K.-J., 688, 689, 690, 950, 951, 952
Malet, G., 153, 263
Malya, P.A.G., 66
Mamaeva, L.S., 539
Manaranche, J.C., 691
Marathay, A.S., 692, 693, 866, 867
Margerum, J.D., 84, 694, 695
Margozzi, A.P., 580
Marlow, F.J., 628
Marsh, D., 696
Marsh, D.G., 818
Martellucci, S., 102, 103
Martin, A.S., 697
Martin, P.C., 364, 698, 699, 700
Martinand, J.L., 382, 639, 701
Martinoty, P., 158, 702, 703, 704, 705, 706, 707
Martins, A.F., 708, 709
Martire, D.E., 201, 202, 203, 1054
Martunand, J.J., 710
Mart'yanova, L.I., 540
Marusic, M., 915, 916
Marvin, D.A., 624
Marzotko, D., 266
Matsuo, T., 2
Maydan, D., 729
Mayer, J., 711
Maze, C., 437
McCaffrey, M.T., 168, 169, 170, 413, 712
McColl, J.R., 713
McDermott, J., 714
McLemore, D.P., 161, 715
McMillan, W.L., 716, 717, 718
McNutt, J.D., 719
Meiboom, S., 720
Meier, G., 82, 311, 312, 433, 434, 435, 436, 697
Meier, H.E., 721
Melamed, L., 722, 723, 724, 725, 726, 727, 728
Melchior, H., 729
Merlin, J.C., 105
Merrill, B.E., 61
Merritt, W.G., 730
Meyer, R.B., 197, 200, 216, 302, 731, 732, 733, 734
Meyer, V., 20
Meyerhofer, D., 735, 736
Michalovcz, A., 710
Michels, H.J., 1011, 1012
Migliori, A., 985
Mignosa, C.P., 1036, 1037, 1038, 1039
Millett, F.S., 737
Mingione, J., 117
Mishra, R.K., 738, 739
Mitra, S.S., 118
Miyashita, K., 771
Mizell, L., 740
Mizuno, H., 741
Moehwald, H., 881, 883
Mondon, F., 125

Moon, F.C., 655
Moravec, F., 742, 743
Morita, Y., 502
Moroi, D.S., 228, 299, 597, 598, 744
Morozov, V.N., 79, 606
Moses, H.A., 204, 361
Mueller, U., 491, 689
Mueller, W.U., 745
Mukamel, D., 422
Muljiani, 302
Mullen, M.E., 746
Muller, U., 953
Munck, E., 830
Murase, K., 604, 747
Murphy, J.A., 297, 748
Murty, C.R.K., 163, 164, 421
Nakagiri, T., 749, 750
Nakamizo, M., 885
Nakamura, K., 1045
Nakatsu, K., 987
Nance, P., 117
Nannichi, Y., 751
Narasimha, R.D.V.G.L., 752
Narasimham, M.A., 753
Nash, J.A., 429
Nassimbene, E.G., 754
Natale, G.G., 755
Nehring, J., 756, 757, 758, 759, 760, 761, 1051, 1052
Nemec, J., 219, 762
Nes, W.R., 66
Nester, E.O., 628
Nicholas, H.J., 48, 49, 50, 51, 586, 587
Nicholas, J.B., 763
Nickerson, M.A., 979, 980
Niessen, A.K., 395
Nikitin, V.V., 79, 606, 1025
Nimoy, J., 694
Nomura, S., 1005
Nordio, P.L., 764
Nosov, V.N., 998
Novak, T.J., 833
Novakovic, L., 765
Oh, C.S., 477, 766
O'Hern, R.J., 767
Ohki, S., 785
Ohnishi, Y., 751
Ohtsuka, T., 768, 769, 770
Okano, K., 499, 500, 501
Olechna, D., 632
Olechna, D.J., 1069
Onnagawa, H., 771
Ooizumi, K., 892, 893
Ootaka, H., 976
Ootake, T., 771
Oron, N., 772
Orsay Liquid Crystal Group, 773, 774, 775, 776, 777, 778, 779, 780, 781
Osredkar, M., 292
Orlov, V.P., 87
Otnes, K., 515, 516, 782
Ozawa, T., 534
Padmini, A.R.K.L., 545, 546, 547
Paleos, C.M., 783, 784, 989
Palikhov, N.A., 1030

Papahadjopoulos, D., 785
Papon, P., 640, 786
Papoular, M., 787, 788, 789
Parker, J.H., 161, 790
Parker, R., 791
Parker, R.S., 300, 301
Parodi, O., 321, 700, 792
Parsons, J.D., 793
Patankar, A.V., 572
Patel, P.R., 362
Patharkar, M.N., 794, 847
Paxton, K.B., 218
Pedulli, G.F., 131
Pelzl, G., 795, 796, 880
Penz, P.A., 797, 798, 799, 800, 801, 802
Perron, R., 485
Pershan, P.S., 215, 364, 654, 698, 699, 700
Petit, J., 485
Petrie, S.E.B., 803
Petrov, A.G., 270, 271, 272, 273, 275, 804
Petscher, E.J., 739
Petty, M.S., 760
Pezolet, M., 805
Pick, P.G., 806, 807
Picot, C., 808
Picot, J.J.C., 794, 846, 847, 848, 849, 1088
Pieranski, P., 129, 211, 438, 439, 809, 810, 811, 1056
Pikin, S.A., 812, 813
Pincus, P., 814
Pintar, M.M., 304, 306, 307, 308, 995
Pirs, J., 108, 109, 110, 111
Pochan, J., 696
Pochan, J.M., 347, 816, 817, 818, 819
Poggi, N.N., 1001
Poggi, Y., 815
Pohl, L., 820
Pohlmann, J.L.W., 336, 337, 338, 339, 340, 341, 821, 822, 823, 824, 825
Poirier, J.C., 610
Poisson, F., 826
Pole, R.V., 461
Pollmann, P., 689, 690, 827
Pontikis, V., 380
Porter, R.S., 235, 236, 237, 828, 863
Pospisil, M., 829
Potasek, M.J., 830
Powell, C.G., 461
Powers, J.C.Jr., 831
Powers, J.V., 832
Poziomek, E.J., 833
Prajapatti, A.P., 231, 233
Prasad, J.S., 179, 183
Price, F.P., 834, 835
Priest, R.G., 836, 837, 838
Prochaska, F.T., 143
Prost, J., 388, 389, 839, 840, 841, 842, 862
Ptak, M., 975
Puyhaubert, J., 843
Pynn, R., 782
Rafuse, M.J., 487, 844, 942
Raich, J.C., 845
Rajan, V.S.V., 794, 846, 847, 848, 849
Rao, D.V.G.L.N., 850
Rapini, A., 788, 851, 852

Rault, J., 209, 210, 853, 854, 855, 856, 857, 858, 859
Raynes, E.P., 19, 32, 512, 579, 860
Raza, M.A., 861
Reeves, L.W., 861
Regaya, B., 388, 862
Rehm, D., 879
Resch, W.A., 1016
Reynolds, R.A., 223, 224
Rhodes, M.B., 863, 864
Ribotta, R., 259, 865
Richards, S.P., 866, 867
Richards, W.C., 347
Rigatti, G., 764
Rigny, P., 194, 195, 196
Riste, T., 515, 516
Roberts, N.N., 1001
Robertson, J.C., 868
Roddier, F., 55
Roncillat, M.,
Rondelez, F., 290, 326, 869, 870, 871, 872
Rosciszewski, K., 873
Rose, P.I., 803
Rosen, H., 874
Rosenberg, H.M., 585, 875
Rubin, D., 722, 723, 724, 725, 726
Runnels, L.K., 876, 877, 878
Rurainski, R., 266, 267
Sabel, J., 559, 560
Sackmann, E., 608, 609, 879, 880, 881, 882, 883
Sackmann, H., 286, 287, 288, 795, 796
Sadagopan, V., 409, 411
Saeva, F.D., 884
Sakagami, S., 885
Sakevich, N.M., 886
Sakurai, Y., 502
Sakusabe, T., 887
Salin, D., 710
Sambucetti, C.J., 23
Sandison, R.D., 387
Sanson, A., 676, 678, 975
Samulski, E.T., 888, 889, 890
Sapriel, J., 53
Sarma, G., 891
Sasabe, H., 892, 893
Sastry, G.S., 545
Sato, S., 894, 895, 896
Saupe, A., 28, 311, 312, 573, 574, 697, 756, 758, 759, 897, 898, 947
Savoie, R., 805
Sawyer, S.P., 568
Scala, L.C., 208, 293, 294, 899
Schadt, M., 116, 146, 476, 900, 901, 902, 911
Schara, M., 903, 920, 921
Scheffer, T.J., 90, 91, 92, 93, 94, 95, 96, 97, 436, 904, 905
Scherrer, H., 116, 906
Scheurle, B., 558, 560
Schiekel, M.F., 907, 908, 909, 910
Schmidt, D., 911
Schmidt, H., 510, 511
Schneider, F., 481, 482
Schnur, J.M., 412, 912, 927

Schultz, T.D., 913, 914
Schwartz, J., 608
Schwerdfeger, C.F., 303, 915, 916
Scudieri, F., 99, 100, 101, 102, 103, 104
Seelig, J., 917, 918
Segre, U., 764
Seiden, P.E., 149
Seki, S., 1, 2
Selzo, C.A., 919
Sensenbaugh, J.D., 285
Sentjurc, M., 903, 920, 921
Serra, A.M., 139
Serve, M.P., 585
Setaka, M., 679, 680
Sethares, J.C., 922
Sette, D., 99, 100, 101, 102, 103, 104
Sexton, M.C., 643
Shashidhar, R., 178, 181, 184, 187, 684, 685
Shaw, D.G., 923, 924
Sheley, C.F., 276
Shelton, J.W., 925, 926
Shen, Y.R., 24, 25, 874, 925, 926, 1063
Sheridan, J.P., 927
Shih, C.-S., 713, 928
Shimojo, T., 596
Shtrikman, S., 422, 929
Shubha, K., 185, 186
Shukla, G.C., 765
Shushinski, M.M., 1091
Silage, D., 331
Silver, D.L., 1008
Simova, P., 578
Simpson, W.A., 930
Skoulios, A., 931
Small, D.M., 236
Smith, C.W., 979
Smith, E.P., 204
Smith, G.W., 932
Smith, H.J., 674
Soeva, F.D., 934
Sobel, A., 933
Sorai, M., 2
Soref, R.A., 174, 844, 935, 936, 937, 938, 939, 940, 941, 942, 943, 944, 945, 946
Speer, R.S., 1048
Spiesecke, H., 947, 948
Spruijt, A.M.J., 400
Stamatoff, J.B., 949
Stanfield, M.K., 150
Stegemeyer, H., 688, 689, 690, 745, 950, 951, 952, 953, 954
Steiger, E.L., 289, 955
Steigner, E., 952
Stein, C.R., 956, 957, 958
Stein, R.S., 808, 863
Steiner, J.W., 236
Steinstrasser, R., 820, 959
Stenschke, H., 960
Stephen, M.J., 349, 350, 746, 961
Stepke, E., 962
Stewart, A., 807
Stewart, G.T., 150
Stieb, A., 82
Stinson, T.W., 659, 963, 964, 965

Straley, J.P., 966, 967
Stranski, I.N., 951
Strnad, R.J., 681
Strzelecki, L., 656
Subramhanyam, H.S., 611, 613
Subramanyam, S.V., 968
Suga, H., 1, 2
Sugiura, Y., 508
Sukharev, S.K., 198, 199
Sung, C.C., 969
Sussman, A., 735, 736, 970, 971, 972, 973
Swift, J., 364, 698, 699
Sy, D., 975
Szymanski, A., 974
Takamatsu, T., 976
Takase, A., 885
Takeuchi, T., 988
Tamamushi, B.-I., 977
Tanaka, S., 741
Tani, C., 978
Tara, N., 178, 181
Tarr, C.E., 979, 980
Tarry, H.A., 414, 416
Taylor, G.W., 981
Taylor, T.R., 29, 30, 687, 982, 983, 984
Teaney, D.T., 985, 986
Tenchov, B., 274
Terauchi, H., 987, 988
Tettamanti, E., 406, 407
Teucher, I., 70, 72, 989, 990, 991, 992, 993
Thiessen, K., 994
Thompson, R.T., 995
Tiddy, G.J.T., 996, 997
Tikhomirova, N.A., 998
Timling, K., 999
Tinius, K., 1000
Tobazeon, N.N., 1001
Tobazeon, R., 356, 616
Tobolsky, A.V., 888
Tolmachev, A.V., 1002, 1002A
Tompkins, E.N., 1003, 1004
Toriyama, K., 1005, 1006, 1007
Treheux, M., 54
Trofimov, A.N., 207, 539, 541, 542
Tseng, C.-C., 461
Tseng, H.C., 1008
Tsenter, M.Ya., 1009
Tsuchiya, M., 770
Tsukamoto, M., 768, 769, 770
Tsunda, J., 596
Tuggey, R.L., 1026, 1028
Tults, J., 628, 1010
Tuyen, L.T., 994
Ubbelohde, A.R., 329, 1011, 1012
Uchida, T., 1013
Uemo, S., 1014
Uhls, D.L., 1015
Uhrich, D.L., 276, 1016, 1059
Vainshtein, B.K., 1017, 1018
Van Boxtel, A.M., 253, 601, 602, 603
Van Der Veen, J., 255, 258A, 1019, 1020, 1021
Van Doorn, C.Z., 1022
Van Meter, J.P., 1023
Vanner, K.C., 1024
Van Zanten, P., 257, 394, 396, 397, 398, 399

Vasil'ev, A.A., 1025
Veracini, C.A., 139, 140, 141
Veracini, L., 77
Verbit, L., 1026, 1027, 1028
Vergotten, G., 105
Veron, D., 643, 1029
Veselago, V.G., 1029A
Veyssie, M., 326, 380, 381, 382, 639
Vigalok, R.V., 1030
Vigdergauz, M.S., 1030
Vilfan, M., 108, 110, 1031
Visintainer, J.J., 297, 1032
Vistin, L.K., 998, 1033, 1034, 1035
Vitek, V., 1057
Vogel, M.J., 1036, 1037, 1038, 1039
Vold, R.L., 1040
Von Planta, C., 1041
Vora, R.A., 230, 231, 232, 233, 234
Voss, J., 882
Wada, M., 894, 895, 896, 1013
Wadati, M., 506, 1042
Wade, C.G., 330, 889, 890
Wagle, S.S., 150
Wagner, P.R., 1043
Wahl, J., 1044
Wako, T., 1045
Walker, W.W., 730
Waluga, T., 711
Wand, Y., 929
Wargocki, F.E., 1046
Watanabe, H., 1013
Watkins, C.L., 1047
Wei, I.Y., 377
Wendorf, J.H., 834, 835
Werchan, R.E., 220
Wetsel, G.C.Jr., 1048
Wheelock, C.E., 1049
White, D.I., 355
White, D.L., 528, 529, 1050
White, J.L., 320
Wild, P.J., 1051, 1052, 1053
Wild, P., 116
Willey, D.G., 1054
Williams, C., 1055, 1056, 1057, 1058
Williams, L., 1080
Williams, R., 736
Williamson, R.C., 333
Wilson, J.H., 1016
Wilson, J.M., 1059
Winsor, P.A., 1060, 1061
Winterscheidt, H., 559, 560
Wohlfarth, E.P., 929
Wolff, U., 432, 575, 576, 1062
Wong, G.K.L., 1063
Wong, S.-Y., 84, 86, 694, 695
Woodard, W.R., 1048
Workman, H., 1064,
Wright, J.J., 1065
Wright, J.R., 582, 583
Wulf, A., 269, 1066, 1067
Wysocki, J.J., 6, 934, 1068, 1069, 1070
Yamada, T., 1071, 1072
Yamaguchi, R.T., 393
Yang, C.C., 1073
Yannoni, C.S., 1074, 1075

Yano, S., 1076
Yeh, C., 335A
Yeh, G.S.Y., 1077
Yim, C.T., 868
Yoshino, K., 484
Young, W.R., 430, 1078, 1079, 1080, 1081, 1082, 1083, 1084, 1085
Yu, H.N., 461

Yun, C.K., 1086, 1087, 1088, 1089, 1090
Zanoni, L.A., 462, 466
Zaschke, H., 264, 265
Zemelman, R., 150
Zhdanova, A.S., 1091
Zolotarev, V.M., 1092
Zupancic, I., 108, 109, 110, 111

LIST OF PERIODICALS

The following list is arranged alphabetically. Only regularly scanned periodicals are listed. The month and year given indicate the latest issue indexed for the present volume...

Periodical	Date
AMERICAN JOURNAL OF PHYSICS (USA) Am J Phys	August, 1973
APPLIED OPTICS (USA) Appl Opt	August, 1973
APPLIED PHYSICS LETTERS (USA) Appl Phys Lett	Aug. 1, 1973
BELL SYSTEM TECHNICAL JOURNAL (USA) Bell Syst Tech J	March, 1973
COMPTES RENDUS HEBDOMADAIRES DES SCEANCES DE L'ACADEMIE DES SCIENCES, Ser. B., (France) Comptes Rendus, Ser. B	June 13, 1973
ELECTRONICS (USA) Electronics	August 2, 1973
ELECTRONICS LETTERS (GB) Electron Lett	July 12, 1973
ELECTRO-OPTICAL SYSTEMS DESIGN (USA) El-opt Syst Des	July, 1973
EXPERIMENTAL MECHANICS (USA) Exp Mech	May, 1973
IBM JOURNAL OF RESEARCH/DEVELOPMENT (USA) IBM J Res/Dev	May, 1973
IBM TECHNICAL DISCLOSURE BULLETIN (USA) IBM Tech Discl Bull	June, 1973
IEEE JOURNAL OF QUANTUM ELECTRONICS (USA) IEEE J Quantum Electron	July, 1973
IEEE, PROCEEDINGS OF, (USA) IEEE Proc	July, 1973
IEEE SPECTRUM (USA) IEEE Spectrum	August, 1973
IEEE TRANSACTIONS ELECTRONIC DEVICES (USA) IEEE Trans Electron Devices	July, 1973
JAPANESE JOURNAL OF APPLIED PHYSICS (Japan) Jap J Appl Phys	April, 1973
JETP LETTERS (USA) JETP Lett	June 5, 1973
JOURNAL DE PHYSIQUE (France) J Phys	May-June, 1973
JOURNAL OF APPLIED PHYSICS (USA) J Appl Phys	July, 1973
JOURNAL OF CHEMICAL PHYSICS (USA) J Chem Phys	July 1, 1973
JOURNAL OF PHYSICS, Pt.D: Applied Physics (GB) J Phys D: Appl Phys	June 11, 1973
JOURNAL OF PHYSICS, Pt.E: Scientific Instruments (GB) J Phys E: Sci Instr	July, 1973
LASER FOCUS (USA) Laser Focus	August, 1973
MOLECULAR CRYSTALS AND LIQUID CRYSTALS (GB) Mol Cryst & Liq Cryst	April, 1973
NATURWISSENSCHAFTEN (Germany) Naturwiss	May, 1973
OPTICA ACTA (GB) Opt Acta	July, 1973
OPTICAL ENGINEERING [formerly SPIE Journal and SPIE Glass] (USA) Opt Eng	Jan.-Feb., 1973
OPTICAL SOCIETY OF AMERICA, JOURNAL OF, (USA) Opt Soc Am J	July, 1973
OPTICAL SPECTRA (USA) Opt Spectra	July, 1973
OPTICS & LASER TECHNOLOGY (GB) Opt & Laser Tech	April, 1973
OPTICS & SPECTROSCOPY (USA) Opt & Spectrosc	December, 1972
OPTICS COMMUNICATIONS (Holland) Opt Commun	May, 1973
OPTIK (Germany) Optik	May, 1973
PHYSICAL REVIEW LETTERS (USA) Phys Rev Lett	July 30, 1973
PHYSICAL REVIEW (A) (USA) Phys Rev (A)	July, 1973
PHYSICS LETTERS (A) (Holland) Phys Lett (A)	June 18, 1973
PHYSICS TODAY (USA) Physics Today	July, 1973
REVIEW OF SCIENTIFIC INSTRUMENTS (USA) Rev Sci Instr	July, 1973
REVUE DE PHYSIQUE APPLIQUEE (France) Rev Phys Appl	February, 1973
SCIENTIFIC AMERICAN (USA) Sci Am	August, 1973
SOLID STATE COMMUNICATIONS (USA) Solid State Commun	July 15, 1973
SOVIET JOURNAL OF OPTICAL TECHNOLOGY (USA) Sov J Opt Tech	December, 1972
SOVIET JOURNAL OF QUANTUM ELECTRONICS (USA) Sov J Quantum Electron	Mch.-Apr., 1973

SOVIET PHYSICS - CRYSTALLOGRAPHY (USA) Sov Phys - Crystallogr   May-June, 1973
SOVIET PHYSICS - DOKLADY (USA) Sov Phys - Doklady   July, 1973
SOVIET PHYSICS - JETP (USA) Sov Phys - JETP   February, 1973
SOVIET PHYSICS - TECHNICAL PHYSICS (USA) Sov Phys - Tech Phys   June, 1973
SOVIET PHYSICS - USPEKHI (USA) Sov Phys - Uspekhi   Nov.-Dec., 1972
U.S. PATENT GAZETTE   July 3, 1973

CATCHWORD TITLE INDEX

Part One: Patent Material

CIRCUITRY: (Drivers; fast turn-on and turn-off; multiplexed operation) 3,503,672   3,503,674
3,519,330   3,532,813   1,219,840   1,220,169   3,575,491   3,575,492   3,575,493
3,637,291   3,653,745   3,654,606   3,725,899   3,740,717

CLOCK: (Liquid crystal display) 3,505,804   3,613,351   1,263,277   3,668,861   3,691,755
3,712,047   1,303,947

DAY/NIGHT MIRROR: 3,614,210

DETECTOR: (Mechanical energy; electric-, magnetic and acoustical energy) 3,597,043   3,693,084
3,713,156   1,316,497

DETECTOR: (Ultrasonic hologrphy)   1,194,544

ELECTRO-OPTICAL DEVICES:
3,499,112   3,551,026   3,569,614   3,572,907   3,578,844   3,588,225
3,597,044   3,600,061   1,246,847   3,622,224   3,622,226   3,623,795
3,625,591   3,627,408   3,642,348   3,645,604   3,647,280   3,650,603
3,652,148   3,655,269   3,655,270   3,661,444   3,663,086   3,666,881
3,674,338   3,674,341   3,674,342   3,675,988   3,675,989   3,687,515
3,690,745   3,694,053   3,697,150   3,700,306   3,700,805   3,702,723
3,703,329   3,703,331   3,705,310   3,707,323   1,302,482   1,304,268
1,304,554   3,716,289   3,716,290   3,716,658   1,306,912   3,718,391
3,718,382   3,718,842   1,308,208   3,722,998   3,723,651   3,725,899
3,726,584   3,727,527   3,728,007   3,728,008   3,730,607   3,731,986
3,732,429   3,734,598   1,318,007   3,736,047   3,737,567

IMAGING SYSTEMS & IMAGE CONVERTERS:
1,251,790   3,655,971   3,663,086   3,666,947   3,666,948
3,680,950   3,704,056   3,707,322   3,711,713   3,718,380
1,309,588

LASER SPECKLE: (Elimination of,)   3,650,608

LIQUID CRYSTAL MIXTURES: (Cholesteric) 3,529,156   3,576,761   3,578,844   3,580,864
3,585,381   3,590,371   3,600,060   3,604,930   3,619,254
3,620,889   3,627,699   3,642,348   3,647,279   3,655,971
3,666,947   3,666,948   3,669,525   3,679,290   3,680,950
3,697,150   3,697,297   3,603,331   3,720,456   3,726,584

LIQUID CRYSTAL MIXTURES: (Nematic)   3,499,702   3,540,796   1,246,847   3,655,270
3,656,834   3,675,987   3,687,515   3,690,745   3,703,329
3,703,331   1,307,809   1,308,237   1,318,011   1,318,012

OPTICAL ELEMENTS: (Passive, employing liquid crystals layers) 3,592,526   3,597,043
3,669,525   3,679,290   3,697,152   3,711,181   3,720,456

REAR PROJECTION SCREEN: 3,674,338   1,316,213

SMECTOGRAPHIC DISPLAYS: 3,524,726   3,716,289   3,723,346

THERMAL IMAGING: 3,527,945   3,569,709   3,576,761   3,590,371   3,604,930   3,617,374
                          3,628,268   3,637,291   3,732,189   3,733,485

THERMOMETRIC ARTICLES: 3,619,254   3,620,889   3,628,268   3,648,280   3,704,625   3,723,346

X-RAY DOSAGE READING: 3,663,390

"""""""""""

Part Two: Article Material

ACOUSTIC FIELD MAPPING: 220, 498, 504, 505

MOLECULAR ALIGNMENT: 98, 225, 318, 325, 367, 451, 480, 517, 533, 625, 626, 627,
                      636, 637, 645, 662, 734, 790, 830, 868, 1043

BIREFRINGENCE: 114, 115, 228, 360 456, 457, 458, 476, 531, 639, 745, 795, 796,
                880, 942, 992, 993

BRAGG REFLECTION: 90, 91, 92, 93, 94

CONDUCTIVE PATTERNS, Fabrication of: 23, 753

DECAY TIME, Reduction of: 20

DIPOLE RELAXATION: 68, 69

CONTROL CIRCUITS (For LC displays):   117, 148, 285, 415, 423, 424, 584, 628, 770, 935,
                                      957, 958, 1025, 1051, 1052, 1053

ELECTRIC-FIELD EFFECTS AND DEVICES: 32, 34, 35, 38, 39, 40, 41, 42, 44, 45, 46, 52,
                                      53, 54, 55, 56, 57, 58, 59, 60, 72, 75, 79, 83,
                                      84, 85, 86, 95, 96, 99, 101, 102, 103, 104, 116,
                                      119, 134, 146, 149, 150, 159, 160, 161, 162, 165,
                                      166, 167, 169, 171, 173, 174, 175, 191, 192, 207,
                                      215, 217, 219, 221, 223, 224, 225, 229, 238, 239,
                                      242, 245, 246, 252, 254, 258, 272, 274, 277, 292,
                                      321, 322, 323, 326, 334, 353, 355, 360, 370, 371,
                                      380, 398, 399, 405, 409, 411, 414, 415, 416, 421,
                                      424, 425, 429, 432, 433, 434, 435, 437, 440, 441,
                                      442, 443, 444, 445, 446, 454, 455, 456, 457, 458,
                                      459, 461, 464, 465, 466, 467, 468, 471, 476, 481,
                                      484, 487, 490, 494, 502, 503, 509, 512, 523, 524,
                                      525, 526, 527, 528, 529, 530, 531, 538, 539, 542,
                                      549, 554, 555, 564, 569, 575, 576, 579, 584, 588,
                                      596, 604, 606, 611, 612, 620, 628, 661, 666, 667,
                                      668, 675, 683, 701, 714, 715, 722, 723, 724, 725,
                                      729, 736, 743, 754, 760, 762, 768, 769, 770, 771,
                                      772, 773, 774, 775, 777, 797, 780, 826, 860, 869,
                                      871, 891, 900, 902, 903, 905, 907, 909, 910, 919,
                                      920, 921, 927, 933, 935, 937, 939, 940, 942, 943,
                                      944, 945, 946, 956, 957, 958, 972, 973, 978, 981,
                                      985, 1005, 1006, 1007, 1010, 1025, 1033, 1034,
                                      1035, 1045, 1046, 1050, 1051, 1052, 1053, 1058,
                                      1062, 1063, 1065, 1069, 1070, 1071, 1085

HYDRODYNAMICS OF LIQUID CRYSTALS: 242, 253, 256, 257, 321, 322, 324, 326, 364, 380,
                                    405, 467, 475, 493, 510, 511, 548, 601, 602, 603,
                                    649, 671, 673, 698, 699, 700, 735, 736, 773, 774,
                                    777, 800, 801, 802, 812, 813, 871, 961, 973, 1001

INFRARED EXPOSURE, Response to: 67, 142, 489, 532, 557, 887, 930, 1002

LASER SPECKLE & COHERENCE, Control of: 27, 741

LIQUID CRYSTALS AND MISCELLANEOUS APPLICATIONS: 37, 47, 54, 75, 119, 132, 133, 134, 151, 167, 171, 357, 358, 437, 454, 455, 459, 464, 503, 556, 617, 618, 620, 647, 677, 714, 751, 778, 803, 908, 933, 938, 939, 959, 962, 1024, 1041

MAGNETIC-FIELD EFFECTS IN LIQUID CRYSTALS: 129, 200, 240, 274, 388, 389, 396, 400, 421, 438, 465, 482, 497, 567, 642, 644, 645, 646, 649, 655, 656, 660, 767, 789, 793, 809, 810, 814, 815, 840, 841, 851, 854, 856, 858, 862, 872, 905, 920, 921, 922, 929, 991, 1022, 1048, 1076, 1087, 1088

MAGNETIC LIQUID CRYSTAL COMPOSITIONS: 126, 767, 1049

MECHANICAL EFFECTS: 102, 103, 208, 216, 238, 239, 251, 252, 260, 261, 333, 347, 461, 480, 498, 504, 505, 540, 541, 545, 546, 547, 562, 563, 568, 651, 655, 664, 665, 687, 696, 702, 727, 728, 746, 755, 787, 788, 817, 818, 852, 893, 998

MOLECULAR MOTION IN LIQUID CRYSTALS: 11, 13, 291, 327, 640, 915, 916

OPTICAL FILTERS, Passive: 7, 40, 317, 318, 319, 447, 488, 527

OPTICAL HARMONIC GENERATION: 412, 925, 926

OPTICAL PROPERTIES OF LIQUID CRYSTALS: 95, 96, 171, 293, 342, 346, 350, 476, 484, 488, 526, 527, 528, 530, 531, 538, 549, 551, 577, 579, 654, 690, 692, 693, 722, 723, 724, 725, 726, 745, 768, 771, 795, 797, 808, 844, 850, 865, 866, 874, 880, 882, 883, 900, 902, 905, 907, 909, 910, 934, 936, 939, 940, 942, 944, 950, 951, 952, 968, 978, 983, 992, 993, 1005, 1033, 1034, 1045

OPTICAL WAVE PROPAGATION IN LIQUID CRYSTALS: 15, 27A, 192

PHASE TRANSITION: 16, 17, 26, 82, 118, 142, 197, 301, 331, 333, 349, 368, 369, 406, 407, 422, 425, 468, 537, 540, 585, 619, 665, 674, 705, 769, 862, 874, 892, 893, 894, 896, 932, 960, 963, 964, 998, 1027, 1030, 1069, 1070, 1073

PITCH CHANGE (Due to various stimuli): 10, 70, 220, 526, 749, 750

PITCH DETERMINATION: 4, 8, 67, 136, 632, 633

ROOM TEMPERATURE NEMATICS: 19, 168, 378, 429, 430, 441, 442, 523, 714, 771, 844, 870

ROTARY POWER: 3, 6, 136, 179, 393, 394, 477, 633, 634, 692, 808, 852, 950, 952, 992, 993

SELECTIVE REFLECTION AND SCATTERING OF LIGHT: 5, 311, 312, 313, 342, 346, 350, 446, 450, 462, 487, 528, 529, 568, 611, 612, 623, 632, 775, 808, 829, 863, 874, 968, 1054, 1073

TESTING OF MATERIALS, Non-destructive:	42, 387, 607, 691, 740, 994, 1015
THERMAL CONDUCTIVITY:	14
THERMO-OPTIC EFFECTS IN LIQUID CRYSTALS:	220, 293, 345, 359, 431, 532, 564, 599, 729, 740, 806, 832, 843, 887, 922, 930, 936, 1002
ULTRAVIOLET EXPOSURE, Response to:	9, 694
VISCOSITY:	14, 127, 213, 331, 378, 472, 544, 562, 563, 622, 703, 704, 779, 839, 977, 1044, 1089
X-RAY (DIFFRACTION) STUDIES:	73, 154, 246, 279, 280, 281, 282, 283, 284, 286, 287, 288, 425, 610, 682, 717, 949, 987, 988

"""""""""""

ANNOUNCEMENTS

VARI-LIGHT CORPORATION

9770 CONKLIN ROAD
CINCINNATI, OHIO 45242
513/791-6330

Vari-Light has been a manufacturer of high purity organic liquid crystalline materials for about ten years, having started with the cholesteryl derivatives and introduced the first room temperature nematic liquid crystal, MBBA (then called VL-462-N) in 1969. The variety and characteristics of our product line have been expanding and reaching new levels of purity since that time. Vari-Light's forte is producing its liquid crystalline products in the highest purity, in order to be able to selectively dope them for use in alternate applications. These materials presently span ranges of, for example, -20°C to +55°C, and are proving themselves in commercial applications.

Among the liquid crystals which we produce are Schiff bases, phenyl benzoate esters, and azoxy-based materials, both in positive and negative anisotropic form. The positive anisotropic materials are presently foremost in our research, being the source for the very promising "field effect" display devices. We might mention that we plan to announce what appears at this writing to be a revolutionary step forward in "nematic field effect displays," and may have a pronounced effect on the MOSLSI-interfaced display market.

In addition to the production of liquid crystals for display purposes, we are also pursuing the use and production of a device called the Acousto-Optical Conversion Cell (AOCC), as covered on our U.S. Patent No. 3,597,043 for the direct visualization of ultrasonic holograms. Separate patent application is in progress on this unique application of a liquid crystal device.

A second product line that has been expanding due to the rate of growth of liquid crystals is that of conductive coatings, both transparent and reflective, and our need for these materials both in research and production has led us into the role of producer of several coatings.

Vari-Light is presently designing and marketing liquid crystal displays of "custom" design for specialized applications involving both dynamic scattering and field effect. These have been produced in prototype and developmental quantities, and should reach low-to-moderate volume in a short time. Our intention is to concentrate on the more unusual requirements of limited quantities and generally avoid the mass confusion of the gargantuan field of alpha-numeric readouts.

One product which is showing strong acceptance in the commercial market is a cholesteric liquid crystal plastic film, THERMAFILM, which is a most practical application of these normally fragile and easily-contaminated materials. THERMAFILM is being used in the non-destructive testing of heating surfaces for de-icing aircraft windows, and in several areas in the electronics inductry.

Vari-Light is growing rapidly within the liquid crystal industry. We will be pleased to receive any comments which you may offer, or be happy to answer any questions we may have raised.

VARI-LIGHT® CHOLESTERIC AND NEMATIC LIQUID CRYSTAL PRODUCTS

ELECTRO-SCIENCE LABORATORIES, INC.
1601 Sherman Ave. • Pennsauken • New Jersey 08110

CONSTRUCTION OF LIQUID CRYSTAL DISPLAYS BY SCREEN PRINTING TECHNOLOGY

Electro-Science Laboratories, Inc. of Pennsauken, N.J. announces improved methods of constructions of LCDs by the use of screen printing technology. Reflective rear mirror coatings, solder glass window frames, metallization borders, solder pastes, spacers, etching resists, and black masking glass layers can all be printed for use in different construction methods. The techniques utilized are similar to those developed and used in Hybrid thick film microcircuit manufacture. ESL, a major thick film paste supplier, has developed a number of new compositions to facilitate display construction.

<u>Reflective Mirrors and Electrodes</u>: Screen Printable reflective coatings #8060 (Silver Color) and #8080 (Gold Color) are specially formulated organo-metallic precious metal compositions. After printing they are fired at 550°C to give thin films of conductive reflective coatings.

<u>Sealing Rings, Spacers or Frames</u>: #4010-B (crystallizing) or #M4011-B (vitreous) pastes are suspensions of fine particle size glass in screen printing vehicles. They are dried at low temperature, then heated to decompose the organics and fuse the glass. The B grades are dull black after firing. The C grades are clear to translucent. By printing 10 to 15 microns thick they serve as plate separation spacers or masking layers.

<u>Metallization Pastes</u>: Solderable, printable coatings, firing at 550°C eg. #590, Silver; #5835, Platinum-Gold; #8835, Gold; and #9635, Palladium-Silver; are available. These bond tenaciously to soda lime glass and can be soldered with suitable alloys.

<u>Solder Pastes</u>: The SP Series of solder compositions are available in paste form which allows solder to be printed on top of the metallizations where joining to a metal cover is needed.

<u>Etching</u>: Tin oxide coated glass can be easily etched, by first printing Etching Resist #1211, drying and then coating with Sprayable Zinc Activator Composition #1212. Hydrochloric Acid in Commercial Spray Etchers may then be used to remove the tin oxide. Indium oxide etching can also be performed after resist coating. In this instance the zinc activator is not required.

Data sheets and further information are obtainable from Electro-Science Labs., Inc. 1601 Sherman Avenue, Pennsauken, New Jersey 08110, U.S.A.

LICRISTAL®

EM LABORATORIES, INC.
affiliate of
E. MERCK, DARMSTADT, GERMANY
500 Executive Boulevard
Elmsford, New York 10523

NEMATIC PHASES FOR ELECTRO-OPTICAL APPLICATIONS

Nematic Phases with Negative Dielectric Anisotropy

Nematic Phase 7 (-10/80) Licristal® (Item #10683)

Pure phase of the azoxy-type--very stable against water and oxidation.

Nematic Range	-10 to +80° C
Dielectric Anisotropy	-0.6
Electrical Resistivity	$\sim 10^{11} \Omega \cdot cm$

Nematic Phase 7A (-10/80) Licristal® (Item #10598)

Special material for the DSM--aligns itself homoeotropically--optimum conductivity for DSM. Same properties as Phase 7 except:

Electrical Resistivity	$5 \cdot 10^8$ to $1.5 \cdot 10^{-9} \Omega \cdot cm$
Threshold Voltage for DSM	7 - 8 Volt
Delay Time + Rise Time	15 - 30 ms
Decay Time	25 - 50 ms

Nematic Phase 7B (-10/80) Licristal® (Item #10605)

Special material for field effect (DAP)--aligns itself homoeotropically--high threshold voltage in DSM--low threshold for DAP-effect. Same properties as Phase 7 except:

Electrical Resistivity	$\sim 10^{10} \Omega \cdot cm$
Threshold Voltage for DAP	5 - 6 Volt

Nematic Phase 8A (-15/60) Licristal^R (Item #10568)

    Mixture of schiff bases--colorless--used only for AC-driving.

Nematic Range	-15 to +60° C
Dielectric Anisotropy	-0.5

<u>NEW</u> Nematic Phase 9A (-20/60) Licristal^R (Item #10763)

    Ester type.

    Completely colorless--high stability against water and oxidation. Same electrical properties as Phase 7A except:

Nematic Range	-20 to +60° C
Dielectric Anisotropy	-0.44

Nematic Phase with Positive Dielectric Anisotropy

<u>NEW</u> Nematic Phase Test Product ZLI 319 (Item #10764)

Nematic Range	<+10 to +58° C
Dielectric Anisotropy	+7
Threshold Voltage in the Twisted Nematic Cell	1.1 Volt
Recommended Driving Voltage	3 to 5 Volt

We are able to supply tailor-made formulations for your specialized needs.

          EM LABORATORIES, INC.
               affiliate of
     E. MERCK, DARMSTADT, GERMANY
        500 Executive Boulevard
         Elmsford, New York 10523

LIQUID CRYSTAL INDUSTRIES, INC.

115 GRANT STREET
TURTLE CREEK, PA. 15145
TELEPHONE (412) 823-4300

Liquid Crystal Industries, Inc. continues to experience rapid growth in liquid crystal technology. The staff of scientists have assisted corporate, university, and government groups in using liquid crystals successfully in a tremendous variety of applications, including the non-destructive testing, electro-optical, and medical fields. LCI is unique for quite a few reasons:

1. Cholesteric testing solutions of ranges between 0-250°C
2. Preblackened liquid crystals for a single-coat process
3. "Memory" liquid crystals with a sensing temperature 0-60°C
4. New sheet form, both reversible and memory types
5. Two room temperature nematics, 0-70°C and 18-45°C
6. Prepared nematic windows and shutters
7. Seminars and consulting services

Our reversible cholesteric liquid crystals are available in bulk or solution form, preblackened or not. Testing can be done with continuous spectrum ranges, 1-50°C broad, accurately detecting temperature differences of only 0.1°C. Formulations can be purchased according to the quantity and range you require. Aerosol applicator and instructions are included. LCI's unique preblackened liquid crystals eliminate the messiness and problems of poor adherence, decreased resistance and resolution encountered when using a conventional two-layer, black paint/liquid crystal system.

In case you want to experiment with various temperature ranges, LCI has prepared two kits. (see photo) Each contains 11 reversible solutions, 25cc each in different ranges, and a memory solution. Application aerosol, indirect testing hoop, and instructions are included. Kit No. 1 ranges from 10°-72°C and Kit No. 2 from 75°-174°C. "Mix and Match" option applies.

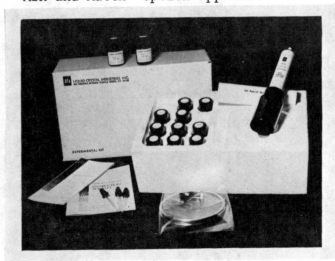

The latest item to be marketed by LCI, reversible liquid crystals in a sheet, has received enthusiastic response from our cutomers. These flexible plastic sheets can be tailor-made to specifications in ranges up to 5°C broad between 5° - 40°C. Adhesive backing is available. Possible applications seem limitless at this point. It is particularly suited to monitoring operations, repeat and uniform surface testing.

The memory liquid crystal, instead of changing reversibly through the spectrum in response to temperature changes, has one irreversible color change at a single selected temperature. The 30°C material, for example, exhibits the characteristic bright green color initially below the sensing temperature. At 30°C the color is lost abruptly; the material becomes black, and remains so with increasing or decreasing temperature. Thus a temperature overshoot leaves a record. The original color is restored by lightly brushing the surface. For indirect testing, the new sheet from memory liquid crystal is ideal. It operates the same and is restored by rubbing the surface. Sensing temperature can be formulated between 0 - 60°C.

With the tremendous interest and growth in the development of nematic materials and technology, LCI has become an important supplier of both. Nematic liquid crystals vary between two states - transparent and translucent. They are sandwiched between two conductive surfaces, at least one of which is transparent. In this condition nematics transmit light, but when 3 to 30 volts AC or DC is applied, the light is scattered and the clear window gradually becomes frosted.

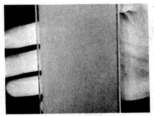

LCI has two nematic materials -- in 0 - 70°C and 18 - 45°C operating ranges. Prepared windows and sample quantities of conductive glass are also available from LCI.

If its quality liquid crystal materials you're looking for, or reliable consultation contact: Liquid Crystal Industries, Inc.
115 Grant Street
Turtle Creek, Pa. 15145
(412) 823-4300

Marconi Communication Systems Limited

A management company for The Marconi Company Limited   A GEC-Marconi Electronics Company

RADFORD CRESCENT
BILLERICAY, ESSEX
ENGLAND

LIQUID CRYSTAL DISPLAYS MANUFACTURED BY THE SPECIALIZED COMPONENTS

DIVISION OF MARCONI COMMUNICATIONS SYSTEMS LIMITED

N. Miller

Two standard devices are in pilot production at present and volume production is expected to begin in early 1974.

The two types available are a single digit seven segment display with a character height of 1.75 inches and four digit seven segment display complete with a polarity sign and decimal points, the character height being 0.6 inches. Both devices are dynamic scattering types incorporating dielectric mirrors, enabling them to be used in either the reflective or transmissive mode depending upon ambient lighting conditions.

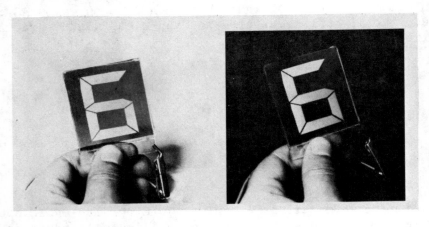

(a)   (b)

Transmission/Reflection Liquid Crystal Numeric Display

(a) against white background
(b) against black background

Under identical lighting conditions.

The cells are constructed of 2mm glass with 25μ spacing, (narrow spacing gives faster operation of lower voltage operation but results in reduced contrast). The transparent electrodes are a mixture of indium and tin oxides deposited on the inside of the cell. The character is formed by photo-etching, at present, though screen printing techniques are being tried. Standard quarter-wave plate optical techniques are employed to achieve the dielectric mirror and after vacuum filling the cells are sealed with epoxy resin.

Transmissive displays have been made and also reflective types employing metallised mirrors but the dielectric mirror displays are considered the best for contrast and clarity. Twisted nematic liquid crystals are being studied with a view to using them in matrix arrays due to the existence of a definite threshold point, thus making multiplexing easier.

It is in this field of matrix arrays that the future is thought to lie and also in large area displays for applications such as clocks and light shutters. These are the fields which best illustrate the capabilities of the Specialized Components Division.

Demonstration Clock Display. Dielectric Mirror Type

Clock Display (Transmissive Mode) Digit Size = 2.4"

DATE DUE